PHENOMENOLOGY AND APPLICATIONS OF HIGH TEMPERATURE SUPERCONDUCTORS

THE LOS ALAMOS SYMPOSIUM—1991
PHENOMENOLOGY AND APPLICATIONS OF HIGH TEMPERATURE SUPERCONDUCTORS
P R O C E E D I N G S

Editors

KEVIN S. BEDELL
Los Alamos National Laboratory

MASAHIKO INUI
Los Alamos National Laboratory

DAVID E. MELTZER
Southeastern Louisiana University
Hammond

J. ROBERT SCHRIEFFER
Florida State University
Tallahassee

SEBASTIAN DONIACH
Stanford University

ADDISON-WESLEY PUBLISHING COMPANY
The Advanced Book Program

Reading, Massachusetts • Menlo Park, California • New York
Don Mills, Ontario • Wokingham, England • Amsterdam • Bonn
Sydney • Singapore • Tokyo • Madrid • San Juan • Paris • Seoul
• Milan • Mexico City • Taipei

This book was typeset by Mary Louise Garcia, using the TEX typesetting language.

ISBN 0-201-56989-2

Copyright © 1992 by Addison-Wesley Publishing Company

All rights reserved. No part of this publication may be reproduced, stored in a retrieval system, or transmitted, in any form, or by any means, electronic, mechanical, photocopying, recording, or otherwise, without the prior written permission of the publisher. Printed in the United States of America. Published simultaneously in Canada

1 2 3 4 5 6 7 8 9 10-MA-95 94 93 92
First printing, March 1992

FOREWORD

This symposium was the second major conference sponsored by the Advanced Studies Program in High-Temperature Superconductivity (HTS). The first one, held in December 1989, focused primarily on the microscopic foundations of HTS. The current symposium focused on fundamental issues of phenomenology and applications of HTS. The style of this meeting was similar to that developed for the previous meeting. The four main ingredients for the success of these meetings were: 1) To bring together a small number of experts in the field; 2) to have some of them present overview talks; 3) allow for extensive informal discussions; and 4) publish the proceedings, including the discussions, in a timely fashion. This type of meeting along with making the proceedings generally available has proven to be most beneficial to the participants as well as the wider scientific community.

To make such a meeting happen, a dedicated and hard-working support staff is needed as well as some benefactors. For this meeting we were fortunate to have both. We begin by thanking Don Parkin, Director of the Center for Materials Science (CMS) and Rod Quinn (past), Ross Lemons (acting), and Kay Adams (present) Directors of the Exploratory Research and Development Center (ERDC), for their generous financial support. We thank the members of CMS, Mike Boring, Bettye McCulla and Brenda Romero who contributed to the success of this meeting. We would like to give special thanks to Laurie Lauer for her invaluable assistance to the conference organizers.

The symposium was structured around five half-day sessions focusing on thin-film and microwave properties, textured materials and Josephson junctions, vortex lattice structure and melting, vortex pinning and creep, a survey of applications and a summary. Each session had three rapporteurs who gave a critical review of a topic, with each review followed by extensive discussions. An excellent reproducing of the discussions was made possible by the tireless efforts of one of our fellow editors, David Meltzer. The speakers and discussants were given the opportunity to review the transcripts of their spoken remarks.

To bring the manuscripts and discussions together using the same format throughout was a formidable task. We would like to give a special thanks to Mary Louise Garcia for the outstanding job she has done in getting this book together. We also would like to thank our fellow editors David Meltzer and Sebastian Doniach for their editorial assistance during the preparation of this book, and Masahiko Inui for his most valuable assistance as co-organizer of the meeting and for his extensive editorial work on this project. We hope this second volume gives insight into the basic aspects of applied HTS as of this date.

Kevin S. Bedell
J. Robert Schrieffer

PREFACE

Phenomenology and Applications of High-Temperature Superconductors, The Los Alamos Symposium: 1991, was sponsored by the Los Alamos National Laboratory, Center for Materials Science, the Advanced Studies Program on High-Temperature Superconductivity Theory (ASP) and the Exploratory Research and Development Center. This is the second symposium in the series. High Temperature Superconductivity, The Los Alamos Symposium: 1989, also published by Addison Wesley, focused on the cutting-edge theoretical and experimental issues in high-temperature superconductors. This symposium, with its focus on the phenomenology and applications of high-temperature superconductors, gives a complementary review of the aspects of the field closely related to the impact of high-temperature superconductors on technology.

The objective of ASP is to advance the field on a broad front with no specific point of view by bringing a team of leading academic theorists into a joint effort with the theoretical and experimental scientists of a major DOE national laboratory. The ASP consisted of fellows led by Robert Schrieffer (UCSB and now FSU) joined by David Pines (University of Illinois), Elihu Abrahams (Rutgers), Sebastian Doniach (Stanford), and Maurice Rice (ETH, Zurich) and theoretical and experimental staff of Los Alamos National Laboratory. This synergism of academic, laboratory, theoretical and experimental research produced a level of interaction and excitement that would not be possible otherwise. This publication and the previous one in the series are just examples of how this approach to advancing science can achieve significant contributions.

Don M. Parkin
Director, Center for Materials Science

Contents

Foreword   v

Preface   vii

1   Thin Films and Microwave Properties
    Chair — J. Robert Schrieffer

Progress in Materials Research and Applications of High-T_c Superconductors
    S. Tanaka   3

Dynamics of the Vortex State in High Temperature Superconductors
    Aharon Kapitulnik   34

High T_c Superconductors at Microwave Frequencies
    G. Grüner   77

2   Textured Materials and Josephson Junctions
    Chair — Seb Doniach

Flux Pinning Enhancement by Y_2BaCuO_5 Inclusions in Melt Processed YBaCuO Superconductors
    M. Murakami   103

High-T_c Josephson Junctions, SQUIDs and Magnetometers
    John Clarke   131

Critical Fields in High-Temperature Superconductors
    D. K. Finnemore   164

3 Vortex Lattice Structure and Melting
Chair — Larry Campbell

Correlations and Transport in Vortex Liquids
    David R. Nelson   187

Direct Images and Unconventional Dynamics of Vortices in the High-T_c Superconductors YBCO (123) and BSCCO (2212)
    *P. L. Gammel, D. Bishop, C. Murray and D. Huse   243

Phase Transition and Transport in Anisotropic Superconductors with Large Thermal Fluctuations
    Daniel S. Fisher   287

4 Vortex Pinning and Creep
Chair — Marty Maley

Muon and Neutron Investigations of Vortex Correlations in High-T_c Superconductors
    G. Aeppli   335

Resistance, Flux Motion and Pinning in High Temperature Superconductors
    B. I. Ivlev   359

Vortex Pinning and Creep Experiments
    P. H. Kes   390

5 Survey of Applications and Summary
Chair — Jim Smith

Survey of Potential Electronic Applications of High Temperature Superconductors
    *Robert B. Hammond and Lincoln C. Bourne   437

Development of High Temperature Superconductors for Magnetic Field Applications
    David C. Larbalestier   463

Conference Summary
    M. Tinkham   499

Contributed Papers   515

*In the case of papers with multiple authorship, the underlined author presented the paper at the conference.

PHENOMENOLOGY AND APPLICATIONS OF HIGH TEMPERATURE SUPERCONDUCTORS

1. Thin Films and Microwave Properties

Chair — J. Robert Schrieffer

S. Tanaka
Tokai University
Kanagawa, Japan

Progress in Materials Research and Applications of High-T_c Superconductors

1. INTRODUCTION: BASIC CONCEPTS

Since the discovery of high-T_c superconductivity in cuprates in 1986, research and development has made rapid progress and applications are expected to appear in the very near future. The international situation has also changed drastically in these five years and the people of the world are thinking about what will happen next and where we are going in the future.

As is evident from our history, technological breakthroughs have a very big impact on the structure of society. In the 1970s and 80s, the information revolution spread especially in the advanced countries, and the society in these countries was greatly influenced. This revolution was based on the very quick development of information processing technology and optical communication technology, both of which were backed by drastic progress of semiconductor technology.

However, it is expected that supercomputers are becoming big users of energy and in the near future several supercomputers will use almost as much electricity as is generated by one nuclear power station. Also the use of many small computers is becoming widely spread and people expect to have more luxuries and a less strenuous life by their use. This seems to suggest that the highly developed information society may be changing to a more highly energy consuming society.

In Fig. 1, the trend of energy consumption in Japan is shown. In the early 1970s, people in Japan believed that by introducing modern information technology they could save labor and energy, and that a more efficient society could be established. Until 1985, this was true. Even though the GNP in Japan increased, energy imports (consumption) decreased greatly. But after 1985, the consumed energy to GNP ratio became almost constant and then the total energy consumption increased rather rapidly. Probably the situation in other developed countries may be the same. Furthermore, environmental problems and the greenhouse effect are becoming serious all over the world. These facts seem to indicate that the energy problem may become quite serious early in the 21st century.

What should we do at present? The answer is clear: We should achieve a new "Energy Revolution" by making some technological breakthroughs.

In that sense, the research and development of high-T_c superconductivity is expected to be very important. Energy saving apparatus, energy storage machines, new energy sources and many other possibilities are opened with high-T_c superconductivity.

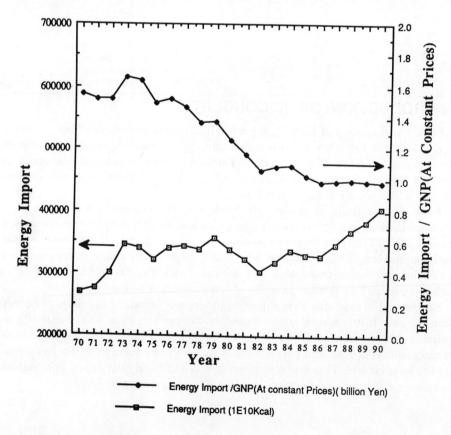

Fig. 1 The trend of energy consumption in Japan.

2. EXPECTATIONS AND THE PRESENT STATUS OF RESEARCH AND DEVELOPMENT

The first phase of R&D of high-T_c superconductivity was from 1986 to 1989. In this phase, the scientists all over the world were shocked by the discovery of high-T_c superconductivity in cuprates and many scientists and engineers started their investigations. As a result, new materials were found one after another, and the critical temperature reached 125K in a thallium compound. Even though some people had doubts about the critical currents in these materials, the critical current actually was increased very rapidly. The critical current in YBCO thin films exceeds 10^6 A/cm^2 even at 77K. Furthermore, in bulk samples it reached 10^5 A/cm^2 in YBCO in identifying and increasing the number of pinning centers.

In the second phase, I expect it will be from 1990 to 1993, the search for new materials having higher T_c will continue steadily. The highest critical temperature was confirmed to be 127K in thallium compounds this year.[1] Even though it is very difficult to find higher T_c in such complicated compounds, it is still hoped to obtain new materials beyond 150K.

Big progress has been made in obtaining high critical currents in YBCO. By introducing the GMG method and then the MPMG method, we obtained YBCO samples in which fine particles of 211-phase are almost uniformly dispersed as is shown in Fig. 2.[2]

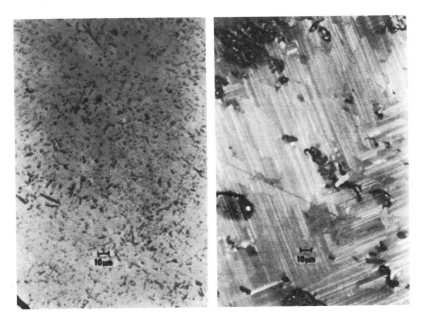

Fig. 2 YBCO samples obtained by the MPMG method. The fine particles of 211-phase are almost uniformly dispersed.

We think that the boundaries between the 123-phase and the 211-phase of YBCO act as pinning centers. It is also found that in Ag-doped MPMG samples, the number of cracks is reduced and this helps increase the critical current. The critical currents of these samples are shown in Table 1. They reach 24 000 A/cm^2 at 77K in a 10T magnetic field, and furthermore reach 2.2×10^5 A/cm^2 at 35T and 31K.[3] This indicates that the pinning force at the boundary between the 123- and 211-phase is much stronger than previously expected. Recently a new method of making such composites was found in our laboratory.[4] YBCO samples doped with Pt or Rh have a similar microstructure to those grown by the MPMG method and fairly high critical currents are observed. By using this method, the quenching process, which is the first step in the QMG and MPMG methods, may become unnecessary.

During the course of development of high critical current in bulk materials, a strong magnetic levitation force was observed, and it has increased, surprisingly, one thousand times in these two years as is shown in Fig. 3.[4] This comes from the trapping of the quantized magnetic flux near the surface of the bulk superconductor because of the existence of strong pinning centers (Fig. 4).

On the other hand, if a bulk superconductor is cooled down in a magnetic field to below the critical temperature, the magnetic flux trapped inside remains even after a permanent magnet, the external magnetic field, is removed. The material then behaves like a magnet and about a 5000-gauss residual magnetic field was observed at 77K. These peculiar features of the MPMG-YBCO bulk material are shown in Figs. 5 and 6.

It must be strongly emphasized that this mysterious magnetic levitation force due to flux pinning is quite a new kind of one, which has been found very recently, and many applications are expected in this field.

The critical current in tapes has also developed very rapidly in the last two years. Bi and Tl compounds are mainly used for the tape, and the strong two-dimensional crystal structure of the materials seems to help in making high quality

Table 1.
Critical Current Density in MPMG YBa$_2$Cu$_3$O$_7$

Temperature	Magnetic Field	Critical Current Density (A/cm^2)
4.2K	10T	880 000
	30T	>330 000
	40T	>100 000
31K	35T	220 000
77K	10T	24 000

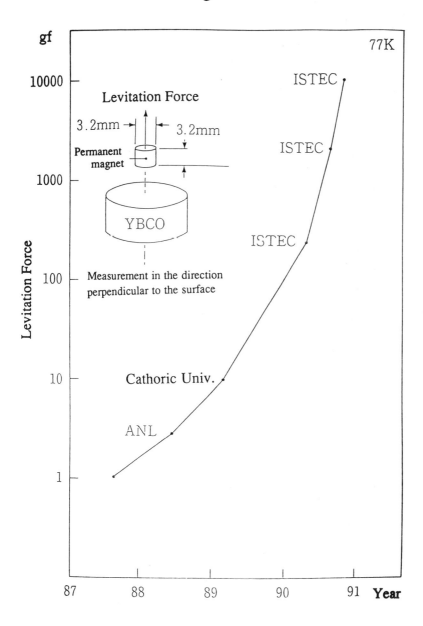

Fig. 3 Increase of magnetic levitation force with time.

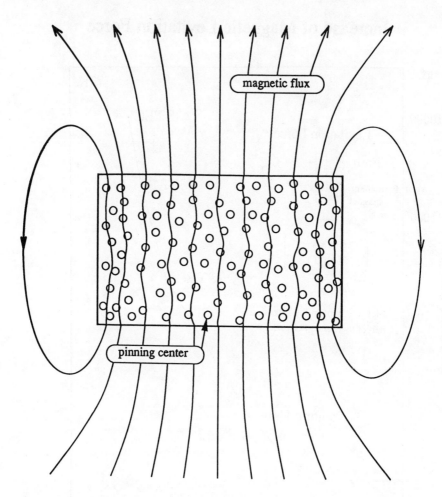

Fig. 4 Schematic illustration of a magnetized superconductor due to the presence of pinning centers.

tapes. An example of progress is shown in Fig. 7. As is seen in this figure, the critical current of the silver sheathed tape of the Bi-compound is very close to application levels.[5]

Fairly good uniformity in the tape was obtained as shown in Fig. 8 and a limited amount of bending does not affect the critical current (Fig. 9).

It must be mentioned here, however, the pinning centers in Bi- and Tl-compounds are still not nearly strong enough at 77K as is the case for MPMG-YBCO. This problem remains to be solved in the very near future.

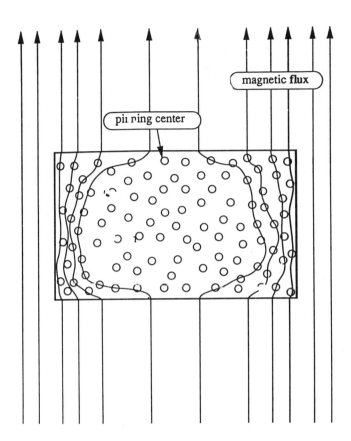

Fig. 5 Schematic illustration of magnetic levitation due to the presence of pinning centers.

As was mentioned above, big progress in the critical currents of bulk and tape form superconductors has been made in the past two years, and they will reach some value suitable for applications at the end of the second phase, 1993. Furthermore, the greatly increased magnetic levitation force developed in the past two years will find a wide field of applications in the near future.

Thin film technology of the high-T_c cuprates also made big progress in these past two years. The critical current of thin films has been greatly increased. Many kinds of preparation methods have been tried to obtain so called "high quality" thin films. Sputtering, laser ablation, ion beam evaporation and MOCVD have

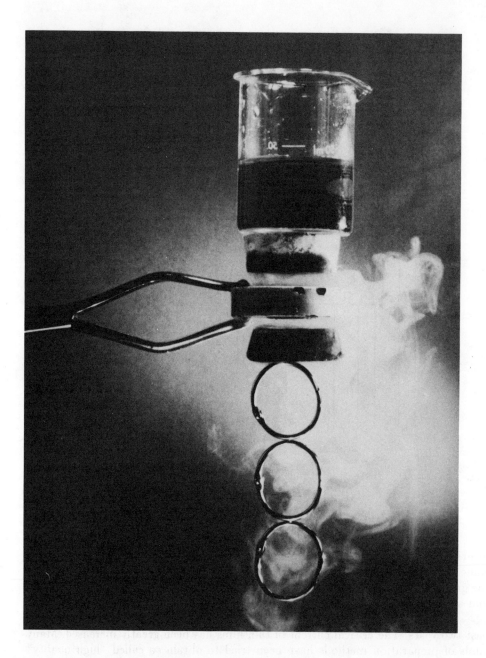

Fig. 6 A demonstration of the effect of flux pinning.

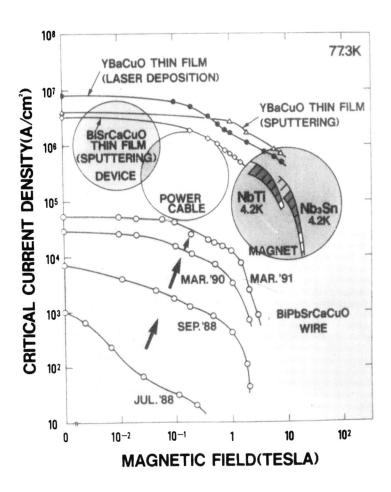

Fig. 7 The critical current density - magnetic field dependence of various superconductors.

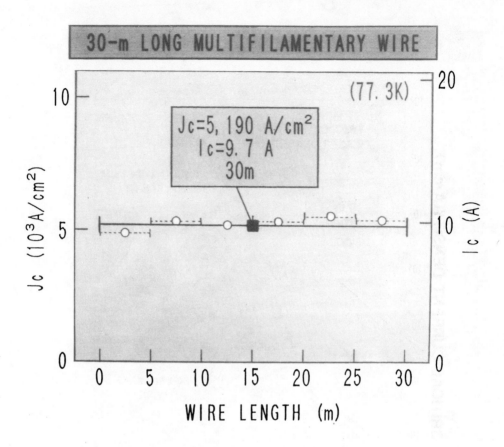

Fig. 8 Dependence of J_c with wire length of a high-T_c wire.

been tried, but it is still not clear what method is the most suitable for getting really "high quality" thin films. We find superconductivity in films less than 100Å in thickness, but nobody is concerned with the roughness of the substrate surface. We must remember how much effort has been expended in VLSI technology in getting a "flat surface" on silicon single crystals. In that sense superconductivity technology is still quite far behind that of semiconductors.

Very recently, it was found from RHEED experiments that the mechanism or process of crystal growth of cuprate thin films is quite different from that of covalent semiconductors, such as Si and GaAs. Also the morphology of the films is very complicated as seen by STM observations. How can we control the growth of truly flat single crystal films? This is the most essential problem to solve. Also it appears to be clear that the c-axis oriented film is preferable to apply SQUID or microwave devices, and to apply future multi-layered active devices, on the other hand, a-axis

Fig. 9 Photograph of a high-T_c wire.

oriented film may be preferable for SQUID or microwave device applications, but for future multi-layered active devices, on the other hand, a-axis oriented films may be preferable. But, how can we control exactly the crystal orientation of the films on the substrate crystal?

At present, so many experiments have been made in obtaining "high quality" films, that I cannot summarize all of them here. Instead I would like to introduce some new approaches, which have been made in our laboratory recently.

One is the oxygen ion beam method. The ion evaporation and molecular beam methods for making cuprate films have conflicting requirements. These processes must take place in an ultra-high vacuum vessel, and the ultra-high vacuum can only be reached by reducing the oxygen content as much as possible in the vessel. But to make superconducting cuprate films we must introduce fairly large amounts of oxygen into the vessel. This means that the stable control of the oxygen pressure in the vessel is almost impossible.

Thus we tried to obtain a high current beam of O^+ ions inside the ultra-high vacuum vessel.[6] As is shown in Fig. 10, the activity of O^+ ions in forming oxide films is higher by several orders of magnitude in comparison to O atoms or O^- ions. The experimental system is shown in Fig. 11. From the ion source we can obtain a 10mA beam of O^+ ions of less than 50eV energy, while the vessel itself maintains an ultra-high vacuum of 10^{-11} Torr. By using this apparatus, we succeeded in obtaining high

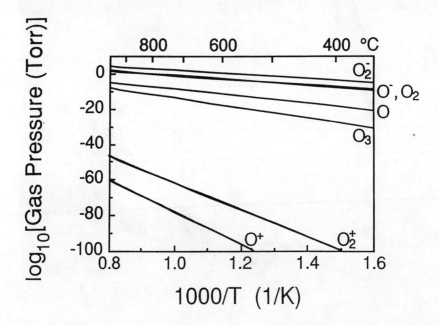

Fig. 10 Temperature dependence of the activity of various oxygen ion species.

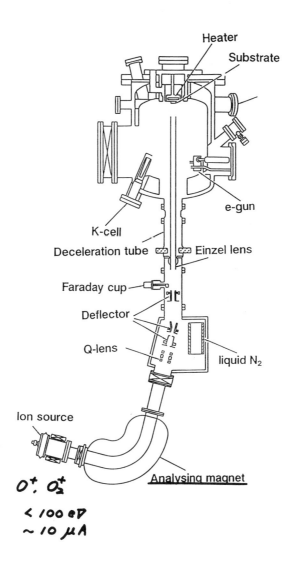

Fig. 11 A diagram of our system equipped with an oxygen ion source generator.

quality CuO films with the substrate at room temperature. The backscattering data of this film is shown in Fig. 12 as an example. The data indicate that the film is almost crystallographically perfect. We are now going to try to grow YBCO films by using this apparatus.

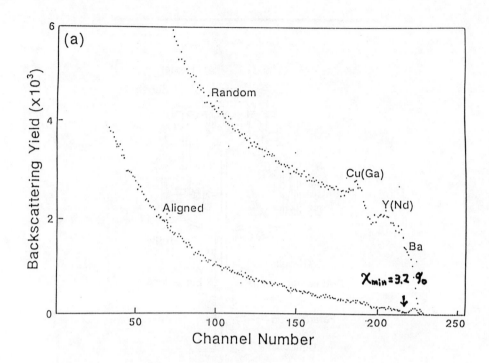

Fig. 12 Data of backscattering from thin films made in the system of Fig. 11.

The other case is magnetron sputtering using the very high frequency of 100 MHz.[7] It is known that by using a high frequency of 45 MHz, we can obtain higher quality films of Al than that made by ordinary 13.5 MHz magnetron sputtering. One of our results of a YBCO film on a substrate of $NdGaO_3$ is shown in Fig. 13. The a-axis oriented film grows beautifully and these structures cover the whole substrate. It is most interesting that this film becomes superconducting or not superconducting as a result of varying the growing conditions.

As mentioned above, thin film technology seems to be still in the very early stages of development and it will take a few more years to be able to target new active devices such as transistors.

Even though we are in a difficult situation in the research and development of thin film technology, a new finding of the proximity effect in a cuprate multi-layer system[8] must be mentioned here. The Hitachi group obtained an interesting result in the $Ho_1Ba_2Cu_3O_7$-$La_{1.5}Ba_{1.5}Cu_3O_y$ system (Fig. 14). Here $Ho_1Ba_2Cu_3O_7$, the superconducting current flows between two electrodes of $Ho_1Ba_2Cu_3O_7$ across a narrow slit of $La_{1.5}Ba_{1.5}Cu_3O_y$. The width of the slit (L in Fig. 14) is of the order of 0.5μm. This may come from the proximity effect on the surface of $La_{1.5}Ba_{1.5}Cu_3O_y$

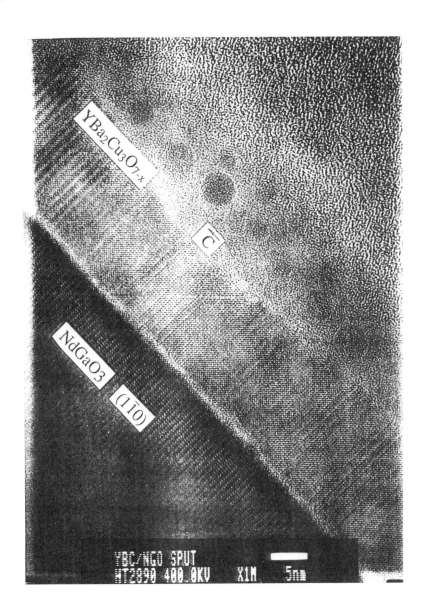

Fig. 13 TEM photograph of an a-axis oriented YBCO film on an NdGaO$_3$ substrate.

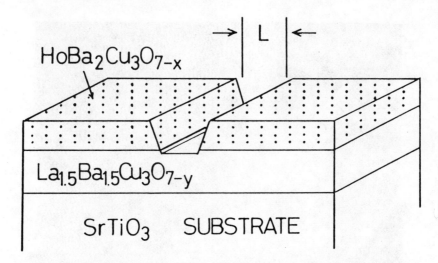

Fig. 14 A multi-layer of $Ho_1Ba_2Cu_3O_7$-$La_{1.5}Ba_{1.5}Cu_3O_y$-$Ho_1Ba_2Cu_3O_7$ system made by the Hitachi group.

crystal. Due to the complexity of a two-layered very fine structure of the system, this result is still not well established. But, if it is true, some applications to new devices may be expected.

As mentioned above, in the second phase of R&D in high-T_c superconductivity, we can see much pronounced progress, and it may be expected that in the third phase, the mid-90's, some small markets may appear. The presence of markets is very important for future developments. In the development of semiconductor technology, new technology opened new markets, and then this accelerated further development. In Section 4, I will discuss what kinds of markets can be expected in the field of high-T_c superconductivity.

3. SUBJECTS FOR FUTURE DEVELOPMENTS

In this section, several subjects for future development of high-T_c superconductivity and its applications will be discussed briefly.

3.1 Search for higher-T_c materials

This is the most puzzling problem at present. In these last five years, many kinds of superconducting materials were discovered, and most of them are oxides including copper as an element. Is it necessary to include copper in future high-T_c materials? Unfortunately we are facing very big difficulties in finding a direction of search for new materials, since the theory of high-T_c superconductivity has not yet been firmly established. We still do not know what kind of carriers to look for in these high-T_c cuprates. Furthermore, high-T_c materials are becoming more and more complicated. A Tl compound having the highest T_c at present includes five elements, and materials consisting of more than six elements have been investigated.

Thinking about how many samples have been prepared all over the world in the past five years (probably millions), the future of the search for higher-T_c materials might not be so bright. Nevertheless, we hopefully may still be able to find new materials by using high-pressure synthesis, low-temperature synthesis and so on. I personally expect that it is not impossible to reach 150K in the near future.

3.2 Introduction of pinning centers into cuprates

We have already succeeded in introducing very strong pinning centers into YBCO as mentioned above. But the introduction of pinning centers, effective at 77K, in other materials has not yet been achieved. It seems to be necessary that such pinning centers must have a size larger than 1μm in diameter as was the case for the 211-phase particle in YBCO. The Bi compound, which is suitable for making tapes, has strong two-dimensionality and this probably prevents the action of large three-dimensional pinning centers. As was shown in Fig. 7, the critical current in the Bi-compound tape reaches almost 10^4 A/cm^2 at 77K in 1T, and this indicates that the critical current can be increased beyond 10^5 A/cm^2 at 77K in 1T, if suitable pinning centers are introduced. Furtheremore, this value is very close to the value for practical applications.

3.3 Preparation of high quality thin films

As was mentioned in the proceeding section, preparations of actually high quality thin films are very important not only for future applications, but also for the basic understanding of high-T_c phenomena, since in most high-T_c cuprates it is very difficult to obtain very pure single crystals of finite size. In thin films, on the other hand, the hope to obtain films of pure single crystals still remains, as they are made in the very clean environment of very high vacuum. Furthermore, the characterization techniques of thin films have been very well developed such as STM, etc.

3.4 Basic research on surfaces and interfaces

At present, we have very little information on the surfaces and interfaces of high-T_c cuprates, except via tunneling phenomena. But, linking future applications to new devices, these investigations are inevitably important. We must remember the invention of transistors in 1947. At that time the knowledge of semiconductors was very limited. But Bardeen and Brattain investigated the surface properties of

Ge and made a very important discovery and invention. The situation in materials research in the high-T_c case at present is much worse compared with the case of semiconductor research in the 1940s. Therefore we must take a new approach to these difficult problems. It may be necessary to develop new equipment, probably very complicated and expensive, for *in situ* observations of many physical quantities during the crystal growing process.

3.5 Improvements of fine pattern lithography

The technology of fine pattern lithography has been developed very well in the semiconductor industry, and can now reach line widths of 0.1-0.2μm. In the case of high-T_c material research, the size is still limited to the order of 1μm. Actually it is not clear at present whether it is necessary to develop such a fine lithography technology of the order of 0.1μm in the case of high-T_c materials in future device research, since future high-T_c devices will be developed under quite different concepts from that of semiconductor devices.

However, it should be possible to obtain some basic knowledge of high-T_c materials by making very thin wires, the width of which is in the submicron region, as the penetration depth in high-T_c cuprates is estimated to be about 0.2μm or so. Also the behavior of quantum magnetic flux can be understood more precisely in such thin wires. Rather it may be necessary to go further to make thin wires in the mesoscopic region as will be attempted in semiconductor technology, if we take into account the coherence length of the order of 10Å. Of course, we must remember that such investigations are closely related to research on surfaces and interfaces mentioned above.

4. EXPECTATIONS OF FUTURE APPLICATIONS

The possibility of applications of superconductivity has long been considered as was shown in Table 2. But as they have to use liquid helium, real applications have been very limited. Josephson devices and superconducting magnets have been used in laboratories and only MRI has been commercialized in the past few years.

After the discovery of high-T_c superconductivity, the situation has changed drastically, since these materials can operate at liquid nitrogen temperature. Furthermore, the development of some devices, magnetic levitation cars, superconducting generators and motors, and SMES, has been greatly accelerated with the hope of using liquid nitrogen, even though these applications still use low-T_c materials.

Here are some practical applications, which will be expected to appear in the market in the near future.

Table 2.
Some Applications of High Temperature Superconductors

---

(1) Transportation
- Mag Lev. Trains
- Spacecraft
- Electromagnetic Ships
- Electric Car

(2) Electricity
- Storage of Electricity
- Superconducting Generator and Motor
- Transmission Lines

(3) Electronics
- Superconducting Wiring in LSI
- One-Wafer Computer
- SQUID Devices
  - Infrared Sensor
  - Magnetic Sensor
- MRI
- Josephson Devices
- Superconducting Transistors
  - MOS Type
  - Bipolar Type

4.1 Microwave devices

Passive devices (like resonators, filters and frequency mixers) made from thin films of high-T_c materials, operating in the microwave range, have been developed very well. Taking into account the future market for mobile transmission systems, the needs for these devices will be increased more and more. Especially, it must be mentioned that the frequency range must be broadened and raised up into the millimeter wave region. Very recently it was shown that by using thin films of high-T_c cuprates many kinds of microwave devices can be made in a very compact size and they work well at 77K.

An example of a strip-line resonator for satellite communications, made in our laboratory, is shown in Fig. 15.

A more simplified 14 GHz straight-line end-coupled resonator and a Chebyshev-type band pass filter are shown in Fig. 16 and Fig. 17, respectively.[8] The Q of the resonator was 2800 and the insertion loss of the filter was surprisingly only 0.1 dB at 77K. Thus we expect that a very sensitive amplifier can be made by combining these superconducting devices and semiconducting HEMT devices, and it will find some market not only in space communication systems, but also in complex multiplex communication systems on the surface of the earth.

Fig. 15 An example of a strip-line resonator for satellite communications made in our laboratory.

4.2 Transportation

The impact of the revolution in mass transportation on society has been quite large in the past. Magnetically levitated linear motor cars with speeds up to 500 km/h may be the next revolutionary technology. In Japan, it has been in development since the 1960s and a test line of 42 km is going to be built near Tokyo.

The first manned ship with superconducting propulsion engines was built in Japan and is shown in Fig. 18. Even though the speed of this ship is not great, we hope that it can achieve higher speeds by some breakthrough in technology, especially the development of tapes or wires that can operate at magnetic fields of more than 15T.

14GHz Straight-line Resonator

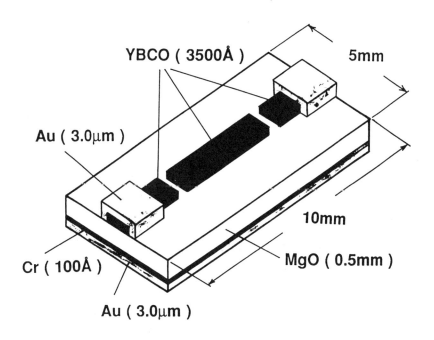

Fig. 16 A 14 GHz straight-line resonator.

4.3 Applications to electric power systems

The development of superconducting generators and superconducting magnetic energy storage systems was started in Japan as is shown in Table 3. Especially, the development of energy storage systems is becoming more and more important. This will be needed in the stabilization of the electric power supply system, and also will be needed as a heavy duty power supply for short periods in the steel industry. It may also be used as a power supply system for high speed trains such as MagLevs. The simplest energy storage system is a big superconducting coil in which magnetic energy is stored. In this system we have to develop very stable superconducting wires and a suitable superconducting switch. What kind of switch will be applicable? There seems to be no clear answer yet.

On the other hand, flywheel storage systems are becoming a very powerful candidate. During the past few years, flywheel systems have been tried for electrical

A Chebyshev Three-pole Band Pass Filter Implemented In Microstripline

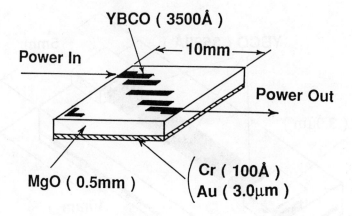

Fig. 17 A Chebyshev three-pole band pass filter.

Table 3.
Plans for Development of Superconducting Generators
and Superconducting Magnetic Energy Storage System in Japan

---

(1) Superconducting Generator
        70 MW
        Now in progress

(2) Superconducting Magnetic Energy Storage System
        (SMES)
        100 kW h
        Started in 1991

  Applications: (1) Stabilization of Electricity Power Supply
              (2) Steel Industry
              (3) MagLev Trains

(3) Energy Storage System with HTCS Fly Wheel

超電導電磁推進船ヤマト1号
(シップアンドオーシャン財団提供)

Fig. 18 The YAMATO-1. The first manned ship with a superconducting propulsion engine.

energy storage. The most famous one is the application to a big nuclear fusion device. But mechanical friction at the central cylindrical bearing prevented further developments. Later, a flywheel system using magnetic levitation and conventional magnets was proposed as a frictionless system.

Now the very strong magnetic levitation force due to flux pinning in YBCO was found, as was mentioned before. A small model of a frictionless energy storage system using MPMG-YBCO at 77K was made in our laboratory as shown in Figs. 19(a) and (b).[9] It must be strongly emphasized that this system is very stable without any position control system. Based on this model, systems having storage power of several amounts were designed as shown in Table 4, and it was shown that a flywheel system of 4m in diameter can have a storage power of 1 MW h. We hope further developments of this system can be made in the very near future.

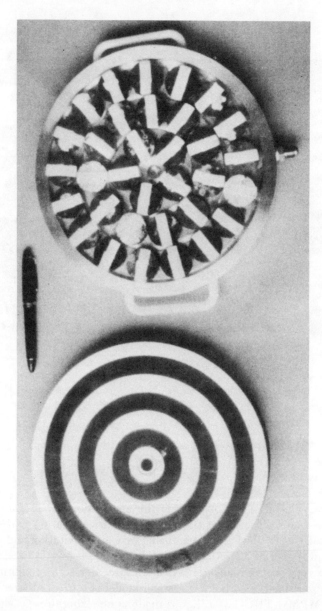

(a)

Fig. 19 (a) and (b) A small model of a frictionless energy storage system using MPMG-YBCO at 77K made in our laboratory.

(b)

Fig. 19 (cont.)

Table 4.
Specifications of an HTCS FlyWheel

Items	1 kW h-class	10 kW h-class	1 MW h-class
Diameter	1 m	2 m	4 m
Thickness	2 cm	5 cm	10 cm × 10
Weight	120 kg	1.2 ton	100 ton
Number of MPMG elements	40	400	30 000
Rotating speed	7200 rpm	3600 rpm	1800 rpm
Stored energy	1.2 kW h	12 kW h	1000 kW h
Energy density	10 W h/kg	10 W h/kg	10 W h/kg

4.4 Magnetic bearings

As an application of the strong magnetic levitation force in MPMG-YBCO, the frictionless magnetic bearing is expected to be very important. This type of bearing can be applied to many kinds of industrial fields. Ultra high speed centrifuges are just one application in the biosciences and biotechnologies. Dust free fabrication machines in semiconductor production lines may also be a big application. We have already succeeded in making a 2.4 kG frictionless magnetic bearing in our laboratory. It can be rotated at speeds of 30 000 rpm at 77K without any external position-control system (Fig. 20).

4.5 Future electronic devices

Since the discovery of high-T_c superconductivity, there has been a great hope to develop new superconducting devices at 77K. At the low temperature of 4.2K, a superconducting transistor using Nb or Pb on silicon has been tried as shown in Fig. 21. But in the field of high-T_c, only weak-coupled devices for SQUID's have been tried. It has shown some appreciable results at 77K, but the fabrication process is still in a very primitive stage.

The final aim of development must be the invention of new active devices operating at 77K, like transistors in the field of semiconductor electronics. Very high speed, very low noise and very low dissipation devices are the most necessary for future electronics.

Fig. 20 A 2.4 kG frictionless magnetic bearing made in our laboratory. It can be rotated at speeds of 30 000 rpm at 77K without any outside position-control system.

The way to such new active devices is still not obvious. Probably we will have to investigate the physics and chemistry of the material itself, in particular surfaces and boundaries, more deeply before obtaining new devices. Throughout the history of semiconductor electronics, we learned that we must be patient.

5. CONCLUDING REMARKS

Research on high-T_c superconductivity covers most of the fields of material science, and therefore, interdisciplinary investigations are necessary by scientists with diverse backgrounds in physics, chemistry, ceramics, metallurgy and so on. At present, after much research on the physical properties of materials, the creation of a theory of high-T_c superconductivity is extremely urgent.

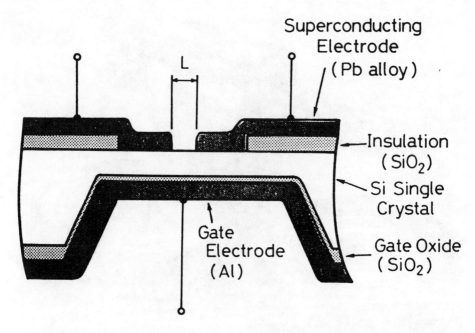

Fig. 21 A schematic cross section of a superconducting transistor using Pb on Si.

If a theory can be successfully established, its effects must be very wide and deep. Solid state physics may be transformed, and the search for new superconducting materials will be accelerated. Furthermore, many applications will be greatly advanced by understanding the phenomena of high-T_c materials, and especially concepts for new electronic devices may be forthcoming.

In the past, interactions between science and technology have been very clear. They sometimes resonate with each other and exhibit rapid progress in a very short period and give a big impact on society. The research and developments of high-T_c superconductivity will hopefully retrace the brilliant history of the great success of the science and technology of semiconductors in the near future. We are very optimistic about this.

ACKNOWLEDGMENTS

I thank all of my colleagues who are working at SRL-ISTEC. Their experimental results excite me greatly and discussions with them are quite fruitful. I also appreciate deeply the membership companies of ISTEC, the Ministry of International Trade and Industry, and New Energy Development Organization for their strong support of our research on superconductivity.

REFERENCES

1. T. Kaneko, H. Yamauchi and S.Tanaka, "Zero-resistance temperature of Tl-based '2223' superconductor increased to 127K," Physica C, Vol. 178 (1991) 377.
2. M. Murakami, S. Gotoh, H. Fujimoto, K. Yamaguchi, N. Koshizuka and S. Tanaka, Supercond. Sci. Technol. 4 (1991) S43.
3. K. Tatsuhara, S. Miura, M. Murakami, N. Koshuka and S. Tanaka, Proc. 2nd ISPP symposium (1991, Elsevier) in press.
4. N. Ogawa, I. Hirabayashi and S. Tanaka, Physica C **177** (1991) 101.
5. By the courtesy of Sumitomo Electric Co.
6. Submitted to Proceedings of 4th International Symposium on Superconductivity (ISS'91), Tokyo.
7. N. Homma, S. Okayama, H. Takahashi, I. Yoshida, T. Morishita and S. Tanaka, Appl. Phys. Lett. **59** (1991) 1383.
8. Submitted to Proceedings of 4th International Symposium on Superconductivity (ISS'91), Tokyo.
9. H. Fukuyama, K. Seki, T. Takizawa, S. Aihara, M. Murakami, H.Takaichi and S. Tanaka, Proc. ISS 91 (1992, Springer-Verlag) in press.

DISCUSSION

M. P. Maley: The results you showed for the BSCCO tape were with the field in the ab plane, parallel to the tape surface. But the anisotropy at 77 K would be orders of magnitude. Do you know what the values are for the field parallel to the c axis?

S. Tanaka: I don't know the tape making [unintelligible], so I can't say anything about that. But anyway, in the sheets, this was compounded just like that, parallel to the ab plane.

F. M. Mueller: In the sequence of applications, are you in Japan considering reexamination of the old classes of superconductors to see if, with modern materials science, they can be made better for some of these applications, in addition to searching for new materials and developing the existing high-temperature materials?

S. Tanaka: Which types of superconductors, niobium-titanium?

F. M. Mueller: The A-15's, the Chevrel phases, and so on.

S. Tanaka: Yes, some people have come to have an interest especially in niobium-tin, and similar materials.

F. M. Mueller: I was particularly thinking of the microwave applications that you were showing.

S. Tanaka: Yes, but you see niobium nitride is a very good material for making the Josephson junctions, so that Japan established how to use it to make Josephson junctions. But the need is not so much a single junction you see, but I am saying to the government, we must try to make a Josephson computer, even if it is very small, however still no result.

A. P. Malozemoff: Can you just confirm, are those two power applications, the superconducting generator and the SMES project in particular, low-temperature or high-temperature superconducting projects?

S. Tanaka: They are going to use niobium-tin, or something like that. But if we succeed to obtain the proper high-T_c materials, then they can change.

V. G. Kogan: Do you have a special reason to call the proximity effect you have observed anomalous?

S. Tanaka: In high-T_c materials the coherence lengths are 20-30 Å, so we expect that in oxide materials the proximity effect must be small, but 500 Å falloff length is very large, it's anomalous, I think.

J. R. Schrieffer: I would think that it has not to do with the superconductor but the material in between, which has its own effective coherence length regardless of what's driving it. So I would be rather surprised if in fact the falloff length in general were very short, if it were a material that gave a long effective coherence length when you put a low-T_c material in contact with it.

S. Tanaka: Yes.

J. R. Schrieffer: I think this is a confusion in the literature which identifies the proximity falloff length with the coherence length of the adjacent superconductor. This is incorrect.

P. L. Gammel: I agree. What's still surprising is that the materials they're putting in have very short normal-metal coherence lengths.

J. R. Schrieffer: How short?

P. L. Gammel: Five-ten Å, I believe.

J. R. Schrieffer: Well, the statement is theoretically true. [Laughter]

P. L. Gammel: Even the best materials, like indium arsenide, have normal-metal coherence lengths which are only 1000 Å, or a few thousand Å, at 4 K.

V. G. Kogan: But it depends on temperature. The lower in temperature it goes, the longer the length, so . . .

P.L. Gammel: Yes, but these are at 4 K, and the temperature dependence comes from the mobility, which is very weak in these strongly disordered materials.

A. Kapitulnik: There is still a possibility that the lanthanum-barium underneath, which was with a stoichiometry of 3-3-6, during the process, could have been superconducting. This material was studied in the past. It is not a stoichiometry which is strictly insulating.

Aharon Kapitulnik
Stanford University
Stanford, CA 94305

Dynamics of the Vortex State in High Temperature Superconductors

The large thermal energy available, the strong anisotropy, and short coherence lengths of high temperature superconductors give rise to new phenomena in the mixed state. We discuss transport and thermodynamic measurements of high-Tc materials and of model systems. In particular, we use experiments on two-dimensional films to compare and isolate two-dimensional effects in the cuprates. By using multilayer systems with similar parameters, we identify decoupling of the superconducting planes in magnetic fields at temperatures much above the irreversibility line. We show that if the irreversibility line is to be considered a melting transition line, it implies melting of the solid state into a liquid of three-dimensional flux lines. We further use Monte Carlo simulations to study the structure of the vortex state as well as melting.

1. INTRODUCTION

The nature of the resistive transition in magnetic fields and the role of the vortex structure and motion in causing dissipation have been major unanswered questions required for the understanding of the phenomenology of high temperature superconductors (HTSC).

The substantial broadening of the resistive transition in an applied magnetic field[1-3] was only observed for thin films of ordinary superconductors,[4-6] hinting at quasi-two-dimensional behavior. A general phenomenology based on giant flux creep was proposed[7-9] to explain these results as well as magnetization decay results. Recent experiments have shown, however, that this scheme does not provide a universal explanation for the anomalous magnetic behavior. Magnetization measurements on $Bi_{2-x}Pb_xSr_2Cu_3O_{10}$[10] have shown that the power law predicted by this model for the irreversibility field, $H_i \propto (T-T_c)^{3/2}$, is followed in this oxide only very near T_c and that instead H_i fits an exponential temperature dependence at lower temperatures. Furthermore, for a set of crystals of $Bi_2Sr_2CaCu_2O_{8+\delta}$ (BSCCO) with a wide range of oxygen stoichiometries, a wide range of exponents were measured indicating their non-universality. Transport measurements on $La_{2-x}Sr_xCuO_4$ single crystals have revealed that the broadening of the resistive transition under applied fields depends only on the direction of the field and not on the direction of the current I, and that it is largest when the field is applied parallel to the c axis.[11] As pointed out by Kitazawa et al.,[11] the considerable broadening of the resistive transition in a geometry where no Lorentz force is applied to the vortices (H ∥ I) rules out an explanation in terms of a flux creep model. Finally, the resistive transition measured along the c-axis of BSCCO crystals shows a continuation of the increasing normal state's increasing resistance well below the ab-plane T_c, before the resistance drops at a much lower temperature.

While there are many competing models for the ab-plane vortex dynamics,[12] experimentally all data are similar. Most of the reported resistive transitions on HTSC exhibit three regimes. As the temperature is lowered below the superconducting transition temperature, a fluctuation dominated drop in the resistance is followed by a weak decrease in resistance, sometimes accompanied by a "knee" or a "hump" and as the temperature is lowered further, the transition ends with a tail due to activated resistance. A typical set of such transitions for a BSCCO single crystal is shown in Fig. 1.[13] (For the procedure of sample preparation see Appendix A). In this paper we will concentrate on the lower part of the transition. We will discuss the emergence of the knee and the nature of the tail of the transition.

HTSC are complicated materials and the fabrication and study of a single, well characterized Cu-O layer is almost impossible. Thus, to get more insight into the importance of two-dimensionality, the short coherence length and the weak interplane coupling, it is instructive to use model systems. We will try to isolate single layer effects by studying superconducting properties of thin films with parameters similar to those of isolated Cu-O layers in the cuprates. We will further show results from measurements on artificially fabricated multilayers with parameters such as coherence lengths, penetration depths and in particular anisotropy that are similar to the oxides. In this review we will emphasize the very weakly coupled systems like $Bi_2Sr_2CaCu_2O_8$. To illustrate the utility of studying model systems, we show in Fig. 2 a series of superconducting transitions for MoGe/Ge multilayer of thicknesses of 60Å and 65Å for the superconducting MoGe and for the insulating Ge respectively.[14] A mass ratio of ~ 500 was found for this system as will be discussed

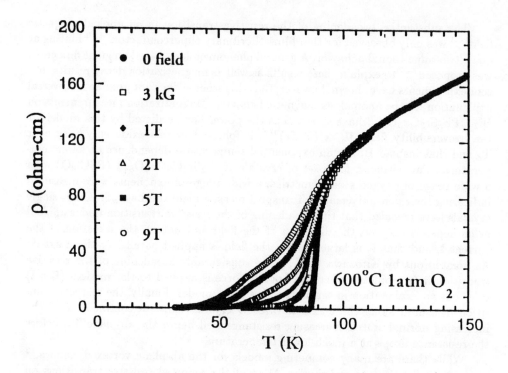

Fig. 1 ab-plane resistive transitions in various perpendicular magnetic fields for a sample annealed in oxygen.[13]

below. It is easy to see the similarity between the results in Figs. 1 and 2. Both systems show the emergence of a kink in the superconducting transition with increasing magnetic field and a tail that extends to very low temperatures. The tail that is seen in all resistive transitions of HTSC as well as for this film is explained in the framework of thermally activated flux motion (flux creep) theory.[3,8,9] In its simplest version by Anderson and Kim,[15] a local Abrikosov lattice is established on length scales λ (a "flux bundle") and can jump between pinning sites in an activated way. If melting of the flux lattice in 2D occurs at T_m, vortex motion is not correlated for $T>T_m$ and individual vortices instead of flux bundles should be considered.

Much work has been devoted to relating the irreversibility in magnetic properties observed in HTSC materials to flux creep.[3] For these models, flux creep determines a crossover temperature for each type of dynamic measurement, implying that no real flux lattice melting is observed in these materials. The irreversibility line is therefore a kinetic line. We will make a few remarks on the irreversibility line at the end of the paper and put thermodynamic constraints on its interpretation as

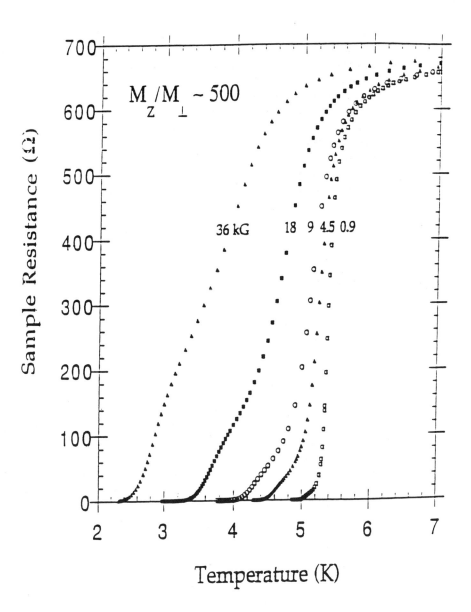

Fig. 2 Sample resistance as a function of temperature for the weakly coupled ($d_i = 65$ Å) MoGe/Ge multilayer sample in different magnetic fields, each perpendicular to the superconducting planes.[14]

a phase transition line. We will also discuss flux creep effects in 2D films and relate them to what is observed in HTSC, displaying the two-dimensional nature of these materials. To illustrate our arguments about the structure of the vortices, we will use the results of Monte Carlo simulations in the vortex state, establishing a phase diagram for the case with no pinning.

In any such study, materials preparation and characterization is an essential aspect that has to be discussed. Thus, we include at the end of the paper an appendix that discusses the materials used in this study.

The organization of the paper will follow the above outline. We first will describe the magnetic properties of a single layer of MoGe where the emphasis here will be on the onset of dissipation or melting. We then will discuss results on the multilayers and compare them to the HTSC crystals.

2. MODEL SYSTEMS

2.1 Melting in Two-Dimensional Films

When a magnetic field is applied to a two-dimensional superconductor below the vortex-antivortex unbinding temperature, $T_{KT} < T_c$, vortices of one type are introduced with vorticity which depends on the direction of the applied magnetic field H (we can assume $H_{c1}=0$). At a temperature below $T_m<T_{KT}$, in the absence of disorder, a triangular lattice will be formed that has a well defined shear modulus. Melting of the vortex lattice occurs[16,17] via a Kosterlitz-Thouless (KT)-type transition[18] where the lattice becomes unstable to the formation of dislocations above a temperature T_m. T_m is calculated to be independent of the magnetic field for fields much smaller than the upper critical field H_{c2}.[17] Since in the universal theory of KT, T_{KT} is determined by the "charge" of the objects only, we expect a melting temperature independent of the magnetic field. This is because the magnetic field determines the lattice constant of the triangular lattice while the "charge" of the dislocation is determined by the type of lattice only. This simple approach that relies on the shear modulus only was corrected by Fisher[17] who introduced a shear modulus which is renormalized by nonlinear lattice vibrations and defects.

To study melting in MoGe thin films, we have performed mutual inductance measurements on a 60Å thin film.[19] This type of film will be shown below to be the basic building block of the multilayer systems,[14] thus providing us with the opportunity to isolate single layer effects and extract the fundamental properties due to interlayer coupling. The technique of mutual inductance has been used successfully by several groups to investigate the properties of thin superconducting films.[20,21] The work we are reporting here[19] uses a configuration consisting of a drive coil and an astatically wound pickup coil placed concentrically on the same side of the film. Briefly, the measurement involves driving an oscillating current

Dynamics of the Vortex State in High Temperature Superconductors

Briefly, the measurement involves driving an oscillating current through the drive coil, which in turn induces shielding currents in the film. These shielding currents then induce an emf in the lower half of the pickup coil. From a measurement of the components of the induced emf that are in phase with the driving signal and in quadrature, we can, using simple linear response theory, calculate the complex conductance, G, of the thin film. The gradiometer configuration of the pickup coil ensures that the drive signal, as well as any constant background noise, is rejected. In addition, the zero field measurements were performed in a shielded dewar to reduce ambient magnetic fields to just a few mG. A typical measurement at zero magnetic field is shown in Fig. 3. Note that the inductive signal starts to increase after the resistive transition shows almost zero resistance, indicating the great sensitivity of the technique to the penetration depth being larger than the sample thickness. From measurements of G we can extract superfluid density.[20]

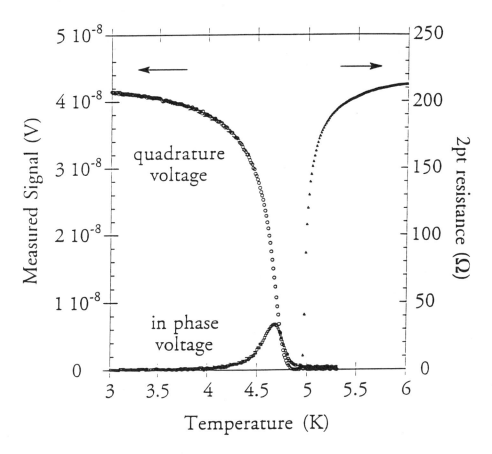

Fig. 3 In-phase and quadrature signals of a mutual inductance measurement of a 60Å MoGe thin film. Note that the signals begin to rise at the same temperature that the 2 point resistance becomes negligible.[22]

By applying a d.c. bias current while performing the measurement we identify a "critical current" above which flux flow dominates the dissipation.[22] For bias currents less than this critical current vortices are still confined to their potential wells, while above it, dissipation dominates the response. This new experimental procedure provides a unique opportunity for investigating vortex dynamics in regions of the H-T plane not easily accessible by other more traditional methods. Our results on MoGe show evidence of a pinning transition at temperatures well below the resistive transition. The films have a critical temperature, determined resistively, of approximately 4.95K. From previous measurements on identical films, the bulk zero temperature magnetic penetration depth is known to be approximately $\lambda(0) = 7700$Å,[23] yielding a perpendicular penetration depth $\lambda_\perp = \lambda^2(0)/d_s = 100\mu$m for a film of thickness $d_s = 60$Å. $\lambda(0)$ was also measured independently as will be discussed below. The Ginzburg-Landau coherence length, as determined by upper critical field slope, is approximately $\xi(0)=55$Å, thus implying strongly type II films. Also, since the film thickness, d_s, is much less than the penetration depth and on the order of ξ, we can assume two-dimensional behavior. R/sqr was measured to be $\approx 250\Omega$, implying a mean free path, $l \approx 3$Å $\ll \xi(0)$. Thus, these films are very dirty superconductors with very large $\kappa \approx 140$. The resistive transition of thin MoGe has been studied previously by Graybeal et al.[6] The zero temperature critical current, as measured resistively, is of the order 10^4 A/cm^2, indicating moderate flux pinning. To apply a uniform d.c. bias current to the film during measurements or to simply measure the films' resistance, we deposited thin gold strips along two opposite edges.

As explained above, applying a d.c. bias current allowed us to identify a critical current for vortex flow. For currents above this critical current, dissipation appears as the inductive signal is reduced. In this way, we were able to apply a known Lorentz force to the vortices while measuring their dynamics. By closely monitoring both the point at which the critical current vanishes and the peak position of the dissipative signal in the mutual inductance measurement, we found that they almost coincide with each other at high magnetic fields but depart from each other at low fields (see inset in Fig. 4). Fig. 4 shows the phase diagram as measured for different frequencies. The fact that the critical current disappears roughly where the peak position is for the high fields indicates that the mutual inductance technique at these fields indeed measures the melting of the vortex lattice only. At higher temperatures and lower fields the kinetic inductance still has a significant temperature dependence which makes the critical current and the dissipation peak position measure different effects.

To compare our result to the theories of Huberman and Doniach[16] and of Fisher,[17] we need to extrapolate to zero frequency. Based on the calculations by Ambegaokar et al.,[24] we extract an $\omega=0$ melting temperature by taking field cuts of $T_m(\omega, H)$ and using the fact that the finite frequency introduces a finite cutoff length above which the motion of the vortex is diffusive. This length: $r_\omega = \sqrt{pD/\omega}$ (D being the dislocation diffusion constant), plays the cutoff role similar to $\lambda_\perp$ for

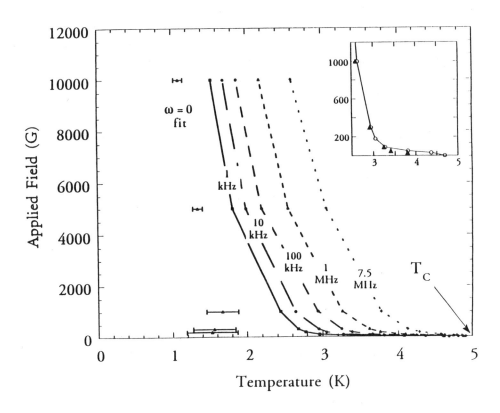

Fig. 4 Mutual inductance dissipation peak location at various measurement frequencies. Bars denote the zero-frequency extrapolation. Inset shows the critical current (see text) together with peak position for 10 kHz.[19]

zero frequency. p is a constant of order 10. Thus, $T_m(\omega)$ is determined from the equation $\xi_D[T_m(\omega)] = r_\omega$, where $\xi_D(T) = a_c \exp\{b(T/T_m-1)^{-\nu}\}$ is the dislocation-antidislocation correlation length. a_c is roughly the core radius and b is a numerical constant. The calculated T_m is found to be in the range 1.7K to 2.9K depending on non-universal constants.[17] Fig. 4 also depicts the zero-frequency extrapolated T_m using the above procedure and $\nu = 1/2$. The choice of this "mean field" ν is used because all $T_m(\omega)$ are well above the extrapolated T_m and hence the renormalized $\nu \cong 0.37$ is not applicable. Note that although $T_m(\omega,H)$ displays strong temperature dependence, the extrapolated result is much less temperature-dependent and in fact shows a tendency to straighten up. Its average value of $T_m \sim 1.2$K is somewhat lower than the predicted temperature but since T_m depends on many corrections and in particular the existence of disorder, this result is quite striking. The full fitting procedure had $T_m(H)$, b and D as adjustable parameters where b is expected to be a constant similar for all fields and D is not expected to depend strongly on the field. We find b $\cong$ 0.02 and $D/\xi^2 \cong 1.10^{12}$. Theory does not give a value for b, but it is expected to be of order unity. The value of $[D/\xi^2]^{-1}$, on the other hand, gives a characteristic time scale for diffusion of dislocations in the system. The value obtained experimentally, which is rather high, is discussed in Ref. 19.

2.2 Flux Creep in Two-Dimensional Films

At finite temperatures, there is always a finite probability for flux creep. Below the melting temperature, flux bundles will be the activated objects.[15] Above T_m, vortex motion is not correlated and individual vortices instead of flux bundles should be considered. This situation will be realized in 2D as well as in the very anisotropic HTSC where the flux lattice is melted in most of the phase diagram. For a temperature T and a current density J applied to the sample which is placed in a perpendicular magnetic field H, the flux creep resistivity can be calculated to be:

$$\rho = \frac{E}{J} = \frac{2f_0 H^2 V L^2}{c^2 kT} \exp\left(-\frac{F_0(T,H)}{kT}\right) \quad (1)$$

E is the electric field, f_0 is some attempt frequency for flux jump, V is a volume of a flux bundle, L is a characteristic hopping distance, c is the speed of light and F_0 is the activation energy.[15] The equation described above can be applied only for low current densities and low-enough temperatures: $F_0 \gg kT$.

In the most general case,[25,5]

$$F_0(T,H) = U(T,H) - TS_{eff}(T,H) \quad (2)$$

where U(T,H) is the barrier energy, and S_{eff} is the effective entropy. S_{eff} is related to the curvature of the pinning potential at the local maxima,[26] but its exact value cannot be known unless detailed knowledge of the pinning potential is available. In the Anderson and Kim model $F_0 \approx U = p\frac{H_c^2}{8\pi}\xi^3$, where H_c is the thermodynamic critical field, ξ is the coherence length and p is the effectiveness factor, typically a small number. This model assumes core pinning in a bulk superconductor with typical size ξ^3. It was proposed that as the field increases and the intervortex distance $a_H > \lambda$, collective effects become important, and hence ξ^3 is replaced by $a_H^2\xi$.[7] At low temperatures, F_0 is usually large for low T_c bulk superconductors and hence the flux creep regime is not easily accessible. Very near T_c, the theory fails because F_0 becomes small due to the temperature dependence of the activation energy.

In the case of two-dimensional superconducting films of thickness d_s, ξ^3 should be replaced by $\pi\xi^2 d_s$ to represent the disc-like shape of the 2D vortices.[5] In addition, the condensation energy should be replaced by the line energy $\varepsilon(d_s)$ to include both core and magnetic energy.[27] Due to the exponential nature of the flux creep phenomenon, it can be measured only when F_0/kT is not too large. In 2D films, the line energy is proportional to the thickness of the film and hence can be arbitrarily small. In the cuprates, this effect combines with the fact that kT is a large number to make flux creep become very important.

Flux creep was studied extensively in ultrathin single crystal Nb films[5] and in thin MoGe films.[6] In both cases, the low-temperature part of the resistive transition showed pure activation ($\ln R \propto -U/kT$ + const.; see Fig. 5 for typical results on Nb films); thus U must either be temperature independent or contain at most a linear correction, i.e. $U(T,H) = U_0(H) - c(H)T$. Since the activation energy is proportional to the core energy plus magnetic energy and both have the same temperature dependence in 2D , $U(T) \propto (\Phi_0/4\pi\lambda)^2 \cdot d_s \propto (1-t)$ near T_c. Thus, the

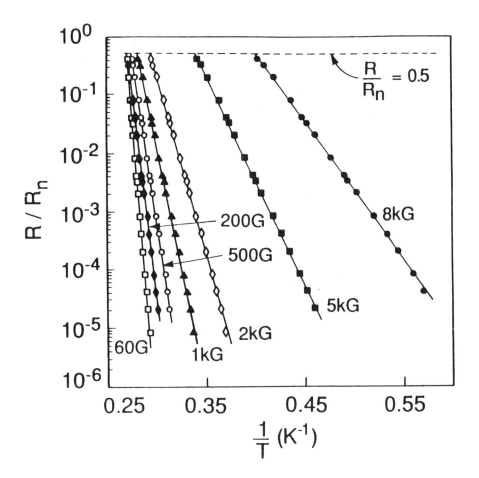

Fig. 5 Normalized sheet resistance (logarithmic scale) vs. 1/T for different applied magnetic fields measured on 20Å Nb film. The dashed horizontal line indicates the midpoint of the transition. The solid lines are fit to Eq. 3.[5]

const. in lnR comes from both the activation energy and the entropy (denote it by K(H)). Therefore, the thermally activated part of the resistive transition can be fitted with:

$$R(H) = R_n f(H) e^{-\frac{U_0(H)}{kT} + K(H)} \qquad (3)$$

f(H) is chosen to give the correct limit in the flux flow regime, i.e. H/H_{c2}. The existence of K(H) is the main result that shows up in all 2D superconducting films.[4,5,6] Using Eq. (3), Hsu and Kapitulnik[5] fitted the resistive transitions of single-crystal Nb films. They have found in the middle field range a power law dependence $U_0(H) \propto H^{-1/3}$, where for stronger fields corresponding to the characteristic pinning distance for their films, a stronger field dependence was found. Graybeal et al.[6] have measured the activation energy in a single layer of MoGe, similar to the one used for the flux lattice melting study. They have found $U_0(H) \propto H^{-2/3}$. We will come back to the field dependence of the activation energy found in this paper when we compare 2D films to flux creep effects in BSCCO crystals.

2.3 Multilayers

The next step was to layer the MoGe films with insulating layers in between.[14] A good choice for the insulator is to use Ge. Varying the Ge thickness (d_i) controls the coupling between the layers. Multilayers were made with d_i varying from 65Å down to no insulating layer at all. The superconducting layer was kept constant with $d_s = 60$Å.

Since the normal state resistivity of MoGe is temperature independent at low temperatures, we may easily extract the fluctuation conductivity above T_c. For each of our samples, we can fit these data with the fluctuation conductivity calculated using the Lawrence-Doniach model[28] considering only the Azlamazov-Larkin contribution[29] (this is justified because of the strong disorder in the films). For the more strongly coupled multilayer with $d_i = 35$ Å, the fit yields $\xi_z(0) = 12.3$ Å. For the weakly coupled system with $d_i = 65$ Å this fit is consistent with a perpendicular coherence length of $\xi_z(0) = 2.5$ Å, the value extrapolated from the more strongly coupled multilayer using the difference in insulator thicknesses and the measured tunneling length of 8.1 Å in amorphous Ge. But ξ_z obviously cannot be determined with great accuracy when the interlayer coupling is so weak. A Ginzburg-Landau mass ratio can be easily derived.[28] We find $M_z/M_\perp \approx 20$ for the sample with $d_i = 35$ Å and $M_z/M_\perp \approx 500$ for the sample with $d_i = 65$. The BCS T_c for these layers also changes, being 6.4K for a thick MoGe film (unity mass ratio), 5.44 for $d_i = 35$Å, 5.30 for $d_i = 65$Å, and 5.08 for a single layer (infinite mass ratio). The fluctuation conductivity of the single layer sample has 2D Azlamazov-Larkin form. All results were checked to confirm independence of measurement current density and frequency below 10kHz. The current densities used were typically of the order of 1A/cm². The resistive transition in a perpendicular magnetic field of one of the multilayer samples is shown in Fig. 2. Unlike bulk MoGe, the resistive transitions of this multilayer sample broaden substantially as the applied field is increased. Note that at the lower part of the transition a kink is developed, in a similar way to what is observed in HTSC. The significance of this kink will be discussed later.

When the resistive transition is plotted Arrheniusly, the kink is manifested in the data as a break below which an activated behavior is recovered (Fig. 6). It is clear that for a given field the steepness of the curves of the strongly coupled multilayers is consistent with a much larger activated energy compared with the more weakly coupled sample. This difference resembles the difference in resistance curves between YBaCuO and BiSrCaCuO samples,[30] as shown in Fig. 7. The resistive transition of Fig. 1 (BSCCO sample) is also plotted Arrheniusly with the break apparent for this system as well.[13]

To study further the importance of the kink, we compare the transition tail for a single layer and the multilayers for a constant field. It is evident that the break indicates the point where the multilayer's resistance departs from that of the single layer (Fig. 8). Therefore, $T^*(H)$ marks a decoupling temperature such that for $T>T^*(H)$ dissipation is governed by vortex "disks" in the single layer and for $T<T^*(H)$, the disks become correlated to form three-dimensional serpentine like vortices. Since at low temperatures the vortices are presumably straight we envision this decoupling temperature as the limit of how wiggly the vortex can be before breaking down to its constituents. Fig. 9 depicts $T^*(H)$ for the sample with $M_z/M_\perp \approx 500$. As expected, $H(T^*)$ deviates from $H_{c2}(T)$ towards much lower fields. Nevertheless, it is not an irreversibility line. This is supported by the fact that the resistance is linear and frequency-independent well below this line.

To study the irreversibility in the multilayer system, we have used the mutual inductance technique[22] as well as SQUID magnetization.[31] The two techniques yield a different irreversibility line as expected from the difference in the measurement frequency. As discussed above with respect to the single layer measurements, the peak position in the mutual inductance measurements is an indication of melting. To use a more general concept, it marks the irreversibility line for the particular measurement frequency (1kHz to 1MHz were used). We mark in Fig. 9 the portion of the dissipation peak in the mutual inductance measurement for a 10kHz measurement frequency. We have also measured the irreversibility line at zero (or rather very low) frequency using a SQUID magnetometer. As expected, T^* was reduced below our capability to measure. Up to $\sim$ 100G we found a shift down of about 0.5 K which will shift the irreversibility line at high fields down to $\sim$ 1.7 K.

Turning now to the flux creep behavior, there are differences between the single layer and the multilayers. The weakly coupled multilayer has a well defined activation energy which is much larger than that of the single layer and of characteristically different field dependence, typically $U_0(H) \propto H^{-0.54}$. The more strongly coupled multilayer displays a much stronger downward curvature, with no simple activation fit possible. We will come back to the origin of the knee in the resistive transitions as well as the nature of the flux creep field dependance when we discuss measurements on BiSrCaCuO crystals.

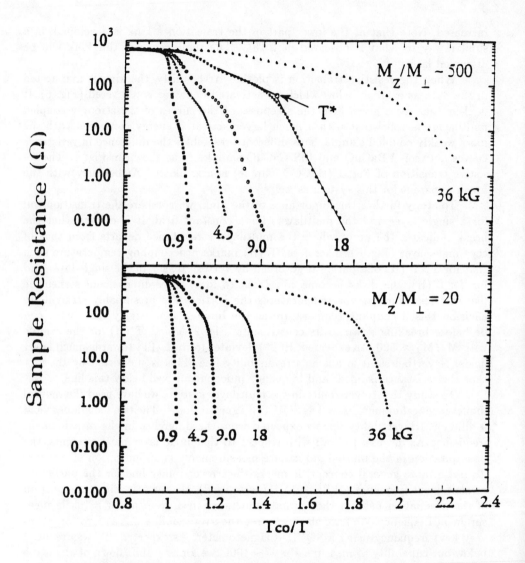

Fig. 6 Arrhenius plots of the sample resistance of two MoGe/Ge multilayers. In the upper plot, of the weakly coupled ($d_i = 65$ Å) multilayer, the slope of each trace defines a roughly constant activation energy at low temperatures. The lower plot, of the more strongly coupled multilayer ($d_i = 35$ Å), shows too much curvature to define a temperature independent activation energy. The upper plot shows an example of the construction which is used to determine $T^*(H)$.[14]

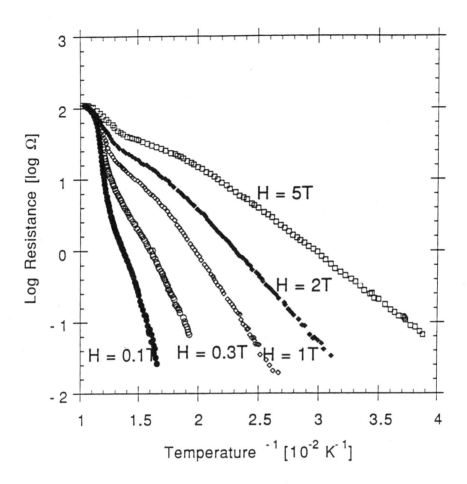

Fig. 7 Arrhenius plots of the resistance of oxygen annealed $Bi_2Sr_2Cu_2O_8$ single crystal shown in Fig. 1.[13]

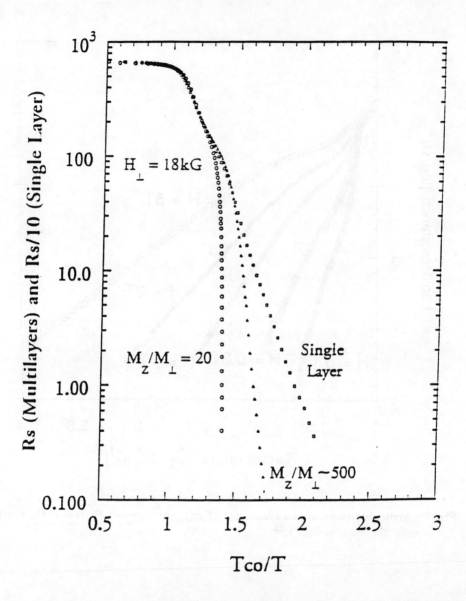

Fig. 8 Arrhenius plots of the sample resistance (scaled by the number of layers in each sample) in the single layer and both multilayers. Note that the vertical axis is normalized only by the number of layers. The inset shows the lines defined in the H-T plane by $T^*(H)$ of the two multilayer samples.[14]

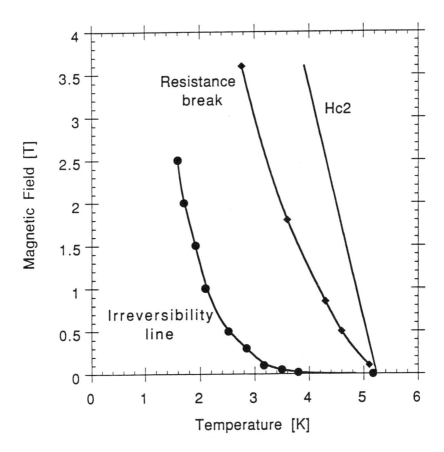

Fig. 9 Phase diagram for MoGe/Ge multilayers displaying H_{c2}, the resistance break ($T^*(H)$) and the irreversibility line determined by mutual inductance measurements.

3. $Bi_2Sr_2CaCu_2O_8$ SINGLE CRYSTALS

3.1 Dissipation in the ab-Plane

While there are many competing models for the ab-plane vortex dynamics, several reported data on $Bi_2Sr_2CaCu_2O_8$[2] and $Tl_2Ba_2CaCu_2O_8$[32] display a prominent excess dissipation at high fields in the form of a resistive "knee" or "hump" at temperatures between 50 to 80 K that, due to the intermediate temperature range involved, is difficult to explain with current models. The most used model of a crossover from flux flow to flux creep[3] (as the temperature is lowered) is difficult to reconcile with the data presented in Fig. 10. We show the resistivity vs. temperature, in fields up to 10.5 T,[33] of a sample annealed at 500°C in a 0.1% H_2/Ar reducing atmosphere.[34] $\rho_n(T)$ is linear, metallic, and of moderately low resistivity, showing that, in the normal state, c-axis contribution to the conduction is negligible. Superconducting fluctuations are evident above 90K at all fields, indicating that the onset temperatures change little with field, consistent with this family of materials. The zero-field transition has no anomalies.[34,35] Upon increasing field, however, a striking resistive peak evolves continuously with field between 55 K and 80K, developing from a "knee" below 2T to a peak above 2T, and exceeding $\rho_n(T_c)$ around 6T. This data strongly suggests that the development of knees in the same temperature range in Refs. 2 and 32 are manifestations of the same phenomenon only with a larger characteristic field.

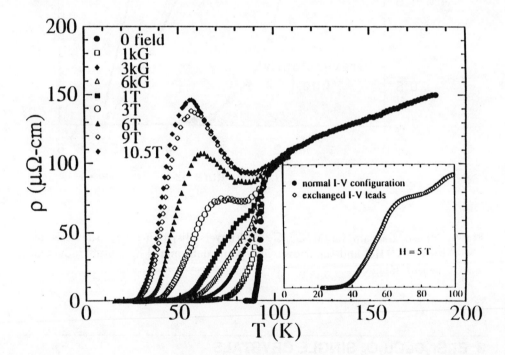

Fig. 10   ab-plane resistive transitions in various perpendicular magnetic fields for a sample annealed in 0.1% H_2/Ar, showing the evolution of the resistance peak. Inset: two sets of data on a similar sample with roles of the current and voltage contacts exchanged, showing the independence of the data from possible contact problems.[33]

In Fig. 11 we show the data at H=3T together with d.c. magnetization both field cooled (Meissner) and zero field cooled (Shielding). One clearly sees that the magnetization is reversible in the temperature range near the peak. Moreover, I-V characteristics show linear behavior with no frequency dependence. Both these results indicate that pinning does not dominate the vortex dynamics in this regime, similar to the behavior near the knee for the multilayers. Samples that showed the most pronounced peaks were annealed in a reducing atmosphere. Samples that were oxygen annealed[34] did not show a pronounced peak but rather only a knee similar to Fig. 1.

The reduced samples are characterized by a T_c of 92 K and the oxygenated samples by a T_c of 83 K. Both were measured by transport and magnetization. The crystals were cut into bars of trapezoidal cross section with typical dimensions

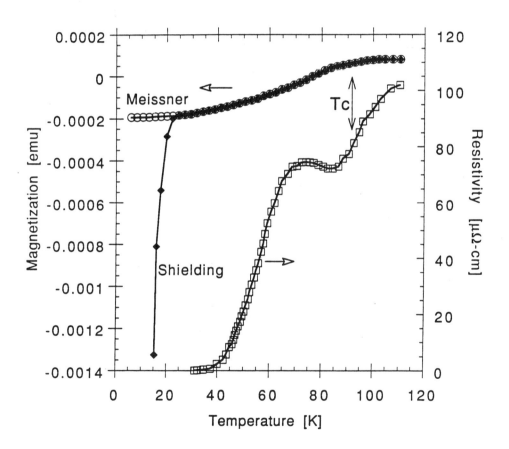

Fig. 11 Magnetization vs temperature at H=3T for the sample of Fig. 1 displayed with the resistive transition.[13]

5 mm × 0.5 mm × .025 mm, cleaved, and then four Au contacts were evaporated down the sides in a linear configuration. The data were carefully checked to be independent of exact contact configuration. Freedom from random contact problems was tested by exchanging the roles of the current and voltage leads on several samples and confirming that the data were unaffected (inset of Fig. 10). The current was directed along the ab-plane; both a.c. and d.c. biases were used with current density $J \leq 50$ A/cm^2, where the current-voltage relation was verified to be ohmic in the temperature range concerned. The magnetic field was oriented in the c-direction and varied from 0 to 10.5 T. We remark that the presence or absence of resistive peaks bore no correlation with the magnitude of ρ_n above T_c or with the $T = 0$ intercept extrapolated from high temperature.

3.2 C-axis Resistance in a Field

Before we continue with the analysis of the data, it is important to discuss the c-axis resistance when a magnetic field is applied along the same axis. Recently, Briceno et al.,[36] and Hess et al.[35] have shown that when the current and field are both in the c-axis direction, the measured resistivity show a maximum. To perform this measurement, four silver pads were evaporated on each side of the crystal surface (8 pads in total). Wires were attached to the pads using silver paint, and annealed at 200°C in O_2. We have measured the temperature dependence of the crystals' resistance in three contact configurations. In each configuration the current is applied through a pair of contacts that are aligned in different crystallographic orientation. The measurements were performed under the application of magnetic fields up to 7T. The direction of the magnetic field was always perpendicular to the ab-plane. In Fig. 12 a set of measurements in zero field and in a field of 1.2T is shown. The known broadening of the transition in the ab-plane is clearly seen. In the c-axis direction, the transition is also broadened but in addition the onset temperature (or the peak temperature) is dramatically reduced. Figure 13 shows a set of measurements in H=0 and H=4T. The broadening of the transition in 4T seems to be similar to the 1.2T one, but the "onset" of the c-direction transition is further reduced. Note that the ab-plane transitions show that the knee appears near the onset temperature of the c-axis transitions. This feature can be seen in both Figs. 12 and 13.

The main feature shown by Figs. 12 and 13 is that the onset temperature for superconductivity along the ab direction is not affected by the applied field, while along the c-axis it is clearly reduced. In fact, $d\rho_c(T)/dT$ measured under an applied field shows no discontinuity at the onset temperature measured in zero field.

Martin et al.[37] measured the perpendicular resistance in zero field and suggested that the behavior of ρ_c above T_c can be explained by either an exponential coupling between planes, or by some distribution of activated conducting paths between them. The effect of a possible application of magnetic field is not considered. In the recently published paper of Briceno et al.,[36] they suggest that a model

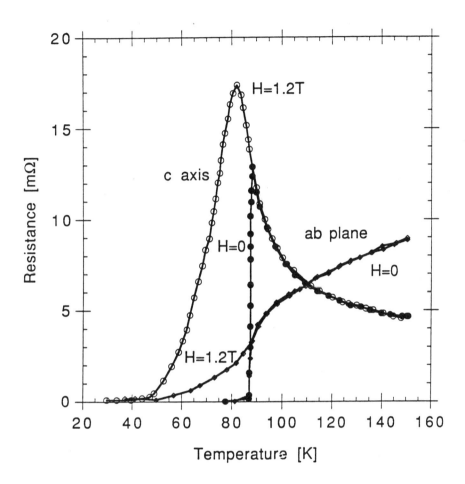

Fig. 12 Resistance along c-axis when the magnetic field is applied in the c direction for H = 0 and H = 1.2 T.[35]

in which a fluctuation-induced dissipation in the mixed state can give rise to the negative slope in ρ_c below T_c. This model cannot be used in the case of a magnetic field smaller than H_{c1} and clearly not for H=0. The above behavior will be interpreted below as due to decoupling of the layers due to the application of a magnetic field. A striking result is that the behavior of $\rho_c(T)$ and that of $\rho_{ab}(T)$ are related, as shown by the correlation between the "knee" in the ab-transition and the peak in the resistance measured along the c-axis.

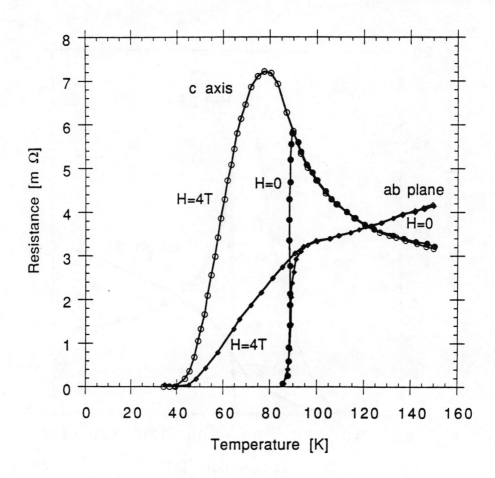

Fig. 13 Resistance along c-axis for H = 0 and H = 4T.[35]

3.3 Flux Creep Effects

Palstra et al.[2] were the first to note the strong flux creep effects in HTSC while measuring activated resistance in applied magnetic fields for BSCCO crystals. They argued that the large magnitude of the dissipative flux motion observed for all HTSC stems from the strong anisotropy and only weakly from the defect structure. To characterize the creep effect in an applied field, they used a power law dependence $U_0 \sim H^{-\alpha}$. α thus may be related to the type of pinning and by itself can vary depending on the magnetic field. When analyzing the activated behavior of BSCCO crystals in magnetic fields from 0.1T to 10T, they found a crossover from $\alpha=1/6$ to $\alpha=1/3$ at high fields. It was not clear however whether this behavior is intrinsic for BSCCO or depends on the type of pinning, current densities etc. This behavior that was reported in Ref. 2 was then reproduced in many laboratories on similar crystals and other HTSC systems. Much work was done since to characterize pinning of either individual vortices or correlated regions of flux lattice for many examples of HTSC in different regions of magnetic fields.

Since in this paper we are interested mainly in the highly anisotropic BSCCO system, we will concentrate on the results of Palstra *et al.* as well as our own results presented in Fig. 7. In the previous chapter we discussed flux creep in some model systems. In particular we emphasized the behavior of 20Å single crystal Nb film as a model system to study flux creep in a single layer of Cu-O. This is because the Nb layer is the closest which has ever been achieved in making ultra-thin films in the clean limit (i.e. $l > \xi$), which is the case for ab conduction in the cuprates. Thus, we will analyze the data in Fig. 5 by plotting $\log(U_0)$ against $\log(H)$ for both the single crystal Nb film and the BSCCO data (see Fig. 14). One immediately notices the similarity between the two systems. In both we find a regime of $\alpha=1/6$ at low fields and a regime of $\alpha=1/3$ at high fields. Nevertheless, we should point out that the data on BSCCO covers only 2 decades in field while the Nb data covers 5 decades. With this wider field range, it is apparent that power law dependences of the activation energy seem unconvincing. Recently, flux creep models using distribution of pinning energies have been proposed to understand the field dependence of $U_0(H)$ in the HTSC.[38] Analyzing our data within such a model produces a reasonable fit for very low fields only (H < 20G). It is clear that this kind of model cannot explain the range of activation energies in the Nb system.

The great similarity between the Nb data presented above and the BSCCO data suggest that indeed the dissipative flux motion observed for all HTSC comes primarily from the quasi-two-dimensional structure of these materials. Since this activation in turn is related to the irreversibility behavior, it is clear that it contains the information about the dynamics of the system. We already noted the change in the exponent α for a similar field range between the single layer MoGe ($\alpha = 0.67$) compared with the weakly coupled multilayers ($\alpha = 0.54$). A similar conclusion comes also from the comparison between the Nb film and the BSCCO crystal. In the field range of 100G to 10 000G, the Nb film exhibits $\alpha = 0.33$ compared to the crystal which represents a multilayer system for which $\alpha = 0.17$. Thus for a similar field range, the interlayer coupling tends to make the activation energy less dependent on the magnetic field. Clearly the limit of a 3D material is obtained where no field dependence is expected.

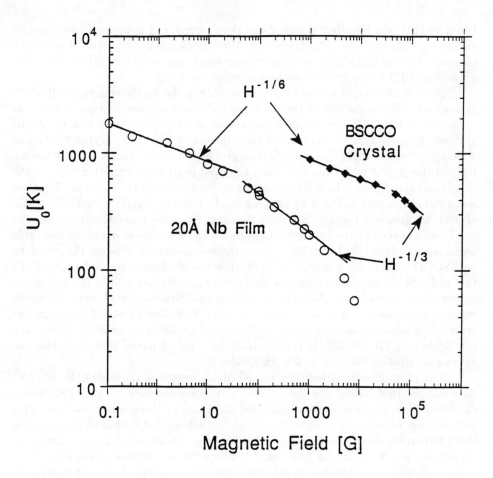

Fig. 14 Temperature-independent part of the activation energy for 20Å Nb film[5] and for BSCCO single crystal[2]. The straight lines show a local fit to a power law $U_0 \sim H^{-a}$.

4. NON-RIGID VORTICES

4.1 Vortex Structure in Monte Carlo Simulations

The fact that vortices in HTSC have a very weak line tension and hence will wiggle and entangle similar to polymers was first discussed by Nelson.[12] He introduced the concept of a liquid of vortex lines to replace the traditional picture of rigid vortices. A Lindemann type of melting was then calculated by Houghton et al.[39] providing insight for the importance of the weak line tension of the vortices. Recently, in a Monte Carlo study, Ryu et al.[40] showed that indeed the entangled phase exists in the melted phase of the vortex state within the simple Lawrence-Doniach (LD) model of Josephson-coupled two-dimensional superconducting sheets.[28] In this model, the free energy is given by:

$$F_{LD} = \frac{H_c^2(0)d}{8\pi} \sum_n \int d^2x \cdot \{|\xi_{ab}(t)(\nabla_{ab} - i2eA_{ab})\Psi_n|^2 + (1-t)|\Psi_n|^2 + \frac{1}{2}\beta|\psi_n|^4$$

$$- g\left|\exp\left(2ie\int_{z_{n+1}}^{z_n} dz \cdot A_{ab}\right)\Psi_{n+1} - \Psi_n\right|^2\} + \frac{1}{8}\int d^3x \cdot h^2 . \tag{4}$$

Ψ is the complex superconducting order parameter, H_c is the thermodynamic critical field, d is the interlayer distance, $t=T/T_c$ is a reduced temperature, β is a constant, n denotes the layer index and A is the vector potential. The subscripts ab denote the ab plane direction. We consider a stack of superconducting layers each of thickness d_s, interlayer spacing d, and dimensionless interlayer coupling strength g which is related to the effective mass ratio via $g=(M_\perp/M_z)$. When applying a magnetic field, vortices appear in the system, made of disk-like vortices in the planes[41,25] and some interplane coupling. A pair of two such segments of a single flux line interact with each other via the magnetic interaction ($\nabla\theta \cdot A$) and the Josephson coupling (the 'g-coupling term') between successive layers. The magnetic coupling for a small separation ($\delta R \ll \lambda$) can be shown to be small and quadratic in the separation.[42,43] The Josephson coupling leads to the pair interaction

$$\delta F_J^{n,n+1} = g\frac{\Phi_0^2}{16\pi^3\lambda^2 d}\int d^2r \cdot (1 - \cos(\Phi_n - \Phi_{n+1})) \tag{5}$$

where Φ_n is the phase of the condensate order parameter at the nth plane. For δR much smaller than a length scale defined by $r_g \equiv \xi_{ab}/\sqrt{g}$ the interaction can be shown to be approximately quadratic in the separation of the cores while for a larger separation, minimization of the LD free energy for a pair of layers leads to

$$-\left|r_g \nabla\Phi\right|^2 + \sin\Phi = 0 \tag{6}$$

where $\Phi(x) \equiv \Phi_{n+1}(x) - \Phi_n(x)$. A more elaborate treatment gives a Sine-Bessel type equation[44] similar to Eq. (6) which has a non-trivial "Sine-Gordon kink-type" solution. This leads to a "string" of vortex with a core running between the planes providing the necessary phase-healing by 2π over a length scale given by r_g. For separation of the vortices beyond $2r_g$, an interplanar core will start to form with energy proportional to the distance. Thus, the interplanar Josephson coupling between two-dimensional vortices may be written in the following way:

$$\frac{\Phi_0^2}{8\pi^3 \lambda_J \lambda} [1 + \ln(\lambda/d)] \cdot \{|R_z - R_{z+1}| - 2r_g\} \quad \text{for } |R_z - R_{z+1}| > 2r_g$$

and

$$\frac{\Phi_0^2}{8\pi^3 \lambda_J \lambda} [1 + \ln(\lambda/d)] \cdot r_g \cdot \left\{ \frac{|R_z - R_{z+1}|^2}{4r_g^2} - 1 \right\} \quad \text{for } |R_z - R_{z+1}| < 2r_g \quad (7)$$

where $\lambda_J = r_g(\lambda/d)$. The vortices in the same plane interact with the usual $K_0(|(r_i,z) - (r_j,z)|/\lambda)$ form[45]

$$\delta F_{in-plane}(R_i - R_j) = \frac{\Phi_0^2 d_s}{8\pi^2 \lambda^2} K_0\left(\frac{|R_i - R_j|}{\lambda}\right). \quad (8)$$

Note that the in plane interaction for the vortex segments is not purely logarithmic,[27] representing an isolated 2D layer, but rather decay for distances larger than λ as a result of the integrity of the vortices in most of the phase diagram.

The above model was simulated by Ryu et al.[40] Two melting lines were identified corresponding to the disappearance of the translational and the hexatic order in the plane. The Lindemann number corresponding to the translational order-disorder transition[39] is found to be about 0.2 but deviated in the extreme density limits.

Fig. 15 is the phase diagram obtained using parameters suitable for the $Bi_2Sr_2CaCu_2O_8$ system ($\xi_{ab}(0) = 15$Å, d = 15Å, $\lambda_{ab}(0) \approx 2000$Å and $M_z/M_\perp \approx 5000$). The three-dimensional boxes display the vortices in the system. The wiggliness of the lines is measured in terms of g. The nonrigidity of the vortices is evident and becomes more pronounced as the temperature is raised. A more detailed analysis of the model and in particular a study of the melting and the hexatic phase boundaries are given in Ref. 40.

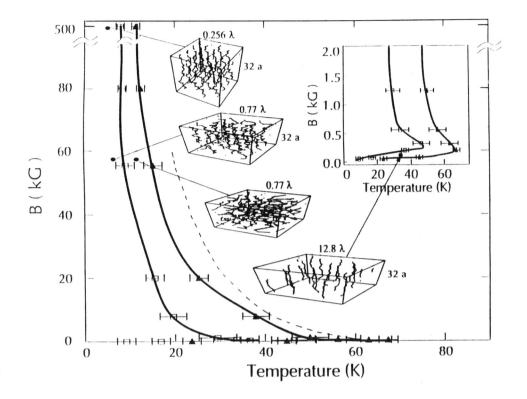

Fig. 15 B vs. T Phase diagram for B = 12.5 G to 50 tesla. Two curves from the simulations are shown. The lower temperature curve (open square) represents the points where the translational order parameter drops to zero. The other curve (black triangle) is for the bond-angle order parameter. The results from elastic continuum theory are also shown (dotted line) for comparison. Some typical configurations are displayed. Note that each picture has a proper size attached to it. The apparent aspect ratio is not exact. The inset shows the very-low-density region.[40]

4.2 Decoupling Fields

It is clear that when the transverse fluctuations of the vortices become too large (of the order of the intervortex distance) two effects may happen. The first is the entanglement of the flux lines and the second is a decoupling of the layers. Nelson suggests that this decoupling procedure comes about when the vertical distance between entangled points reaches the interlayer separation and hence the flux lines lose their identity.[12] It is easy to notice however that even in the absence of entanglement (i.e., a model that allows flux lines cutting), decoupling must occur.[46]

Using Eq. (4), we can calculate the mean square displacement of one vortex disk from the one "connected" to it in the next layer:

$$\langle r^2 \rangle = \frac{\int_0^\infty r^2 d^2 r \cdot \exp\left(-\frac{U(r)}{kT}\right)}{\int_0^\infty d^2 r \cdot \exp\left(-\frac{U(r)}{kT}\right)}. \tag{9}$$

We denote $L_T = (\langle r^2 \rangle)^{1/2}$ as the averaged thermal length of the string. For temperatures above melting but below $T(H_{c2})$, decoupling will occur when

$$L_T = c\left(\frac{\Phi_0}{H}\right)^{1/2} \tag{10}$$

and c is a numerical constant. A decoupling field is therefore estimated as:

$$H_{decoupling} = c^2 \frac{\Phi_0}{L_T^2}. \tag{11}$$

For intermediate fields, above the reentrance regime shown in Fig. 15, the thermal energy is large enough to support the formation of vortex strings in the liquid state. L_T can be evaluated from Eqs. (7) and (9), and combined with Eq. (11) it yields a decoupling field H_{D1}. Near T_c this decoupling field is:

$$H_{D1} = c_1 \cdot g \cdot \frac{\Phi_0^5}{(kT)^2 \lambda^4(0)} \left(1 - \frac{T}{T_c}\right)^2; \tag{12}$$

$c_1 \sim 3 \times 10^{-6}$ is a constant which sensitively depends on the choice of c. At lower temperatures $(1-T/T_c)^2$ is replaced by the square of the temperature dependence of the order parameters. For $Bi_2Sr_2CaCu_2O_8$ crystals, we expect near T_c, $H_{D1} \sim 0.1 \times (1-T/T_c)^2$ given in Tesla. For $YBa_2Cu_3O_7$ $H_{D1} \sim 600 \cdot (1 - T/T_c)^2$. This latter result agrees remarkably in both the exponent and the prefactor with the results of Farrell et al.[47] showing experimental evidence for flux lattice melting (they find H_0=800T with the same temperature dependence). It could very well be that the damping of the mechanical oscillator was indicative of the decoupling between the planes. We can also try to fit this formula to the data of Fig. 9. In this case T_c of the multilayers is much smaller and the anisotropy is larger than for YBCO. Fitting the data to Eq. (10), we find an excellent agreement with a $(1-T/T_c)^2$ temperature dependence and H_0=12T. A calculated H_0 based on the MoGe parameters yields 25T. In view of the arbitrariness of the decoupling factor c (in a way similar to a Lindemann factor,[46]) the above agreements are remarkable, suggesting that the decoupling of the planes in layered superconductors is a well-defined process. If we try however to fit the data for BSCCO, with the peak positions marking the decoupling, the agreement is not so good. In this very anisotropic case the measured H_0 is 10 times larger than the calculated one.

As mentioned above, a decoupling field based on entanglement was also calculated by Nelson.[12] His calculation agrees with a field H_{D2} where L_T is calculated using the quadratic potential, i.e., distance smaller than r_g in our language. In this case the average energy is equal to $(1/2)kT$ and if we use Eq. (8) and substitute the relations for r_g and λ_J we find:

$$H_{D2} = g \cdot \frac{\Phi_0}{2\pi^2 kT} \left(\frac{\Phi_0}{4\pi\lambda}\right)^2 . \qquad (13)$$

We feel this result will be more appropriate for higher fields where the decoupling temperature is lower and the distances between vortices are much smaller than λ. In fact, for BSCCO crystals, this second decoupling field makes a better fit, possibly because when the anisotropy is that large, r_g is very large (~ 1500Å), of order of $\lambda(0)$. Thus, this reduces the H-T regime where $L_T > 2r_g$ which is the string formation regime. For such high anisotropy we may need to identify the decoupling field closer to T_c to agree with H_{D1}. At very low fields, the magnetic energy may be dominant as suggested by Nelson.[12] Since this interaction is quadratic, similar to the Josephson coupling, H_{D2} becomes the relevant decoupling field. We note that a similar result was also obtained by Glazman and Koshelev[48] who suggested a sharp phase transition at H_{D2}. We believe, based on the experimental data discussed here, that a gradual crossover is more representative of the data. This is more apparent when we include the results on the BSCCO single crystals where no effect was seen in the magnetization but a sharp effect appears in the resistance data.

4.3 Dissipation Due to Nonrigid Vortices

To understand the effect of the ab-plane resistance peaking up and even exceeding the normal state resistance, we have to incorporate the c-axis resistance. This is because $\rho_n(T)$ decreases with T, and whatever model is used for the ab-plane vortex motion it cannot result in a vortex-induced resistance that exceeds $\rho_n(T_c)$. As emphasized by Briceno et al.,[36] this is true in the Bardeen-Stephen model[49] even if $\rho_n(T)$ is semiconducting. On the other hand, the correlation between the appearance of the knee with the c-axis resistance peak is evident. Thus, we propose[33] that the motion of distorted vortices can produce an excess dissipation by coupling to the c-axis current paths in addition to the ab-plane resistance. This model accounts for at least the qualitative features of the data in Fig. 10. We assume that below T_c (onset) the layers can be thought of as Josephson coupled,[41,25] and that this weak coupling allows distortions and non-rigid movement of vortex lines. This model and its implications for the transport and magnetic properties of the superconductor are discussed in the next chapter. When the distorted vortex line moves, the time-dependent interlayer phase difference arising from the motion results in resistive interlayer current jumps, coupling in the larger c-axis dissipation to current flow nominally in the ab-plane.

Consider a vortex line which is primarily in the c-direction but with a kink in the b-direction, shown schematically in Fig. 16. The local vector potentials $A_n(x,y,t)$ and $A_{n+1}(x,y,t)$ in the n and n+1 layer will then be unequal and will have some c-axis component, resulting in an interlayer local phase difference:

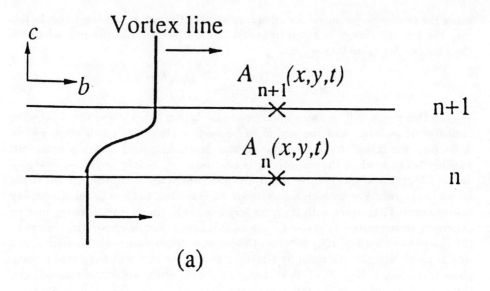

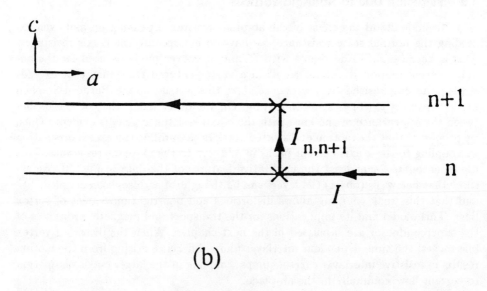

Fig. 16 Schematic representations of: a.) a distorted vortex line and the local vector potentials, b.) the local interlayer current jump caused by motion of the distorted vortex line.[33]

$\gamma_{n,n+1}(x, y, t)$. If the vortex moves in response to a small applied current ($J \ll J_c$) passed in the a-direction, the local vector potentials will change with time, resulting in an interlayer voltage:

$$V_{n,n+1}(x,y) = \frac{\hbar}{2e} \frac{\partial \gamma_{n,n+1}}{\partial t} = \frac{\hbar}{e} \frac{\pi}{\Phi_0} d \left| \frac{\partial A_{n+1}}{\partial t} - \frac{\partial A_n}{\partial t} \right|. \quad (14)$$

If the vortex is a classical straight rigid rod, by symmetry the time derivatives cancel and so no voltage develops. However, distortions in the structure or motion of the vortex line prevent this cancellation and an interlayer voltage appears. This voltage will cause some current to locally jump from the n to the n+1 layer as shown schematically in Fig. 16, with $I_{n,n+1}(x,y) = V_{n,n+1}(x,y)/R_c$, where R_c is the resistance in the c-direction and d is the interlayer distance. The total voltage will be calculated as an average over all interlayer current paths in the volume of the sample. If we apply now a bias current I in the ab-plane, the total resistance is then given by:

$$R(T, H) = R_{ab}(T, H) + \frac{I_c(T, H)}{I} \cdot R_c(T). \quad (15)$$

Here R(T,H) is the measured resistance, R_{ab} is the resistance due to the vortex motion in the ab-plane only, R_c is the c-axis resistance obtained from the data (e.g. as shown above) and $I_c(T,H) = L^{-2} \sum_n \int\int I_{n,n+1} dxdy$ is the c-direction current that can be calculated if the details of the vortex structure at each temperature are known. To illustrate that a set of resistive curves similar to those in Fig. 10 can be obtained, we simulated Eq. (11) using a simple form: $I_c = aH \cdot \exp\{-\sigma(T-T^*)^2\}$. As expected I_c is vanishingly small near T_c and near $T = 0$ and thus has to peak at some temperature T^* in between. Using the experimental data for $\rho_c(T)$ we show a typical simulation in Fig. 17. It is clearly seen that the simulation reproduces all the essential features of the data. It is important to note that the simulation argues against a c-axis contribution in the normal state if such a procedure is tried with the calculated percentage of I_c/I (see legend in Fig. 17). Using now the data of Fig. 10 together with the irreversibility line measured magnetically on similar crystals, we can establish the phase diagram depicted in Fig. 18. Indeed it looks remarkably similar to the one we measured for MoGe/Ge multilayers (Fig. 9). We therefore suggest that the knee in the resistive transition at finite magnetic field (that may develop into a peak if the normal state c-axis resistance has a strong semiconducting behavior) is a universal signature of decoupling of the planes in the vortex liquid state. This claim that was proven experimentally for the multilayers case is easily transferred to the BSCCO single crystals case. T^*, that marks the maximum in $I_c(T,H)$, will have to be very close to the decoupling temperature. Starting from low temperatures, the flux lines are rigid and most likely pinned and thus very little I_c occurs. As the temperature is raised, melting occurs and flux lines move as individuals. As will be discussed in the next section, the wiggling of the flux lines grows with temperature causing I_c to grow. For a fixed field, decoupling

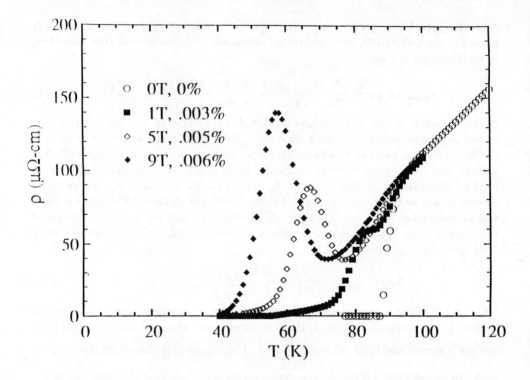

Fig. 17 Simulated $\rho(T,H) = R(T,H) \times$ (area/length) using Eq. (5) with $\sigma^{-1}=60$, showing the evolution of a resistive peak. The percentages shown are the fitted magnitudes of $I_c(T,H)/I$. $R_{ab}(T,H)$ and $R_c(T)$ contributions are obtained from measured data.[33]

will occur at some temperature, typically when the distortion of the vortices is of the order of the intervortex distance. At this point, I_c reaches its maximum and will only decrease as T increases towards T_c. Physically, the Lorentz force on the vortices acts only on the in-plane disks which, above T*, do not drag with them the interplane segments. Thus, the question of whether a peak will be observed, or only a knee, depends on what is the exact form of $R_c(T)$. For MoGe/Ge multilayers the c-axis resistance is roughly constant (it is governed by tunneling) and thus only a knee appears. In a similar way, in YBCO crystals, the c-axis resistance is either decreasing with temperature or very slightly increasing, hence causing at most a knee.

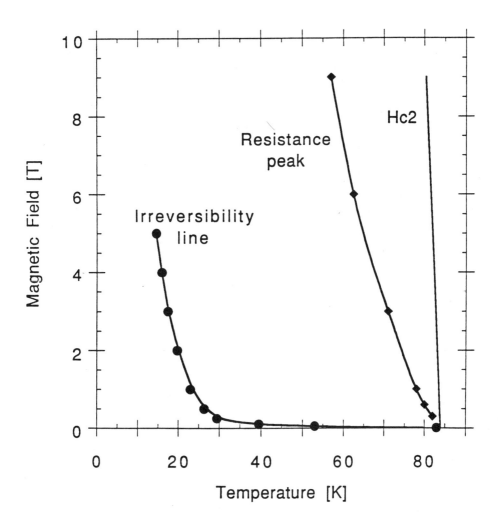

Fig. 18 Similar to Fig. 9, a phase diagram for $Bi_2Sr_2Cu_2O_8$ single crystals displaying H_{c2}, the resistance peak (Fig. 10) and the irreversibility line measured magnetically.

5. THE IRREVERSIBILITY LINE

Irreversibility between zero field cooled and field cooled measurements on HTSC materials were first noted in magnetization measurements by Muller et al.[50] They found that the Meissner (field cooled) and the shielding (zero field cooled) magnetization curves overlapped down to a temperature $T_i(H)$ which is much lower than the expected $T_c(H)$ temperature. Below $T_i(H)$ the two curves diverged. Mapping this irreversibility onset temperature as a function of field they obtained the irreversibility line in the H-T plane. The irreversibility line has also been studied in other superconducting cuprates using a variety of techniques. In many cases it was reported to obey the power law relation $H_i = H_0(1-T_i/T_c)^\mu$ with $\mu \sim 3/2$.[50,7,8] It was originally proposed that this behavior resulted from a superconductive glass state in analogy to spin glasses.[50] Alternately, phenomenological models have been developed to explain this power law and generally rely on thermally activated flux-motion processes.[7,8] Mechanical oscillator studies have been interpreted as evidence for vortex lattice melting similar to the analysis of data taken on two-dimensional films.[51] Real melting involves the loss of translational and orientational order. Thus, it seems that the measured translational correlations of up to 20 lattice constants at low fields on single crystals of $Bi_2Sr_2CaCu_2O_8$ by Grier et al.[52] is the first observation of the development of crystalline order. In fact, in that experiment, the correlations seem to grow steadily with increasing applied field. Orientational correlations persist to even longer distances,[52,53] typically thousands of lattice constants exhibiting hexatic order.[53] It is not clear, however, if this observation is the evidence of the competition between the disorder and the repulsive intervortex interaction, or a nonequilibrium state quenched from the "hexatic liquid" state. Probably one of the key factors in real materials is pinning. The random pinning severely modifies the nature of the long-range order of the Abrikosov lattice as was pointed out by Larkin.[54]

A theory that inherently relies on the existence of disorder, describing the irreversibility line as melting of a vortex glass, has been suggested[55,56] based on the critical behavior of I-V curves.[57] Theoretical studies of such a system yield $\mu \sim 4/3$ for the vortex glass transition and such a power law has been found by several groups. However, this low-field power law was found to be not universal and thus might point to difficulties in its unique interpretation. In particular, recent experiments on $Bi_2Sr_2CaCu_2O_8$ single crystals with different oxygen stoichiometries showed a large range of this exponent. Moreover, proton irradiated samples gave an irreversibility line which is exponential with temperature.[58]

If indeed the irreversibility line marks the melting phase transition, it has to satisfy a few thermodynamic constraints. Since we believe the transition is three-dimensional, it should develop long-range order via either a first- or a second-order phase transition. We will see below that with the accuracy of existing measurements, a true phase transition is not ruled out.

If the transition is first order, the line in the H-T plane has to satisfy a Clausius-Clapeyron equation: $\Delta S = \Delta M(dH/dT)$, where ΔM denotes the jump in the magnetization as we cross the transition point (we expect the magnetization to be more negative below the melting point) and ΔS is the change in entropy through the transition. When careful Meissner measurements are taken on BSCCO crystals[59] with minimal effect of hysteresis due to irreversibility, no jump is apparent for fields as high as 5T. Assuming that we are limited by the sensitivity of our measurement which is 5×10^{-5} emu, using a sample size of $(L^2 \times b) \approx 7mm^2 \times 10\mu m$, and a slope of $\sim -0.7\text{T/K}$ of the H-T phase boundary which is typical at high fields (H > 0.5T), we calculate: $\Delta S = \Delta M (\Phi_0/HL^2)(d/b)(dH/dT) = 10$ k/vortex-CuO_2-plane. This is a rather high number that indicates there may still be a jump hidden. Note that a simple argument of ordinary melting where uncorrelated 2D disks freeze into a three-dimensional vortex lattice would give $\Delta S \approx 2k$/vortex-CuO_2-plane. If, as expected from theory and from the simulations, a hexatic phase exists before freezing, even less entropy change is expected through the transition due to the strong correlations. Moreover, three dimensional correlations that connect the disks into line vortices will cause a dramatic decrease in the entropy similar to polymers. To investigate further the possibility of a sharp transition, we have performed heat capacity measurements in magnetic fields of up to 8T[60] (at 8T we expect the melting to occur at $\sim 10K$). The sensitivity of the measurement was 0.1% of the total background (for which $C_p \sim 250$ erg/K) which was mostly the BSCCO crystal (of size ~ 10mg). In these units, a sharp jump of width 1K that contains $\sim 2k$/vortex-CuO_2-plane was calculated to give a signal which is $\sim 10\%$ of the background, much above our sensitivity. However, no effect was observed and this result puts a bound on the entropy change through the transition at high fields of less than $\sim 0.02k$/vortex-CuO_2-plane (a sharper peak will result in a much bigger expected signal and hence a much smaller bound). This is a much more stringent bound than the one obtained from magnetization measurements.

If a second-order phase transition is assumed, one can use the more general Maxwell relations for a magnetic system. Measuring M(T) for many fields, we can use this set of data to calculate $\Delta S = -\int (\partial M/\partial T)_H \cdot dH \approx -\int (\partial M/\partial T)_H \cdot (dH/dT).dT$ and the integration extends from T=0 to $T=T_i$. Again, looking at the magnetization data, we see very little temperature dependence for high fields near and below the irreversibility line. In fact, the 5T data is almost flat on the scale of the sensitivity. Thus for high fields, $(\partial M/\partial T)_H \approx 0$ and very small entropy has to be involved with this transition ($\Delta S < 0.05$ k/vortex-CuO_2-plane). At low fields, there is some slight curvature to M(T). We have modeled this curvature with a power law, and using the almost power law behavior of the irreversibility line near T_c (H = $H_0(1-T_i/T_c)^\mu$) we integrated the above eqauation. The result is $\Delta S \approx 0.03k$/vortex-CuO_2-plane. Coming back to the heat capacity data, we again try to look for an accumulation of entropy under the specific heat curve for the possibility of a second-order phase transition. In this case we expect $c_p \propto (T_i-T)^{-\alpha}$. If $\alpha \geq 0$, we expect a singularity that should show in the measurements. For this case it is easy to put a bound of $\sim 0.02k$/vortex-CuO_2-plane similar to the first-order phase transition case. If $\alpha < 0$, no singularity is expected and the entropy will be spread over a large temperature range. For this last case, if the entropy is spread over ~ 10 K, we may argue that $\Delta S < 0.1$ k/vortex-CuO_2-plane.

The bounds that we have found all agree that the entropy associated with the melting transition, if described by the irreversibility line, is very small per vortex-disk. This is a very important result because it gives an insight into the correlations of the vortices prior to freezing. One possibility that has been suggested by Fisher et al.[56] is that the correlations may first occur in the planes as a result of strong two-dimensional vortex-glass correlations and that a freezing transition will occur when three-dimensional correlations develop.[63] A second possibility that is strongly supported by the Monte Carlo simulations is that the correlations develop in the single vortex. In this case, the data suggest that in the liquid state just above freezing, disks are not free but rather correlated, presumably via interlayer coupling, to form three-dimensional objects. Freezing into a flux lattice or even a vortex glass will take place with nonrigid line vortices. Decoupling has to occur at much higher temperatures. Thus, entropy will be released with the decoupling in the vortex liquid state and may cause anomalies in transport measurements, in particular energy transport. As already mentioned, no anomaly is found in magnetization measurements near what we identified as the decoupling temperature (Fig. 7). Thus the above analysis is consistent with decoupling of the planes that is only a crossover effect and it happens at temperatures much above the irreversibility line.

ACKOWLEDGMENTS

This work was supported by NSF Grant DMR-86-57658, by AFOSR grant F49620-89-C-0001 and by NSF through Stanford Center for Materials Research. Results described in this paper were obtained in collaboration with many people in our KGB group at Stanford and outside. In particular, Mike Hahn, Julia Hsu, Mark Lee, Anthony Leone, Lou Lombardo, Seungoh Ryu, Jeff Urbach, Whitney White, Ali Yazdani, Mac Beasley, Ted Geballe, Guy Deutscher and Seb Doniach. I am grateful for all these collaborations as well as for iluminating discussions with David Nelson and Daniel Fisher. Finally I would like to thank Mike Hahn, Lou Lombardo and Kathryn Moler for critical reading of the manuscript.

APPENDIX A: Materials preparation and characterization

A. MoGe/Ge films and Multilayers

All of the Mo-Ge films were grown by multitarget magnetron sputtering.[61] During sample growth, the quarter-inch square substrates are mounted on a room-temperature table which rotates at 300 rpm to ensure film uniformity. The two elemental targets, Mo and Ge, can be independently shuttered to grow either Ge insulating layers or Mo-Ge superconducting layers. The substrates must be kept below 110°C to prevent crystallization of the a-Mo-Ge.

The single-layer films were deposited on sapphire substrates. A 200Å amorphous Ge buffer layer was deposited first to smooth out any surface roughness. Then 60Å of a-Mo-Ge was deposited. Finally, the film was capped with a Ge protective layer, typically between 15 - 30Å. Multilayers were grown on a Si_3N_4/Si substrate with a 200Å a-Ge buffer layer.

B. $Bi_2Sr_2CaCu_2O_8$ Single Crystals

The $Bi_2Sr_2CaCu_2O_{8+\delta}$ crystals studied in this work were grown using a directional solidification method.[34] The direction of the crystal growth was controlled by the temperature gradient across the alumina crucible, typically 5 to 10°C/cm. The growth rate was determined by the rate of the temperature change, 0.4 to 1.0°C/h from 900°C down to 725°C. The crystals grown by this method were plate-like with the c-axis normal to the plane; areas of the crystals could be as large as 2 cm². The stoichiometry of the as-made crystals was $Bi_{2.10}Sr_{1.94}Ca_{0.88}Cu_{2.07}O_{8+\delta}$. The oxygen contents of the crystals can be varied by annealing the crystals in different environments (discussed below). According to microprobe analysis, the cation stoichiometry given above was not altered by the oxygen annealing. XPS core level spectroscopy results indicated that some addition/removal of oxygen occurs on the Bi-O planes.

To change the oxygen content (i.e. change δ), the crystals were annealed at 500°C in 0.1% H_2 in Ar, Ar, 1 atm O_2, and 12 atm O_2 for 24 to 60 hours. Two crystals were also annealed at 600°C in 1 atm of O_2. After annealing, the crystals were quenched by cooling in liquid N_2 or water. The changes in δ were determined by thermogravimetric analysis (TGA) during the annealing periods. More detailed information on the crystal growth and the annealing procedures can be found in Ref. 34. Recently, we described an improved method for growing high-quality single crystals.[62] By using magnesia crucibles which are far less reactive with the melt than alumina, more precise control was allowed. Samples using both methods are discussed in this study interchangeably.

REFERENCES

1. B. Oh, K. Char, A. D. Kent, M. Naito, M. R. Beasley, T. H. Geballe, R. H. Hammond, A. Kapitulnik and J. M. Graybeal, Phys. Rev. **B37**, 7861 (1988).
2. T. T. M. Palstra, B. Batlogg, R. B. Van Dover, L. F. Schneemeyer and J. V. Waszczak, Phys. Rev. Lett. **61**, 1662 (1988).
3. A. P. Malozemoff, in "Physical properties of High Temperature Superconductors -I," editor D.M. Ginsberg, (World Scientific Singapore, 1989) pp. 71-150.
4. P. M. Horn and R. D. Parks, Phys. Rev. **B4**, 2178 (1971).
5. J. W. P. Hsu and A. Kapitulnik, Appl. Phys. Lett. **57**, 1061 (1990).
6. J. M. Graybeal and M. R. Beasley, Phys. Rev. Lett. **56**, 173 (1986).
7. Y. Yeshurun and A. P. Malozemoff, Phys. Rev. Lett. **60**, 2202 (1988).
8. M. Tinkham, Phys. Rev. Lett. **61**, 1658 (1988).

9. M. Inui, P. B. Littlewood and S. N. Coppersmith, Phys. Rev. Lett. **63**, 2421. (1989).
10. P. de Rango, B. Giordanengo, R. Tournier, A. Sulpice, J. Chaussy, G. Deutscher, J. Genicon, P. LeJay, R. Retoux and B. Raveau, J. de Phys. **50**, 2857 (1989).
11. K. Kitazawa, S. Kambe and M. Naito, in "Strong Correlation and Superconductivity," edited by H. Fukuyama, S. Maekawa and A. P. Malozemoff, (Springer Verlag, Berlin, 1989).
12. For a recent review see e.g. D. R. Nelson, this volume.
13. J. W. P. Hsu and A. Kapitulnik, unpublished data.
14. W. R. White, A. Kapitulnik, and M. R. Beasley, Phys. Rev. Lett. **66**, 2826 (1991).
15. P. W. Anderson and Y. B. Kim, Rev. Mod. Phys. **36**, 39 (1964).
16. B. A. Huberman and S. Doniach, Phys. Rev. Lett. **43**, 950 (1979).
17. D. S. Fisher, Phys. Rev. **B22**, 1190 (1980).
18. J. M. Kosterlitz and D. J. Thouless, J. Phys. **C6**, 1181 (1973).
19. M. R. Hahn, W. White, P. Lerch, M. R. Beasley and A. Kapitulnik (1991), preprint.
20. A. T. Fiory and A. F. Hebard, AIP Conf. Proc. **58**, 293 (1980).
21. B. Jeanneret, Ch. Leemann, and P. Martinoli, Jap. J. Appl. Phys. **26**, 1417 (1987).
22. M. R. Hahn, Ph.D Thesis, Stanford University, (1991).
23. N. Missert, Ph.D Thesis, Stanford University, (1990).
24. V. Ambegaokar, B. I. Halperin, D. R. Nelson, and E. D. Siggia, Phys. Rev. **B21**, 1806 (1980).
25. S. Doniach in "High Temperature Superconductivity "Proceedings, edited by K. S. Bedell, D. Coffey, D. E. Meltzer, D. Pines and J. R. Schrieffer (Addison-Wesley, Redwood City, CA 1989) p. 406.
26. H. A. Kramers, Physica **7**, 284 (1940).
27. J. Pearl, J. Appl. Phys. **37**, 4139, (1966).
28. W. E. Lawrence and S. Doniach in Proc. 12th Int. Conf. on Low Temp. Phys., Kyoto (1970), edited by E. Kanada (Keigaku, Tokyo 1971), p. 67.
29. L. G. Azlamazov and A. I. Larkin, Phys. Lett. **26A**, 238 (1968).
30. T. T. M. Palstra, B. Batlogg, R. B. Van Dover, L. F. Schneemeyer and J. V. Waszczak, Phys. Rev. **B41**, 6621 (1990).
31. L. W. Lombardo, W. White, M. R. Beasley and A. Kapitulnik, to be published.
32. K. Togano et al., Jap. J. App. Phys. **28**, L907 (1989); H. Mukaida et al., Phys. Rev. **B42**, 2659 (1990).
33. J. W. P. Hsu, D. B. Mitzi, A. Kapitulnik, and M. Lee, (1991) preprint.
34. D. B. Mitzi, L. W. Lombardo, A. Kapitulnik, S. S. Laderman, and R. D. Jacowitz, Phys. Rev. **B41**, 6564 (1990).
35. N. Hess, G. Deutscher, L. Lombardo and A. Kapitulnik, Proceedings of M^2S, Japan, July 1991, to appear in Physica C.
36. G. Briceno, M. F. Crommie, and A. Zettl, Phys. Rev. Lett. **66**, 2164 (1991).
37. S. Martin, A. T. Fiory, R. M. Fleming, L. F. Schneemeyer, and J. V. Warzczak, Phys. Rev. Lett. **60**, 2194 (1988).

38. C. W. Hagen and R. Griessen, Phys. Rev. Lett. **62**, 2587 (1989).
39. A. Houghton, R. A. Pelcovits, and S. Sudbo, Phys. Rev. **B40**, 4763 (1989).
40. S. Ryu, S. Doniach, G. Deutscher and A. Kapitulnik, "Monte Carlo Simulation of Flux Lattice Melting in High-Tc Superconductors," submitted for publication (1991).
41. G. Deutscher and A. Kapitulnik, Physica A **168**, 338 (1990).
42. A. Buzdin and D. Feinberg, J. de Phys. **51**, 1971 (1990).
43. John R. Clem, Phys. Rev. **B12**, 1742 (1975).
44. J. P. Carton, J. de Phys., (1991).
45. P. G. deGennes, "Superconductivity of Metals and Alloys" (W.A. Benjamin, New York, 1966), p. 227.
46. A. Kapitulnik, G. Deutscher and S. Doniach, Stanford Report (1989), unpublished.
47. D. E. Farrell, J. P. Rice, and D. M. Ginzberg, (1991), Case Western Reserve preprint.
48. L. I. Glazman and A. E. Koshelev, Phys. Rev. **B43**, 2835 (1991).
49. J. Bardeen and M. J. Stephen, Phys. Rev. **140**, 1197 (1965).
50. K. A. Muller, M. Takashige, and J. G. Bednorz, Phys. Rev. Lett. **58**, 1143 (1987).
51. P. L. Gammel, L. F. Schneemeyer, J. V. Waszczak, and D. J. Bishop, Phys. Rev. Lett. **61**, 1666 (1988); P. L. Gammel, J. Appl. Phys. **67**, 4676 (1990).
52. D. G. Grier, C. A. Murray, C. A. Bolle, P. L. Gammel, D. J. Bishop, D. B. Mitzi, and A. Kapitulnik, Phys. Rev. Lett. **66**, 2270 (1991).
53. C. A. Murray, P. L. Gammel, D. J. Bishop, D. M. Mitzi and A. Kapitulnik, Phys. Rev. Lett. **64**, 2312, (1990).
54. A. I. Larkin, Zh. Eksp. Teor. Fiz. 58, 1466 (1970) [Soviet Phys. - JETP 31, 784 (1970)]; A. I. Larkin and Yu. N. Ovchinnikov, J. Low Temp. Phys. **34**, 409 (1979).
55. M. P. A. Fisher, Phys. Rev. Lett. **62**, 1415 (1989).
56. D. S. Fisher, M. P. A. Fisher, and D. A Huse, Phys. Rev. **B43**, 130 (1991).
57. R. H. Koch, V. Foglietti, W. J. Gallagher, G. Koren, A. Gupta, and M. P. A. Fisher, Phys. Rev. Lett. **63**, 1511 (1989).
58. L. W. Lombardo, A. Kapitulnik and A. Leone, IEEE Magnetism, (1991), to appear.
59. L. W. Lombardo and A. Kapitulnik (1990), unpublished results.
60. J. S. Urbach, L. W. Lombardo and A. Kapitulnik (1991), to be published.
61. J. M. Graybeal, Ph.D. thesis, Stanford University, 1985 (unpublished).
62. L. W. Lombardo and A. Kapitulnik, J. of Crystal Growth, (1991), to appear.
63. D. S. Fisher, this volume.

DISCUSSION

D. R. Nelson: There is some decoupling in your beautiful artificial layers in the bismuth; I wonder about the YBCO. You can do various estimates of this decoupling field; some of them are as large as 50 tesla. I'm concerned in particular, though, about identifying Farrell's signal with the decoupling point. If in fact there is also an irreversibility temperature below that line he should see two peaks. My own view is that probably at the decoupling there isn't a great peak in the kind of torsional oscillator measurements that he does, anyway, so I'm a little bit worried about your interpretation.

A. Kapitulnik: It could be that at the lower melting temperature there is another peak. It could also be that these two are very close to each other – that's exactly what I would assume for YBCO, they are very close to each other – and maybe the peaks will extend. But still, for the moly-germanium multilayers I can show exactly where decoupling occurs. I can do the fit, and I find that it agrees with the prefactor and the exponent. For the YBCO, since I don't have a single layer, I cannot show it. It's just by analogy I'm inferring this result. It's just peculiar that again it agrees with both the prefactor and exponent.

D. S. Fisher: Just two comments, one on this question of the decoupling field. It seems to me, if I understand it, this calculation that you've done is very similar to the calculation which is done for the Lindemann criteria for melting in the limit that the dominant wave vector of the excitations in the z direction is of order of an interplane spacing. But you've assumed not-very-high fields. I'm not sure if it's a separate calculation. I think there is a $[1-(T/T_c)]^2$ that you have there. Probably by the time that this calculation is relevant, fluctuations are large enough that the penetration length will no longer have the simple BCS temperature dependence, and what you'll really see is something more like a 4/3 power again there.

A. Kapitulnik: I didn't talk about the irreversibility line in this range of fields, and the fit to the power law. But by now we have an enormous amount of data on many crystals – that are supposed to be very good crystals – that show a whole variety of exponents going from 3/2, 4/3, 1.1, and sometimes even a much smaller exponent. So I would say that for our crystals, I don't see any 4/3 at this range of fields.

D. S. Fisher: Well it probably was in an intermediate crossover regime.

The other point, just on this last point of yours, about how much coupling there is between the layers, I may not have quite understood what you meant. But if there is a large amount of correlations in the layers – which there is definite evidence for – and the correlations within the layers build up to be quite long range before the coupling between the layers becomes important – which will certainly happen at very high fields, and I think there's a good indication that it's happening at fields of a few tesla . . . if that happens, then the amount of entropy left for anything going on down to zero temperature is very little. So even though eventually the irreversibility line will involve coupling between the layers, it doesn't mean that anything looks like a vortex line. If you took a picture of it, it could still be vortex points that would look as if they were randomly positioned.

A. Kapitulnik: I agree that, for example, if you first will have hexatic order in each layer – which occurs above the melting point – and now you go to low temperature, you will have much less entropy.

D. S. Fisher: The thing is, you're talking about everything as if the pinning were very weak. There's rather good evidence from the μSR and other data that in these regimes the effective strength of pinning is very strong, so there are probably only short-range lattice correlations even within the layer. So one really has to think of what's going on within the layers already in terms of vortex glass behavior.

A. Kapitulnik: Again I don't have time to talk about it, but I have a study of the irreversibility line as a function of irradiation, that presumably will tell you with how much pinning you started. Because, you can see if a small amount is doing anything or not. The effect of irradiation is dramatic, so I would assume that the pristine crystals are much cleaner than . . .

D. S. Fisher: I think that μSR suggests the opposite.

P. H. Kes: If your argument is correct about the irreversibility line in the multilayer where the vortices are decoupled, the irreversibility line actually would mimic the resistive transition of the single layer. Did you check that? If the vortices are decoupled, you would expect the peak . . .

A. Kapitulnik: No, but what I'm saying is that they are coupled.

P. H. Kes: No, they are decoupled at T*, right?

A. Kapitulnik: Yes, but as you go to lower temperatures where activation exists, then they are coupled. And, of course, the activation energy is different, as I showed.

D. R. Nelson: They always couple before there's any phase transition?

A. Kapitulnik: Exactly. My message is that the coupling between the layers occurs before the irreversibility line, as I reduce the temperature.

S. Doniach: I'd just like to make a comment that I think Aharon is presenting a rather black and white picture. One can imagine that as one goes through T* you start to develop a persistence length along the vortices so that you could define an effective persistence length which could be a few layers, or many layers, depending where you are in the phase diagram. That means that between the melting line and this T* it's something like a polymer liquid, that the vortices are not infinite objects but they are objects that have a persistence length of a number of layers. So I think there's a rather rich physics in that intermediate liquid, that's something like a polymer.

A. P. Malozemoff: I want to turn the discussion to the data that you showed on the resistivity of the bismuth with that remarkable peak. Is the difference from, for example, Tom Palstra's earlier data the field range that you studied, or is it the quality of the materials involved?

A. Kapitulnik: It's the same field range with different crystals.

A. P. Malozemoff: Different in what way?

A. Kapitulnik: Well first, the starting crystals are different, and I'm sure Peter will talk about results on these starting crystals. Then we learned how to reduce these crystals, extract oxygen out, in order to change their T_c. This behavior occurs always when there is almost no oxygen in the bismuth-oxygen layers. In other words, the δ is as much as you can extract out.

A. P. Malozemoff: I see. So the coupling between the layers in these crystals may be very different from those that Tom first studied.

A. Kapitulnik: There is one comment about all this behavior. If you look, and you say, based on our very simple model, what will really determine whether you'll have a peak or just a knee, it's basically what is the perpendicular resistance. Now for good YBCO, it's either very shallow sometimes, or most of the times for 1-2-3 O_7 it goes down. I don't think anybody measured the perpendicular resistance for oxygen-reduced samples. So I don't know whether they peak.

For the Mo-Ge, the resistance in the c direction inferred from some tunneling measurement (I don't want to go into it) seems to be constant. So therefore it seems that if the resistance in the c direction as you reduce the temperature goes up very sharp, you will get a peak; when it has much weaker temperature dependence you will get a hump, and for even weaker temperature dependence you may not see it at all.

A. P. Malozemoff: So now about the interpretation, in terms of bent vortices, of your resistance which intermixes a component of R_c: I was wondering whether this requires a net tilting of the vortices, because aren't there going to be cancellations if you have an *equal* number of vortex kinks tilting to the right and to the left.

A. Kapitulnik: Yes, normally you'll have a tilt. And when you average all the layers, some will contribute positive, some negative voltage, but locally it looks like a tilt.

A. P. Malozemoff: But then when you get the net signal over the sample are you requiring that there be a net tilt of vortices? If not, won't the signal average out?

A. Kapitulnik: Well, yes. I assume that yes.

M. P. Maley: Would you say something more about your irradiation results on the irreversibility lines?

A. Kapitulnik: Can I give you a preprint instead?

M. P. Maley: Sounds great to me.

A. Kapitulnik: I have a piece of a talk about it, but it's rather long.

P. L. Gammel: I was wondering if you really think there is a change in the coupling between vortices in different layers, whether or not you could use some of these techniques that were used in DC flux transformers, to actually probe the coupling between vortices. Say, flow countercurrents on the top and bottom and look for drag effects.

A. Kapitulnik: I heard that things like this are being done at Berkeley. Is it true?

J. Clarke: It's true.

F. M. Mueller: Successfully?

P.L. Gammel: Within these polymer-type models, if you drag vortices using a current on the top then – absent pinning, of course; then the vortices are dragged on the bottom – those types of measurements should be a much more direct probe of the layer coupling, I would think. If there's really a pretty good correlation in the c axis, the flux transformer should be a much more direct probe than these indirect measurements.

A. Kapitulnik: We didn't talk yet, but my students at least know about the results. I wasn't around, so correct me . . . . You did noise measurements above and below, and you saw a difference, right?

J. Clarke: Yes.

A. Kapitulnik: Above and below, and you see a difference between the bismuth and the other high-T_c materials. Well, you should talk about it, not me. But this is a good way to do it, to look for correlations between motion above and motion below.

P. L. Gammel: Yes, right.

V. Vinokur: I would like to draw attention to the fact that in the physical picture suggested this decoupling energy which determines the position of the kink or the peak in fact should not be obliged to be the energy of the coupling. In fact this is nothing but the characteristic energy of plastic deformation in vortex structure, and it's not necessary that this temperature is the real decoupling energy between layers.

Another detail of this proposed explanation: am I to understand that you need this quasi-Lindemann factor in your estimates just to account for the fact that your activation energy at the temperature of the kink is larger than the temperature?

A. Kapitulnik: Yes, but when it appeared it's just in the right place. I would say that if we consider the pinning in the vortex liquid then this factor would appear in a natural way, and it will be some big factor which accounts for influence of pinning on the motion of vortex structure. And then at this temperature you should have a real transition from viscous flow to the activation region of motion of the vortex liquid.

G. Grüner
University of California, Los Angeles
Los Angeles, CA 90024

High T_c Superconductors at Microwave Frequencies

I discuss various experiments conducted in the micro- and millimeter wave spectral range on thin-film and single-crystal specimens of the high-temperature oxide superconductors. For high-quality films the surface resistance R_s is, except at low temperatures, due to thermally excited carriers, with extrinsic effects playing only a secondary role. Because of the low loss, various passive microwave components such as resonators, delay lines and filters, with performance far superior to those made of normal metals, can be fabricated. The conductivity measured at millimeter wave frequencies displays a peak below T_c. Whether this is due to coherence factors or due to the change of the relaxation rate when the materials enter the superconducting state remains to be seen.

1. INTRODUCTION

The electrodynamics of the superconducting state is markedly different from that of normal metals, and exploring the frequency-dependent response over a broad spectral range may lead to information on the symmetry of the ground state, and on the magnitude of the gap. Experiments conducted about thirty years ago played a primary role in establishing the basic features of low-temperature superconductors and gave strong evidence for the BCS description. A significant amount of work has also been done recently on the high temperature superconductors, and the activity is, to a large extent, driven by expectations that these materials can be used for various applications at micro- and millimeter wave frequencies.

The parameter which is usually measured at these frequencies is the so-called surface impedance $Z_s = R_s + iX_s$ with R_s the surface resistance and X_s the surface reactance. The former determines the loss, and the latter the resonance frequency when the material forms part of a resonant circuit — the usual configuration for the evaluation of R_s and X_s. These parameters have been calculated, first by Mattis and Bardeen for certain limits of the ratio of the coherence length ξ and mean free path ℓ. Subsequently numerous calculations have been made for different superconducting ground states or for different ratios of ξ and ℓ. While the experiments do not give direct information on the complex conductivity $\sigma(\omega) = \sigma_1(\omega) + i\sigma_2(\omega)$, both components can be evaluated if *both* R_s and X_s are measured, as will be discussed later. The temperature dependence of σ_2 is related to the temperature dependence of the single particle gap Δ. For singlet superconductors σ_1 develops a peak below T_c. This is due to case-II coherence factors, and the behavior expected is similar to the celebrated Hebel-Slichter peak for the nuclear spin-lattice relaxation rate.

In this short review, I first discuss the fundamental parameters of the superconducting state which are measured at micro- or millimeter wave frequencies, usually well below the gap frequency $\omega_g = \Delta/\hbar$. This is followed by a short description of the experimental techniques and a discussion of results on conventional, low-temperature superconductors. Experiments on high-T_c superconducting films and on single crystals will be discussed next with a short summary of the remaining questions.

2. THE SURFACE IMPEDANCE AND CONDUCTIVITY

The surface impedance Z_s is one of the parameters which reflects the electrodynamics of the superconducting state. At the same time it is perhaps the most important technical parameter which determines the application potential of the superconductors for high-frequency passive components and devices.

The surface impedance is defined as[1]

$$Z_s = \frac{E_o}{\int_o^\infty j\,dz} \qquad (1)$$

where E_o is the electric field of the surface, j the ac current in the sample and the x direction is perpendicular to the surface. Z_s is given, in terms of the complex conductivity $\sigma = \sigma_1 - i\sigma_2$, by

$$Z_s = \left(\frac{i\mu_o\omega}{\sigma_1 - i\sigma_2}\right)^{1/2} = R_s + iX_s \qquad (2)$$

where μ_o is the permeability of free space, R_s and X_s are the surface resistance and surface reactance, and ω the measuring frequency.

The surface reactance is given, in the superconducting state well below T_c, by

$$X_s(T) = \mu_o \omega \lambda(t) \qquad (3)$$

with λ the penetration depth. The measured $X_s(T)$ can be compared with various models of the superconducting state.[2]

The surface resistance has been calculated in the local limit by Mattis and Bardeen,[3] and $R_s(T)$ is in good agreement with experiments conducted on classical superconductors, as will be discussed later.

In the limit where the real part of the conductivity σ_1 far exceeds the imaginary part σ_2 (typically for low temperatures and for $\hbar\omega \ll k_B T_c$) the surface resistance is approximately given by

$$R_s = \frac{1}{2}\left(\frac{\omega\mu_o}{\sigma_2}\right)\sigma_1/\sigma_2 \qquad (4)$$

where $\sigma_2 = 1/(\omega\mu_o^2)$. Both σ_1 and σ_2 can be evaluated[3,4] using weak coupling BCS theory for an isotropic gap Δ at various temperatures. The overall temperature and frequency dependence is given at low temperatures $k_B T \ll \Delta$ by

$$R_s = A\frac{\Delta}{T}\left(\frac{\omega^2}{\Delta}\right)\ln\left(\frac{\Delta}{\hbar\omega}\right)\exp\left(-\frac{\Delta}{kT}\right) \qquad (5)$$

where A is a numerical factor, reflecting the properties of the normal state at $T > T_c$. The exponential behavior reflects the appearance of an isotropic BCS gap, while the ω^2 dependence of R_s reflects the inductive response of the superfluid. In contrast, for a normal metal R_s is proportional to $\omega^{1/2}$.

$R_s(\omega)$, calculated using the Mattis-Bardeen theory for a material with $T_c = 92K$, is displayed in Fig. 1 together with R_s calculated for metallic Cu at $T \cong 300K$ and $T = 77K$. At 300K the normal skin effect regime is appropriate, and $R_s \sim \omega^{1/2}$, while at $T = 77K$, the anomalous skin effect limit applies, and $R_s \sim \omega^{2/3}$ and is also independent of the mean free path.[1] The single-particle gap clearly shows up in $R_s(\omega)$, and the surface resistance is that of a normal metal when ω is above the gap frequency. For $\omega \ll \omega_g$, R_s is proportional to ω^2 in the superconducting state. The calculations have been made for parameters appropriate for $YBa_2Cu_3O_7$, and it is evident that R_s at $T = 0.5T_c$ is significantly less than that of copper, implying superior device performance.

As will be discussed later, Eq. (2) can be used to extract the components of the complex conductivity σ_1 and σ_2 from the measured parameters R_s and X_s, when *both* are evaluated. These have also been calculated, and the temperature dependences reflect the so-called coherence factors associated with the BCS ground state. The coupling to the electromagnetic field leads to case-II coherence similar to the case of the nuclear spin-relaxation time. Consequently, it is expected that $\sigma_1(T)$ develops a peak below T_c, with features similar to the Hebel-Slichter peak for the relaxation rate.

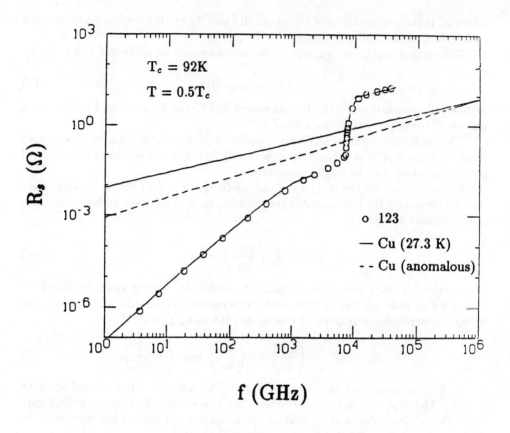

Fig. 1 Calculated frequency dependence of the surface resistance R_s for parameters characteristic of YBa$_2$Cu$_3$O$_7$, of temperature $T = 0.5T_c$.

$\sigma_1(T)$ and $\sigma_2(T)$ calculated using the Mattis-Bardeen theory with parameters $T = 9.3$K (appropriate for Nb) for a measurement frequency of f = 60 GHz and for various gap values is displayed in Fig. 2.

3. EXPERIMENTAL TECHNIQUES

The majority of the experiments are conducted by using resonant structures, with the superconducting material forming part of the structure.[5-9] Both cavity end wall[8] and cavity perturbation[9] techniques have been utilized. In all cases R_s and X_s are determined using a configuration where these parameters refer to

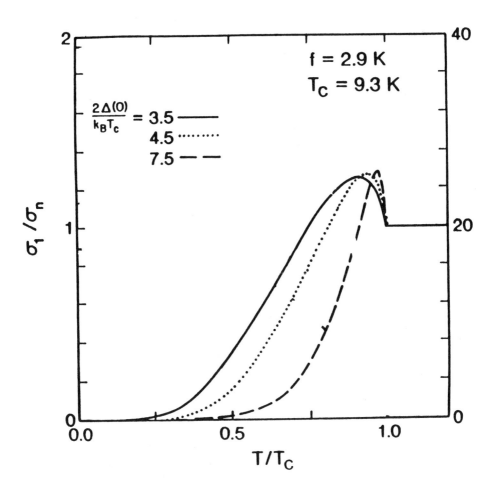

Fig. 2 Calculated temperature dependence of σ_1 for parameters given on the figure.

electric currents flowing in the planes (if thin films are investigated). Alternatively, microstrip and transmission line[11] configurations can be used with films acting as either the central conductor or groundplate of the resonator. The latter techniques are usually employed at high rf or low microwave frequencies, while cavity methods are most advantageous at micro- and millimeter wave frequencies.

The various radio frequency and microwave techniques are usually applied to bulk or thin film samples, and depend on measuring relatively small changes in the resonance frequency f_o and quality factor Q as a function of the temperature.

The changes normally represent a small perturbation on the resonance, and can be linearly related to the changes in the surface impedance of the sample:

$$\Delta\left(\frac{1}{Q}\right) - 2j\left(\frac{\Delta\omega_o}{\omega_o}\right) = \gamma\Delta Z_s = \gamma\left(\Delta R_s + j\Delta X_s\right). \qquad (6)$$

γ is a geometric factor associated with the dimensions of the resonator and the sample.

Nonresonant methods[11] have also been employed but these techniques are far less widespread. Their advantage lies in the possibility of measuring the surface impedance at several frequencies, but the evaluation of R_s and X_s from the measured parameters is not entirely straightforward, and the sensitivity is less than that of the resonant techniques.

4. EXPERIMENTAL RESULTS

At first I briefly discuss some experiments which have been conducted on low-temperature superconductors, in order to demonstrate the essential features of the experimental results. The discussion of the experiments on high temperature superconductors is in two parts: experiments on thin films where the surface resistance (or loss) has been investigated in detail, and experiments on single crystals where attempts have been made to evaluate both R_s and X_s and consequently σ_1 and σ_2.

4.1 Low-temperature Superconductors

A significant amount of work was performed several decades ago on conventional superconductors and various techniques have been employed to measure the surface reactance (proportional to the penetration depth well below T_c) and the surface resistance. The experiments have been contrasted with theory, and they have been accounted for using the BCS theory.[13-15] As an example, in Fig. 3 I display the temperature dependence of the surface resistance measured[16] on Nb at $\omega = 60$GHz. The full line is a calculation using the Mattis-Bardeen theory.[3] The excellent agreement between theory and experiment is evident from the figure, and a detailed analysis can also lead to the magnitude of the single particle gap from the temperature dependence of R_s well below T_c, see Eq. (5). The temperature dependence of X_s leads at $T \ll T_c$ to the evaluation of the temperature dependence of the penetration depth λ, see Eq. (3). R_s and X_s can be combined, using Eq. (2) to evaluate the temperature dependence of σ_1 and σ_2. The former is displayed in Fig. 4 together with calculations again based on the local limit with $\ell < \xi_o$.[3] The peak of σ_1 below T_c is here due to case-II coherence factors and is well accounted for by the BCS theory.[17] It is important to note that a two fluid model does not give a coherence peak, which also becomes smaller as the frequency increases and approaches ω_g. Also in the clean limit, $\ell > \xi_o$, the coherence peak is suppressed.

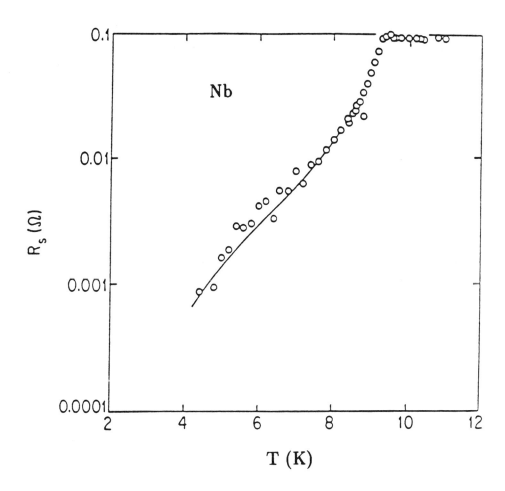

Fig. 3 Temperature dependence of R_s for Nb. The full line is a calculation based on Mattis-Bardeen theory.[3]

4.2 High T_c superconductors

$R_s(T)$ measured[20] on a ceramic, sputtered thin film and laser ablated thin films of $YBa_2Cu_3O_7$ is displayed in Fig. 5. One observes a largely reduced surface resistance when $T < T_c$, and one also finds a finite residual resistance as $T \to 0$. $R_s(T)$ depends also strongly on the film quality, with laser ablated films — the highest overall quality — giving the lowest $R_s(T)$ values. Also displayed on the figure are the Mattis-Bardeen calculations with parameters appropriate for $YBa_2Cu_3O_7$.

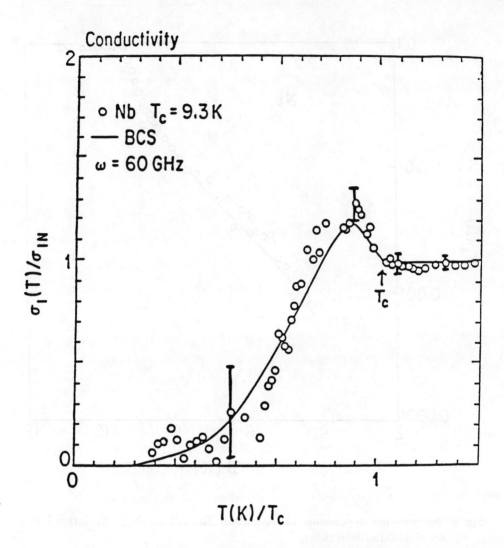

Fig. 4 Temperature dependence of σ_1 for Nb, as evaluated from the measured R_s and X_s.

Because of finite film thickness d, leakage through the films, which are placed at the end of the cavity may be important, in particular at high temperatures where the penetration depth λ or the skin depth δ is comparable with d. Such a leakage can be modeled[21] and in the normal state the behavior can be described in detail, and the ohmic and radiation losses can be separated. A similar procedure also leads to significant corrections near T_c in the superconducting state, and the ohmic losses

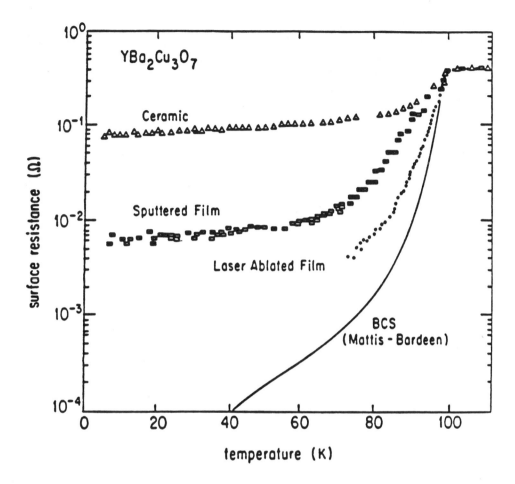

Fig. 5 Temperature dependence of R_s for ceramic, sputtered and laser ablated $YBa_2Cu_3O_7$. Ref. 20 and references cited therein.

can reliably be evaluated. The experimental results to be discussed later have been obtained by performing such analysis and consequently they represent the ohmic losses in the materials. A similar procedure can be applied to evaluate the intrinsic surface reactance, but $X_s(T)$ has not been analyzed in detail to date.

The finite residual surface resistance is evident on the ceramic material[18] and on the sputtered films.[19] Laser ablated films give small residual surface resistance, which often is within the experimental uncertainty.[20,21]

The large R_s at low temperatures for the majority of materials is due to a second phase[22] or due to grain boundaries[23] which act like Josephson coupling between the grains. The effect of nonsuperconducting second phase has been considered in detail utilizing the effective medium theory,[22] and comparisons have been made with experiments conducted on ceramic materials. A good agreement between theory and experiments has been found. Based on such a comparison it has been suggested that ceramic materials may have a significant amount of second phase. For sputtered and laser ablated films, the amount of second phase is small, and the films are also highly c-axis oriented. It is expected that the main source of the low-temperature residual loss is due to grain boundaries. In a simple model[23] of grain boundary effects on the surface impedance of thin films, the material is modeled as a network of superconducting grains coupled by Josephson junctions, which are described using the standard resistively shunted junction model with negligible capacitance. The parameters of the model are the junction $J_c R$ product (J_c is the junction critical current and R the junction resistance) and the effective grain size a. The parameters which enter are $L_G = \mu_0 \lambda_{ab}^2$, the kinetic inductivity of the grain for current flow in the ab plane (penetration depth λ_{ab}), and $L_j = \hbar/2eJ_c$, the unit areal inductance of the junction.

For such a description of the grain boundary, the surface impedance is given by

$$Z_s = \omega \mu_0 \lambda_{eff} \left[\lambda^2{}_{ab} + \lambda_j^2\right]^{1/2}, \quad \lambda_j = \left[\frac{\hbar}{a2eJ_c\mu_0}\right]^{1/2}, \quad (7)$$

with

$$\lambda_{eff} = \left[\lambda_{ab}^2 + \lambda_j^2\right]^{1/2}, \quad \lambda_j = \left[\frac{\hbar}{a2eJ_c\mu_0}\right]^{1/2}, \quad (8)$$

and λ_{ab} the penetration depth in the planes. The model accounts well for the increased penetration depth and increased surface of sputtered superconducting films. It also leads to R_s proportional to ω^2 and this has also been found by experiments conducted in the millimeter wave spectral range.[24]

For laser ablated films grain boundary effects are not important, at least at temperatures not too far below T_c, and this leads to a largely reduced surface resistance as shown in Fig. 5. Experiments conducted at various laboratories[20,25] on different laser ablated films indicate that R_s is - - except well below T_c - - determined by intrinsic effects, i.e. by the losses due to thermally excited carriers. This is further demonstrated[27] on Fig. 6 where R_s measured using the same method is displayed for samples prepared at Bellcore and at Superconductor Technologies, Inc. Also included are data on two laser ablated $Tl_2Ba_2CaCu_2O_8$ films with $T_c = 105K$ and $T_c = 103K$, respectively.

Within experimental error the temperature dependence of the normalized surface resistance R_s/R_N is the same for all films, and it is also comparable to the surface resistance measured by other groups.[26]

The frequency dependence of R_s was measured by various techniques on films prepared by the same group, and the experimental results are displayed in

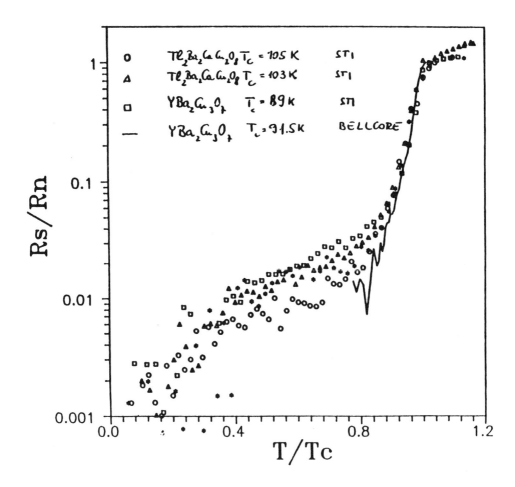

Fig. 6 Temperature dependence of R_s for various laser ablated $YBa_2Cu_3O_7$ and Tl films. Ref. 27.

Fig. 7. The various frequency ranges have been covered by various techniques. A meander-line configuration was used at low frequencies[28] while at millimeter wave frequencies, a cavity endplate configuration was employed. The full line represents the quadratic frequency dependence expected for a superconductor at frequencies $\omega \ll \omega_g$.

The paper by Mattis and Bardeen[3] is used in general to compare experimental results with expressions of R_s based on the weak-coupling BCS ground state. For a detailed comparison with experiments, the theory has to be extended to included finite mean free path effects, and also the possibility of gap values which

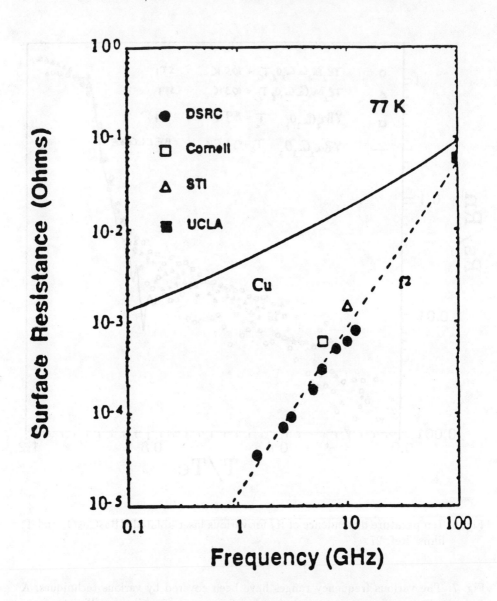

Fig. 7 Frequency dependence of R_s measured on laser ablated films prepared at Bellcore. For details see Ref. 24.

are different from the BCS weak coupling limit $2\Delta/k_BT_c = 3.52$. This, together with consequences of the two fluid model, has been discussed in detail, and numerical results for $Z_s(\omega)$ within the BCS framework have been given by Chang and

Scalapino.[30] The major assumptions underlying these calculations are the following. For a c-axis oriented film, one neglects the electron transfer between the planes, so that only the component of the wave vector of the electromagnetic field parallel to the layers enters. It is effectively zero for the micro-and millimeter waves of interest, so that the electromagnetic kernel relating the current density j to the vector potential A is evaluated with q = 0. Thus the electromagnetic response is described with σ depending on $\ell/\pi\xi_o$, $2\Delta(0)/k_BT_c, \omega/\Delta(0)$, and T/T_c. Here ℓ is the mean free path, which is limited by elastic scattering due to impurities, and ξ_o is the zero-temperature coherence length $\hbar v_F/\pi\Delta(0)$. It is of course an open question as to whether such a BCS analysis is appropriate and in particular, whether the scattering lifetime, which arises from the underlying dynamics, can be approximated by an elastic scattering mean free path even over the relatively narrow temperature range of these experiments.

Within the BCS approach, R_s depends also on both the ratio $\ell/\pi\xi_o$ and on the ratio $2\Delta/k_BT_c$. An estimate of $\ell/\pi\xi_o$ can be obtained from

$$\sigma_n \lambda_L^2(0) = \frac{T}{\mu_o} = \frac{\ell}{\pi\xi_o} \left(\frac{k_BT_c}{2\Delta(0)}\right) \frac{2\hbar}{\mu_o k_B T_c} \qquad (9)$$

Taking $\sigma_n^{-1} = 65\mu\Omega$ cm, $T_c = 90$ K, $2\Delta(0)/k_BT_c = 3.52$, and $\lambda_L(0) \simeq 1500$Å gives $\ell/\pi\xi_o \simeq 0.9$. The ratio R_s/R_N has been calculated for various values of $2\Delta(0)/k_BT_c$. In Fig. 8, the normalized surface resistance is compared with calculations using the above assumptions. The two sets of experimental curves correspond to two different extrapolations of the T → 0 residual resistance, and have been discussed in detail.[20] The dependence is displayed for $\ell/\pi\xi_o = 1$ in Fig. 8 along with the data corrected for leakage effects. As can be seen from Fig. 8, the experimental results fall below the calculated values for the weak-coupling $2\Delta = 3.52k_BT_c$ limit. This clearly shows that $2\Delta(0)/k_BT_c$ exceeds the weak-coupling limit, and a gap value of $2\Delta(0) \cong 5k_BT_c$ appears to describe our findings. A large single particle gap, as inferred from the surface resistance studies, is in broad agreement with a host of experimental results which consistently lead to large single particle gaps, well exceeding the weak coupling limit.

As discussed earlier, when both R_s and X_s are evaluated the components of the optical conductivity, σ_1 and σ_2, can be evaluated. Due to finite film thickness, which leads to leakage effects in particular close to T_c where the penetration depth is large, R_s and X_s can not be evaluated to high accuracy. Consequently, the conductivity can not be evaluated. This is however possible by conducting experiments on single crystals, with the specimens placed inside a resonant structure, and evaluating and analyzing the resultant changes in the quality factor and frequency shift.

Our experiments on $Bi_2Sr_2CaCu_2O_8$[31] and on $YBa_2Cu_3O_7$[32] single crystals were conducted at f = 60GHz using a resonant cavity operating in the TE_{011} mode, with the specimen placed inside the cavity in a maximum ac magnetic field H_{ac} (with the electric field E_{ac}, ideally zero) or in a maximum electric field E_{ac} (for which H_{ac} is zero). For both configurations the single crystal platelets were

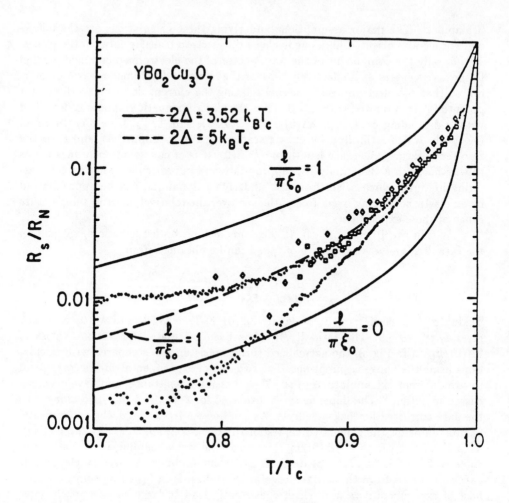

Fig. 8 Temperature dependence of the surface resistance R_s together with calculated $R_s(T)$ for two different gap values, and including finite mean free path effects. Ref. 31.

oriented to result in the shielding current flow (for ideal orientation) parallel to the conducting layers, and consequently the in-plane electrodynamics is examined. For the maximum H_{ac} configuration, the applied field is perpendicular to the layers, while for the maximum E_{ac} configuration it is parallel.

In Fig. 9, I display the temperature dependence of R_s and X_s for the Bi compound with both parameters normalized to their T = 100K value. The temperature independent numerical constant X_s^o, which reflects an unknown geometrical factor,

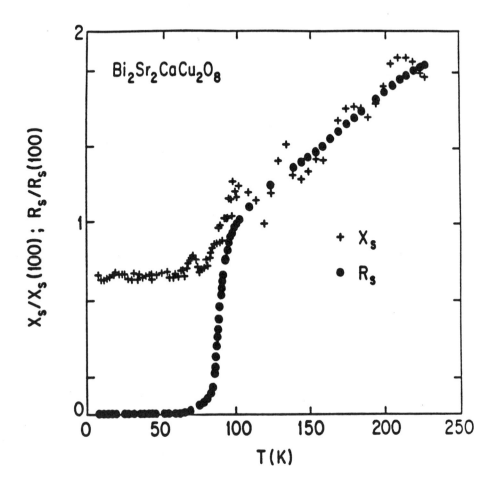

Fig. 9 Temperature dependence of the surface resistance and surface reactance for $Bi_2Sr_2CaCu_2O_8$ for a maximum H_{ac} configuration. Ref. 31.

has been evaluated in the following way. In the normal state above T_c, and in the so-called Hagen-Rubens limit ($\omega\tau < 1$ with τ the single particle relaxation time), Eq. (1) leads to

$$R_s = X_s = \left(\frac{\mu_0\omega}{2\sigma_1}\right)^{1/2} \qquad (10)$$

In other words, both R_s and X_s are expected to have the same value at all temperatures in the normal state. Optical experiments give a relaxation time $\frac{1}{\tau} \sim 500\text{cm}^{-1}$ at T=100K and larger $\frac{1}{\tau}$ values at higher temperatures, and consequently, the

Hagen-Rubens limit applies at our measurement frequency. We can then use Eq. (10) to establish the additive frequency shift Δf_o^o, or alternatively, X_s^o by requiring that $X_s = R_s$ in the normal state, and data presented in Fig. 9 have been obtained in this way. The observation that X_s is equal to R_s in the entire measured temperature range $T > T_c$ clearly demonstrates that the Hagen-Rubens limit applies, and strongly supports our procedure for evaluating X_s^o.

With the R_s and X_s values shown in Fig. 9 we can use Eq. (2) to evaluate σ_1 and σ_2. As discussed earlier, it follows from Eq. (2) that we can evaluate these components only up to a numerical factor A. This factor, however, depends only on the dimensions of the specimens and is the same for conductivity values both below and above T_c. Consequently, this factor is eliminated if $\frac{R_s}{R_N}$, $\frac{X_s}{X_N}$ or $\frac{\sigma_1}{\sigma_N}$ is derived. In Fig. 10 I display $\frac{\sigma_1}{\sigma_N}$ as a function of temperature using two different geometrical configurations, where the specimen was placed in the maximum electric or in a maximum magnetic field. The close similarity of the two results strongly suggests that demagnetization effects, and finite magnetic field induced loss mechanisms (due to losses and increased penetration into the superconductor due to frozen in flux, for example), do not play a role, and the results displayed on the figure reflect the electrodynamics of the London state.

The sharp peak observed slightly below T_c could be interpreted as evidence for case-II coherence. There are however several questions concerning the interpretation. First, because of the rather large anisotropy of the electronic structure, fluctuation effects are important in $Bi_2Sr_2CaCu_2O_8$ and the conductivity strongly increases above T_c due to superconducting fluctuations. In the Figure it was assumed[31] that $\sigma_{1N} = AT$, i.e. the conductivity has only the usual linear term, and consequently the strong increase may be largely due to fluctuation effects,[34] and not due to the enhanced density of states below T_c. Also displayed on the figure is $\sigma_1(T)$ calculated assuming the weak-coupling BCS limit; it is evident that it leads to a feature significantly broader than the experimental result.

Due to the increased dimensionality, fluctuation effects are suppressed in $YBa_2Cu_3O_7$, and consequently such effects are expected to play only a minor role in the electrodynamics. In Fig. 11 the normalized conductivity measured on single crystals is displayed as a function of temperature. We do not find an enhancement above T_c and a detailed analysis clearly indicates that the peak of $\sigma_1(T)$ occurs well below the superconducting transition where R_s is already significantly reduced from the normal state value. Also displayed on the figure is the nuclear spin-lattice relaxation rate,[35] and the temperature dependence of the optical conductivity,[36] neither showing a peak below T_c. A conductivity which increases below T_c has been found by several other experiments using different techniques. Utilizing a coherent domain spectrometer,[12] σ was measured as a function of frequency in the upper millimeter wave spectral range and a broad peak is recovered at low frequencies. Optical experiments at 60GHz,[37] and experiments at microwave frequencies on fine powders,[38] also lead to a peak; in both cases $\sigma_1(T)$ is rather similar to that displayed on Fig. 10.

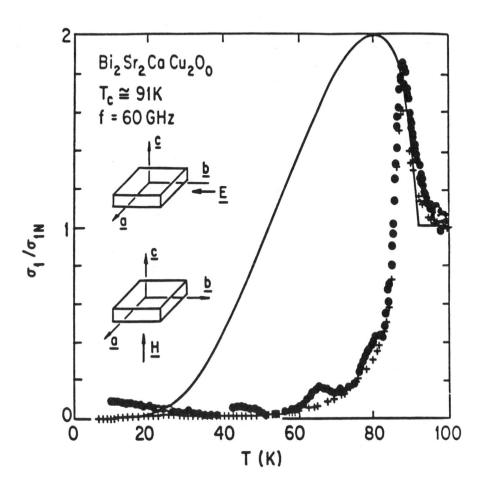

Fig. 10 Temperature dependence of σ_1 for $Bi_2Sr_2CaCu_2O_7$, as evaluated using a maximum E_{ac} and maximum H_{ac} configuration. The measurement configurations are also displayed on the Figure. The full line is calculated assuming $2\Delta = 3.52 k_B T_c$. Ref. 31.

Several interpretations have been advanced to account for the behavior of the conductivity and the spin lattice relaxation rate. The conductivity can be modeled by assuming a large gap, but neglecting other strong-coupling effects. Such a calculation describes $\sigma_1(T)$ with $2\Delta = 10 k_B T_c$, and if such a description is appropriate, it suggests a gap opening near T_c which is significantly more dramatic than that given by the weak coupling limit. This is also in accord with conclusions reached

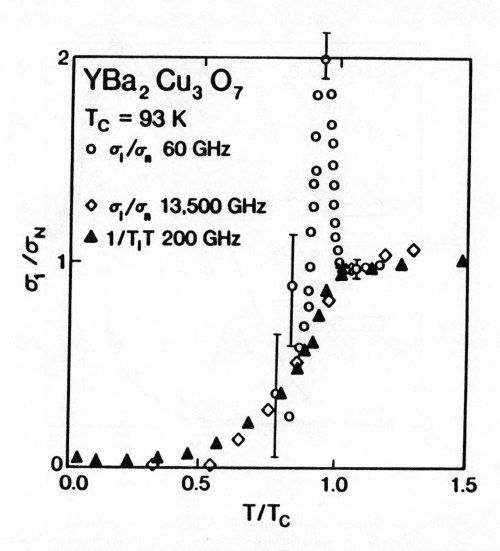

Fig. 11 Temperature dependence of σ_1 for $YBa_2Cu_3O_7$, together with the temperature dependence of the nuclear relaxation rate $1/T_1$ and σ measured at optical frequencies.

by examining the temperature dependence of R_s on films, as discussed earlier. The description also leads to a peak which rapidly diminishes as the measurement frequency is increased. Consequently, the fact that R_s decreases below T_c at optical frequencies can not by itself be regarded as evidence for unusual pairing as suggested.[36] Such an interpretation, however, does not account for the absence of a coherence peak in $1/T_1$. It has been suggested therefore that the increased conductivity reflects an increased relaxation time τ, which arises as the consequence of a freeze-out of low-frequency scattering processes. Such a model[12] also leads to a peak in $\sigma_1(T)$ below T_c, which also gets more pronounced at low frequencies.

5. CONCLUSION

The surface impedance of high temperature superconductors has been thoroughly explored both on thin films and on crystalline specimens. Both the temperature- and frequency-dependent losses measured in various thin films, prepared mainly by laser ablation, indicate that these losses arise, at least at temperatures above approximately 50K, from intrinsic mechanisms: they are due to the thermally excited carriers.

Experiments where the conductivity σ_1 is evaluated indicate an increase below T_c with respect to the conductivity in the normal state. This increase of σ_1 below T_c has been observed by several groups in $YBa_2Cu_3O_7$ and the behavior is by now well established — with some differences between $\sigma_1(T)$ observed by different techniques. Whether this is due to case-II coherence factors — i.e. due to the pileup of the density of states near the gap — or due to the freeze-out of scattering processes in the superconducting state, remains to be seen, and experiments at various frequencies may answer this question. Both $\sigma_1(T)$ and $R_s(T)$ imply the rapid opening of the single particle gap below T_c, with $\Delta(T)$ having a temperature dependence significantly stronger than that given by the weak coupling BCS limit. A larger $2\Delta(0)/k_B T_c$ ratio is in agreement with estimates based on recent optical and photoemission[39] studies, and results for the temperature dependence of the penetration depth.[40] Our results are also consistent with a nodeless gap, with no evidence for any significant density of states in the region below the superconducting gap.

In comparing various experiments which involve current flow, it is important to keep in mind the large anisotropy of these materials. For our configuration, the currents flow within the planes, and consequently R_s values referring to current flow within the planes are obtained. In contrast, for experiments in which the currents flow perpendicular to the planes, one may observe different superconducting characteristics, owing to the high anisotropy of these materials.

Due to the low loss, superior performance is expected at frequencies below about 100 GHz, with respect to passive devices made of normal metals. This has been clearly demonstrated by various groups, and by now resonators, delay lines and filters with unmatched performance are available[41] in the market.

With the feasibility of applications clearly demonstrated, it is believed that issues other than the loss mechanisms in the superconductor will determine the further development of the field. Low dielectric constant and low loss substrates, compatible with microwave applications are highly desirable, and questions on the convenience of various cooling schemes also play an important role.

ACKNOWLEDGMENTS

This research was partly supported by the California Competitive Technology program and partly by the University of California within the framework of the INCOR program. Many of the experiments discussed here were performed by L. Drabeck, O. Klein and K. Holczer. I enjoyed useful discussions with D. Scalapino, J. J. Chang, P. Littlewood and C. Varma. The samples which were investigated were provided by V. Venkatesan (Bellcore), J. R. Hammond (STI), L. Mihály (Stony-Brook) and by I. F. Shchegolev (Chernogolovka).

REFERENCES

1. C. Kittel, Quantum Theory of Solids, Wiley Publishing Co. (1963). M. Tinkham, Introduction to Superconductivity, (McGraw Hill, New York 1975).
2. For the early results on high T_c material see, J. Carini, L. Drabeck and G. Grüner, Mod. Phys. Lett. **3**, 5 (1989).
3. D. C. Mattis and J. Bardeen, Phys. Rev. **111**, 412 (1958).
4. A. A. Abrikosov.
5. J. Carini *et al.*, Solid State Comm. **67**, 373 (1988). A. N. Awasthi *et al.*, Solid State Comm. **67**, 373 (1988).
6. N. Klein *et al.*, App. Phys. Lett. **54**, 757 (1989). G. Müller *et al.*, IEEE Trans. Magn. **25**, 2402 (1989).
7. S. Sridhar and W. L. Kennedy, Rev. Sci. Instru. **59**, 571 (1988).
8. The cavity methods have been summarized by A. M. Portis, D. W. Cooke and E. R. Gray, Journal of Superconductivity **3**, 297 (1990). All high T_c cavities have also been fabricated. C. Zakopoulos, W. L. Kennedy and S. Sridhar, Appl. Phys. Lett. **52**, 2168 E. Minehara, R. Nagai and M. Takenchiu, J. Appl. Phys. **28**, 2100 (1989). W. J. Ratcliffe *et al.*, IEEE Trans. Magn. **25**, 990 (1989). V. F. Gantmakher *et al.*, Sov. Phys. JETP **68**, 833 (1989).
9. Z. L. Buranov and I. F. Shchegolev. Prib. Tekh. Eksp. **14**, 171 (1971).
10. S.M. Anlage *et al.*, Journal of Superconductivity **3**, 311 (1990). E. Belohoubek *et al.*, Proc. SPIE **1187**, 348 (1989).
11. A. A. Valenzuela and P. Rensel, Appl. Phys. Lett. **55**, 1029 (1989). D. E. Oates, A. C. Anderson and P. M. Mankiewich, Journal of Superconductivity **3**, 251 (1990). S. M. Anlage *et al.*, Appl. Phys. Lett. **54**, 2710 (1989). A. Fathy *et al.*, Microwave J. **3**, 75 (1988).
12. M. C. Nuss *et al.*, Phys. Rev. Lett. **66**, 3305 (1991).

13. J. R. Waldram, Adv. Phys. **13**, 1 (1964).
14. M. A. Biondi et al., Rev. Mod. Phys. **30**, 1109 (1958). M. S. Khaikin, Soviet Phys. JETP **35**, 961 (1958).
15. A. B. Pippard, Proc. R. Soc. (London) **A216**, 547 (1953).
16. O. Klein et al., to be published.
17. Experiments on Pb (K. Holczer, O. Klein and G. Grüner, Solid State Comm. **78**, 875 (1991) also lead to a coherence peak.
18. J. Carini et al., Phys. Rev. **B37**, 9726 (1988).
19. L. Drabeck et al., Phys. Rev. **B39**, 785 (1989), ibid IEEE Transactions on Magnetics **25**, 810 (1989).
20. L. Drabeck et al., Phys. Rev. **B40**, 7350 (1989). L. Drabeck et al., Phys. Rev. B. **39**, 785 (1989).
21. L. Drabeck et al., J. Appl. Phys. **68**, 892 (1990).
22. K. Scharnberg and D. Walker, Journal of Superconductivity, **3**, 269 (1990). D. Walker and K. Scharnberg, Phys. Rev. **B42**, 2211 (1990).
23. T. L. Hylton et al., Appl. Phys. Lett. **53**, 1343 (1988). T. L. Hylton and M. R. Beasley, Phys. Rev. **B39**, 789 (1989). A somewhat different description is given by J. Halbrittes, J. Appl. Phys. **68**, (1990).
24. A. Inam et al., Appl. Phys. Lett. **56**, 1178 (1990).
25. L. Drabeck et al., Journal of Superconductivity **3**, 317 (1990).
26. M. Hein et al., , J. Appl. Phys. **66**, 5940 (1989).
27. H. Eddy et al., J. Appl. Phys. (to be published). D. W. Cooke et al., J. Superconductivity **3**, 3 (1990) pp. 261-267.
28. D. Kalokities et al., J. Electron. Mat. **19**, (1990).
29. Several other studies also clearly establish that R_s is proportional to ω^2 both for intrinsic and extrinisc (i.e. grain boundary, etc. determined) surface resistance, see for example W. L. Kennedy, C. Zahopoulos and S. Sridhar, Solid State Comm. **70**, 741 (1989). D. W. Cooke et al., Solid State Comm. **73**, 297 (1990).
30. J. J. Chang and D. Scalapino, Phys. Rev. **B40**, 4299 (1989).
31. K. Holczer et al., Phys. Rev. Lett. **67**, 152 (1991).
32. O. Klein et al., to be published.
33. L. Forro et al., Phys. Rev. **B42**, 8704 (1990). L. Forro et al., Phys. Rev. Lett. **65**, 1941 (1990).
34. M. L. Horbach et al., Phys. Rev. Lett. (submitted).
35. M. Takigawa et al., Phys. Rev. **B39**, 7371 (1989). S. E. Barrett et al., Phys. Rev. **B41**, 6283 (1990).
36. R. T. Collins et al., Phys. Rev. **B43**, 8701 (1991).
37. P. H. Kobrin et al., Physica **C176**, 121 (1991).
38. H. M. Cheak, A. Porcha nd J. R. Waldram, Physica **B165**, 1195 (1990).
39. C. G. Olson, Physica **B169**, 112 (1991).
40. S. Sridhar et al., Phys. Rev. Lett. **63**, 1873 (1989).
41. R. B. Hammond, This conference.

DISCUSSION

M. Tinkham: I had a question about one of your graphs showing scatter of data on the losses in films with temperature. You said it was pretty good, yet it looked to me like it bottomed out and then it continued to go down at still lower temperatures.

G. Grüner: Yes.

M. Tinkham: Is that serious? That's unusual, isn't it?

G. Grüner: Yes. Now, what happens is that we usually find this behavior to be very much magnetic-field-dependent. My feeling is that somehow it reflects frozen-in flux. We systematically find this bumpy behavior, and we did a fair amount of work on studying it. You can enhance this effect with a magnetic field, and that's why I think that it's an extrinsic effect.

F. M. Mueller: George, you're being very pessimistic based on your present experience about the quality of the films and whether there could be any improvement. Now, in the first talk we heard today I saw some stunning material which no one, and certainly not me, would have ever thought could have been made. Are you really that pessimistic? Has it been long enough, and is this a realistic estimate, as to whether these things can be applied or not?

G. Grüner: No, I'm not a pessimist, I'm a realist, and . . .

F. M. Mueller: Well I'm hinting that . . .

G. Grüner: No, no, no, no, no, wait, wait, wait. Wait. If you think that you can make better and better material and the surface resistance will be smaller and smaller and smaller and smaller – that's not true. I mean, at a certain temperature, okay, there is going to be a fundamental limit which is determined by the thermally excited electrons. And you cannot suck them out. So, the fact that you cannot lower the losses . . .

F. M. Mueller: Without limit? I agree with that statement.

G. Grüner: Okay, and at relatively high temperatures. If you go to low temperatures then you can improve it, and if you go to niobium the surface resistance is still five orders of magnitude smaller than what I am talking about. My feeling is that if you want to limit your applications here, around 77 K for example, you are not going to see an improvement. You will see an improvement as far as high-power device performance is concerned if you solve these flux-pinning problems and so on. The other thing is that these materials are really very good, okay. We are talking about, one or two orders of magnitude better than Cu for the conditions I discussed, which is not bad.

F. M. Mueller: Suppose we imagine that Tanaka's 150 K superconductor were developed. Would you have the same opinions?

G. Grüner: Right. There are two improvements, and for example that's why the thallium R_s we saw was much better than the 1-2-3. I had scaled transition temperature, and the thallium has significantly higher transition temperatures, so at the same real temperature you will have a lower loss. So you have to go up to higher transition temperature. There will be two improvements; one is T/T_c will be smaller at a particular temperature. Second, the gap is going to be larger and therefore you will get a larger reduction.

D. Scalapino: Following up on what you were saying. I agree with what you say, that it looks like we're certainly approaching a limit. If one weren't afraid of altering the T_c of a material, or its critical currents, by putting in some type of impurity to give a shorter, elastic mean free path, that would be a direction to go. But of course anybody who's done that knows that's usually disastrous for the main effect so it's hardly a way around what George says. My question to you though is, what's the possibility of getting good material for, say, a single-layer cuprate film that goes superconducting at about ten degrees? Is there anybody who might supply you that where you once again look particularly at the coherence factors for a material that has a lower T_c just to see what it's doing?

G. Grüner: This is a question to the audience, or to me?

D. Scalapino: I'm wondering whether you would be willing to do that, and whether you can get the materials?

G. Grüner: Yes, I'm willing to do that, if somebody gives me a good material.

A. Kapitulnik: The group at Varian are making very good films now of the single-layer material.

M. P. Maley: What's the maximum surface field in your power measurements at which the losses begin to increase dramatically? . . . I've done this same kind of measurement on single crystals to see the effect of grain boundaries and that sort of thing.

G. Grüner: We really did not focus on that too much, because we just wanted to eliminate all of these kinds of effects. Our experiments were conducted in the low-ac-field limit. We operated at low power levels and changed the power by an order of magnitude to see whether there is any change in R_s. You can calculate it, but I don't have a figure on top of my head. Bob Hammond may have a figure related to it.

R. Hammond: Yes, I'll say more about that on Saturday.

S. A. Wolf: I have a comment. These materials are improving. I can quote you for a film at four degrees. It's actually better than the single crystals at four degrees. This is a film that was made in England, and the latest film actually broke down precipitously at a surface magnetic field of 200 gauss. Just to put that in perspective that's about where niobium was only six or seven years ago, and that corresponds to building a cavity which would have an accelerating field of about 6 MeV/m, which is quite considerable. Now the figures at 77 K of course are worse, but it shows that these materials are becoming quite reasonable, and in fact the surface resistance is beginning to show more or less an exponential decay all the way down in temperature without some of these bumps and wiggles, if you're careful to shield the field from the cavity.

[Question:] What gap do you get?

S. A. Wolf: You get about 2.5 $k_B T_c$.

J. R. Schrieffer: Doug Scalapino, Nijat Bulut, and I have been thinking about existence of the Hebel-Slichter peak in the spin channel and the conductivity peak in the charge channel of high-T_c materials.

Since BCS is a very simple theory, it does not distinguish between spin-flip and non-spin-flip processes because there are no exchange interactions included.

When we include exchange interactions we find the Hebel-Slichter peak in $1/T_1$ is suppressed while the corresponding peak in σ_1 is not.

To see this you consider an up-spin incoming quasiparticle and the down-spin outgoing quasiparticle as sort of a spin-one electron-hole pair; then the electron attracts the hole by the exchange interaction, producing a low-energy spin fluctuation or collective mode rather than an incoherent quasiparticle scattering process. For strong binding the ultimate decay of the collective mode into two quasiparticles has little effect.

However, since the current operator entering σ_1 does not flip the spin, then according to the Hubbard model the particle and hole do not interact and no collective mode is formed and the pairing theory is adequate.

2. Textured Materials and Josephson Junctions

Chair — Seb Doniach

M. Murakami
ISTEC, Superconductivity Research Laboratory
1-10-13, Shinonome, Koto-ku, Tokyo 135 Japan

Flux Pinning Enhancement by Y_2BaCuO_5 Inclusions in Melt Processed YBaCuO Superconductors

While nonsuperconducting particles are known to serve as effective pinning centers in conventional superconductors, their effect in high T_c superconductors is still controversial. In this paper, we give evidence that nonsuperconducting Y_2BaCuO_5 (2-1-1) inclusions can act as pinning centers in melt processed YBaCuO superconductors even when their size is orders of magnitude larger than the coherence length.

1. INTRODUCTION

Flux pinning enhancement in bulk high temperature superconductors is critical for achieving high J_c values required for practical applications. It has been reported that neutron irradiation is effective in raising the flux pinning force both in the Bi and the Y systems,[1,2] although the kind of defects responsible for this flux pinning enhancement is not clear.

It is also proposed that Y_2BaCuO_5 (2-1-1) inclusions can also enhance flux pinning in the YBaCuO system.[3] Other nonsuperconducting particles such as Ag[4] and $BaSnO_3$[5] have also been reported to be effective in enhancing flux pinning. The beneficial points of the pinning enhancement due to these nonsuperconducting particles is that they are stable and can be dispersed artificially. However, it has been suggested that since the size of these nonsuperconducting particles is relatively large compared to the coherence length, their contribution should be minimal.

It is known that interface pinning is essential in the case of large nonsuperconducting particles. A comparison of J_c values for YBaCuO samples containing the 2-1-1 inclusions with different sizes supports the fact that the interface can contribute to flux pinning.

In this paper, I give evidence of flux pinning enhancement by the presence of 2-1-1 inclusions.

2. SAMPLE PREPARATION

In the YBaCuO system, the superconducting phase $YBa_2Cu_3O_x$ (1-2-3) is produced by the following reaction:

$$Y_2BaCuO_5 + L(3BaCuO_2 + 2CuO) \rightarrow 2YBa_2Cu_3O_x$$

Using this reaction it is possible to disperse Y_2BaCuO_5 (2-1-1) inclusions in the 1-2-3 matrix. It has also been found that the size and the content of the 2-1-1 inclusions can be changed by controlling the starting compositions and processing conditions. In this study, YBaCuO samples containing 2-1-1 inclusions with different amounts and sizes were prepared by several melt processes such as MTG, QMG, and MPMG.[6] Precursor powders were prepared by melt-quenching (MPMG), mixing of 2-1-1 and BaCuO, mixing of Y_2O_3 and BaCuO, and sintering (MTG). The powders were pressed into pellets, reheated to temperatures above the peritectic reaction and then slowly cooled through the peritectic temperature at a rate of 1°C/h. Fig. 1. shows typical optical microstructures of melt processed YBaCuO.

3. COMPARISON OF CRITICAL CURRENTS

A comparison of critical current density was performed only in the condition where the magnetic field is parallel to the c axis. Although little attention has been paid to this, it is very critical for comparison of flux pinning.

Critical currents which flow in the ab plane strongly depend on the field direction. When the field is applied perpendicular to the c axis, J_c values are usually very high and are extremely difficult to measure by direct transport measurements due to heating at the current contacts[7] (see Fig. 2). In this field direction, the upper critical field (H_{c2}) is at least three times larger than that with the field parallel to the c axis, for which case J_c drops rapidly in a magnetic field.

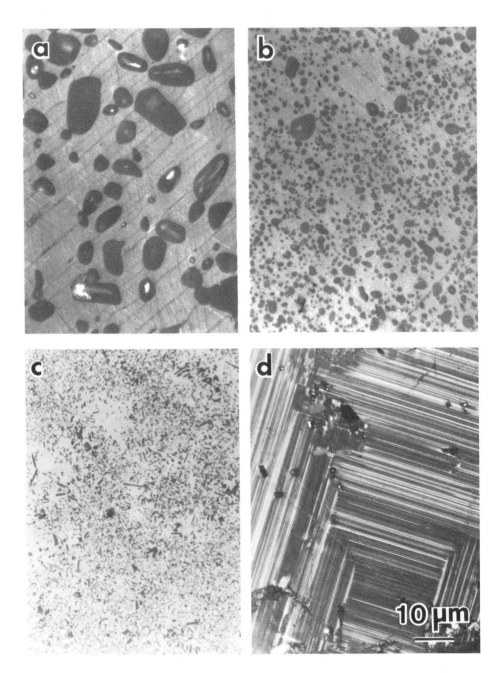

Fig. 1 Optical micrographs for YBaCuO samples prepared by different melt processes: (a) MTG; (b) modified MTG; (c) MPMG (Y:Ba:Cu=1.5:2.25:3.25); (d) MPMG (Y:Ba:Cu=1:2:3). Note that the size of and the amount of the 2-1-1 inclusions can be changed by controlling the processing conditions.

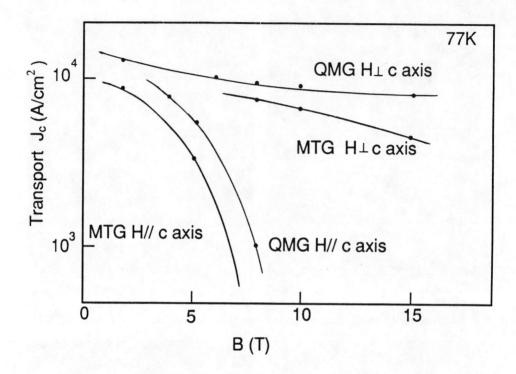

Fig. 2 Transport J_c versus B for YBaCuO samples prepared by the QMG and the MTG processes.[7] It is notable that large J_c values are obtained even at 77K and 15T for both samples when the magnetic field is applied perpendicular to the c axis. J_c values at low fields are difficult to measure. In contrast, the J_c values for the field parallel to the c axis drop rapidly in a magnetic field. The amount of the 2-1-1 inclusions is larger in a QMG sample than in an MTG sample.

We have also observed many stacking faults perpendicular to the c axis independent of the presence of nonsuperconducting particles in melt grown YBaCuO crystals[8] as shown in Fig. 3. These stacking faults can serve as pinning centers and increase the J_c values effectively in this field direction.[9]

It should also be noted that a number of cracks are formed along the ab planes as shown in Fig. 4 and they can also serve as strong pinning centers when the field is in this direction (see Fig. 5). Therefore, the transport J_c values for this field direction result from the combined flux pinning effects due to the stacking faults, the cracks and the other defects including nonsuperconducting particles. Such combined effects complicate the analysis and will smear the effect of the nonsuperconducting particles.

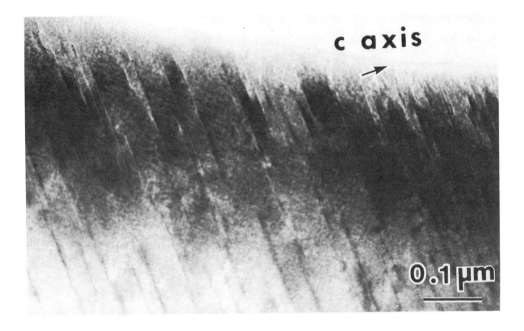

Fig. 3 Transmission electron micrograph of melt processed YBaCuO without 2-1-1 inclusions. Note that a number of stacking faults are present along the *ab* plane. Such defects will contribute to flux pinning when the field is applied parallel to the *ab* plane.

Magnetization measurements can be used as a contactless method to obtain J_c values. However, in contrast to transport measurements, magnetization J_c values with the field perpendicular to the c axis are usually much lower than those with the field parallel to the c axis. This causes great confusion when a comparison of flux pinning is made. The source of such a large difference can be easily understood by considering the fact that for the field perpendicular to the c axis, the shielding currents must flow along the c axis in magnetization measurements (see Fig. 6). It is important to note that cracks do not aid in increasing the flux pinning force for such currents. Rather they will allow field penetration and reduce the path for the currents and thereby result in a significant decrease in magnetization or estimated J_c values. It should also be noted that an anisotropy in the elementary pinning force (f_p) exists due to the elliptical vortex structure (see Fig. 7). Even when the pinning sites are isotropic, flux pinning is intrinsically smaller for the currents along the c axis, which also reduces the J_c value.

In this paper, I used J_c values obtained from magnetization measurements with the field parallel to the c axis for the comparison of flux pinning. By this technique: (1) we can avoid the reduction of J_c due to heat generation in a transport

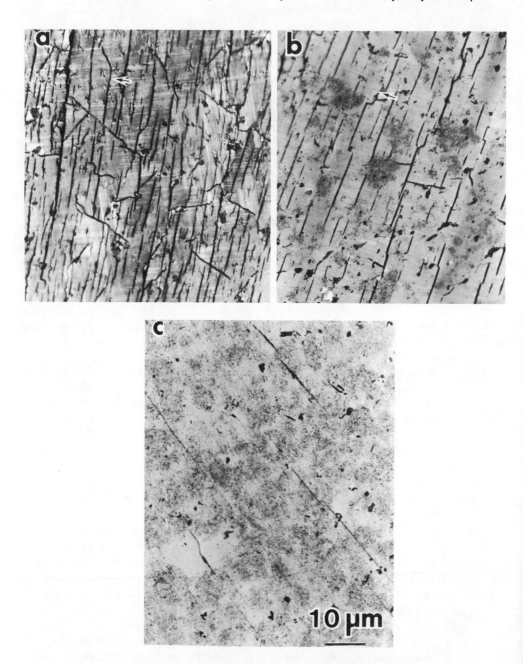

Fig. 4 Optical micrographs for MPMG processed YBaCuO samples containing (a) 0% 2-1-1; (b) 10% 2-1-1; and (c) 25% 2-1-1 viewed from a direction perpendicular to the c axis. The arrow indicates the direction of the c axis. Note that a number of cracks are observed along the ab plane. The amount of cracking can be reduced by dispersing 2-1-1 inclusions.

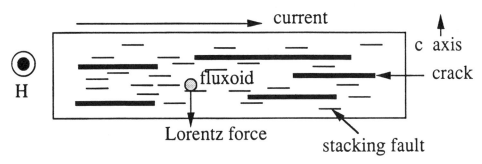

Fig. 5 Schematic illustration showing that the stacking faults and cracks along the ab plane contribute to flux pinning in transport J_c measurements when the field is perpendicular to the c axis. In this condition, the currents flow along the ab plane and are not impeded by the stacking faults and the cracks. Furthermore, these defects effectively pin the fluxoids against the Lorentz force.

measurement; (2) we do not need to take the anisotropy into account (see Fig. 8); (3) we can minimize the effects of the stacking faults and the cracks (see Fig. 9). It then becomes possible to clarify the effects of nonsuperconducting particles on flux pinning.

It has also been confirmed that magnetization hysteresis scales with sample thickness for every sample (see Fig. 10), which guarantees that the critical state is established in the whole sample[10] and thereby J_c is obtainable from the following relation:[11]

$$J_c = (M^+ - M^-)/d$$

where M^+ and M^- are the values of the magnetization (A/m) measured during application of increasing and decreasing field, and d is the thickness (m). A typical dimension of the sample was $4 \times 3 \times 1$ mm^3 with its long axis parallel to the c direction.

It was also confirmed from magneto-optic observation that the critical state is really established in the melt processed YBaCuO samples as will be shown later.

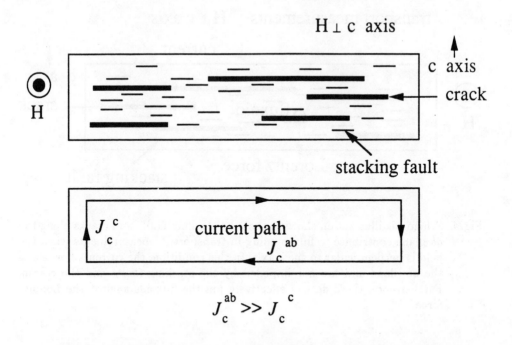

Fig. 6 Schematic illustration showing how the currents flow in magnetization measurements with the field perpendicular to the c axis. Although the field direction is the same as in the transport measurements, the currents must flow along the c axis. In this case, the stacking faults and the cracks will reduce J_c significantly. As presented in Fig. 4, the introduction of the 2-1-1 inclusions is effective in reducing the amount of cracking, and thereby enhances magnetic hysteresis in this field direction. Therefore, magnetization measurements should not be used for comparison of flux pinning in this field direction, especially when the effects of nonsuperconducting particles are to be clarified.

3.1 Flux creep

Flux creep is caused by the escape of the pinned fluxoid from the pinning center with the help of thermal activation,[12] and therefore is strongly correlated with flux pinning. We measured flux creep for YBaCuO samples with different 2-1-1 contents using a SQUID magnetometer at 77K in various magnetic fields by detecting magnetization decay in the course of increasing and decreasing the field.

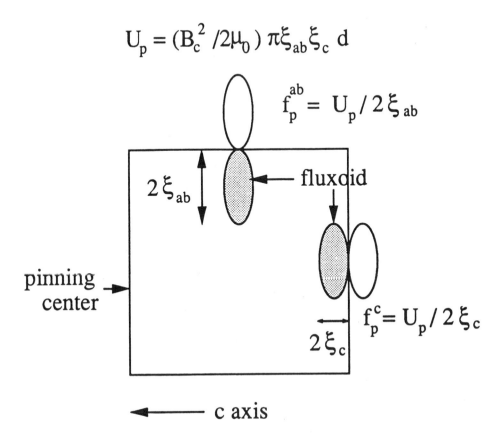

Fig. 7 Anisotropy in the elementary pinning force (f_p) when the fluxoids are aligned perpendicular to the c axis. Since the fluxoid core is elliptical, the pinning force depends on the direction. The anisotropy in f_p is ξ_{ab}/ξ_c.

3.2 Irreversibility line

The irreversibility line was determined from magnetization loops as the field where the magnetization hysteresis disappears.

3.3 Observation of flux pinning sites

The most direct way to observe flux pinning sites is a decoration of ferromagnetic particles on a superconductor in the mixed state.[13] We deposited Ni particles on a YBaCuO sample after it was field-cooled at 10Oe down to 10K and the external field was removed.[14]

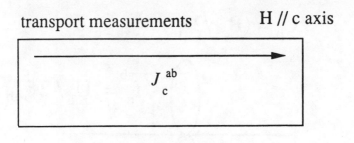

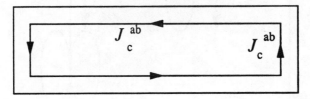

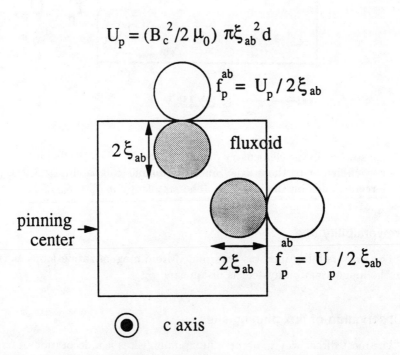

Fig. 8 Schematic illustration showing that the flux pinning is isotropic when the magnetic field is applied parallel to the c axis. In this field direction, the transport J_c equals to magnetization J_c.

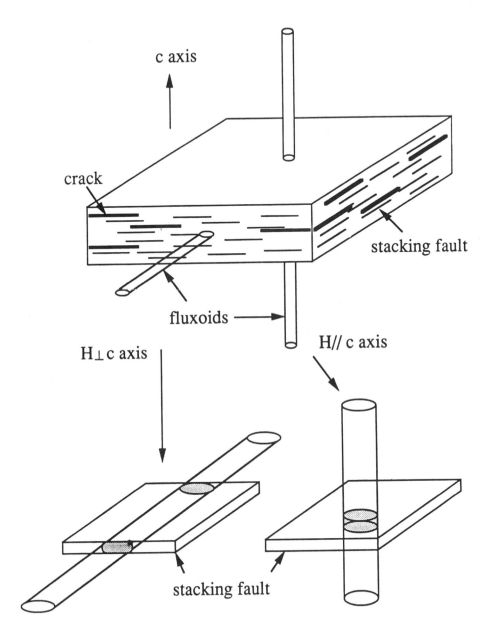

Fig. 9 Schematic illustration of the interaction between the fluxoid and the planar defects along the ab plane for the field parallel and perpendicular to the c axis. Flux pinning by stacking faults and cracks is very effective for the field perpendicular to the c axis, while it is minimal for the field parallel to the c axis.

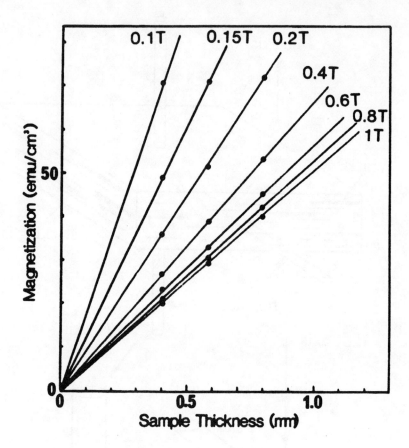

Fig. 10 Magnetization hysteresis $\Delta M(M^+ - M^-)$ versus sample thickness (d). Note that ΔM scales with d, which guarantees that the critical state is established in the entire sample and J_c can be calculated from the magnetization hysteresis.

We also used the Faraday effect of an iron garnet film with perpendicular magnetization.[15] The domain structure observed optically in the iron garnet film placed on the YBaCuO samples reveals the magnetic field distribution.

4. RESULTS OF COMPARISON

4.1 Critical currents

Fig. 11 shows the J_c-B properties for three different YBaCuO samples with different 2-1-1 contents. Here the size of the 2-1-1 inclusions was maintained at about 1 micron, since their size also affects flux pinning. It is clear that J_c can be increased by increasing the 2-1-1 content, supporting the fact that the 2-1-1 inclusions contribute to flux pinning.

4.2 Flux creep

Fig. 12 shows flux creep for YBaCuO samples with and without 2-1-1 inclusions. It is clear that the flux creep rate can be reduced by introducing the 2-1-1 inclusions, which also supports the fact that the 2-1-1 inclusions are pinning centers.

4.3 Irreversibility line

Fig. 13 shows irreversibility lines for a YBaCuO sample with 2-1-1 inclusions 10% by volume and for a single crystal. It is obvious that the irreversibility line can be shifted toward the higher H-T region by introducing the 2-1-1 inclusions. This result also supports the fact that the dispersion of the 2-1-1 inclusions is effective in enhancing flux pinning.

4.4 Direct observation of flux pinning

4.4.1 Magneto-optical effects

Fig. 14 shows how the flux penetrates into a superconductor. Here the white region corresponds to a large single domain of an iron garnet film. For the superconductor it is the region where the flux penetrates. In the case of the sample in which the 2-1-1 inclusions are distributed homogeneously, the field enters the sample from the edge with an almost uniform front. The concentration of magnetic field due to demagnetizing effects is also observed at the edge of the sample at low fields. It is also notable that the field penetrates into the superconductor in the manner predicted by the critical state model. The field gradient is related to J_c by $dH/dx = J_c$ and the J_c values obtained from this relation agreed very well with the J_c values calculated from magnetization hysteresis. These results clearly demonstrate that the critical state is established even in high temperature superconductors.

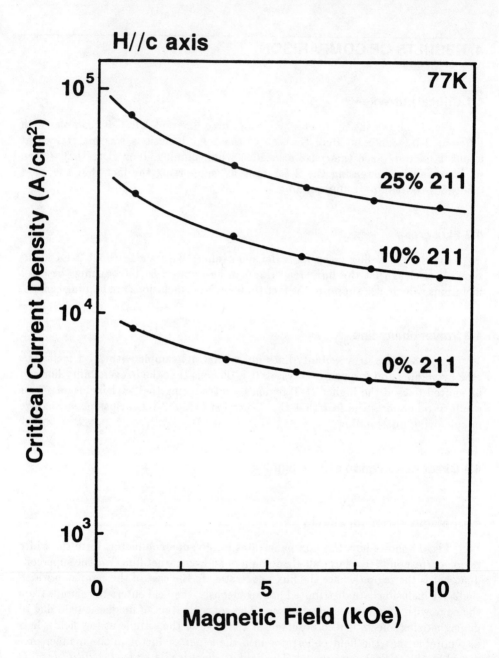

Fig. 11 Magnetic field dependence of J_c values for YBaCuO samples with different 2-1-1 contents with the field parallel to the c axis. Note that J_c is improved by increasing the 2-1-1 content.

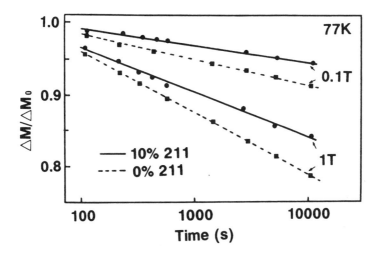

Fig. 12 Time decay of magnetic hysteresis (M^+-M^-) for two YBaCuO samples with and without 2-1-1 inclusions. Note that the flux creep rate can be reduced by introducing 2-1-1 inclusions.

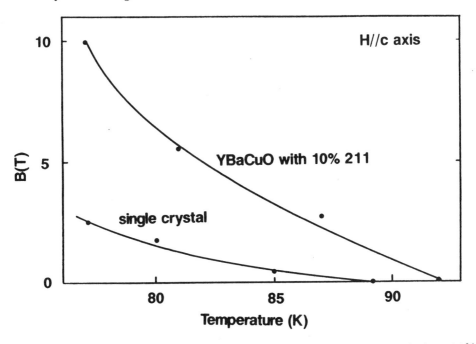

Fig. 13 Irreversibility lines for a YBaCuO sample containing 2-1-1 inclusions 10% by volume and for a single crystal. Note that the irreversibility line can be shifted to higher H-T region by introducing the 2-1-1 inclusions.

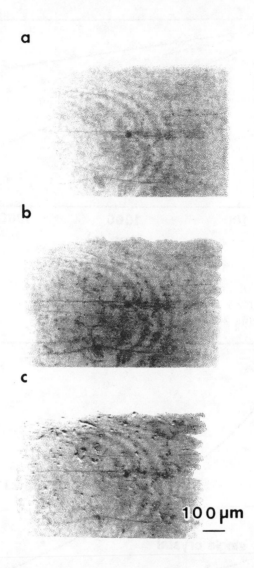

Fig. 14 Observation of flux distribution for a YBaCuO sample with 2-1-1 inclusions distributed uniformly. The observation was performed at 10K in magnetic fields of (a) 84Oe; (b) 209Oe; and (c) 753Oe in the increasing field process. At 84Oe, the magnetic field is completely shielded by the superconductor and concentration of the field is observed at the edge due to the demagnetizing effect. At 209Oe and 753Oe, the magnetic flux penetrates into the superconductor with a smooth flux front, which agrees very well with the behavior of strongly pinned type-II superconductors predicted by the critical state model.

However, in the sample where the distribution of the 2-1-1 inclusions is inhomogeneous, the field penetration is not uniform as shown in Fig. 15. The dark region corresponds to a very fine domain structure. For the superconductor it is the region where magnetic flux does not penetrate, indicating that the flux pinning force is larger in such a region than in the white region where the magnetic field has penetrated. Optical microscopic observation revealed that the density of the 2-1-1 inclusions is higher in the dark region and thus shows higher pinning force, which also supports the fact that the 2-1-1 inclusions are pinning centers.

Fig. 15 Observation of flux distribution for a YBaCuO sample with 2-1-1 inclusions distributed inhomogeneously at 77K and 850Oe. The region where the flux cannot penetrate is observed, indicating that the flux pinning force is larger in such a region. Optical microscopic observation revealed that the density of the 2-1-1 inclusions is large in this region, which supports the fact that the 2-1-1 inclusions provide flux pinning centers.

4.4.2 Decoration with ferromagnetic particles

Fig. 16 shows the distribution of Ni particles deposited on a YBaCuO sample containing the 2-1-1 inclusions. It is interesting to note that Ni particles are present only on the 2-1-1 or the interface of the 2-1-1/1-2-3. This result strongly supports the fact that the 2-1-1 inclusions provide flux pinning sites in melt processed YBaCuO.

5. DISCUSSION

5.1 The size of pinning centers

There seems to be a misunderstanding that the size of pinning centers must be of the order of the coherence length. There is much evidence that large precipitates contribute to flux pinning in conventional superconductors.[16,17] Fig. 17 shows a schematic illustration of the pinning potential for a large and a small pinning site. In the case of the large pinning site, there is no energy difference for flux motion inside and therefore flux pinning occurs at the interface.[18] Because of this, the effectiveness of flux pinning per unit volume of the precipitates is not as high as that of the small pinning centers (see Fig. 18). However, it should be noted that the pinning potential is deeper than for the small pinning site. For example, the pinning potentials are $(B_c^2/\mu_0)\pi\xi^2 d$ and $(B_c^2/\mu_0)(4/3)\pi\xi^3$ for a large and a small pinning site, respectively. Here, B_c is the thermodynamical critical induction, ξ is the coherence length and d is the size of the pinning site. When the size of the pinning site (d) is 1 micron as in the case of the 2-1-1 inclusions, the pinning potential becomes at least two orders of magnitudes larger than a pinning site of the order of the coherence length. The latter gives a very small pinning energy of around 0.01eV, which is almost comparable to the thermal energy at 77K. This difference seems to be very important for flux pinning of oxide superconductors at higher temperatures.

The thermal energy becomes larger with increasing temperature, and when it becomes equal to the depth of the pinning potential well, such pinning sites become ineffective. Of course, many pinning sites may contribute to the pinning of one fluxoid collectively. But as suggested by many researchers, the elastic constant to bend a fluxoid, C_{44}, seems to be very small in high T_c superconductors,[19] which will result in small $l_{44}=(C_{44}/a_L)^{1/2}$, the characteristic length for the bending where a_L is the Labusch parameter. Since two pinning centers separated by a distance larger than l_{44} cannot pin a fluxoid collectively as shown in Fig. 19, small C_{44} disfavors the collective contribution of pinning centers. The relatively large dissipation of high J_c thin films may be attributed to the combination of this small C_{44} and the

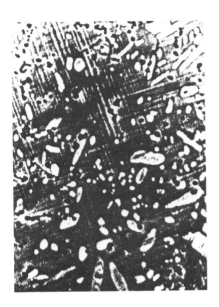

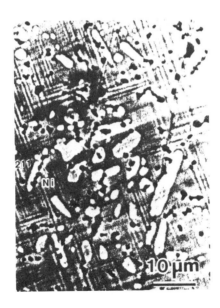

Fig. 16 Observation of flux distribution for a YBaCuO sample containing the 2-1-1 inclusions using the Bitter pattern technique. Note that the FLL has no long range order and the fluxoids are pinned by the 2-1-1 or the 2-1-1/123 interfaces. This result also strongly supports that the 2-1-1 inclusions are pinning centers.

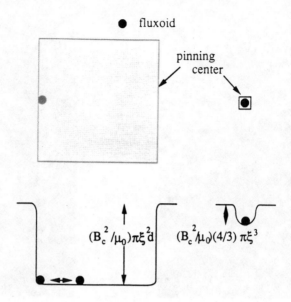

Fig. 17 Schematic illustration of the flux pinning potential for a large and a small pinning site. The pinning potential is much deeper in the large pinning site

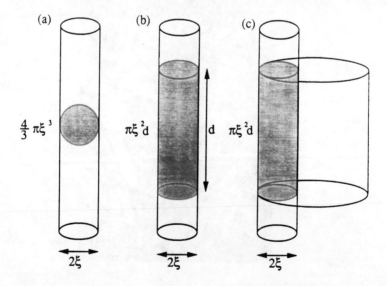

Fig. 18 Schematic illustration of the interaction between the fluxoid and the pinning centers with different sizes. In the case of a small pinning site (a), the whole volume contributes to flux pinning, while only the interface contributes to flux pinning in the case of a large pinning site (c), therefore, the effectiveness of flux pinning per unit volume is larger in the smaller pinning site.

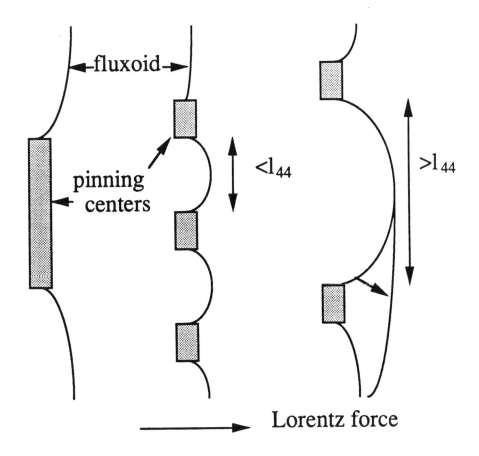

Fig. 19 When the elastic constant for the bending, C_{44}, is small, a fluxoid segment smaller than l_{44} can escape from the pinning center with the help of thermal activation.

relatively small size of the pinning sites, while the pinning energy of a YBaCuO sample containing 2-1-1 inclusions of about 1 micron is much higher despite smaller J_c values, which may support the assertion that the depth of the pinning potential well should be large to overcome thermal activation at high temperature.

On the other hand, it is also true that the size of 2-1-1 particles is still not optimized. Before we discuss the size effects, we briefly review the estimation of the pinning force due to large particles.

5.2 The bulk pinning force and J_c

While the elementary pinning force (f_p) can easily be obtained, the statistical summation of f_p is not simple except for a very simple case, because the elasticity of the flux line lattice (FLL) must be taken into account. When the FLL is absolutely rigid, the total pinning force (F_p) is zero. In the contrary limit, F_p can be simply obtained by $F_p = Nf_p$ when the FLL is plastic or when the elastic constant for the FLL shear, C_{66}, is very small. Here N is the number of the interactions between a pinning center and a fluxoid. In the elastic lattice, F_p is described as a power series in f_p and also as a function of the elastic constants of the FLL.

In the case of high temperature superconductors, it is known that the FLL is very soft, or that C_{66} is very small due to a large $\kappa(=\lambda/\xi)$. This indicates that each fluxoid will be placed at an energy minimum, which is supported by the direct observation of the FLL presented in Fig. 15, where the FLL has no long range order and the fluxoids are pinned by the 2-1-1 inclusions. In such a case, a direct summation provides a good approximation for obtaining the bulk pinning force.

Let us assume that the sample contains 2-1-1 inclusions N_p per unit volume and d in size. Then the number of interactions N per unit volume becomes

$$N = N_p(d/a_f)$$

where a_f is the FLL spacing and $a_f = 1.075(\phi_0/B)^{1/2}$ where ϕ_0 is the flux quantum and B is the magnetic induction. From the relation $F_p = J_c B = N f_p$, J_c is obtained as

$$J_c = \pi B_c^2 \xi N_p d^2 / 4\mu_0 \phi_0^{1/2} B^{1/2}$$

The estimation of J_c using the above relation agrees very well with the experimental results presented in Fig. 11. It is also notable that this relation suggests that J_c can be increased by decreasing d. Fig. 20 shows J_c (77K and 1T) as a function of the grain size of the 2-1-1 inclusions. It is clear that J_c is proportional to $1/d$, which also supports the fact that the 2-1-1/1-2-3 interface provides flux pinning (see Fig. 21). It is also important to realize that in order to improve J_c, we need to reduce the size of the 2-1-1 inclusions.

6. SUMMARY

In this paper, I have shown that nonsuperconducting Y_2BaCuO_5(2-1-1) particles can enhance flux pinning in $YBa_2Cu_3O_7$, even when their size is larger than the coherence length and the FLL spacing. In such a case, the interface provides pinning. Theoretical estimates based on direct summation agree very well with the experimental results. The application of direct summation for obtaining the bulk pinning force is justified by the direct observation of the FLL, where the FLL has no long-range order and the fluxoids are pinned by the 2-1-1 inclusions.

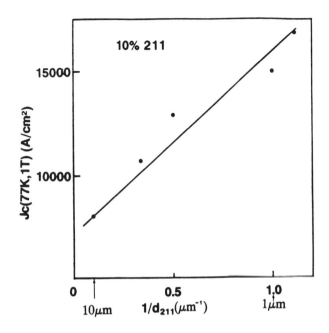

Fig. 20 J_c versus the reciprocal of the average grain size of 2-1-1 inclusions. Note that J_c is increased by reducing the size of the 2-1-1.

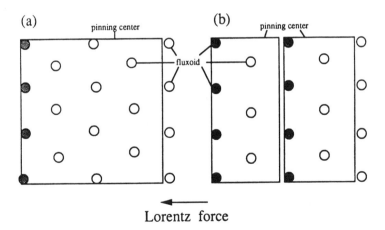

Fig. 21 Schematic illustration of the interaction between the fluxoids and large pinning centers. The shaded fluxoids are pinned at the interface. The number of pinned fluxoids can be increased by reducing the size of the pinning center provided that the volume fraction of pinning centers is maintained.

ACKNOWLEDGMENTS

I am grateful to Mr. K. Yamaguchi, H. Fujimoto, S. Gotoh, Dr. N. Koshizuka and Dr. S. Tanaka of Superconductivity Research Laboratory for valuable discussions. I would like to thank Dr. Y. Higashida and Dr. Y. Kubo of Japan Fine Ceramics Center for their observation of the FLL using the decoration of ferromagnetic particles technique.

REFERENCES

1. H. W. Weber and G. W. Crabtree, "Studies of High Temperature Superconductors," (A. V. Narlikar ed., Nova Science Publishers, New York, 1991) vol. 9 in press.
2. H. Kupfer *et al.*, Cryogenics **29** (1989) 268.
3. M. Murakami *et al.*, Supercond. Sci. Technol. **4** (1991) S43.
4. C. Y. Huang, H. H. Tai and M. K. Wu, Mod. Phys. Lett. B3 (1989) 525.
5. J. Shimoyama *et al.*, Jpn. J. Appl. Phys. **29** (1990) L1999; P. J. McGinn *et al.*, Appl. Phys. Lett. **59** (1991) 120.
6. For MTG, S. Jin *et al.*, Appl. Phys. Lett. **52** (1988) 2974. For modified MTG, K. Salama *et al.*, Appl. Phys. Lett. **54** (1989) 2352. For QMG, M. Murakami *et al.*, Jpn. J. Appl. Phys. Lett. **28** (1989) L1125. For MPMG, M. Murakami *et al.*, High-Temperature Superconductors Materials Aspects (ed. H. C. Freyhardt, R. Flugiker and M. Peuckert, DGM, Oberursel, 1991) 13; Z. Lian *et al.*, Supercond. Sci. Technol. **3** (1990) 490; D. Shi *et al.*, J. Appl. Phys. **68** (1990) 228; H. Hojaji *et al.*, Mat. Res. Bull. **25** (1990) 765.
7. J. W. Ekin *et al.*, Appl. Phys. Lett. **59** (1991) 360; K. Kimura *et al.*, Paper presented at JIM annual spring meeting (1991).
8. K. Yamaguchi *et al.*, J. Mat. Res. **6** (1991) 1404.
9. K. Watanabe *et al.*, Appl. Phys. Lett. **56** (1990) 1490.
10. M. Murakami *et al.*, Mod. Phys. Lett. **4** (1990) 163.
11. C. P. Bean, Phys. Rev. Lett. **8** (1962) 250.
12. P. W. Anderson, Phys. Rev. Lett. **9** (1962) 309.
13. H. Trauble and U. Essmann, J. Appl. Phys. **39** (1968) 4052.
14. Y. Higashida *et al.*, Physica C to be published.
15. S. Gotoh *et al.*, Jpn. J. Appl. Phys. **29** (1990) L1083.
16. R. I. Coote, J. E. Evetts and A. M. Campbell, Canad. J. Phys. **50** (1972) 421.
17. K. Yamafuji *et al.*, Cryogenics **31** (1991) 431.
18. T. Matsushita, Jpn. J. Appl. Phys. **20** (1981) 1955.
19. P. H. Kes and J. van den Berg, "Studies of High Temperature Superconductors," (A. V. Narlikar ed., Nova Science Publishers, New York, 1991) vol. 8.

DISCUSSION

N. V. Coppa: Rather than considering the 2-1-1 as causing a square-well potential in the pinning potential, why not, or do you, consider there to be a pseudo-third-phase at the interface and thereby causing a local pinning-potential minimum in that region?

M. Murakami: Well I don't think so.

N. V. Coppa: Why not?

M. Murakami: Well, one reason is that we observed the interface very carefully, and we tried to find some – well, not a pseudo-third phase but some kind of other phase at the interface. But we could not find any other phase; the interface is very sharp.

N. V. Coppa: Okay. By virtue of the fact that you have two different materials bonded together, that interface can be looked at as a third phase.

A.P. Malozemoff: The 2-1-1 is already normal . . .

N. V. Coppa: Yes, I understand that.

A. P. Malozemoff: . . . so the other one would be normal too.

M. Murakami: Well, 2-1-1 is an insulator . . .

N. V. Coppa: I understand that.

M. Murakami: We're not looking at superconducting properties . . . superconducting at the insulating interface.

N. V. Coppa: And you model that with a square potential, with an infinite slope at the interface.

M. Murakami: Yes.

N. V. Coppa: Well, actually at the interface what you have is an arrangment of atoms that are unlike the bulk of the 2-1-1, or the bulk of the 1-2-3, correct? Assuming that, that may cause a local minimum in that region, in the pinning potential.

J. R. Schrieffer: It's essential to notice that there's a non-locality in the problem, of order the coherence length, and the physics is averaged over this size. I think everything in principle you say is fine, but I don't think you would ever observe it because the averaging is over a size large compared to the thickness of the third phase.

A. P. Malozemoff: On the same general point: it struck me that in your micrograph showing the nickel particles, there were many 2-1-1 grains where there were no nickel particles. Why is that?

M. Murakami: Well, the reason is we applied only 10 oersted, so the density of flux lines is not high enough to cover every 2-1-1.

V. Vinokur: The distance between stacking faults should be large enough. That means that there shouldn't be a significant difference between pinning by stacking faults for the case when magnetic field is parallel to the c axis and when it is perpendicular to the c axis. Because in that case, when vortices are moving perpendicular to the stacking faults immediately after leaving this particular trap, vortex cannot feel it any more, and so in both cases the effective averaged pinning force is proportional to the density of stacking faults. And then maybe the more probable scenario for this process is the role of intrinsic pinning, because their density is much higher.

M. Murakami: You mean intrinsic pinning is more effective than stacking faults in traps?

V. Vinokur: Yes, I believe so.

M. Murakami: It's difficult to separate the effect of stacking faults and intrinsic pinning, I believe, in that field configuration.

V. Vinokur: What is the effective concentration of cuprum-oxide planes?

M. Murakami: You mean if the concentration is much higher . . .

V. Vinokur: . . . the amplitude of pinning faults is approximately the same. And that critical current should be close to the pair-breaking current. But the concentration of stacking faults, and the concentration of this kind of defects . . . I mean cuprum-oxide planes is very different.

M. Murakami: Yeah, Yeah, it's a good point.

M. P. Maley: When you did the experiments of looking at how the hysteretic measurement of the magnetization change was scaled with the dimension of the field along the c axis, did you try the same experiment with the field along the *ab* plane?

M. Murakami: Yes.

M. P. Maley: Does that scale with the dimension? That is, do the currents sense the granularity and flow around the grains, or do they tend to flow around the sample.

M. Murakami: Well, I would say it scales with dimension along the c-axis, the direction where the flux-pinning force is weaker.

M. P. Maley: If that's true ... you were saying basically you have some kind of a rectangle where the c axis is up this way, and ...

M. Murakami: The field penetration is very small from this side, but very large from the other side.

M. P. Maley: Yeah. If it scales with the thickness along the c axis, then eventually ...

M. Murakami: No, No. The other direction.

M. P. Maley: Oh, the other direction.

M. Murakami: But in order to look at that effect, you need the sample which has the long axis along the c direction, so it's very difficult to prepare that kind of sample.

J. R. Schrieffer: I was fascinated with this question of what happens when you rotate a magnetized sphere. If you rotate a uniform magnetic dipole about its symmetry axis, B does not change and there is no induced E field as long as it's a perfect dipole. But the earth is not a perfect dipole, and Greenland does not have a time-varying field in the rotating frame on the period of 24 hours. Thus rotating magnetic sources produce rotating B fields.

D. C. Larbalestier: I think that in the system you are working with you have some very interesting potentialities that it doesn't seem to me are in any other system of high-T_c materials in which pinning centers are being deliberately manipulated, yet. You've shown that you can change the percentage of 2-1-1 particles that you put in, and it is reasonable to think that they are basically spherical particles.

M. Murakami: Well, it's not perfect spheres.

D. C. Larbalestier: Well let's accept that for the moment. But the second thing is that you change the density of cracks, in particular. You don't know what you do to the stacking faults, do you?

M. Murakami: There's no change in the density of stacking faults.

D. C. Larbalestier: If you measure now the critical current density not by measuring the magnetization, but actually by measuring transport current so that you measure for H parallel to the ab planes and H parallel to c, then you might be able to extract out the relative pinning strengths of the 2-1-1 particles in the two different directions, and to play that off against the pinning effect produced by cracks and stacking faults . . .

M. Murakami: Oh, I see your point.

D. C. Larbalestier: . . . which should be strong, as you suggested, in the configuration H parallel to ab.

M. Murakami: Actually this (Fig. 2) is the result of transport J_c in different field directions. The MTG sample contains 2-1-1, but the size is very large. So they will not be effective. It's really difficult to measure transport J_c at low fields. Probably the limit is around 30 000 A/cm. We don't know the real J_c at lower fields, but it's clear that we can find some effect due to the presence of 2-1-1 inclusions, from transport J_c. But it's really difficult to tell which should be strong.

D.C. Larbalestier: How much 2-1-1 was there?

M. Murakami: In the case of QMG?

D. C. Larbalestier: Yes.

M. Murakami: I'm not sure, this is data from Nippon Steel. Probably 20%.

John Clarke
University of California
Berkeley, CA 94720

High-T_c Josephson Junctions, SQUIDs and Magnetometers

1. INTRODUCTION

There has recently been considerable progress in the state-of-the-art of high-T_c magnetometers based on dc SQUIDs (Superconducting Quantum Interference Devices). This progress is due partly to the development of more manufacturable Josephson junctions, making SQUIDs easier to fabricate, and partly to the development of multiturn flux transformers that convert the high sensitivity of SQUIDs to magnetic flux to a correspondingly high sensitivity to magnetic field. Needless to say, today's high-T_c SQUIDs are still considerably less sensitive than their low-T_c counterparts, particularly at low frequencies (f) where their level of 1/f noise remains high. Nonetheless, the performance of the high-T_c devices has now reached the point where they are adequate for a number of the less demanding applications; furthermore, as we shall see, at least modest improvements in performance are expected in the near future. In this article, I outline these various developments. This is far from a comprehensive review of the field, however, and, apart from Sec. II, I shall describe largely our own work. I begin in Sec. II with an overview of the various types of Josephson junctions that have been investigated, and in Sec. III, I describe some of the SQUIDs that have been tested, and assess their performance. Section IV discusses the development of the multilayer structures essential for an interconnect technology, and, in particular, for "crossovers" and "vias." Section

V shows how this technology enables one to fabricate multiturn flux transformers which, in turn, can be coupled to SQUIDs to make magnetometers. The performance and possible future improvements in these magnetometers are assessed, and some applications mentioned. The last section is a brief summary.

2. JOSEPHSON JUNCTIONS

The development of a reproducible and reliable junction technology has been a major challenge in the creation of a successful high-T_c thin-film technology. A variety of structures have been made, all exhibiting a nonhysteretic current-voltage (I-V) characteristic, and for convenience of reviewing, I have divided them into three groups: "grain boundary" junctions, which may be natural or engineered, "artificial barrier" junctions, where one deliberately introduces a barrier between two superconducting films, and "weakened superconductor" junctions in which a region of the film is weakened in a controlled manner.

2.1 Natural Grain-Boundary Junctions

To my knowledge, the first thin-film Josephson junction was made by Koch et al.[1] [Fig. 1(a)]. These workers deposited a polycrystalline film of YBaCuO (YBCO) which consisted of randomly oriented grains with typical dimensions of the order of 1μm. Using a photomask, they ion-implanted a region of the film to make it nonsuperconducting, leaving a bridge of superconducting material crossed by one or possibly several grain boundaries. Subsequently, many groups have made similar junctions, usually by removing regions of the film with acid etching or ion-beam milling, and extending the work to other materials, notably BiSrCaCuO (BSCCO)[2] and TlBaCaCuO (TBCCO).[3] In the case of YBCO films, one can grow 45° grain boundary junctions by depositing c-axis films on (100) MgO substrates, where the lattice mismatch is relatively large. Although the a- and b-axes align preferentially along the cubic axis of the substrate, some grains are rotated by 45°; the fraction of such grains can be controlled by the growth conditions.[4,5] Grain boundary junctions with relatively low critical current density ($\leq 10^5$ A cm^{-2}) can exhibit[4] almost ideal resistively shunted junction[6] (RSJ) behavior, while those with higher critical current densities tend to exhibit characteristics consistent with flux flow.[4] As we shall see, the relatively well-behaved nature of junctions with lower critical current has resulted in SQUIDs with good performance in the white noise region. However, junctions of this kind have no real technological future because they have more or less random properties and one has no control at all of their location on the substrate.

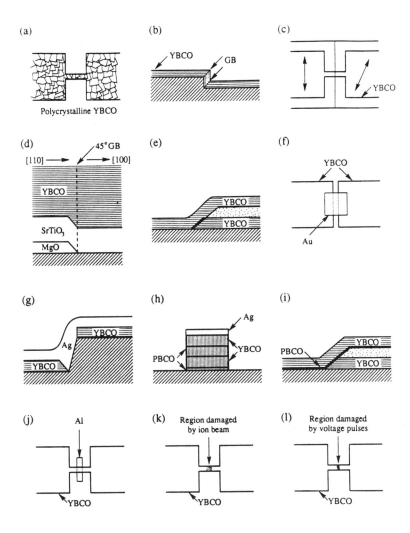

Fig. 1 Schematic representation of nine types of high-T_c thin-film Josephson junction: (a) naturally occurring grain boundary junction (Ref. 1), (b) step-edge grain boundary (GB) junction (Ref. 7), (c) bicrystal grain boundary junction (arrows indicate [100] axes, Ref. 9), (d) bi-epitaxial grain boundary junction (Ref. 10), (e) edge junction (Ref. 12), (f) YBCO-Au-YBCO c-axis junction (Ref. 14), (g) YBCO-Au-YBCO a-axis junction (Ref. 16), (h) trilayer YBCO-PBCO-YBCO junction (Ref. 18), (i) trilayer YBCO-PBCO-YBCO edge junction (Ref. 19), (j) poison stripe junction (Ref. 20), (k) ion beam damage junction (Refs. 21, 22), (l) electroshocked junction (Ref. 23).

2.2 Grain Boundary Junctions on Step Edges

Simon et al.[7] have fabricated grain boundary junctions by etching a step in the substrate and depositing a YBCO film across the step with a thickness less than the step height. The film is subsequently patterned to form a bridge across the step [Fig. 1(b)]. Provided the step edge is sharp enough, a grain boundary junction is formed at the top and bottom. The degree of sharpness required has been quantified by some beautiful transmission electron microscopy at Jülich,[8] which showed that if the step angle is shallow, less than 45°, a c-axis film grows across the entire step, whereas if the step angle is steeper than 45° the film has an a-axis orientation on the step, thus producing two grain boundaries. These junctions have the obvious advantage that they can be placed at will on the substrate, and they have been used successfully to make SQUIDs.

2.3 Grain Boundary Junctions on Bicrystal Substrates

Dimos et al.[9] fabricated grain boundary junctions with controllable angles by depositing YBCO films on $SrTiO_3$ bicrystal substrates. In this technique, a crystal is cut in a direction perpendicular to the (100) plane and the faces are polished so that the axes have a known misorientation when the two parts of the crystal are subsequently fused together. The crystal is then diced to form (100) substrates. When the YBCO is deposited with an *in situ* process, the ab-axes mimic the orientation of the SrTiO3, producing a grain boundary of known orientation θ [Fig. 1(c)]. It seems necessary to have a very sharp interface between the two $SrTiO_3$ regions to obtain a sharply defined grain boundary in the YBCO. The critical current density of microbridges patterned across the grain boundary decreases with increasing θ in a reproducible way. Some of the best SQUIDs have been made with this technique. This technique is admirable for devices requiring a small number of devices, but it is probably unrealistic to extend it to circuits requiring a large number of junctions on a chip.

2.4 Bi-epitaxial Grain Boundary Junctions

This method derives from the bicrystal technique, but induces a controlled grain boundary at the edge of a seed layer[10] [Fig. 1(d)]. In the first version, a seed layer of MgO, perhaps 100Å thick, is grown epitaxially by laser deposition onto an r-plane sapphire substrate. Using a photo-mask and an ion mill, one removes the MgO from part of the substrate, and then laser deposits a $SrTiO_3$ buffer layer followed immediately by a YBCO film. The YBCO film grows epitaxially on the SrTiO3, but reorients its (100) axis by 45° where it crosses the edge of the seed layer. Thus, a microbridge patterned across this edge contains a single 45° grain boundary junction. This technique has the advantage that junctions can be placed at will on a chip by means of two straightforward photolithographic processes. Subsequently,

Char et al.[11] demonstrated alternative structures for producing bi-epitaxial growth. These junctions have been used very successfully in SQUIDs.

In concluding this brief discussion of grain boundary junctions, one might ask "what are they?" and "why are they so good?" The answer to the first question will very likely answer the second; but the nature of the interface remains obscure, and a continuing effort to understand it is very much in order.

2.5 Edge Junctions

This is the first of the junctions I shall describe in which one deliberately inserts a "barrier" between two superconducting films. Laibowitz et al.[12] produced edge junctions in which they first deposit a YBCO film and a nonepitaxial insulating film and then ion-mill an edge [Fig. 1(e)]. They expose the edge to a plasma in an oxygen-fluorine mixture in an attempt to mimic the growth of insulating barriers on low-T_c barriers. Subsequently, they deposit a YBCO film to form an edge junction. In this structure, the supercurrent flows along the ab-planes of the films. Wang et al.[13] have made junctions with a similar technique, omitting the oxyfluorination. Laibowitz et al.[12] report that both junctions and SQUIDs have been made with this technique, but that the yield is low and the junctions' parameters are unpredictable. Although this approach is appealing, one would have to improve reproducibility and yield if these junctions are to be technologically useful.

2.6 Superconductor–Normal Metal–Superconductor Junctions

This class of junctions makes use of the proximity effect. To my knowledge, the first junction was made by Schwarz and co-workers,[14] who deposited a YBCO film, made a very thin slit in it with electron-beam lithography and evaporated a gold film across the slit [Fig. 1(f)]. This junction exhibited supercurrents at temperatures up to 16K. In this structure, supercurrents flow into and out of the YBCO along the c-axis, a direction in which the proximity effect is weak,[15] so that, in hindsight, this is not the preferred configuration.

This drawback was overcome by DiIorio et al.,[16] who fabricated structures in the configuration of Fig. 1(g). A step is milled in a LaAlO$_3$ substrate, and YBCO films are sputtered at an angle to the substrate so that they do not connect along the step edge. A silver film is deposited to contact the two YBCO films in the ab-plane. These junctions have a high yield and almost ideal RSJ characteristics, and should be capable of at least moderate levels of integration on a chip.

Another class of proximity effect junctions, pioneered by Rogers et al.,[17] involves PrBaCuO (PBCO) as the barrier. In their early work, these authors grew an epitaxial c-axis trilayer of YBCO-PBCO-YBCO, with a PBCO thickness of typically 500Å and patterned it to form junctions. Some of these junctions maintained a supercurrent at temperatures as high as 65K. However, given the weak proximity effect of c-axis films,[15] it seems likely that the observed supercurrent arose from superconducting shorts through the PBCO. Subsequently, Barner et al.[18] fabricated epitaxial a-axis trilayers which they patterned into junctions that exhibited a supercurrent at 80K or higher [Fig. 1(h)]. The critical currents scaled approximately with junction area, and the critical current density could be varied over the range from 10^2 to 10^4 A cm^{-2} by varying the thickness of the PBCO. Although this reproducibility is very promising, the I-V characteristics exhibit excess current, suggesting that the current distribution may be nonuniform.

An alternative approach to a-axis junctions was adopted by Gao et al.[19] [Fig. 1(i)], who fabricated edge junctions with PBCO barriers. After depositing a YBCO film and a nonepitaxial insulating layer, they ion-milled an edge and deposited PBCO and YBCO. Junctions and SQUIDs made in this way have flux-flow-like characteristics, exhibiting supercurrents up to liquid nitrogen temperatures. Although this is an attractive approach to making junctions, as in the case of the trilayer junctions it is unclear why the I-V characteristics are not more RSJ-like.

2.7 "Weakened" Structures

Several groups have made YBCO microbridges in which they deliberately weaken a region. Simon et al.[20] deposited a YBCO film across a thin aluminum strip which "poisons" the superconductor locally, reducing its transition temperature drastically [Fig. 1(j)]. In another method, one focuses[21] high-energy ions onto a microbridge or uses a mask patterned with electron beam lithography[22] to define the area exposed to an unfocused ion beam [Fig. 1(k)]. It is likely that an appropriate dose of ions destroys superconductivity in most of the irradiated film, leaving narrow superconducting filaments connecting the banks of unirradiated film. In a third technique,[23] controlled electrical pulses were applied to a patterned microbridge at 77K, producing a weakened region that, again, possibly consisted of narrow superconducting threads [Fig. 1(l)]. These various techniques all produce weak-link structures that are reasonably controllable. However, it appears that this class of junctions generally exhibits flux-flow characteristics, probably because the weakened region is long compared to the superconducting coherence length in the direction of current flow, and is thus less desirable for most applications.

2.8 Final Comments

Which of these various junction types look most promising for the long term? A great deal of science has been learned from natural grain-boundary junctions, but they are obviously unsuitable for manufacturing. The bicrystal approach yields remarkably consistent junctions and could well be used for circuits requiring a few junctions. Of the grain boundary devices the bi-epitaxial junctions look particularly promising: they can be placed at will by means of two photolithographic steps, and appear to be reasonably reproducible. The other junctions to develop further would appear to be those with artificial barriers. Silver barriers seem quite reproducible and reasonably straightforward to fabricate. Junctions with PBCO barriers in either a-axis trilayers or with an edge configuration between c-axis films are appealing in having an all-perovskite structure, but there is no clear-cut evidence that the supercurrent arises from a proximity effect rather than superconducting filaments. Since the proximity effect is of considerable fundamental importance and this type of junction may still turn out to be very useful, one hopes that work in this area will be pursued.

3. SQUIDS

3.1 Introduction

Numerous workers have made dc and rf SQUIDs from high-T_c superconductors, and I shall make no attempt to survey them all. Indeed, since the best performance has been obtained from dc SQUIDs, I shall neglect rf SQUIDs entirely.

The dc SQUID consists of two junctions with nonhysteretic I-V characteristics connected in parallel to make a superconducting loop of inductance L. Ideally, each junction has the same critical current I_o and resistance R. When the magnetic flux Φ threading the SQUID is changed, the critical current oscillates with a period of one flux quantum, $\Phi_0 = h/2e \approx 2 \times 10^{-15}$ Wb (h is Planck's constant and e the electronic charge). One usually operates the SQUID at a constant bias current greater than the critical current so that the voltage is also periodic in the flux. By detecting small changes in voltage in the region where $(\partial V/\partial \Phi)_I$ is near a maximum, one can measure changes in flux of much less than Φ_0: the best low-T_c SQUIDs have a resolution approaching $10^{-6}\Phi_0$ Hz$^{-1/2}$ at frequencies above the 1/f noise region.

There are certain criteria to be met in the design of SQUIDs. First, to observe quantum interference effects, we require the modulation depth of the critical current, Φ_0/L, to be much greater than the rms noise current around the loop, $(k_B T/L)^{1/2}$. A computer analysis[24] indicates that this requirement leads to $L \lesssim \Phi_o^2/5k_B T \approx 800$pH at 77K. In practice, one usually designs SQUIDs well within this constraint, and inductances of 50-100pH are typical for high-T_c SQUIDs. Second, optimum performance is achieved[25] when $2LI_o \approx \Phi_0$; for L = 100pH one finds $I_o \approx 10\mu$A. Needless to say, given the difficulty in controlling I_o and its strong temperature dependence at temperatures around 77K, few high-T_c SQUIDs have been operated under optimum conditions.

It is convenient to characterize the noise of SQUIDs in terms of the flux noise energy

$$\varepsilon(f) = S_\Phi(f)/2L , \qquad (3.1)$$

where $S_\Phi(f) = S_v(f)/(\partial V/\partial\Phi)_I^2$ is the spectral density of the flux noise and $S_v(f)$ is the spectral density of the voltage noise. At frequencies above the 1/f region, the spectral density of the noise is white, and originates in Nyquist noise in the junctions. For an optimized SQUID with resistively shunted junctions, computer simulations[25] have shown that

$$\varepsilon(f) \approx 9k_BTL/R .\qquad(3.2)$$

Thus, the noise energy scales with T implying that the resolution of SQUIDs at liquid nitrogen temperatures can never be as good as their liquid-helium-cooled counterparts. At present, as we shall see, the potential white noise performance of high-T_c SQUIDs is moot at low frequencies because of the high level of 1/f noise.

3.2 Practical Devices

To my knowledge, the first thin-film high-T_c SQUID was made by Koch et al.[1] from YBCO films with naturally occurring grain boundaries. They patterned the film using a photomask, ion-implanting portions to make them insulating, to produce a superconducting square washer with an inside length of 50μm interrupted by two microbridges [Fig. 2]. These early devices exhibited a well-defined voltage modulation at 4.2K, and a discernible albeit rather aperiodic response at 68K. This design was the prototype for a large number of SQUIDs made subsequently, most of them patterned photolithographically with acid-etching or ion-milling. The vast majority of these devices have been made from YBCO, although a few have been made from BiSrCaCuO or TlBaCaCuO. Most of the junction technologies described in Sec. II have been incorporated into SQUIDs, and, indeed the magnitude of the SQUID response is sometimes taken as a measure of junction quality.

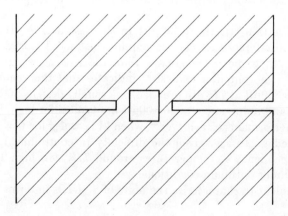

Fig. 2 Prototype planar dc SQUID fabricated from thin film of YBCO on a sapphire substrate (redrawn from Ref. 1).

To illustrate the current generation of SQUIDs, I briefly describe some of the YBCO devices made at Conductus and characterized at Berkeley. The configuration of the SQUID is shown in Fig. 3; the outer dimension of the square washer is 250 μm. Two bi-epitaxial junctions are grown along the line indicated by the arrow. Current and voltage leads are attached to two separate contacts on each of the arms at top and bottom of the photograph. A representative I-V characteristic of a single junction is shown in Fig. 4, and is reasonably close to the predictions of the resistively shunted junction model. Figure 5 shows the modulation of the voltage across the SQUID by a magnetic field; the amplitude ranges from about 100μV at 4.2K to about 0.5μV at 77K. The noise is measured in a conventional flux-locked loop (Fig. 6) in which a 100kHz modulating flux with a peak-to-peak amplitude of $\Phi_0/2$ is applied to the current-biased SQUID, and the resulting alternating voltage

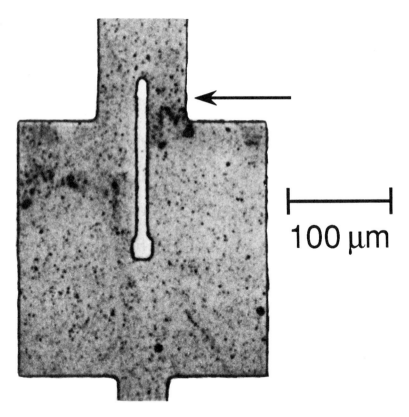

Fig. 3 Photograph of dc SQUID with bi-epitaxial junctions. The two junctions are along the line (marked with arrow) demarking the edge of the MgO seed layer which is in the region above the arrow (from Ref. 46).

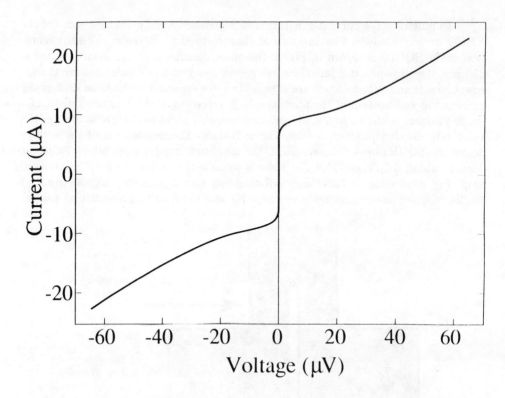

Fig. 4 Typical I-V characteristic of Conductus bi-epitaxial junction at 4.2K (courtesy A. H. Miklich).

across the device is amplified and lock-in detected. The voltage from the lock-in detector is connected via a resistor to an inductor coupled to the SQUID so that the flux fed back cancels any applied flux $\delta\Phi$. The output voltage across the resistor is thus proportional to $\delta\Phi$. This circuit enables one to measure flux changes $\delta\Phi \ll \Phi_0$ while extending the dynamic range to many flux quanta. One determines the flux noise by connecting the output to a spectrum analyzer, with no input signal applied to the SQUID.

To give a perspective on the performance of high-T_c SQUIDs, we plot $\varepsilon(f)$ vs. f in Fig. 7; it is emphasized that these are representative values of a small number of devices only. Near the bottom of the figure is the noise of a thin-film Nb-based SQUID made at Berkeley and operated at 4.2K. Several other groups have achieved comparable or better performance; in particular, lower levels of l/f noise have been achieved.[26,27] Also shown are the performances of commercially available bulk-Nb

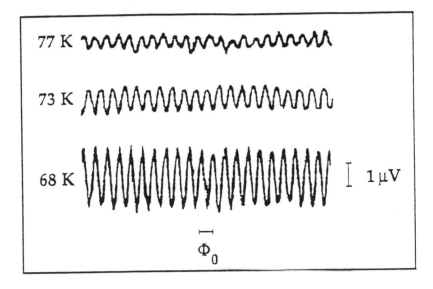

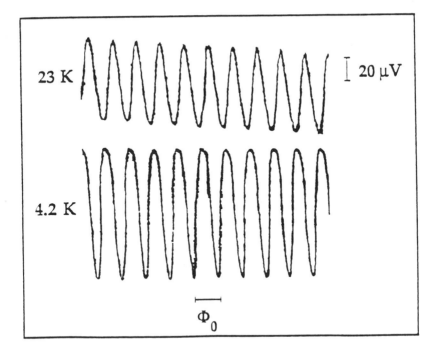

Fig. 5 V vs. Φ for Conductus bi-epitaxial dc SQUID at 5 temperatures (from Ref. 10).

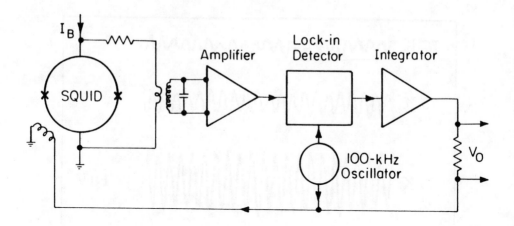

Fig. 6 Flux-locked loop operation of dc SQUID.

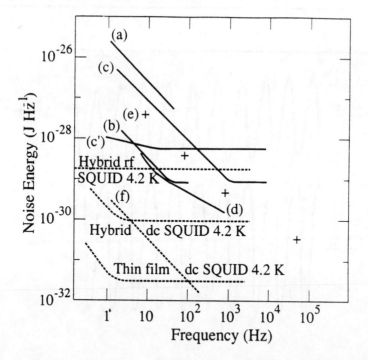

Fig. 7 Noise energies of three Nb SQUIDs at 4.2K and six high-T_c SQUIDs at or near 77K: (a) YBCO grain boundary SQUID (Ref. 29), (b) YBCO edge junction SQUID (Ref. 13), (c), (c') bi-epitaxial YBCO SQUID at 56K, 82.5K (Ref. 46), (d) TlBaCaCuO grain boundary SQUID (Ref. 3), (e) bicrystal YBCO SQUID (Ref. 32), (f) flux noise of YBCO film (ref. 38).

dc and rf SQUIDs.[28] Although clearly higher in noise than the thin-film SQUIDs, the bulk-Nb dc SQUID has been adequate for many applications, including magnetoencephalography, and the rf SQUID has been adequate for applications less demanding in sensitivity such as susceptometers.

The highest noise energy shown is that of an early YBCO grain boundary dc SQUID.[29] From roughly 1Hz to 100Hz, the noise energy scales as 1/f. A much lower noise energy was obtained by Koch et al.[3] in a TlBaCaCuO grain boundary dc SQUID; between 10Hz and 100Hz the slope of the noise energy is significantly shallower than -1. The results from a Conductus/Berkeley bi-epitaxial YBCO dc SQUID are shown at 56K and 82.5K. At the lower temperature the noise energy scales as 1/f at frequencies below 1kHz, becoming white at higher frequencies. When the temperature is raised to 82.5K, the white noise increases by almost an order of magnitude, reflecting the appreciable reduction in the signal from the SQUID as the critical current becomes small. However, the 1/f noise is greatly reduced. The reason for the significant reduction is not known, but a similar effect has been reported by other workers.[30,31] The spectrum from a YBCO SQUID with 2μm-wide junctions grown on a bicrystal[32] exhibits 1/f noise at frequencies up to several tens of kilohertz. At 71kHz where the noise is white the noise energy is 3×10^{-31} J Hz^{-1}, in good agreement with the predictions of Eq. (3.2) and to my knowledge the best value yet reported at 77K.

Encouraging results have been obtained[16] at 4.2K on SQUIDs with YBCO-Ag-YBCO junctions, which exhibited a white noise energy of less than 10^{-30} J Hz^{-1} at frequencies above 2kHz. It will be of considerable interest to measure the performance at 77K. Metal barriers presumably do not contain electron traps, and it is possible that the 1/f noise due to critical current fluctuations will be much less than in other junctions.

The general picture that emerges from Fig. 7 is that 1/f noise dominates at frequencies up to a few hundred Hz. This is a crucial limitation, particularly for biomagnetic applications which demand high resolution at frequencies as low as 0.1Hz. Thus, it is of considerable importance to understand and if possible reduce the 1/f noise level. In low-T_c SQUIDs two independent sources of 1/f noise have been identified.[33] The first is so-called flux noise, which arises from the random motion of flux lines pinned in the body of the SQUID, changing the flux linked to the SQUID and producing noise. It should be emphasized that this is an irreducible noise in the sense that it induces a flux in the SQUID loop in the same way as an external signal. Thus, any modulation technique that reduces this noise reduces the signal by the same amount. The second noise source arises from fluctuations in the critical currents of the junctions. In low-T_c tunnel junctions, these fluctuations arise from the trapping and release of electrons in the barrier, a process that locally modulates the tunnel barrier and thus the critical current and tunneling resistance.[34] Each trap has a characteristic trapping time τ that produces a Lorentzian power spectrum of the form $\tau/[1 + (2\pi f \tau)^2]$. The summation of the processes from a number of statistically independent traps yields a 1/f power spectrum.[35] In the case of the dc SQUID, the two independently fluctuating sources produce a voltage spectral density given approximately by[33]

$$S_v(f) \approx \frac{1}{2} \left[(\partial V/\partial I_o)_I^2 + L^2 (\partial V/\partial \Phi)_I^2 \right] S_{I_o}(f) , \qquad (3.3)$$

where $S_{I_o}(f)$ is the spectral density of the l/f current noise in each junction. The first term on the right hand side of Eq. (3.3) is the "in-phase mode" in which each of the two junctions produces a fluctuation of the same polarity. This noise source is (ideally) eliminated by the flux modulation scheme shown in Fig. 6 for frequencies below the modulation frequency. The second term on the right-hand side of Eq. (3.3) is the "out-of-phase mode" in which the fluctuations are of opposite polarity and produce a current around the SQUID loop and hence a flux. Fortunately, this term can be greatly reduced by any of several double-modulation schemes.[26,28,33]

Our picture of l/f noise in high-T_c SQUIDs is still emerging but is becoming clearer. In some early experiments, Ferrari et al.[36] made direct measurements of the l/f flux noise in YBCO films and BSCCO flakes by mounting each of them in turn parallel to and about 100μm away from an Nb-based SQUID. The apparatus was mounted in a vacuum can immersed in liquid ^{4}He so that the temperature of the high-T_c sample could be raised to above its transition temperature while the SQUID remained at 4.2K. All samples exhibited l/f noise; in the case of YBCO, the magnitude dropped dramatically as the quality of the films was improved. A detailed model[37] explained the data in terms of the thermally activated hopping of vortices between pinning sites, each hopping process yielding a Lorentzian power spectrum. This process can contribute to the l/f noise in SQUIDs. However, in the case of high quality *in situ* YBCO films[38] the magnitude of the noise can be low, as we see from Fig. 7. Of course, it may not be possible to maintain such high film quality and concomitant low noise in a SQUID that has been through several processing steps, but this result is very encouraging and suggests that flux noise may not be dominant in SQUIDs made from *in situ* films.

An obvious possible implication of this result is that the l/f noise arises from critical current fluctuations. Koch and co-workers[31] have used a double modulation scheme in an attempt to reduce the l/f noise in a series of SQUIDs. In many devices there was no reduction in the noise level, but in some the noise power was reduced by as much as an order of magnitude. The latter observation is most encouraging, but the lack of noise reduction in most of the SQUIDs remains puzzling. Much effort will undoubtedly be expended in this area in the near future, and one can be optimistic about the outcome.

This concludes our brief review of SQUIDs, except for one observation. Although SQUIDs are very sensitive to changes in magnetic flux, their relatively small pick-up area, particularly in the case of high-T_c SQUIDs, implies that they are not actually very sensitive to changes in magnetic field. As an example, consider a square-washer high-T_c SQUID with an inner dimension $d_1 = 25\mu$m and an outer dimension $d_2 = 200\mu$m. The body of the SQUID focuses[39] the applied flux through the central hole, producing an effective pick-up area $d_1 d_2 \approx 5000\mu$m^2. If we take an optimistic flux noise of $10^{-4}\Phi_0$ Hz$^{-1/2}$ at 1Hz the corresponding magnetic field sensitivity is a meager 40 pT Hz$^{-1/2}$, a value that compares unfavorably with off-the-shelf flux-gate magnetometers. This result emphasizes an extremely important point: no high-T_c SQUID is likely to be useful as a magnetometer unless it is coupled to a high-T_c flux transformer that provides a substantial gain. Now for a given size of pick-up loop of inductance L_p the gain of the flux transformer is optimized when the inductance L_i of the input coil coupled to the SQUID is approximately equal[40] to L_p. Since the area of the SQUID is very much less than that of a useful pick-up loop, the input coil is necessarily multiturn. In a thin-film technology, the need to make a superconducting contact to the innermost turn of the coil inevitably leads us to multilayer films. In particular, we require *crossovers* – two high-T_c films separated by an insulating layer, and *vias* – a superconducting contact between the two films through a window in the intervening insulator. These are the two essential components of *interconnects* – the wiring between individual devices on a chip.

4. INTERCONNECTS

The requirements for a high-T_c interconnect technology are severe.[41] To minimize processing temperatures it is essential to have *in situ* processes for all layers so that the films grow epitaxially and do not require subsequent annealing at higher temperatures. At present, to my knowledge only YBCO has been used successfully in these structures. The insulator must be able to grow on both the substrate material and on YBCO and to support epitaxial growth of YBCO on its upper surface. Furthermore, the insulator must have low interdiffusion and chemical reactivity with both the substrate and YBCO at temperatures up to the maximum processing temperature, say 750°C, and good coverage of and adhesion to both the substrate and YBCO, particularly at the film edges. Finally, of course, the resistivity must be high at device operating temperatures, say 77K. Faced with these requirements, one is quickly led to the same materials options for the insulator as those used for substrates, for example, $SrTiO_3$, YSZ and $LaAlO_3$.

In early work, Kingston *et al.*[41] used shadow masks to define the geometry of the films; this procedure enables one to test different deposition procedures rapidly, and avoids possible problems with the wet chemistry associated with photolithography. The films were laser deposited on cleaved and polished MgO substrates clamped to a heater block that was coated with silver paste to ensure good thermal contact. The excimer laser was operated at 248 nm with an 18 ns pulse length. The first YBCO layer, about 3000Å thick, was deposited at a substrate temperature of about 730°C in an O_2 pressure of about 200mTorr. After the deposition was completed, the chamber was filled with 1 atm O_2 and the heater block cooled to 450°C. After 15 min at this temperature, the sample was cooled. The insulating layer, $SrTiO_3$ in the first structures, was deposited in a similar manner with a substrate temperature of 660°C, and the same cooling procedure was used to allow O_2 to diffuse through the $SrTiO_3$ into the YBCO. The second YBCO film was deposited under the same conditions as the first. The two YBCO films were patterned to produce overlapping strips crossing at right angles, with the SrTiO$_3$ film intervening. Under the correct growth conditions, each film grows epitaxially, as was demonstrated by x-rays,[41] and is shown by transmission electron microscopy[42] in Fig. 8 for a SrTiO$_3$ film grown on YBCO. This image shows a clean interface with no evidence of second phase formation. In both YBCO and SrTiO$_3$ layers, crystallinity is maintained up to the interface which is very sharply defined.

Crossovers made in this manner exhibited excellent electrical characteristics.[41] The two YBCO films had sharp resistive transitions, with zero resistance at typically 88K. The resistivity of the SrTiO$_3$ increased rapidly as the temperature was lowered from room temperature, with typical values of well over $10^9 \Omega$ cm at 77K.

Although successful multiturn coils and flux transformers were constructed using shadow masks for the first two films and a photomask followed by an etch for the upper YBCO layer,[43,44] it is clear that a manufacturable technology requires all-photolithographic processing. Such processing poses new challenges, for example, in the growth of an epitaxial film over the steep edges of an underlying film patterned photolithographically and in the complete removal of all chemicals

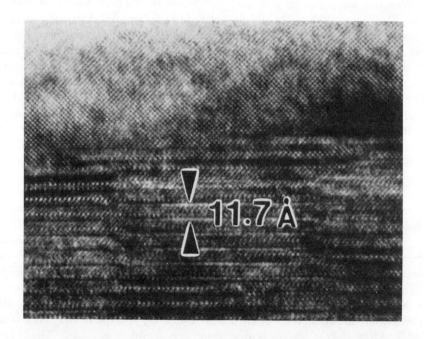

Fig. 8 Transmission electron micrograph of SrTiO$_3$ film grown on YBCO film (from Ref. 42).

used to pattern one film prior to the deposition of the next. At this juncture, several groups[45-49] have developed successful albeit somewhat different processes. In the Berkeley process, the first YBCO film is patterned by etching in 0.1% nitric acid, the resist is stripped and the sample is immersed in a 1% solution of Br in methanol to etch clean the YBCO surface. The etch also produces some rounding of the edges of the film to assist the epitaxial growth of the subsequent layer. To make both crossovers and vias the YBCO is covered with $SrTiO_3$, which is patterned to leave an intact area over the YBCO to form a crossover and to open a window to the YBCO to form a via. To make a window the $SrTiO_3$ is patterned with an ion mill with the beam at an angle of 30° to the substrate. This procedure produces an edge along one side of the window beveled at about 8° to the substrate, allowing epitaxial growth of the YBCO film. The ion mill is allowed to remove typically 100 nm of the YBCO in the window, to ensure the complete removal of the $SrTiO_3$ and to enable the upper and lower YBCO films to make contact in the ab-plane as well as along the c-axis. After the resist is stripped from the $SrTiO_3$, the upper YBCO film is deposited and patterned with the acid etch. In one variant of these procedures, Eidelloth et al.[48] etched both YBCO layers with ethylenediaminetetracetic acid (ETDA) and the window in the $SrTiO_3$ film with HF, which stops at YBCO. The surface of the first YBCO was restored by a short immersion in ETDA.

The processing technology for interconnects continues to evolve, particularly with regard to the insulating material; for example, Lee et al.[49] have used a $LaAlO_3$ film sandwiched between two $SrTiO_3$ layers to reduce the overall dielectric constant of the insulator. Although refinements to the technology will probably continue for a long time, the essential process is established.

5. FLUX TRANSFORMERS AND MAGNETOMETERS

A number of groups have used the interconnect technology[45-49] to make flux transformers suitable for coupling to high-T_c SQUIDs. A photograph of a flux transformer made at Berkeley appears in Figs. 9(a) and (b). In this device the first YBCO film is patterned to form the "crossunder." The $SrTiO_3$ layer has two windows: one provides the contact to the innermost turn of the coil and the other the contact to the single turn pick-up loop. The upper YBCO film is patterned to make the five-turn spiral coil and the much larger pick-up loop shown in Fig. 9(b). Using their wet etch techniques, Eidelloth et al.[48] have made 5-, 10- and 20-coils all with zero resistance at 89K, 87K and 79K respectively.

The Berkeley/Conductus group[46] has made magnetometers by clamping the input coil to the SQUID shown in Fig. 3 with a 3μm mylar spacer between the two chips. The magnetometers were mounted on a variable temperature insert in a liquid ^{4}He dewar or immersed directly in liquid N_2. The critical current of the flux transformer was determined by applying an increasing magnetic field and noting

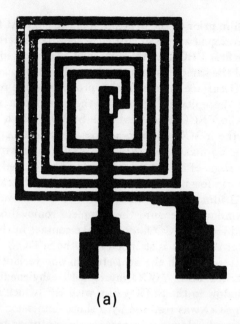

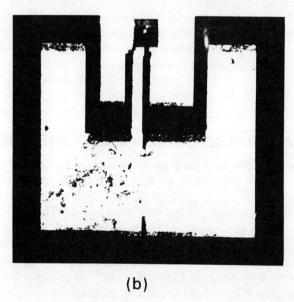

Fig. 9 (a) Photograph of 5-turn spiral input coil of flux transformer (from Ref. 46). The two leads at the lower edge of the figure connect to the single-turn pick-up loop shown in (b) (from Ref. 44). (The complete flux transformer shown in (b) in fact contains a 10-turn input coil.)

the value at which the period of the oscillations of the voltage across the SQUID abruptly changed. In one particular transformer the critical current was about 30 mA at 15K, dropping to about 2 mA at 77K; both the SQUID and the magnetometer continued to operate at temperatures up to 86K. At 77K the critical current density referred to the cross-sectional area of the turns of the input coil was about 5×10^4 A cm^{-2}. This value was probably determined by steps in the upper YBCO film that occur where the SrTiO$_3$ crosses the edges of the lower YBCO film.

The gain g of the transformer is defined as the factor by which it increases the magnetic field response over that of the SQUID alone. In the Berkeley transformer the sense of the windings produces a negative value of g. The gain was determined from the periodic response of the SQUID with and without the transformer, and is plotted versus temperature in Fig. 10. The gain was -83 ± 3 over the range from 4.2K to 80K.

One can use the measured gain and estimated parameters of the flux transformer to obtain the mutual inductance between the input coil and the SQUID. For the bare SQUID, we define the *effective* area (which includes the effects of flux focusing[39]) as

$$A_s = \Phi_0/B_0 , \qquad (5.1)$$

where B_0 is the magnetic field required to induce one flux quantum. The rms magnetic field sensitivity is given by

$$S_{BO}^{1/2}(f) = S_\Phi^{1/2}(f)/A_s . \qquad (5.2)$$

When the flux transformer is coupled to the SQUID, an applied field B induces a supercurrent

$$J = -BA_p/(L_p + L_i) \qquad (5.3)$$

in the transformer and hence a flux

$$\Phi_s = B[A_s - A_p M_i/(L_p + L_i)] \qquad (5.4)$$

in the SQUID. Here, A_p is the effective area of the pick-up loop and $M_i = \alpha(L_i L_0)^{1/2}$ is the mutual inductance between the SQUID and the input coil; we have neglected the area of the input coil relative to A_p. Equation (5.4) yields the effective magnetometer area, $A_m = \Phi_s/B$, and hence the gain

$$g \equiv \frac{A_m}{A_s} = 1 - \frac{A_p}{A_s} \frac{M_I}{L_i + L_p} . \qquad (5.5)$$

The second term on the right-hand side of Eq. (5.5) is much greater than unity. Finally, provided the transformer contributes negligible noise, the rms noise of the magnetometer is

$$S_B^{1/2}(f) = S_{BO}^{1/2}(f)/|g| . \qquad (5.6)$$

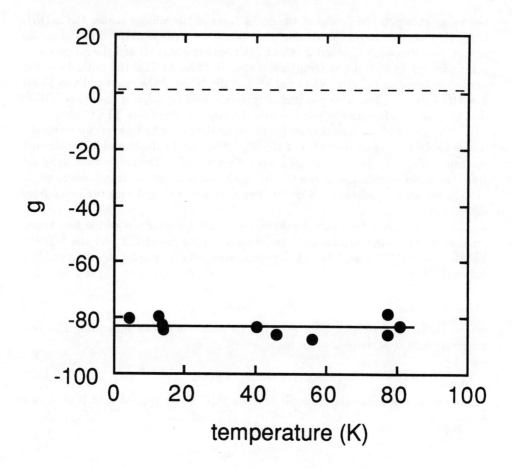

Fig. 10 Magnetic field gain g of flux transformer vs. temperature. Dashed line at g = +1 represents gain of bare SQUID (from Ref. 46).

Estimated or measured values of various parameters are: $A_p \approx 81$ mm^2, $A_s \approx 1.2 \times 10^{-2}$ mm^2, $L_p \approx 20$nH, and $L \approx 0.11$ nH. It is difficult to estimate L_i reliably because of the uncertainty in the separation between the input coil and the SQUID and hence in the reduction of L_i by the screening action of the body of the SQUID. If we assume $L_i \ll L_p$ (implying that the flux transformer was not optimized), Eq. (5.5) yields $M_i \approx 0.25$ nH. The value of α is known poorly because of the uncertainty in L_i; for example, if we estimate $L_i \approx 1.5$ nH, we find $\alpha \approx 0.6$.

To determine the noise of the magnetometer, it was operated in a flux-locked loop. Figure 11(a) shows $S_B^{1/2}(f)$ at 4.2K and 77K, with the SQUID immersed in

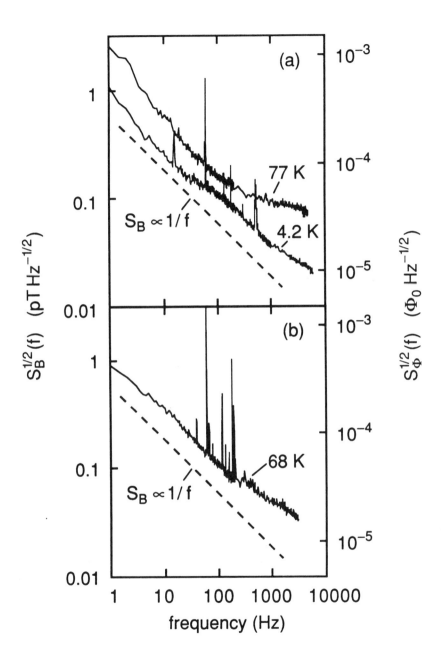

Fig. 11 (a) Rms magnetic field noise $S_B^{1/2}(f)$ of magnetometer and flux noise $S_\Phi^{1/2}(f)$ of SQUID at 4.2K and 77K. Dashed line represents $S_B(f) \propto 1/f$. (b) $S_B^{1/2}(f)$ for magnetometer at 68K (from Ref. 46).

liquid ^{4}He and N_2, respectively. The sharp spikes are due to pick-up of 60Hz and its harmonics and to microphonics. At 4.2K the rms noise scales approximately as $1/f^{1/2}$ over most of the frequency range shown, while at 77K the noise begins to flatten out above 200Hz indicating that white noise is beginning to dominate in this frequency range. In the $1/f^{1/2}$ region, the rms noise increases by a factor of only 2 when the temperature is raised from 4.2K to 77K. At 77K the observed magnetic field resolution is 0.6pT Hz$^{-1/2}$ at 10Hz and 0.09 pT Hz$^{-1/2}$ at 1kHz; to my knowledge these represent the best values yet reported for a thin-film high-T_c magnetometer operating at 77K.[50] At both 4.2K and 77K the measured flux noise in the SQUID [right-hand axis of Fig. 11(a)] was very close to that observed in the SQUID in the absence of the transformer, implying that the noise contribution of the transformer was negligible at both temperatures.

The performance of the magnetometer suspended in helium gas at 68K is shown in Fig. 11(b). The rms noise was a factor of about 2 lower than at 77K: whether the noise at 77K was enhanced by bubbling of the liquid nitrogen or whether the improvement represents a run-to-run variability in the performance, for example, due to the trapping of different amounts of magnetic flux, is unclear. At 68K, the rms noise at 10Hz and 1kHz was 0.35 pT Hz$^{-1/2}$ and 0.05 pT Hz$^{-1/2}$, respectively.

Modest improvements in the sensitivity of the magnetometer, say by a factor of 2 to 3, should be possible by increasing the number of turns on the input coil so that $L_i \approx L_p$ and by increasing the coupling coefficient α. For a given SQUID performance, further improvements will necessitate increasing the size of the pick-up loop; provided one maintains L_i and L_p approximately equal, the sensitivity should improve as $A_p^{-3/4}$. For example, increasing the dimension of the pick-up loop from 10mm to 30mm should improve the sensitivity by a factor of about 5. Furthermore, as discussed in Sec. III.B, one may realistically expect the level of 1/f noise in SQUIDs to be reduced by double modulation; such reductions translate directly into improved magnetometer performance provided the 1/f noise of the flux transformer remains negligible.[51]

Oh et al.[47] have also fabricated flip-chip magnetometers operating at 77K, with bicrystal SQUIDs and 5-turn input coils. The magnetic field sensitivity of one device with a 0.5 × 0.5 mm^2 pick-up loop was 3.8 pT Hz$^{-1/2}$ at 77K and 1kHz, and of another with a 1 × 1 mm^2 pick-up loop was 1.6 pT Hz$^{-1/2}$ at 43K and 1kHz.

A very recent development has been the fabrication of an integrated SQUID magnetometer by Lee et al.[49] in which the flux transformer and SQUID are deposited on the same chip. In this process, the flux transformer is deposited first, and the SQUID, with bi-epitaxial junctions, is deposited subsequently. This process involved seven epitaxial layers, and the magnetometer worked at temperatures up to 76K. At the time of writing, no performance figures are available, but this very considerable achievement demonstrates that relatively complicated structures are possible with the high-T_c technology.

How useful are the high-T_c thin-film magnetometers made to date? The best sensitivity, 1pT Hz$^{-1/2}$ at 1Hz improving to 0.05 pT Hz$^{-1/2}$ in the white noise region, is adequate for some of the less demanding applications such as nondestructive evaluation[52] and some geophysical measurements.[53] This sensitivity is also sufficient to obtain a magnetocardiogram, as was recently demonstrated by Miklich et al.[54] The magnetometer was immersed in liquid nitrogen contained in a thin-walled glass dewar extracted from a household thermos flask. The flask was rigidly supported in a shielded room with the plane of the magnetometer parallel to the chest of the standing subject and within 25 mm. The dewar and subject were surrounded by two concentric high permeability cylinders which attenuated the ambient magnetic field by a factor of 300 (see Fig. 12). Magnetocardiograms were obtained from three healthy male subjects; a representative example is shown in Fig. 13. Although an improvement of at least one order of magnitude would be desirable for serious clinical measurements, this result does demonstrate that high-T_c magnetometers are within reach of practical applications.

6. CONCLUDING REMARKS

A great deal of progress has been made in the last year. The technology for junctions, at least for those with nonhysteretic I-V characteristics, is now beginning to look manufacturable, for example, with the bi-epitaxial process or with normal barriers. Many groups have made thin-film SQUIDs capable of operating in liquid nitrogen. At frequencies up to at least a few hundred hertz 1/f noise is dominant, and high enough in magnitude to pose a serious limitation to applications, such as magnetoencephalography, which require high resolution at frequencies below 1Hz. Fortunately, there is evidence that the 1/f noise level can be reduced by double modulation schemes, and one may be optimistic about significant reductions in the near future. The essential technology for interconnects -- crossovers and vias -- has been established, although there will undoubtedly be refinements in future work. This technology has been used to fabricate flux transformers that have been coupled to SQUIDs in a flip-chip arrangement to make magnetometers. The most sensitive of these magnetometers is adequate for the less-demanding applications, and even to obtain a mangetocardiogram. One can expect at least an order-of-magnitude improvement in the sensitivity on a reasonable time scale. Finally, the first integrated SQUID magnetometer, involving seven epitaxial layers, has recently been demonstrated. Thus, the technology for high-T_c SQUIDs is rapidly coming together. This technology, needless to say, is equally applicable to other kinds of high-T_c electronic circuitry, and prototype digital circuits seem likely to appear in the not-too-distant future.

Fig. 12 Photograph of apparatus used to obtain magnetocardiogram; M. S. Colclough is on left, A. H. Miklich on right.

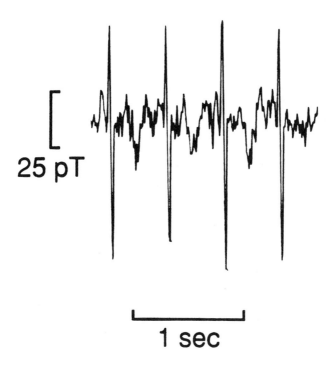

Fig. 13 Magnetocardiogram (not of Miklich) obtained with apparatus in Fig. 12 (from Ref. 54).

ACKNOWLEDGMENTS

The parts of this work carried out at Berkeley are due to the strenuous efforts of M. J. Ferrari, J. J. Kingston, Ph. Lerch, A. H. Miklich and F. C. Wellstood to whom I express my thanks; R. Gronsky and M. E. Tidjani performed expert transmission electron microscopy. I thank K. Char, M. S. Colclough, L. P. Lee and G. Zaharchuk, of Conductus, Inc., with whom much of this work was carried out. I wish to thank R. H. Koch for informative discussions. This work was supported by the California Competitive Technology Program and by the Director, Office of Energy Research, Office of Basic Energy Sciences, Materials Sciences Division of the U.S. Department of Energy under contract number DE-AC03-76SF00098.

REFERENCES

1. R. H. Koch, C. P. Umbach, R. J. Clark, P. Chaudhari, and R. B. Laibowitz, Appl. Phys. Lett. **51**, 200 (1987).
2. D. W. Face, J. M. Graybeal, T.P. Orlando, and D. A. Rudman, Appl. Phys. Lett. **56**, 1493 (1990).
3. R. H. Koch, W. J. Gallagher, B. Bumble, and W. Y. Lee, Appl. Phys. Lett. **54**, 951 (1989).
4. S. E. Russek, D. K. Lathrop, B. H. Moeckly, R. A. Buhrman, D. H. Shin, and J. Silcox, Appl. Phys. Lett. **57**, 1155 (1990).
5. S. M. Garrison, N. Newman, B. F. Cole, K. Char, and R. W. Barton, Appl. Phys. Lett. **58**, 2168 (1991).
6. D. E. McCumber, Appl. Phys. Lett. **39**, 3113 (1968); W. C. Stewart, Appl. Phys. Lett. **12**, 277 (1968).
7. R. W. Simon, J. B. Bulman, J. F. Burch, S. B. Coons, K. P. Daly, W. D. Dozier, R. Hu, A. E. Lee, J. A. Luine, C. E. Platt, S. M. Schwarzbek, M. S. Wire, and M. J. Zani, IEEE Trans. Magn. **MAG-27**, 3209 (1991).
8. C. L. Jia, B. Kabius, K. Urban, K. Herrman, G. J. Cui, J. Schubert, W. Zander, A. I. Braginski and C. Heiden, Physica C **175**, 545 (1991).
9. D. Dimos, P. Chaudhari, J. Mannhart, F. K. LeGoues, Phys. Rev. Lett. **61**, 219 (1988).
10. K. Char, M. S. Colclough, S. M. Garrison, N. Newman, and G. Zaharchuk, Appl. Phys. Lett. **59**, 733 (1991).
11. K. Char, M. S. Colclough, L. P. Lee, G. Zaharchuk, Appl. Phys. Lett. **59**, 2177 (1991).
12. R. B. Laibowitz, R. H. Koch, A. Gupta, G. Koren, W. J. Gallagher, V. Foglietti, B. Oh, J. M. Viggiano, Appl. Phys. Lett. **56**, 686 (1990).
13. S. Wang, X. Zeng, Y. Dai, Y. Hu, H. Jiang, R. Liu, and J. Li, SQUID '91, IV International Conference on Superconducting and Quantum Effect Devices and their Applications, Berlin, Germany, 18-21 June 1991.
14. D. B. Schwarz, P. M. Mankiewich, R. E. Howard, L. D. Jackel, B. L. Straughn, E. G. Burhart, A. H. Dayem, IEEE Trans. Magn. **MAG-25**, 1298 (1989).
15. M. Yu Kupriyanov and K. K. Likharev, IEEE Trans. Magn. **MAG-27**, 2460 (1991); M. R. Beasley in *Proceedings of Conference on Materials and Mechanisms of Superconductivity 1991*, Japan (to be published).
16. M. S. DiIorio, S. Yoshizumi, K-Y Yang, J. Yang and M. Maung, Appl. Phys. Lett. **58**, 2552 (1991).
17. C. T. Rogers, A. Inam, M. S. Hedge, B. Dutta, X. D. Wu, and T. Venkatesan, Appl. Phys. Lett. **55**, 2032 (1989).
18. J. B. Barner, C. T. Rogers, A. Inam, R. Ramesh, and S. Bersey, Appl. Phys. Lett. **59**, 742 (1991).
19. J. Gao, W. A. M. Aarnink, G. J. Gerritsma, D. Veldhuis, and H. Rogalla, IEEE Trans. Magn. **MAG-27**, 3062 (1991).

20. R. W. Simon, J. F. Burch, K. P. Daly, W. D. Dozier, R. Hu, A. E. Lee, J. A. Luine, H. M. Manasevit, C. E. Platt, S. M. Schwarzbek, D. St. John, M. S. Wire, and M. J. Zani in *Proceedings of the Conference on the Science and Technology of Thin Film Superconductors 1990* (to be published).
21. M. J. Zani, J. A. Luine, R. W. Simon, R. A. Davidheiser, Appl. Phys. Lett. **59**, 234 (1991).
22. G. Cui, Y. Zhang, K. Hermann, Ch. Buchal, J. Schubert, W. Zander, A. E. Braginski, and C. Heiden, LT-19 Satellite Conference on High Temperature Superconductors, Cambridge, England, August 1990 (unpublished).
23. D. Robbes, A. H. Miklich, J. J. Kingston, Ph. Lerch, F. C. Wellstood, and J. Clarke, Appl. Phys. Lett. **56**, 2240 (1990).
24. J. Clarke and R. H. Koch, Science, **242**, 217 (1988).
25. C. D. Tesche, and J. Clarke, J. Low Temp. Phys. **29**, 301 (1977).
26. V. Foglietti, W. J. Gallagher, M. B. Ketchen, A. W. Kleinsasser, R. H. Koch, S. I. Raider, and R. L. Sandstrom, Appl. Phys. Lett. **49**, 1393 (1986).
27. C. D. Tesche, R. H. Brown, A. C. Callegari, M. M. Chen, J. H. Greiner, H. C. Jones, M. B. Ketchen, K. K. Kim, A. W. Kleinsasser, H. A. Notarys, G. Proto, R. H. Wang, and T. Yogi, in *Proceedings 17th International Conference on Low Temperature Physics*, LT-17 (North-Holland, Amsterdam, 1984) p. 263.
28. Biomagnetic Technologies Incorporated, San Diego, California.
29. R. L. Sandstrom, W. J. Gallagher, T. R. Dinger, R. H. Koch, R. B. Laibowitz, A. W. Kleinsasser, R. J. Gambino, B. Bumble, and M. F. Chisholm, Appl. Phys. Lett. **53**, 444 (1988).
30. R. Gross, private communication.
31. R. H. Koch, at SQUID '91, IV International Conference on Superconducting and Quantum Effect Devices and their Applications, Berlin, Germany, 18-21 June 1991.
32. M. Kawasaki, P. Chaudhari, T. Newman, and A. Gupta, Appl. Phys. Lett. **58**, 2555 (1991).
33. R. H. Koch, J. Clarke, W. M. Goubau, J. M. Martinis, C. M. Pegrum, and D. J. Van Harlingen, J. Low Temp. Phys. **51**, 207 (1983).
34. C. T. Rogers and R. A. Buhrman, Phys. Rev. Lett. **53**, 1272 (1984).
35. P. Dutta and P. M. Horn, Rev. Mod. Phys. **53**, 497 (1981).
36. M. J. Ferrari, M. Johnson, F. C. Wellstood, J. Clarke, P. A. Rosenthal, R. H. Hammond, and M. R. Beasley, Appl. Phys. Lett. **53**, 695 (1988).
37. M. J. Ferrari, M. Johnson, F. C. Wellstood, J. Clarke, D. Mitzi, P. A. Rosenthal, C. B. Eom, T. H. Geballe, A. Kapitulnik, and M. R. Beasley, Phys. Rev. Lett. **64**, 72 (1990).
38. M. J. Ferrari, M. Johnson, F. C. Wellstood, J. Clarke, A. Inam, X. D. Wu, L. Nazar, and T. Venkatesan, Nature **341**, 723 (1989).
39. M. B. Ketchen, W. J. Gallagher, A. W. Kleinsasser, S. Murphy, and J. Clem. in *SQUID '85, Superconducting Quantum Interference Devices and their Applications* (Walter de Gruyter, Berlin 1985) p. 865.
40. For a discussion, see J. Clarke, Proc. IEEE **77**, 1208 (1989).

41. J. J. Kingston, F. C. Wellstood, Ph. Lerch, A. H. Miklich, and J. Clarke, Appl. Phys. Lett. **56**, 189 (1990).
42. M. E. Tidjani, R. Gronsky, J. J. Kingston, F. C. Wellstood, and J. Clarke, Appl. Phys. Lett. **58**, 765 (1991).
43. F. C. Wellstood, J. J. Kingston, and J. Clarke, Appl. Phys. Lett. **56**, 2336 (1990).
44. F. C. Wellstood, J. J. Kingston, M. J. Ferrari, and J. Clarke, Appl. Phys. Lett. **57**, 1930 (1990).
45. J. J. Kingston, F. C. Wellstood, Du Quan, and J. Clarke, IEEE Trans. Magn. **MAG-27**, 974 (1991); F. C. Wellstood, J. J. Kingston, M. J. Ferrari, and J. Clarke, ibid, 2569 (1991).
46. A. H. Miklich, J. J. Kingston, F. C. Wellstood, J. Clarke, M. S. Colclough, K. Char, and G. Zaharchuk, Appl. Phys. Lett. **59**, 988 (1991).
47. B. Oh, R. H. Koch, W. J. Gallagher, R. P. Robertazzi, and W. Eidelloth, Appl. Phys. Lett. **59**, 123 (1991).
48. W. Eidelloth, W. J. Gallagher, R. P. Robertazzi, R. H. Koch, and B. Oh, Appl. Phys. Lett. **59**, 1257 (1991).
49. L. P. Lee, K. Char, M. S. Colclough, and G. Zaharchuk, to be published in Appl. Phys. Lett.
50. We note that SQUIDs made from bulk YBCO are much more sensitive to magnetic field than their thin film counterparts with comparable flux noise and inductance. The bulk SQUIDs involve a cylindrical hole of length ℓ and radius r with an inductance $\mu_o \pi r^2/\ell \approx$ 100pH for ℓ = 10mm and r = 0.5mm. Thin film devices of comparable inductance have a hole of dimension $\approx 60\mu$m. Thus, the rf SQUIDs of J. E. Zimmerman, J. A. Beall, M. W. Cromar and R. H. Ono [Appl. Phys. Lett. **51**, 617 (1987)] and of A. G. Likhachev, V. N. Polushkin, S. V. Uchaikin and B. V. Vasiliev [Superconductor Science and Technology **3**, 148 (1990)] had estimated magnetic field sensitivities of aobut 1 pT Hz$^{-1/2}$ and 0.3 pT Hz$^{-1/2}$, respectively, at 1Hz.
51. M. J. Ferrari, J. J. Kingston, F. C. Wellstood, and J. Clarke, Appl. Phys. Lett. **58**, 1106 (1991).
52. See, for example, H. Weinstock, IEEE Trans. Magn. **MAG-27**, 3231 (1991).
53. See, for example, J. Clarke, IEEE Trans. Magn. **MAG-19**, 288 (1983).
54. A. H. Miklich, F. C. Wellstood, J. J. Kingston, J. Clarke, M. S. Colclough, K. Char, and G. Zaharchuk, Nature **352**, 482 (1991).

DISCUSSION

K. E. Gray: John, what do you think are the advantages, if any, of having a hysteretic junction instead of the non-hysteretic?

J. Clarke: Well, that's obviously a good question. There are certain applications where you really would like hysteresis. One is the voltage standard which, as you know, at present involves about 20 000 niobium junctions. If you could make the array out of high-T_c, that would be great. Another possible application would be in developing computer devices, although I'm personally not so convinced that that's viable. I think the real reason we'd all like to see hysteretic junctions however is that we could see an energy gap!

D. S. Fisher: You said that you could make devices that work, but are there any devices yet that could work better than low-temperature superconducting devices?

J. Clarke: You're talking about SQUIDS?

D. S. Fisher: Yes.

J. Clarke: There will never be a SQUID at 77 K better than SQUIDs at 4 K using low-T_c superconductors in terms of noise performance.

D. S. Fisher: Intrinsic noise limitations?

J. Clarke: The white-noise power is intrinsically higher at 77 K because it scales as the temperature. I think the reason for developing the high-T_c technology is because we can all think of applications where you'd much rather have liquid nitrogen than liquid helium. The one that I'm personally most fond of is geophysical applications, where you'd like to be able to put SQUIDs out for years without having to go and fill up the dewar. That would make geophysics with SQUIDs very practical.

D. S. Fisher: How much higher in temperature would we have to go before we could do it with solid state cooling?

P. L. Gammel: To 100 K.

J. Clarke: Yes, I would say 120 K, but . . .

P. L. Gammel: I read a claim of 100 K using Bi-Te in *Laser Focus World* recently.

J. Clarke: OK, 100 K. All right. And that's just too high, I think, for any of the devices, even the thallium ones. Obviously for YBCO and BSCCO it's too high.

L. J. Campbell: Are the large spikes in the noise at certain frequencies in the diagram you showed well understood?

J. Clarke: Yes. They arise from two things: 60 cycles and microphonics.

P. L. Gammel: I had two questions. First, what's the yield, approximately, on your crossovers and vias?

J. Clarke: On good days, 100%.

J. Wilkins: How many good days?

D. S. Fisher: What's the yield of good days?

J. Clarke: It's a little hard to say, because we've actually recently gone away from strontium titanate because of its undesirably high dielectric constant. We're now using cerium oxide. I think the yield on that is now pretty high. On any one chip, either they all work or they all don't work, and we've probably had four or five chips that work.

P. L. Gammel: I had a second question. On junctions which are fairly wide, 50 microns or so, do you think the junction is over the entire width?

J. Clarke: Of course, I don't really know that, but I suspect that the junction length is certainly much greater than the coherence length and most of them, including the ones we made with these electroshock techniques, have characteristics that are much more flux-flow-like than RSJ-like. It's very likely that the weak links are filaments of high-T_c material.

P. L. Gammel: I don't remember specifically; don't those types of junctions tend to be noisier?

J. Clarke: Oh yes. They're not very useful. These junctions give a small signal when you use them in a SQUID, and so tend to produce high noise levels. So all I'm saying is I don't think this is a profitable way to go.

R. Lemons: Why do you think it's been so difficult to make hysteretic junctions out of the high-T_c materials?

J. Clarke: Well, there are lots of theories about that. One of them of course is that you always get a disordered layer on the interface, and this is being looked at very carefully by people at Stanford, Phillips and other places where they're trying to make proximity-effect structures. They start with, say, YBCO, followed by the normal layer, and put, say, a lead layer on the top. And what they find is that the proximity effect is *extremely* small, maybe zero, in the c-axis direction. And it's small, but non-zero, in the a-axis direction.

Now, there's been some theoretical work by Kupriyanov and Likharev where they try to explain this in terms of scattering at the interface because you have a very different density of states on the two sides of the interface. And to some extent, maybe not entirely, this explains the result – I don't think to everyone's satisfaction, but it's going in the right direction.

A. P. Malozemoff: Why does the magnesium oxide buffer lead to a rotation by 45° of the yttrium barium copper oxide?

J. Clarke: As I understand it, all of these rotations are sort of black magic. No one can predict *a priori* what's going to happen. But you put down MgO on sapphire, and then strontium titanate on top of that, and the mismatch in the lattice constants causes the strontium titanate to rotate by 45°. The YBCO on top of it follows suit. And there's a large class of materials which Kookrin Char and co-workers at Conductus have been exploring very vigorously, which will give you a rotation. You can look at all the lattice parameters, and have a good guess at whether or not something will work, but I don't think you actually know until you try. You just pattern a whole lot of combinations and look for a 45° rotation with x rays, and you very quickly find out whether or not any given combination gives a rotation.

The rotation is easy; the thing that's very hard is getting a good interface along the boundary between the two grains.

D. C. Larbalestier: John, I was very intrigued in your question, "Why are grain boundary junctions so good?" Part of this was stimulated by the June *Physics Today*. I opened Randy Simon's article and he had this nice high-resolution picture, taken at Jülich, of the step-edge junction showing the two grains coming in like this, and then you have the planes doing this, and they meet at 45°, along an (013) plane. And this, apparently, makes a great Josephson junction. Now, in my *Physics Today* article, I showed a micrograph of Gao and Merkle's with exactly the same crystallography as exemplifying the fact that that same connection in the bulk crystal we found to have a flux-pinning character. So the interesting question is that apparently the same grain boundary faceting can give both Josephson and flux-pinning properties. So I wondered, does it consistently give Josephson junction behavior in the step-edge junctions, or do some of them not work as Josephson junctions?

J. Clarke: I think the fact is a lot of them do not work very well in the sense of having good RSJ-like characteristics. You can pretty well be sure that you'll get a weak link of some kind, in our experience. But we've made lots of these structures, and many of them do not have RSJ-like characteristics.

D. C. Larbalestier: The follow-up I would say though is that in fact as one is beginning to get microscopy on exactly the same junctions one is looking at, in fact, there's no smoking gun that's evident in the microscopy. I mean, really, they are extremely perfect, it's very hard to see the disorder. Of course, you don't see oxygen.

J. Clarke: My question still stands.

D. C. Larbalestier: Yes.

M. P. Maley: What are your feelings about the origin of the $1/f$ noise?

J. Clarke: First of all I think flux noise, so called, we understand pretty well. You have flux lines stuck on pinning sites; they hop around under thermal activation, and we have a model that fits that pretty well. Let me comment, however, that if you look at the temperature dependence the noise doesn't go up very rapidly with temperature, so that you can operate close to T_c. This is well understood in terms of the theory, and it relates to the fact that the distribution of energy of the flux-pinning sites is somewhat narrow. You don't have a very broad distribution.

Now, if you have critical current fluctuations, in the low-T_c junctions, the origin is very well understood. You have an electron trap in the barrier, an electron tunnels in, and locally there is a change in the barrier height, and a corresponding change in the critical current. Subsequently, the electron tunnels out, and the barrier height pops back again. That gives you a Lorentzian behavior, and if you have a distribution of trapping times, adding together these Lorentzians, you get $1/f$ noise.

That could be the origin in the high-T_c junctions. It is possible that you have traps in the barrier, but I don't think there's any direct evidence for that yet. But there are other mechanisms you could come up with; for example, Roger Koch suggested that if you imagine there are flux lines pinned on either side of the barrier the flux lines stop the supercurrent going through the junction if it's close enough, because you have little normal regions. So you could come up with a scenario where these flux lines jump around, and again they cause a deviation in the critical current.

M. P. Maley: If you were to look at BSCCO and thallium films, you might expect that the $1/f$ noise would be much worse.

J. Clarke: Generally speaking, that's absolutely true. We've looked at both. In thallium, the noise was high enough you could practically hear it with your unaided ears.

D. S. Fisher: What is the typical density of the flux lines?

J. Clarke: Well, that's a very good question, Dan. We have measured the dependence of the noise on the magnetic field in which we cool the film. If you cool in something very close to zero field, you still have noise, which is to say that there are residual flux lines in the film. And presumably these are flux lines generated near T_c and as you cool down most of them annihilate, but some of them get stuck on the flux lines. Now, in the better quality YBCO films, the density of flux lines is roughly equivalent to cooling in one gauss, just to give you an order of magnitude.

X-D Wu: I just want to make a comment about the 45° rotation. It's not magic. The reason you have a 45° rotation is because you have a cubic system. And also you get 45° because there's a good lattice match along 45° rotation. The MgO on sapphire, MgO rotated 45°. $SrTiO_3$ on MgO followed the MgO direction. On the other hand, $SrTiO_3$ on the sapphire itself did not rotate; that's why you get a 45° rotation. Of course sapphire is not the only choice for the substrates.

J. Clarke: I believe that you don't know until you try that you're going to get a 45° rotation as opposed to no rotation. I agree you don't get an intermediate angle.

X-D Wu: You only can get a 45° rotation, you can not make it 25°.

J. Clarke: Exactly. It might be zero.

R. Lemons: If the $1/f$ noise were due to charge trapping at the interface, how would it depend on the area of the junction?

J. Clarke: Well, all things being equal, you'd expect that the density of traps would scale linearly with the area.

R. Lemons: The statistics of the fluctuation ought to be proved.

J. Clarke: And indeed, in the low-T_c junctions, that's been done. If you make a series of junctions with the same barrier height, but increase the area, the critical current scales with the area, but the noise goes up as the square root of the area. That's quite well established.

R. Lemons: Has any one been following this approach?

J. Clarke: There just isn't enough control yet in the junction technology that one can address that issue. If you were to make a whole bunch of junctions of the same width, there would be a quite wide variation in noise anyway, even though you made them all at one time. It'll come: all of these are the right questions. The answers are just not in yet.

D. C. Larbalestier: John, I wonder, can you not change the trap density just by sucking a little bit of oxygen out of the grain boundary?

J. Clarke: Maybe. Actually Kawasaki and co-workers at IBM showed that annealing in ozone decreased the $1/f$ noise.

D. K. Finnemore
Ames Laboratory
Ames, IA 50011

Critical Fields in High-Temperature Superconductors

An analysis of various methods to obtain the critical fields of the high-temperature superconductors from experimental data is undertaken in order to find definitions of these variables that are consistent with the models used to define them. Characteristic critical fields of H_{c1}, H_{c2} and H_c that occur in the Ginzburg-Landau theory are difficult to determine experimentally in the high temperature superconductors because there are additional physical phenomena that obscure the results. The lower critical field is difficult to measure because there are flux pinning and surface barrier effects to flux entry; the upper critical field is difficult because fluctuation effects are large at this phase boundary; the thermodynamic critical field is difficult because fluctuations make it difficult to know the field where the magnetization integral should be terminated. In addition to these critical fields there are at least two other cross-over fields. There is the so-called irreversibility line where the vortices transform from a rigid flux line lattice to a fluid lattice and there is a second cross-over field associated with the transition from the fluctuation to the Abrikosov vortex regime. The presence of these new physical effects may require new vocabulary.

1. INTRODUCTION

There is an enormous variation in the published values for the characteristic critical fields of the high-temperature superconductors and there is no general agreement on procedures to extract these quantities from measured values. The range of values can be as large as a factor of ten. Indeed, a linear extrapolation of magnetization vs temperature, M vs T, data can lead to a positive slope, dH_{c2}/dT, for the temperature dependence of the upper critical field, H_{c2}.[1] The only variable for which there is general agreement is the measured value of the H_{c2}, for $Y_1Ba_2Cu_3O_{7-\delta}$, Y(123), close to the critical temperature, T_c. The excellent measurements by Welp and co-workers,[2] by Hao and co-workers[3] and by Welp et al.[4,5] give results that agree from sample to sample and also fit the model used to define this critical field.[6] By choosing different assumptions for the analysis, values of dH_{c2}/dT can range from -1.6 T/K to -1.9 T/K but these differences are not very large.

For the lower critical field, H_{c1}, all of the measurements determine the flux entry field rather than the field where a vortex is first thermodynamically stable. The papers assume that there are no barriers to the first flux entry or that somehow the barriers do not matter. These measurements only approach H_{c1} to the extent that the barriers are small compared to H_{c1}. Even in a classical material such as Nb or V, the measurement of H_{c1} is a subtle enterprise. This confusion about the definition of H_{c1} and resulting analysis can easily lead to a difference of a factor of two. Because it is difficult to get around this barrier problem, it may be easier to determine H_{c1} by using muon rotation to determine the penetration depth, λ, and substitute this value into

$$H_{c1} = \frac{\phi_o}{4\pi\lambda^2} \ln \kappa . \tag{1}$$

where ϕ_o is the flux quantum and κ is approximately the ratio of the penetration depth to coherence distance, ξ. Because κ appears in the logarithm, modest errors in the coherence distance will not affect H_{c1} very much. For penetration depths in the range of 0.14 μm and κ of 52, Eq. (1) gives H_{c1} of 30 mT at 4.2K for a material such as Y(123).

In the measurements of the cross-over field from the thermodynamically reversible range to the regime of irreversibility, the so-called irreversible line, H_{irr} vs T, most of the values are relatively close to one another. There are, however, subtle differences because the measurements probe the sample on a different time scale or with different forces and thus get somewhat different results. For example, a determination of H_{irr} via a megahertz susceptibility measurement will give a result different from a dc magnetization measurement because the time scale of the two probes is so different. These two quantities should have different names because they are probing different phenomena.

Beyond H_{c1}, H_{c2} and H_{irr} lines in the H-T plane, there is the very difficult problem of describing the magnetic-field dependence of the critical fluctuations along the superconductor-normal metal phase boundary.[6,7] In a classical superconductor,[7] the jump in C_p follows the H_{c2} vs T line because the jump results from a sharp onset of the superconducting state from the normal state. The jump primarily reflects the condensation energy as modified by the presence of the Abrikosov lattice. In the high-T_c materials the specific heat from the fluctuations dominates over a temperature interval about 4K wide so the jump in C_p has a form different from the classical materials. The position of the jump does not follow the H_{c2} line.[6]

In simplified concept, the critical fields can be derived from a straightforward measurement of a magnetization curve such as shown in the sketch of the ideal case in Fig. 1a. The magnetization drops with rising magnetic field at the Meissner slope, and the first abrupt change in slope occurs when flux enters at H_{c1}. At higher fields, there is another abrupt change in slope when the vortices disappears at H_{c2}. The real case, illustrated in Fig. 1b, is not so simple. There are flux-entry barriers that give rise to hysteresis near H_{c1}. There are fluctuations that give rounding near H_{c2} as shown in Fig. 1b. Both effects complicate the measurements.

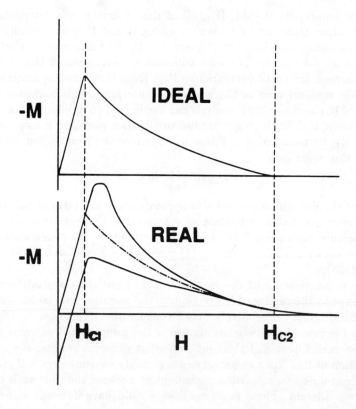

Fig. 1 Comparison of the ideal and real magnetization curves for a high-T_c material.

The purpose of this paper is to describe some of the most elementary aspects of the measurement of these critical fields and cross-over phenomena that divide the H-T plane into distinct sectors. In this work there are two types of uncertainty that must be considered. First, there are the experimental problems of sample quality and the accuracy of the measurements; second, there are the conceptual problems of defining the quantity to be measured and dealing with other phenomena that obscure the measurement. In the early experiments, sample quality was a major contributor to the errors. More recently, however, the quality of materials such as Y(123) has improved a great deal and the reproducibility of the measurements is good. The central issue now is the fact that the Ginzburg-Landau approximations originally used to define the characteristic fields do not completely describe the data and more sophisticated vocabulary and models are needed.

2. UPPER CRITICAL FIELD

The definition of H_{c2} arises from the Ginzburg-Landau-Abrikosov-Gor'kov (GLAG) theory, and values of H_{c2} should be derived from data that obey this theory. More specifically, H_{c2} is defined to be the magnetic field where the Abrikosov lattice first appears and it is important to remember that this definition of H_{c2} does not contain the complications of fluctuations. In a careful study of the magnetization, M, vs temperature, T, curves near T_c, for a single crystal of $Y_1Ba_2Cu_3O_{7-\delta}$, Y(123), Welp and co-workers[2] took data below the fluctuation regime (i.e. below 89K) and extrapolated the data to M = 0 to define H_{c2}. They assumed that data from roughly 82 to 89K were linear in T as predicted by GLAG and fit this limited range of data to obtain H_{c2} vs T and a value of $dH_{c2}/dT = -1.9$ T/K.

These ideas are illustrated by very similar data for a grain-aligned sample[9] of Y(123) having a grain size of about 35 μm ss shown in Fig. 2. These data are essentially the same as the Welp *et al.*[2] results. Analyzing these data the same way as Welp *et al.* analyzed the single crystal data gives $dH_{c2} = -1.9$ T/K which is in good agreement with the single-crystal data. Indeed, as shown in Ref. 9, the rounding in the M vs T data of Fig. 2 quantitatively agrees with the fluctuation C_p data to an accuracy of about 30%. The rounding is not an accident of sample inhomogeneity but a true reflection of fluctuations.

There is a problem with the linear extrapolation analysis to get H_{c2} that was recognized by Hao *et al.*[3] The data of Welp *et al.*, and the data of Fig. 2 do not really fit the GLAG theory between 82 and 89K because the slope, dM/dT, decreases by more than a factor of two as the magnetic field increases form 0 to 5 T, as shown in Fig. 3. The GLAG theory, however, predicts that all of these slopes should be the same. This discrepancy arises because the GLAG regime extends only from T_c to about 88K for these materials.[3,5,12] The slope, dH_{c2}/dT, is so steep that the region of GLAG applicability is only about 4K wide and modifications to the theory are

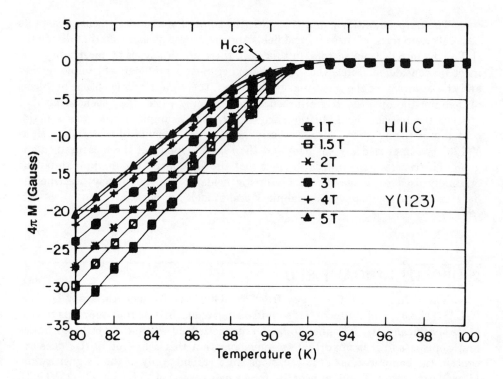

Fig. 2 Magnetization data for grain-aligned Y(123)

needed. As Hao and Clem[3] point out, the London theory[12] that might be used over a wide range of temperatures, also is inadequate because it ignores the free energy of the cores. This is contrary to the usual expectation for high-κ materials. Generally it is thought that the area of the core is tiny compared to the total area of the vortex so the core energy can be ignored. This is not the case. The circulating supercurrents in the vortex and the resulting excluded flux also are tiny so that the magnetic and core energies are comparable. Stated another way, the core has large free energy per unit volume but a tiny volume; the magnetic energy has a tiny free energy per unit volume but a large volume; the total free energies are comparable. Working out the full theory gives the observed magnetic-field-dependent slopes and it also puts some curvature in the M vs T plots.

There is a rather narrow temperature range in which the theory[3] can be used to fit the data. Below 80K, irreversibility prevents reliable equilibrium magnetization measurements; above 88K, the fluctuations make the theory inapplicable. There is, however a window between 80 and 88K where comparison with the model is applicable.

Critical Fields in High-Temperature Superconductors

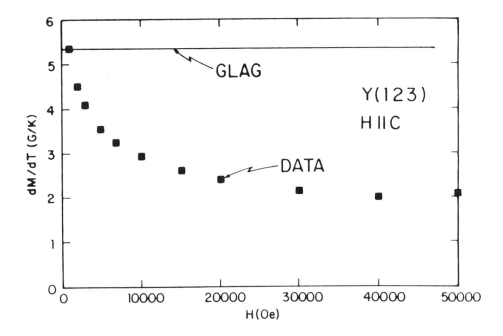

Fig. 3 Change in slope of the M vs T data in reversible range.

It should also be noted that above 84K the data are reversible to the accuracy of the measurement and from 80 to 84K the hysteresis is only a few percent. At every temperature on Fig. 2 there is a temperature increasing as well as a temperature decreasing data point and there is a small amount of hysteresis. Taking the average of the field-increasing and field-decreasing magnetization as suggested by the critical state model permits data accurate to about 1%.

With this more sophisticated theory[5] and with the proper limitations on the range of data used, the data are seen to fit the theory rather well and for the case of Y(123), at least, there is a self-consistent definition for H_{c2}. The correction caused by using the Hao et al.[5] theory rather than the Ginzburg-Landau theory is on the order of 20%.

Several other methods that have been used to determine H_{c2} are unsatisfactory. For example, the attempts to use electrical resistivity as a probe of H_{c2} are at best a broad indication of the location of H_{c2}. Resistivity data probe the depinning and motion of vortices rather than thermodynamic variables so they do not measure H_{c2}. As the sample warms, the point where vortices first begin to move and show dissipation is not generally the point where they first exist, H_{c2}, in the high-T_c materials. As pointed out by Tinkham,[11] the first onset of a decrease in the R vs T curve gives the closest approach to H_{c2} but dissipation measurements are not an appropriate tool to determine a thermodynamic variable in these systems where the vortex lattice is so mobile.

3. LOWER CRITICAL FIELD

The lower critical field, H_{c1}, is defined to be the lowest field at which vortices are thermodynamically stable in the interior of a superconductor. As pointed out by Wollan et al.,[13] all studies of the phase transition at H_{c1} unfortunately must contend with the basic irreversibilities in the transition which arise because there are free-energy barriers which inhibit the entry of fluxoids. A major contributing factor is the image force on the vortex at the vacuum-superconductor interface[12] that is illustrated in Fig. 4. This alone guarantees hysteretic effects even for a perfect interface. In addition, there are other imperfections that lead to surface pinning and other types of bulk pinning. Even for the well documented cases of Nb and V,[13] the determination of H_{c1} is difficult. It is necessary to develop a model for the hysteresis, measure the flux entry and flux exit fields, show that the data fit the model and then extract the value of H_{c1} from the field increasing and field decreasing magnetization data. The first deviation from the straight line portion of the initial slope of the magnetization curve is the flux entry field. It is not H_{c1}. As the hysteresis and pinning effects get smaller and smaller, the entry field approaches H_{c1}. As discussed by deGennes,[12] however, the image barrier is a substantial effect.

There are many ways of determining the flux entry field with very high precision. Umezawa and coworkers[14] made high precision magnetization measurements over a wide temperature range and find by plotting the deviation for M from a linear fit to the low field data that flux enters rather abruptly at a well defined field and measure this entry to be approximately $11 (1 - T/T_c)$ mT. Other methods that depend on rf and audio frequency loss processes also give the flux entry field to high precision. A number of measurements have shown flux entry fields that have the shape shown in Fig. 5.[15] The abrupt change in slope of the H_{en} vs T curve at about 25K almost surely arises from a change in the temperature dependence of the barrier field, not the change in H_{c1} vs T. Although these values can be measured extremely precisely, it is important to remember that precision of measurement is not the same as accuracy of value.

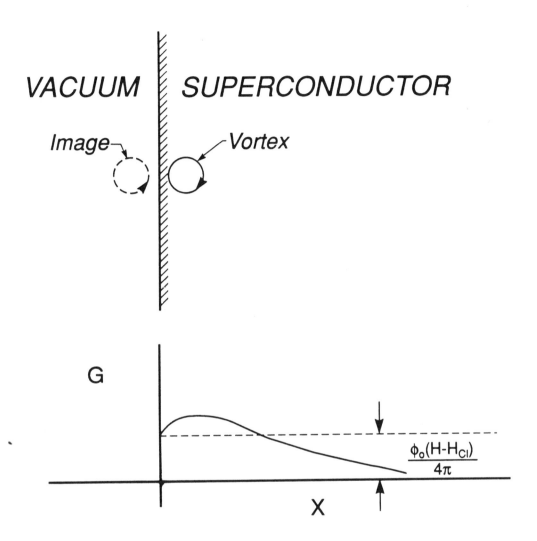

Fig. 4 Free energy barrier to flux entry due to image forces.

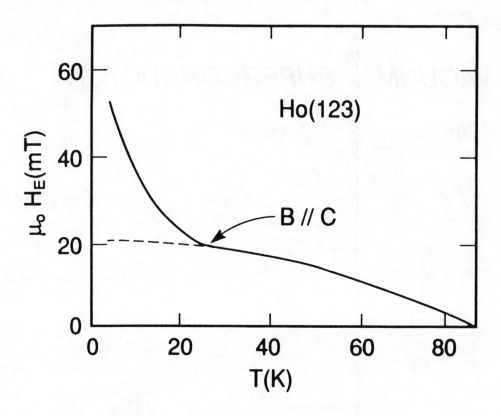

Fig. 5 Flux entry fields for high temperature superconductors.

4. THERMODYNAMIC CRITICAL FIELDS

As the name implies, the thermodynamic critical field, H_c, can be determined from thermodynamically reversible magnetization data and the free energy difference between the superconducting and normal states. As discussed by Pippard,[16] the area under the magnetization curve gives the free energy difference and the thermodynamic critical field curve usually is given by

$$G_N - G_S = \frac{H_c^2}{4\pi} = \int_0^{H_{c2}} M\,dH \quad . \tag{2}$$

The upper limit on this equation, however, must be modified for the high T_c materials because there is a lot of flux exclusion associated with the fluctuations and at H_{c2} as defined by Hao and Clem,[5] the sample is far from normal. Magnetization curves shown in Fig. 6 illustrate that there is a great deal of area under the curve, and hence free energy, left above 5 T even at 87K. There is no reliable way to know where to truncate the integral by extrapolating the data of Fig. 6. Because there are fluctuations present, the determination of the thermodynamic critical field requires both reversible magnetization data and the ability to measure experimentally at fields well above H_{c2}.

4.1 H_{c1}, H_{c2} and H_c as Fitting Parameters

The method of determining the critical fields discussed by Hao and Clem[5] and carried out in Refs. 3 and 4 is worth further examination. In their view, the way to proceed in the determination of the critical field curves is to give up the direct measurement of these quantities by probing the magnetic field regions where they occur. There are too many complicating phenomena going on in those regimes. Instead,

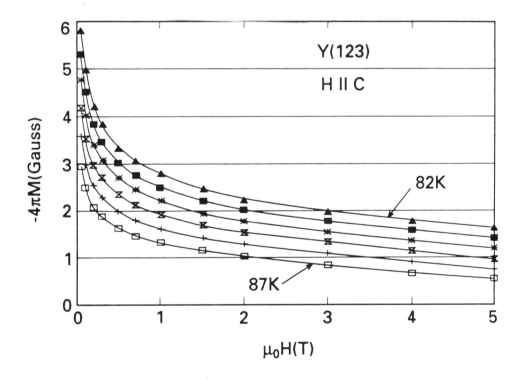

Fig. 6 Magnetization curves for Y(123). Data are at 1K intervals between 82K and 87K.

one can make measurements of the magnetization at fields well away from these critical fields in magnetic field regions where clean reversible measurements can be made, and then infer the values of H_{c1}, H_{c2} and H_c by fitting the magnetization vs field and temperature data to a theory that contains these quantities.

Hao and Clem[3] developed a theory for the magnetic field dependence of the magnetization that applies at fields and temperatures outside the Ginzburg-Landau regime and Welp *et al.* have fit extensive data for two single crystals of Y(123) to determine H_{c1}, H_{c2} and H_c self-consistently. The theory applies in the intermediate field regime, $H_{c1} \ll H \ll H_{c2}$, where the magnetization is reversible and $\xi \ll a \ll \lambda$. Here ξ is the coherence distance and $a = [\phi_o/H]^{1/2}$ is the average spacing between vortices. For Y(123), $\xi = 1.5$ nm, $\lambda = 150$ nm and a ranges from 63 nm at 0.5 T to 20 nm at 5 T. The basic relation connecting H_{c1}, H_{c2} and H_c to the magnetization data is

$$4\pi M = \frac{\alpha H_{c1}}{2 \ln \kappa} \ln \frac{H}{\beta H_{c2}} . \tag{3}$$

where α and β are determined from the theory and are constants for fairly wide ranges of the field. In this equation, H_{c1} and H_{c2} are replaced by the variables H_c and κ. These variables are much more convenient because H_c depends only on temperature and not magnetic field; $\kappa = \lambda/\xi$ is assumed to be independent of both field and temperature. The relations are:

$$H_{c1} = H_c \frac{\ln \kappa}{2\kappa} , \tag{4}$$

and

$$H_{c2} = H_c (2)^{1/2} \kappa . \tag{5}$$

Eq. 3 then becomes

$$4\pi M = \frac{\alpha H_c}{4\kappa} \ln \frac{H}{\beta \sqrt{2} \kappa H_c} . \tag{6}$$

In the range from 82 to 89K and for fields from 1 to 5 T, the theory predicts that α and β are essentially constant and equal to $\alpha = 0.77$ and $\beta = 1.44$ for Y(123). The data are then fit to Eq. (6) to find the best H_c at each temperature subject to the condition that κ is the the same at all temperatures.

By fitting data between 1 and 5 T and temperatures between 85 and 91.5K, they find for H along the c-axis that $\kappa^c = 52$ and for H along the ab-plane the $\kappa^{ab} = 370$; $dH_c/dT = 20.2$ mT/K for both field directions; $dH_{c2}/dT = -1.64$ T/K for H along c and -12.2 T/K for H along ab. These results are very consistent with other measurements. It is a bit surprising that Eq. (6) fits the 91.5K data quite so well as they do in light of the fluctuation data of the Illinois group,[6] but this analysis probably represent the best values to date.

5. FLUCTUATION REGION

The jump in specific heat at T_c is a feature of the high-temperature superconductors that is clearly different from the behavior of classical superconductors.[6] Classical superconductors are typified by the results sketched in Fig. 7a for $PbMo_6S_8$ from the work of the Geneva group.[7] As a magnetic field is applied, the temperature where the jump in C_p occurs is steadily suppressed and follows the H_{c2} line. Hence the jump in C_p can be used as a probe to plot the H_{c2} vs T curve. For this classical case, presumably the jump in C_p arises from the condensation energy of the electron gas in the presence of the flux line lattice.

For the high T_c materials, the behavior is quite different. As a magnetic field is applied, the jump in C_p continues to occur near the same temperature and the only change is in the magnitude of the C_p jump.[6,9] As shown for the case of Y(123), in Fig. 7b, the application of fields up to 7 T suppresses the peak of the C_p jump by a factor of 3 but the location of the initial deviation from the normal state shifts relatively little compared to a classical superconductor. The H_{c2} line is shown by the arrows at the top of Fig. 7b. It is clear that the initial deviation from the normal state occurs at temperatures well above the H_{c2} vs T line. The feature of the C_p data that corresponds most closely to the H_{c2} vs T curve is the temperature where the C_p data become independent of magnetic field.

Flux expulsion does not begin at the H_{c2} line either. As shown by an expanded view of the magnetization in Fig. 8, There is a cross-over in the magnetization data for Y(123) at 91.5K just as was observed for the Bi(2212).[1] The H_{c2} vs T line shown by the arrows on Fig. 8 seems to be unrelated to the temperature where massive flux exclusion begins. There is, of course, flux expulsion on the part per million scale at 200K due to fluctuations and this reaches the part per ten thousand scale at 91.5K for all fields up to 5 T. There is as yet no satisfactory theory for flux exclusion in this cross-over regime near T_c.

6. CONCLUSIONS

Values of H_{c1} and H_{c2} can not be determined by direct measurements of the magnetization in the regions of H_{c1} and H_{c2}. The vortex image barrier and flux pinning preclude direct measurement and a rather complicated study of the barriers and hysteresis must be used in conjunction with an appropriate theory for the hysteresis to derive a value of H_{c1}. This has not yet been done for the high-T_c materials. Fluctuations preclude the direct measurement of H_{c2} as a change in slope of a magnetization curve. Both H_{c1} and H_{c2} must be viewed as fitting parameters in an appropriate theory that accurately describes magnetization data between H_{c1} and H_{c2} in the region where the data satisfy thermodynamic reversibility. This type of modeling gives κ and H_c as well as H_{c1} and H_{c2}. For high-T_c materials, the

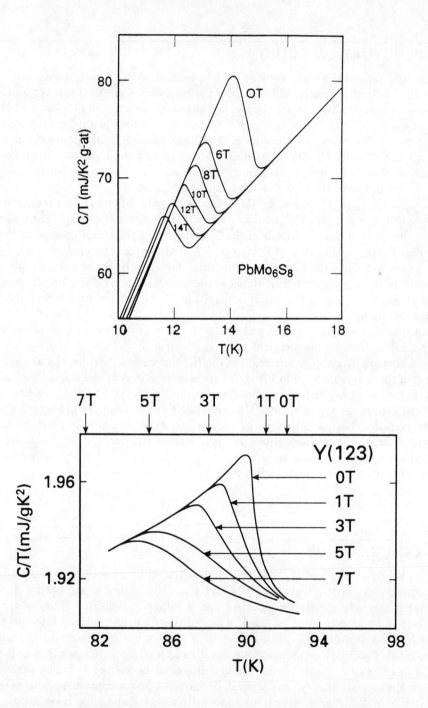

Fig. 7 Comparison of the jump in C_p for the classical superconductors with the behavior of high-T_c materials.

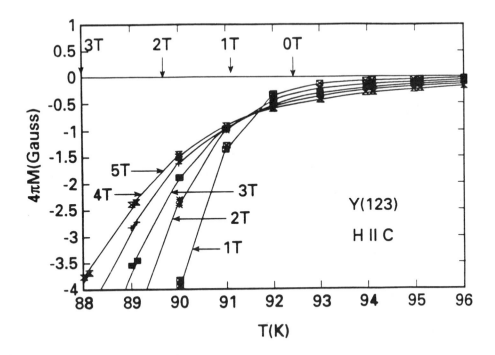

Fig. 8 Expanded view of magnetization in the fluctuation regime showing the cross-over at 91.5K. Flux expulsion does not begin at the H_{c2} line indicated by the arrows.

jump in C_p is controlled by fluctuations and hence can not be used to determine the H_{c2} vs T line. There is no obvious signature of H_{c2} in C_p data. The cross-over in the magnetization data observed for both Y(123) and Bi(2212) is a distinctive feature of fluctuations for both of these systems.

ACKNOWLEDGMENTS

We thank M. Tinkham for helpful comments on this work. Ames Laboratory is operated for the U.S. Department of Energy by Iowa State University under Contract No. W-7405-ENG-82. This work was supported by the Office of Basic Energy Sciences.

REFERENCES

1. P. H. Kes, C. J. van der Beek, M. P. Maley and M. E. McHenry, D. A. Huse, M. J. V. Menken and A. A. Menkovsky, (Preprint).
2. U. Welp, W. K. Kwok, G. W. Crabtree, K. G. Vandervoort, J. Z. Liu, Phys. Rev. Lett. **62**, 1908 (1989).
3. Z. Hao, J. R. Clem, M. W. McElfresh, L. Civale, A. P. Malozemoff and F. Holtzberg, Phys. Rev. B **43**, 2844 (1991).
4. U. Welp, S. Fleshler, W. K. Kwok, K. G. Vandervoort, J. Downey, B. Veal and G. W. Crabtree, Physical Phgenomena at High Magnetic Field Conference at Florida State University.
5. Z. Hao and J. R. Clem, Phys. Rev. B **43**, 2844 (1991).
6. S. E. Inderhees, M. B. Salamon, J. P. Rice, D. M. Ginsberg, Phys. Rev. Lett. **66**, 232 (1991).
7. J. Cors, D. Cattani, M. Decroux, A. Stettler and O. Fischer, Physica B, 165 (1990); J. Y. Genoud, T. Graf, A. Junod, D. Sanchez, G. Triscone and J. Muller, Physics C (1991).
8. K. S. Athreya, O. B. Hyun, J. E. Ostenson, J. R. Clem and D. K. Finnemore, Phys. Rev. B **38**, 11 846 (1988); M. M. Fang, J. E. Ostenson, D. K. Finnemore, D. E. Farrell and N. P. Bansal, Phys. Rev. B **39**, 222 (1989).
9. Junho Gohng, and D. K. Finnemore, Phys. Rev. B **42**, 7946 (1990).
10. W. C. Lee, R. A. Klemm, D. C. Johnson, Phys. Rev. Lett. **63**, 1012 (1989).
11. M. Tinkham, Phys. Rev. Lett. **61**, 1658 (1988).
12. V. G. Kogan, M. M. Fang, and S. Mitra, Phys. Rev. B**38**, 11 958 (1988); P. G. de Gennes, Superconductivity of Metals and Alloys (Benjamin, New York, 1966), p. 83.
13. J. J. Wollan, K. W. Haas, J. R. Clem and D. K. Finnemore, Phys. Rev. B, 1874 (1974).
14. A. Umezawa, G. W. Crabtree, U. Welp, W. K. Kwok, K. G. Vandervoort, Phys. Rev. B **42**, 8744 (1990).
15. M. Wacenovsky, H. W. Weber, O. B. Hyun, D. K. Finnemore, and K. Mereiter, Physica C **160**, 55 (1989).
16. A. B. Pippard, Classical Thermodynamics, (Cambridge University Press, London, 1957), p. 129.

DISCUSSION

D. S. Fisher: Going back to the first part of your talk, and the magnetization versus field, temperature and so on, another way to look at this is that there seems to be quite a lot of experimental and also theoretical evidence for believing that within a few degrees around the transition one actually has a lot of fluctuation effects. A lot of the characteristic fields that are measured – irreversibility lines and so on – show a lot of curvature away from linear slopes. So that another way that

Critical Fields in High-Temperature Superconductors

one can try to analyze the magnetization data, and I think that at least qualitatively it's consistent with the things you're seeing, is to analyze it with the fluctuations in it and this will give all sorts of scaling forms. You might try to look at that.

D. K. Finnemore: Yes, George Crabtree has a paper on that.

G. W. Crabtree: Can I make a couple of comments. First of all, to expand a little bit on what you said, the intermediate field regime seems to work very well experimentally if you take a log H form. We're able to fit it, and you are too, very nicely with very reasonable parameters that give you a value for H_{c2} and H_{c1} that would be, presumably, the mean field result if you could measure it. But you need to measure it in a region where the fluctuations don't exist.

D. S. Fisher: At these temperatures close enough to T_c fluctuations do exist, and they're quite big.

G. W. Crabtree: It depends on the magnitude of the field. If you then look very close to where M goes to zero – what you would normally call H_{c2} – there are large fluctuations and you're right, those have to be analyzed in a very special way. You can do this according to something we just did, and I don't think Doug has the paper yet . . .

D. K. Finnemore: Unless I'm the referee, and I'm not the referee.

G. W. Crabtree: At least we know one person who is not the referee. One way to do it is to use the Ginzburg-Landau model and scale them according to a three-dimensional form that should work for any Ginzburg-Landau kind of approach to the fluctuation problem. You find that if you take data let's say at five different fields, and have very different M versus H curves, and if you scale them this way, both M and H, they collapse onto one curve.

D. S. Fisher: How do you mean, scale them with a Ginzburg-Landau form?

G. W. Crabtree: It's a scaling form – we can talk about it later – it's a very simple form, $(T-T_c)/(HT)^{2/3}$. That should be the scaling variable that one gets. Okay, that works. But to do that you have to have a $T_c(H)$. So what we have done . . .

D. S. Fisher: No, you can scale everything with the zero-field transition temperature.

G. W. Crabtree: No, no. $T_c(H)$ occurs in that scaling form. It's $[T-T_c(H)]$. So you need $T_c(H)$. You can't test this scaling form without $T_c(H)$.

D. S. Fisher: I think the right scaling form to use is with T_c at zero field, and then you scale things with respect to that and the magnetic field . . .

G. W. Crabtree: I disagree with that. But let me finish my story, we can talk about that in a moment. We took the G-L scaling form and adjusted the function $T_c(H)$ to get the best scaling, that is, to get the M(T) data to fall on top of one another for different values of the field. We found that a linear $T_c(H)$ seemed to be best. Then if you *assume* a form of $T_c(H)$ which is linear, and let only the slope change, and change the slope until you get the best fit, you get a slope which is very close to the numbers that Doug and I have gotten in the past. So I think that the point is, you now have three ways to analyze the data: The first way, which is admittedly wrong, assumes that there is an Abrikosov form to the magnetization and you simply do a linear extrapolation to M = 0. You get a number that is 1.9 T/K for 1-2-3. Second, you can do a sort of mean-field approximation in the intermediate-field regime, $H_{c1} \ll H \ll H_{c2}$, where M(H) is nonlinear, which is based on either Ginzburg-Landau or London, and do it in a range where the field is low enough, far enough away from H_{c2} so that the fluctuations are not strong, and you get a number that is 1.6 T/K. Then you can do it a third way, which is to look explicitly at the fluctuation regime, and do this $T_c(H)$ analysis that I mentioned, and you get a number that is something like 1.7 T/K. So all three of these ways give numbers that are very close to one another.

D. S. Fisher: If I actually ask what do I expect the form to look like: what should M(H) look like in this regime where you're within a few degrees of the zero-field T_c and in not-enormously-large fields. Then one does expect to have a scaling form scaled with the zero-field transition. And there is no well-defined $T_c(H)$ until you go down much further but this is going down to where the melting curve is or where there's an irreversibility line.

G. W. Crabtree: But this $T_c(H)$ is a number that would exist in mean-field theory if you could extract that part of the magnetization out. Of course, you have to add other terms which take care of the fluctuations. That's the way this thing is formulated.

D. S. Fisher: It's a somewhat inconsistent way of doing it.

D. K. Finnemore: Let me try to phrase Fisher's point slightly differently, George. The coherence distance determines the fluctuation volume, and the coherence distance is $\hbar v_F / k_B T_c$, and so the coherence distance doesn't change as you increase the field. Isn't that your point?

D.S. Fisher: [Nods affirmatively.]

D. R. Nelson: The issue also comes up in two-dimensional films in *zero* field where there is a kind of bulk T_c, that is roughly defined, namely the T_c of a three-dimensional material, somewhat above the true Kosterlitz-Thouless transition. And here also we could get rid of these fluctuations in principle if we had an analogous convenient way, like going to four dimensions or 17 dimensions to suppress them. If you try to pull this number out, it's only going to be accurate, you know, to 5% or something, but it's certainly a worthy thing to try to do. Very close to zero-field T_c I agree that it would be fun to try to fit these scaling forms, but from an experimentalist's point of view it's perfectly sensible to determine an underlying mean-field $T_c(H)$, even though it could never be determined with absolute precision.

D. S. Fisher: I'm not even sure that it's even well-defined in principle.

D. R. Nelson: No it's not, but it's fine within a fluctuation parameter.

D. S. Fisher: It seems to me one ought to try to make contact between the real experiments and real theory, not sort of fake theory which is fudged in different ways to force it to fit. [Laughter]

F. M. Mueller: Do you have an estimation as to an extrapolated H_{c2} in the low-temperature regime?

D. K. Finnemore: I wouldn't do that on a bet. When somebody uses some theory to do that all they're telling you is the slope.

F. M. Mueller: So you're perfectly happy with 1.6 T/K?

D. K. Finnemore: No, it's just that I don't want to tell you something I don't know. I don't know what it is at $T = 0$. I haven't measured it. Nobody can measure it with the present techniques, there's no way to do it. Well, the way you could do it is if you could make a sample that was so perfect, and it was thermodynamically reversible over the whole range, then you could extract it from these kind of data. So it's not in principle impossible.

F. M. Mueller: There are techniques for doing it, you believe, but . . .

D. K. Finnemore: Oh yes, well. Let me tell you, picosecond, nanosecond and microsecond measurements in these things are hard.

F. M. Mueller: But would it be worthwhile to do?

D. K. Finnemore: Yes.

B. R. Coles: Doug, did I understand you to say that flux expulsion remains at T_c independent of H, and if so how do you reconcile that with a $T_c(H)$?

D. K. Finnemore: It's a very qualitative statement, that when you cool down you get a part-per-million fluctuation expulsion at 150 K and 100 K. And when you get close to T_c things diverge, and the flux expulsion goes up by orders of magnitude in tenths of kelvin. And, if you increase field, H_{c2} will drop by 5 K, but that abrupt rise (which I think is associated with fluctuations) takes place right at T_c. And it doesn't change much. That's what it means when the jump in specific heat takes that rise there. That's just a measure of the flux expulsion. And the way to view it is that you've got these bubbles of superconductivity that come and go, and you don't really get a flux-line lattice that you can see in the specific heat until you're 3 or 4 K farther down.

D. C. Larbalestier: I guess I'm really perturbed because one thing for example, as somebody who tries to worry about pinning situations, we need things like κ and H_c, we need values that at least there's some agreement on them . . .

D. K. Finnemore: Oh, all this stuff is within 30%. No matter what you do you get it within 30%.

D. C. Larbalestier: Okay. Now the question is, are people here collectively happy with this idea that dH_{c2}/dT is 1.6 T/K. The community needs standard values of dH_{c2}/dT and H_c that we can work with.

Tom Palstra is sitting here. You did some entropy transport measurements, for example, for H parallel to c, and in a very respectable fashion get 7 or 8 T/K, as I recall. Now this makes an enormous difference to . . .

D. K. Finnemore: I mean, if you take the flux expulsion – we've tried to measure it, and sometimes it backbends like Peter Kes showed, but usually it's about 10 T/K, if you sort of look at it. So that an initial flux expulsion has about his slope.

T. T. M. Palstra: I did do the comparison between the magnetization experiments and the entropy transport, because according to the Maki theory the magnetization results should be completely similar to the entropy transport, and if you looked to the change of slopes, for instance, you find that they are completely different for the entropy transport and the magnetization results. Part of the reason might be that for the entropy transport determination of T_c you assume that all the resistance that you're measuring is due to flux motion, and that underlying assumption might be wrong. Those kind of problems are not true for magnetization measurements, and I don't know yet exactly where that will lead you but it's certainly a complication.

D. C. Larbalestier: Can I just follow up. The most complete data, it seems to me, has come out of George Crabtree's group. We now have from the measurements you made the H_{c2}/T, κ's, H_c's, and all the rest of it . . . Is there really a collective feeling that those are the numbers we should use for the moment, or are people unhappy about that. Because when we think about pinning energies, which are going to go as H_c^2, we need to quantify what uncertainties really exist, and there are values in the literature ranging from 0.8 to 1.6 tesla for $H_c(0)$. It would be nice to converge on some numbers and agree that we were going to use these, or at least that we were going to fight to get some better ones.

S. Doniach: If I could just bring it back to the critical phenomena, because I think that's a very intriguing piece of the physics, and I ask how big the Ginzburg region is for these materials. I mean, is it several degrees, or is it a fraction of a degree?

D. S. Fisher: It actually depends on what condition one uses. If you just use a straight Ginzburg criterion you obtain a fraction of a degree, if I remember correctly. If you ask where does the penetration length start to significantly deviate from its Ginzburg-Landau form, i.e., where it will start to deviate from a $(T-T_c)^{-1/2}$, then it actually turns out to be about ten degrees. And if you ask where the resistance at zero field gets rounded, it's of order a tenth of a degree. So it depends on the condition, and that's why we're in the process of trying to work through some of the coefficients and things more carefully, so one can actually try to fit things. But the thing which seems to have the widest regime is how things depend on magnetic field – in reasonably large fields, like irreversibility lines, and so on, and I think this should also apply for these M(H)'s. So there YBCO seems to be such that the critical regime has some definition which yields of order ten degrees width.

G. W. Crabtree: May I comment. In this paper – which we should have a copy of, because I'm sure it would be interesting to look at all these details – we fit on the same crystal, measurements of the conductivity and the magnetization. And from data out of the literature – actually, Tom Palstra's data of the Ettinghausen effect, and the data of Inderhees and Salamon from Champaign on the specific heat – and we find that they're all described by the same type of scaling laws for each quantity. There's a slightly different ordinate, but the abscissa's always the same. And, the Ginzburg number that you get is something like 1/100. That describes all of it; all four of those quantities are described by a single Ginzburg number, and the range is a few degrees. It makes perfect self-consistent sense, which I think is a new feature. That wasn't clear to us when we started, but surprisingly that's what we found.

V. Vinokur: The Ginzburg parameter depends on . . .

G. W. Crabtree: It depends on field. At higher fields the fluctuation regime is longer than it is at lower fields. We're in the high-field regime altogether anyway, of course. That is, we're limited by the lowest Landau level for the order parameter. That length is what limits you; that's why it works.

3. Vortex Lattice Structure and Melting
Chair — Larry Campbell

David R. Nelson
Harvard University
Cambridge, Massachusetts 02138

Correlations and Transport in Vortex Liquids

The theory of the vortex-line liquids which arise in the copper oxide high temperature superconductors is described. We discuss correlations in the presence of weak disorder, and the "viscous electricity" which results when entangled flux liquids attempt to flow past a few strong pins. We show, using an analogy with single particle quantum mechanics, that thermal fluctuations lead to a large renormalization of the binding energy of an isolated line to planar or linear pins. A related mapping of the statistical mechanics of many flux lines onto the physics of two-dimensional boson superfluids is reviewed, with an emphasis on the physical meaning of phase coherence. We argue that "boson localization" provides an appropriate description of flux lines when many planar or linear pins are present. A number of experimental tests of the theory are proposed.

1. INTRODUCTION

In 1957, Abrikosov presented his remarkable mean field theory of the mixed state of type-II superconductors.[1] For applied fields H such that $H_{c1} < H < H_{c2}$, the magnetic field penetrates in the form of quantized flux tubes which, in the absence of disorder, form a regular lattice of parallel defect lines. In conventional low temperature superconductors, this line lattice was believed to exist at essentially all temperatures up to $H_{c2}(T)$.[2] Disorder and pinning (which play a crucial role in transport experiments) were described by the extent to which they broke up the translational order embodied in the underlying Abrikosov flux lattice.[3] Flux flow in the presence of an externally imposed current was assumed to take place via the motion of blobs of flux crystal, with dimensions given by disorder-induced translational correlation lengths parallel and perpendicular to the field direction.[3]

The new high temperature (HTC) superconductors can apparently be described by the same s-wave Ginzburg-Landau functional used by Abrikosov for conventional materials. Striking differences arise, however, because the phenomenological coupling constants have very unusual values, and because the theory must be solved in a qualitatively different, high temperature regime. It has been argued, in particular, that a melted vortex liquid replaces the conventional Abrikosov flux lattice over large regions of the phase diagram because of the weak interplanar couplings, high critical temperatures and short coherence lengths characteristic of the new HTC materials.[4-7]

These theoretical suggestions were inspired by early flux decoration experiments on $YBa_2Cu_3O_7$ (YBCO).[8] For simplicity, we shall confine our attention throughout this review to results for magnetic fields along the z-axis, perpendicular to the CuO_2 planes. Although flux quanta (decorated via the Bitter technique) were observed emerging from a single crystal sample at $T = 4.2$ K, no flux patterns have ever been discerned at $T = 77$ K, possibly due to time-dependent flux wandering in an equilibrated flux liquid. Regions of extensive crystallinity were subsequently observed at low temperatures in YBCO by Dolan et al.,[9] suggesting that melting into a vortex liquid could proceed from a highly ordered state, as well as from a disorder dominated "vortex glass."[10] Extensive decoration studies of $Bi_2Sr_2CaCu_2O_8$ (BSCCO) have recently revealed a transition from disorderly flux arrays at very low fields ($H \lesssim 15$ Oe) to "hexatic glass" configurations with very long-range orientational order at higher fields.[11,12] A broken orientational symmetry — even if translational order is absent — is sufficient to insure a thermally driven phase transition (hexatic "melting") to an isotropic vortex liquid at higher temperatures.

Because the magnetic field becomes almost uniform when vortex lines are much closer than the in-plane London penetration depth λ_{ab}, decoration experiments are usually restricted to fields less than a few hundred gauss. The first experimental claim for melting at high fields ($H \gtrsim 10^3$ Oe) was based on a peak in vibrating reed experiments on YBCO and BSSCO by Gammel et al.[13] The very sharp peak found in recent low-frequency torsional oscillator experiments by Farrell et al.[14] supports the hypothesis that this effect is an actual phase transition, and not merely a thermal enhancement of the vortex mobility.[15] Resistance measurements in very clean YBCO crystals by Worthington et al.[16] have also been interpreted as evidence for an underlying melting transition. The vanishing shear modulus associated with melting would allow the vortex liquid to flow freely around macroscopic pinning centers, leading to an abrupt increase in the resistivity.[17]

In real HTC superconductors, any putative melting must take place in the presence of disorder, such as oxygen vacancies or twin boundaries. One can certainly question the interpretation of experiments like those above when the disorder is strong. Strong disorder, varying on scales comparable to the intervortex spacing, can undoubtedly destroy the translational and orientational order associated with the Abrikosov flux lattice, and completely wash out a sharp melting transition. We know, moreover, that arbitrarily *weak* disorder will *always* destroy translational correlations at sufficiently large length scales.[3]

The operative question, however, is how large the translational and orientational lengths actually become below the "melting" transition. If the disorder is predominately due to oxygen vacancies, which pin very weakly at liquid nitrogen temperatures,[18] translational correlation lengths of hundreds, even thousands of lattice constants may be possible at high fields. The measured translational correlation lengths at low fields in Ref. 12 become as large as 20 lattice constants when $B \approx 100$ G and are in fact steadily *increasing* with applied field. The orientational correlation range is *hundreds* of lattice constants and displays the same trend with field. Under these circumstances, large regions of vortex lines at high fields may appear to be crystalline, disrupted only by widely spaced macroscopic defects like twin boundaries. "Melting" would then be an appropriate description for the destruction of the translational and orientational order within these crystalline "grains," even if different physics arises on scales large compared to the grain spacing. Note that macroscopic planar defects like twin boundaries are an example of *correlated* disorder. Theoretical treatments based on microscopic Gaussian disorder with only short-range correlations[3] are *not* appropriate at these large length scales, especially if the twins pass completely through the sample.

In this paper, we shall be interested primarily in vortex liquids, regardless of whether they form a "polycrystalline" array or a disorder-dominated glass at low temperature. A vortex liquid has no translational order, and can flow appreciably on experimental time scales. To have some idea of when vortex liquids can occur, we first discuss where they appear in the temperature-field phase diagram, both for relatively clean HTC superconductors and for superconductors dominated by disorder.

1.1 Phase Diagrams

A phase diagram for very pure materials, with the applied field H perpendicular to the copper-oxide planes, is shown in Fig. 1a. The line $H_{c2}^0(T)$ marks the onset of a significant flux expulsion from the sample and is not a sharp phase transition. The melting transition near H_{c1} at low fields was first predicted[4] by mapping the statistical mechanics of thermally excited vortex lines onto the physics of two-dimensional boson superfluids.[19] Melting occurs from this point of view because bosons with a purely repulsive pair potential (decreasing faster than $1/r^2$) are always melted by zero-point motion at sufficiently low density. When applied to flux lines with a repulsive exponential interaction,[20] this argument simply means that entropy (i.e., "zero-point motion" in the boson language) always favors an

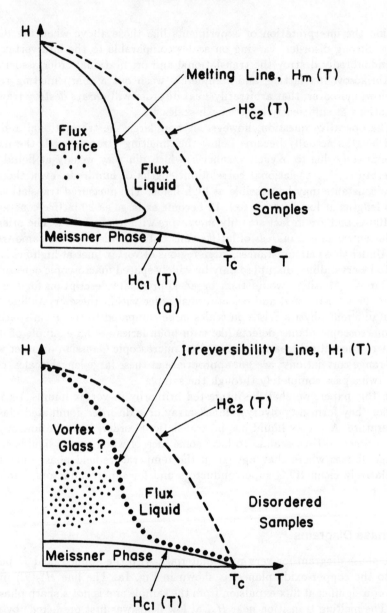

Fig. 1 Phase diagrams for (a) clean and (b) highly disordered high temperature superconductors. The dashed lines mark the onset of the Meissner effect, which is not a sharp phase transition.

entangled flux liquid over an ordered vortex crystal at low densities. When the flux lines are very dense, the interaction between neighboring lines is a logarithmic, rather than exponential repulsion.[20] The boson analogy then predicts that zero-point motion always melts the crystal at sufficiently *high* densities which accounts for the reentrant upper branch of the melting curve in Fig. 1a.[21]

A more precise determination of the melting curve for a particular material follows from the Lindemann criterion. As pointed out by Nelson and Seung,[22] the Lindemann integrals are dominated by wave vectors near the zone boundary, and wave vector-dependent (i.e., nonlocal) elastic constants should be used to obtain quantitatively accurate estimates. This has been done by Houghton *et al.*,[7] who produced reasonable fits to the high field melting lines proposed by Gammel *et al.*[13] for YBCO and BSCCO materials with a single adjustable Lindemann parameter. The importance of wave vector-dependent elastic constants in superconductors has also been emphasized by Brandt.[23] The tilt and bulk moduli are soft near the zone boundary, which *enhances* the fluctuation-induced melting discussed in Ref. 5.

The Lindemann condition is a *criterion*, and not a *theory* of flux lattice melting. As will be discussed further in Section 2, an exact theory of the melting of line crystals would be equivalent to understanding quantum melting of two-dimensional bosons as a function of pressure or $\hbar$ at $T = 0$. To the best of my knowledge, an analytical theory does not exist, although there are interesting efforts in this direction by Sengupta *et al.*[24] and by Chui.[25] Dislocation mediated melting of the flux lattice would lead to a three-dimensional *hexatic* line liquid, as pointed out in Ref. 26. It can be shown that amplitude fluctuations cause the continuous mean field theory transition of Abrikosov at H_{c2} to become first order just below six dimensions.[27] These fluctuations suppress the transition below the mean field H_{c2} line, so that there is a nonzero local order parameter amplitude, even in the normal state. One can then invoke a famous argument due to Landau[28] to argue that melting probably remains first order in $d = 3$.[27] A weakly first-order transition, possibly smeared by macroscopic disorder like twinning planes, is the best theoretical guess at the moment.

Figure 1b shows an alternative phase diagram,[29] for HTC materials in which extrinsic impurity-induced disorder dominates over much of the phase diagram. According to Ref. 29, the Abrikosov flux lattice may be replaced by a thermodynamically distinct vortex-glass phase, separated by a sharp transition line from a high-temperature vortex liquid. Although the vortex glass theory is almost entirely phenomenological at the moment, this scaling approach has produced a striking fit to resistivity experiments by Koch *et al.*[30] Even if there is no true vortex glass phase, there will still be a gradual crossover from a low-temperature regime dominated by fluctuations in the impurity potential, to a high-temperature region dominated by thermal fluctuations. In this case, the dotted line in Fig. 1 would simply represent a locus of crossover temperatures.[15] If the disorder is weak, there should actually be a deep notch of flux liquid, stable to both crystallization and disorder, cut into this line. See Eqs. (2.59) and (2.61) below.

To provide a point of reference for vortex liquids in three dimensions, it is worth summarizing what is known about *point* vortices in two dimensions, where there is a well developed melting theory.[31] Liquid phases of point vortices were proposed in conventional low temperature superconducting films over a decade ago.[32,33] A phase diagram predicted (for clean samples) using the dislocation/disclination theory of two-dimensional melting is shown in Fig. 2.[33] Fluctuations are particularly strong in two dimensions, so a substantial region of the mean field Abrikosov phase diagram is occupied by a flux liquid, just as for three-dimensional HTC materials. The estimate of the melting line is based on the Kosterlitz-Thouless dislocation criterion.[31-33] This condition is an *upper bound* on the true melting temperature, which insures that there is indeed a large region of melted flux liquid, even if the melting transition occurred via some other mechanism. In addition to an isotropic liquid of point vortices, there is also a region of hexatic order, to be discussed further in Section 3.

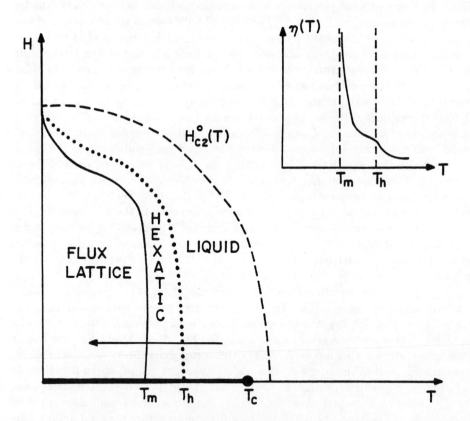

Fig. 2 Schematic phase diagram for a thin film superconductor.[33] H_{c1} is zero in thin films. Inset shows the expected temperature dependence of the viscosity along the path indicated by the arrow.[31]

The inset to Fig. 2 shows the shear viscosity[34] of the vortex liquid on the path indicated by an arrow, as predicted by the theory of dislocation mediated melting.[31] The hexatic-to-liquid transition is continuous and characterized by a strongly diverging intervortex viscosity. As we shall see, the viscosity of a flux liquid in three dimensions can also be large, not only near the melting temperature but over much of the temperature field phase diagram.[5] As will be discussed in Section 4, these large viscosities allow the effects of a few strong pins to propagate large distances, with important implications for transport experiments.[35]

Strong disorder will, of course, eliminate the crystalline phase in Fig. 2, just as it does in three dimensions.[3] No vortex glass transition is expected in two dimensions although sluggish dynamics should accompany a diverging vortex glass correlation length at zero temperature.[29] As pointed out by Chudnovsky,[36] hexatic orientational order is *preserved* in both two and three dimensions for the model originally considered by Larkin and Ovchinnikov.[3] Although hexatic order may eventually die off at *very* large length scales in more realistic systems,[26] the range of orientational correlations can greatly exceed the translational correlation lengths computed in Ref. 3, in qualitative agreement with Chudnovsky's ideas. A striking confirmation of this point of view appears in recent experiments on two-dimensional magnetic bubble arrays[37] which, like vortices, are subject to an underlying random substrate potential.

1.2 Correlations and Transport

Liquids of vortex *lines*, in contrast to liquids of point vortices in two dimensions, are a new type of matter and are of some interest in their own right. An important physical characteristic of such liquids is the "entanglement length" ℓ_z,[4,5] i.e., the spacing along the field direction between collisions or close encounters between flux lines.[38] This length describes the thermal wandering of a vortex line transverse to its average direction as it meanders through a superconducting sample. For flux lines with line tension $\epsilon_1 = (\phi_0/4\pi\lambda_{ab})^2 \ln \kappa_{ab}$ and mass anisotropy $M_\perp/M_z \ll 1$, ℓ_z is given approximately by[39]

$$\ell_z \simeq (M_\perp/M_z)(\phi_0 \epsilon_1/B\, k_B T) \quad . \tag{1.1}$$

Note that ℓ_z decreases with increasing magnetic field and mass anisotropy.

Consider a vortex liquid confined to a slab of thickness L along the field direction with average copper oxide plane spacing c_0. Vortex liquids should then display the three qualitatively different regimes shown in Fig. 3, depending on the value of the magnetic field relative to two important crossover fields. The first field is obtained by equating ℓ_z to L[39]

$$B_{x1} = (M_\perp/M_z)(\phi_0 \epsilon_1/L\, k_B T) \quad . \tag{1.2}$$

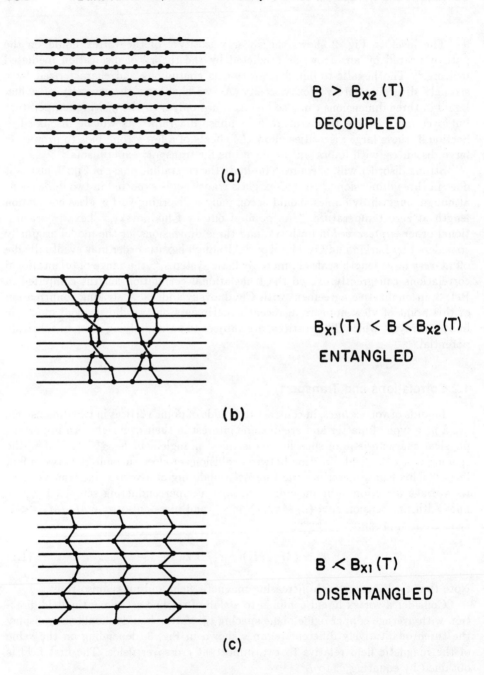

Fig. 3 Schematic of flux lines relative to the CuO_2 planes in HTC materials, viewed edge on. Three regimes are shown: (a) decoupled, (b) entangled and (c) disentangled.

For $B \lesssim B_{x1}$, we have a *disentangled* flux liquid, whose properties should be qualitatively similar to the liquid of point vortices indicated in Fig. 2. For $L = 1$ mm, this field is about 100 gauss.

Above $B = B_{x1}$, we have an entangled flux liquid. Entanglement continues to increase until B exceeds B_{x2}, which is the field such that $\ell_z = c_0$,

$$B_{x2} = (M_\perp/M_z)(\phi_0\epsilon_1/c_0\, k_BT) \quad . \tag{1.3}$$

When $B > B_{x2}$, flux lines are "superentangled," in the sense that vortices hop more than an intervortex spacing when passing from one CuO_2 plane to the next. It then makes little sense to draw lines connecting them. Above B_{x2} (about 50 tesla for YBCO but only $\sim 1\,T$ for BSSCO), the liquid degenerates into decoupled planes of point vortices. A crossover at $B_{x2}(T)$ appears to be the explanation for the resistivity changeover observed recently in artificial multilayers by W. R. White et al.[40] An identical criterion for decoupling vortices in neighboring copper oxide planes has been derived recently by Glazman and Koshelev.[41] A related condition for decoupling in the *crystalline* phase was also derived by these authors, as well as by Fisher et al.[29]

Three-dimensional equal-time correlations in flux line liquids may be measurable via neutron diffraction.[42] An interesting alternative would be to compare (using magnetic decoration or tunneling microscopy) the configuration of flux lines entering a sample with that which emerges on the other side. Remarkably, this conceptually simple experiment does not appear to have been attempted during the 25-year history of the flux decoration procedure. A basic theoretical result which bears on such measurements is that in-plane vortex line density fluctuations decay exponentially along the z-axis.[5] Consider, in particular, a set N of vortex lines which wander along the z axis with trajectories $\{\mathbf{r}_j(z) = [x_j(z), y_j(z)]\}$, as shown in Fig. 4. We assume that $B_{x1} < B < B_{x2}$, so that the vortex lines are both entangled and well defined. The Fourier-transformed density in a constant-z cross section,

$$\rho_{\mathbf{q}_\perp}(z) = \frac{1}{\sqrt{N}} \sum_{j=1}^{N} e^{i\mathbf{q}_\perp \cdot \mathbf{r}_j(z)} \tag{1.4}$$

then decays exponentially to zero,

$$\begin{aligned}\tilde{S}(q_\perp, z) &\equiv \langle \rho_{\mathbf{q}_\perp}(z) \rho^*_{\mathbf{q}_\perp}(0) \rangle \\ &\approx S_2(q_\perp) e^{-|z|/\ell_z(q_\perp)} \quad .\end{aligned} \tag{1.5}$$

The function $S_2(q_\perp)$ is the vortex structure function in a constant-z cross section. As will be discussed in Section 3, the decay length $\ell_z(q_\perp)$ diverges as $q_\perp \to 0$, and is simply related to the phonon-roton spectrum of the equivalent two-dimensional boson superfluid. We expect that $\ell_z(q_\perp)$ is of order the entanglement length ℓ_z discussed above when $q_\perp \approx \pi/d$, where d is the intervortex separation. Equation (1.5)

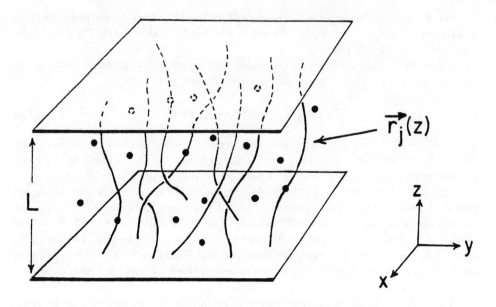

Fig. 4 Schematic of flux lines wandering through a sample of thickness L in presence of random impurities.

is directly applicable to the two-sided decoration experiment suggested above. For vortex liquids in superconducting slabs of thickness L, the reduced dimensionality will only become evident for fluctuations with wave vectors $q_\perp$ satisfying

$$\ell_z(q_\perp) \gtrsim L \; . \tag{1.6}$$

The above results were first derived in the absence of the pinning disorder shown in Fig. 4. Dense flux liquids, however, are *stable* to weak disorder, and will persist even if the Abrikosov flux lattice is replaced by a glassy phase at low temperatures.[43] Weak disorder merely produces "Lorentzian squared" corrections to the Fourier transform of Eq. (1.5) — see Section 2. This insensitivity of vortex liquids to disorder arises because flux lines collide and "restart" their random walks along the z-axis before the effects of randomness can significantly alter their trajectories.

The explicit microscopic derivation of these results[5,43] assumes simplified models which ignore nonlocal interactions between vortex lines.[44] Such interactions produce quantitative changes in phase boundaries at high fields.[7,23] Nonlocal interactions, however, are easily incorporated into a long wavelength "hydrodynamic" version of the theory,[43] with essentially no changes in the basic conclusions.

There are also important issues related to the *dynamics* of flux line liquids. As was suggested in Refs. 5 and 35, these liquids can become very viscous, allowing the effects of a few strong pinning centers to propagate over long distances. A large viscosity will arise provided (1) the lines are entangled; and (2) there is a large barrier to flux line crossing. The entanglement criterion simply means that B must exceed $B_{x1} \approx 100$ G. A barrier to line crossing arises because two singly quantized vortices behave like a doubly quantized one in the region where the crossing takes place. The energy stored in the circulating supercurrents which extend a distance λ_{ab} perpendicular to the lines is proportional to the *square* of the effective flux quantum. If ϵ_1 is line tension, and the crossing takes place over a distance ℓ_0 along the z-axis, the crossing energy is then roughly[5]

$$U_x \approx (2^2 - 1^2 - 1^2)\epsilon_1 \ell_0$$
$$= 2\epsilon_1 \ell_0 \qquad (1.7)$$

when the lines are dilute. For flux lines at arbitrary densities, this formula becomes[45]

$$U_x = 2\epsilon_1 \ell_0 \ln[H_{c2}^0/H]/\ln[H_{c2}^0/H_{c1}] \quad . \qquad (1.8)$$

Although the crossing energy vanishes near the mean field upper critical field H_{c2}^0, it is more than 20 times $k_B T$ (with $\ell_0 = 10$ Å) at liquid nitrogen temperatures for $B = 1$ tesla in YBCO, consistent with a very high viscosity.[35]

In isotropic superconductors, some of this interaction energy can be reduced by forcing the lines to cross at a 90 degree angle.[46] Something like this surely takes place above the decoupling field B_{x2}, where the requisite large crossing angles are produced by thermal fluctuations and the continuum picture breaks down. At lower fields, however, crossing at large angles requires additional line length and the net crossing energy remains of the order given by Eq. (1.8).[47] Because the currents prefer to flow in the anisotropic CuO_2 planes even when vortex lines tilt, the critical angle for easy flux cutting is in any case much *larger* than the 90 degrees required for isotropic materials.

Suppose we cool a HTC superconductor from the normal state at fixed magnetic field, $B_{x1} < B < B_{x2}$. The viscosity is always small near $H_{c2}^0(T)$, where vortex lines are little more than loci of zeros of the order parameter. The flux liquid should become significantly more viscous, however, once $k_B T < U_x(T)$, as the vortex lines become more tangible and resist crossing. If this intervortex viscosity becomes large well before the equilibrium freezing line, one might expect regimes where a vortex liquid drops out of equilibrium at a polymer-like glass transition well before it ever freezes into a flux crystal.[5,45] Such a "polymeric glass" state has little to do with oxygen vacancy disorder, although it will greatly influence the response of the system to a few, strong pins. The polymer glass idea provides an interesting alternative to the "vortex glass" conjecture[29] as an explanation for experimentally observed "irreversibility lines."[48]

How can we understand transport measurements in vortex line liquids? Because the underlying Abrikosov flux lattice is melted, it does not make sense to invoke the traditional picture[3] of collective motion of crystalline flux bundles.[49] At the opposite extreme is the "thermally assisted flux flow" (TAFF) idea,[50] which essentially assumes an ideal gas of disconnected flux bits moving in a tilted washboard potential. This approach allows thermally activated depinning to be incorporated into the usual[20] Bardeen-Stephen flux flow resistivity formulas. The TAFF picture, however, takes little account of intervortex interactions and is not well-suited for dealing with phase transitions. To describe transport in strongly interacting vortex liquids, it is essential to include the intervortex viscosity.

To see how this viscosity affects flux flow experiments, consider the flow of an isotropic flux liquid in a channel contained between two flat twin boundaries in the xz plane at $y = -W/2$ and $y = W/2$, shown in Fig. 5. The magnetic field is in

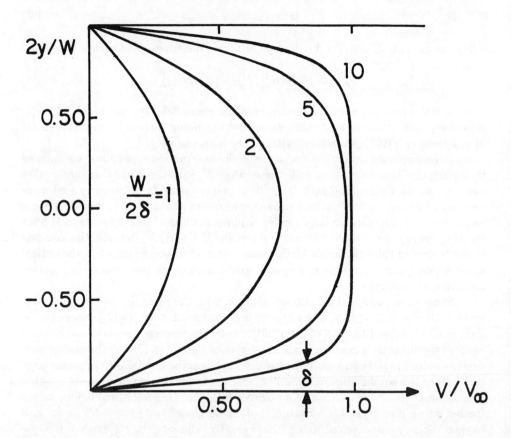

Fig. 5 Flux line velocity profiles for flow in a channel between two twin boundaries. We have assumed for simplicity that the velocity is forced to vanish on the boundary. The viscous length δ is indicated for the case $W/2\delta = 10$.

the z-direction. The transport current is applied along $\hat{y}$, $\mathbf{j}_T = j_T\hat{y}$, and yields a constant driving Lorentz force[20]

$$\mathbf{f}_T = \frac{1}{c} n_0 \phi_0 \hat{z} \times \mathbf{j}_T \qquad (1.9)$$

in the x direction, where n_0 is the areal vortex density. The equation of motion which describes the vortex liquid velocity field is then[35]

$$-\gamma \mathbf{v} + \eta \nabla_\perp^2 \mathbf{v} + \mathbf{f}_T = 0 \quad . \qquad (1.10)$$

Equation (1.10) is a hydrodynamic description of flux flow, valid on scales large compared to the intervortex spacing. The combination of the drag and the viscous term introduces an important new length scale into the problem,

$$\delta = \sqrt{\eta/\gamma} \quad . \qquad (1.11)$$

Here γ is the vortex "friction" coefficient of Bardeen and Stephen[20] augmented by the effects of weak microscopic disorder such as oxygen vacancies. The parameter η is the intervortex viscosity discussed above. We assume for simplicity that $\mathbf{v} = 0$ at the walls, due to strong pinning by the twin boundaries, which are believed to contain exceptionally high concentrations of point disorder. The solution of (1.10) is then

$$v_x(y) = v_\infty \left[1 - \frac{\cosh(y/\delta)}{\cosh(W/2\delta)} \right] \quad . \qquad (1.12)$$

Here $v_\infty = f_T/\gamma$ is the usual Bardeen-Stephen limiting flux line velocity obtained far from any strong pins. The flow velocity drops to zero in a boundary layer of width δ. If a polymer-like glass transition or a nearly continuous melting transition is responsible for the irreversibility line, the viscosity $\eta(T)$ should rise dramatically with decreasing temperature. The length $\delta(T)$ can easily become a micron or more.[35] The friction $\gamma(T)$ remains finite within these scenarios. It is clear from Fig. 5 that the resulting increase in $\delta(T)$ causes the influence of the boundaries to propagate into the interior of the channel eventually to choke off the flow.

Upon computing the resistivity ρ associated with the flux line velocity profile in Fig. 4, one finds[35]

$$\rho = \rho_0 \left[1 - \frac{2\delta}{W} \tanh(W/2\delta) \right] \quad , \qquad (1.13)$$

where $\rho_0 = (\phi_0 n_0/c)^2/\gamma$ is the limiting Bardeen-Stephen flux flow resistivity for a twin-free sample. An identical formula holds for a periodic array of parallel twins with spacing W. Note that if $\delta(T)$ diverges, we have

$$\rho(T) \approx \frac{1}{12} \left(\frac{\phi_0 n_0}{c} \right)^2 W^2/\eta(T) \quad , \qquad (1.14)$$

so that $\rho(T)$ is *independent* of $\gamma(T)$ and vanishes like the reciprocal of the viscosity in this limit.

It is very important to determine the parameters $\gamma(T)$ and $\eta(T)$ separately. Although the bulk resistivity can be interpreted in terms of a diverging intervortex viscosity $\eta(T)$,[35] it is hard to rule out an even stronger divergence in $\gamma(T)$ (as predicted by the vortex glass hypothesis[29]) to explain the vanishing resistivity in flux flow experiments. In Section 4, we discuss how to make an independent determination of $\gamma(T)$ and $\eta(T)$, and provide a clean experimental test of the different explanations of the irreversibility line.

1.3 Contents

In the remainder of this paper, we present a more technical review of the ideas sketched above. In Section 2, we discuss the statistical mechanics of flexible vortex lines at finite temperature. We describe the thermal wandering of a single line, and show that this produces a large, temperature-dependent renormalization of the binding energy of a flux line to planar and linear defects. The boson analogy, which allows us to gain intuition about flux liquids from two-dimensional many-particle quantum mechanics, is presented in detail. This Section concludes with a discussion of when disorder has a significant effect on the results for pure systems.

In Section 3, we discuss equilibrium correlations in flux liquids, using a hydrodynamic approach which is more general than the microscopic boson theories. We also discuss the intermediate hexatic phase, which may appear in both glassy form[36] and as an equilibrated flux liquid.[26] We conclude with some comments about vortex liquids when the dominant source of disorder is random planar twin boundaries or linear defects which pass completely through the sample. We argue that conventional models which include only microscopic point disorder[3,29] are then *inappropriate*, and that a better description of the vortex line configurations and dynamics results from an analogy with boson localization.

Flux liquid dynamics is discussed in Section 4. We treat flux motion in the presence of a few strong pins and describe how to distinguish between the four currently popular explanations of the irreversibility line via an experiment which probes the physics of "viscous electricity."

2. STATISTICAL MECHANICS OF FLEXIBLE LINES

2.1 Model Free Energy

Consider the Gibbs free energy for the N flux lines shown in Fig. 4, in a slab of thickness L, whose positions with a field $\mathbf{H}$ along the z direction (perpendicular to the CuO_2 planes) in a sample length L are given by $\mathbf{r}_j(z) = [x_j(z), y_j(z)]$, $j = 1, \ldots, N$. Since the ratio of the penetration depth λ_{ab} to the coherence length ξ_{ab} is typically quite large, $\kappa_{ab} = \lambda_{ab}/\xi_{ab} \approx 10^2$, we shall often work in the London limit. If $\epsilon_1 = \left(\frac{\phi_0}{4\pi\lambda_{ab}}\right)^2 \ln \kappa_{ab}$ is the energy per unit length of a single flux line, and $\phi_0 = 2\pi\hbar c/2e = 2.07 \times 10^{-7}$ gauss-cm^2 is the flux quantum, the energy reads

$$G = \left(\epsilon_1 - \frac{H\phi_0}{4\pi}\right) NL + \sum_{i>j} \int_0^L V[\mathbf{r}_{ij}(z)]dz + \sum_{j=1}^N \int_0^L dz\, V_D[\mathbf{r}_j(z), z]$$

$$+ \frac{1}{2}\tilde{\epsilon}_1 \sum_{j=1}^N \int_0^L \left|\frac{d\mathbf{r}_j(z)}{dz}\right|^2 dz \qquad (2.1)$$

where $\mathbf{r}_{ij} = \mathbf{r}_i - \mathbf{r}_j$, $V(\mathbf{r}_{ij})$ is a vortex line pair potential, which in the London approximation takes the form[20]

$$V(r_{ij}) = \frac{\phi_0^2}{8\pi^2\lambda_{ab}^2} K_0(r_{ij}/\lambda_{ab}) \quad , \qquad (2.2)$$

where $K_0(x)$ is the modified Bessel function, $K_0(x) \approx (\pi/2x)^{1/2}e^{-x}$ for large x. Although we have neglected interactions along the z-axis, this simplified model is still quite nontrivial and informative. Qualitatively similar results, moreover, emerge when these "nonlocal" interactions are taken into account using the hydrodynamic approach of Sec. 3.

The calculations of Refs. 5 and 43 are in fact easily carried out for *arbitrary* in-plane pair potentials. Some of the calculations in Ref. 5, for example, used a cutoff delta function pseudopotential,

$$V(r_{ij}) = v_0 \int_{k<\lambda} \frac{d^2k}{(2\pi)^2} e^{i\mathbf{k}\cdot[\mathbf{r}_i(z)-\mathbf{r}_j(z)]}$$
$$\equiv v_0 \delta_\lambda(\mathbf{r}_i(z) - \mathbf{r}_j(z)) \quad , \qquad (2.3)$$

where $v_0 = \phi_0^2/4\pi$. With this latter potential, it is easy to show that the crossing energy for the lines inclined at angle θ in Fig. 6 is finite and given by[44]

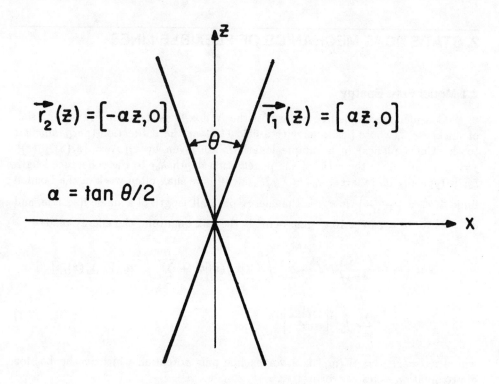

Fig. 6 Two flux lines inclined symmetrically about the field direction $\hat{H}\|\hat{z}$, and crossing at an angle θ.

$$U_x = \frac{v_0}{2\pi} \int_0^\lambda q\,dq \int_{-\infty}^\infty dz\, J_0(2\alpha q z)$$
$$= \frac{v_0 \lambda}{2\pi} \cot(\theta/2) \ . \tag{2.4}$$

The renormalization group calculation in Ref. 5 is just a resummation of perturbation theory in U_x.

The function $V_D[\mathbf{r}_j(z), z]$ is a Gaussian pinning potential for individual vortex lines, representing the effects of impurities. If the defects are randomly distributed, as in the case of oxygen vacancies, the quenched fluctuations in the impurity potential will obey

$$\overline{V_D(\mathbf{r}, z) V_D(\mathbf{r}', z')} = \Delta \delta(\mathbf{r} - \mathbf{r}')\delta(z - z') \ , \tag{2.5}$$

with $\overline{V_D(\mathbf{r}, z)} = 0$. The overbar represents an average over the impurity disorder. An approximate formula for the coefficient Δ is given by Fisher, Fisher, and Huse,[29] namely,

$$\Delta \approx \frac{1}{4} \gamma_I^2 v_c n_I [T_c/(T_c - T)]^2 (\phi_0^4/16\pi^2 \lambda_{ab}^2) \ , \tag{2.6}$$

where v_c is the volume of the unit cell in the underlying crystal, n_I is the fractional occupation number of impurities in this cell, and

$$\gamma_I = d[\ln T_c(n_I)]/dn_I \qquad (2.7)$$

is a dimensionless impurity coupling constant (typically of order unity). Note that disorder due to a frozen distribution of grain or twin boundaries is qualitatively different. The randomness is *correlated* by the planar nature of the disorder in this case.

For an isotropic superconductor, the last term in Eq. (2.1) comes from the expansion of the total line energy, $E_i = \epsilon_1 \int_0^L (1 + |d\mathbf{r}_i/dz|^2)^{1/2} dz$. In this case, $\tilde{\epsilon}_1 = \epsilon_1$. For the anisotropic layered compounds under consideration here, $\tilde{\epsilon}_1$ is considerably smaller,

$$\tilde{\epsilon}_1 = \frac{M_\perp}{M_z} \epsilon_1 \quad , \qquad (2.8)$$

where $M_\perp$ is the in-plane effective mass, and $M_z \approx 10^2 M_\perp$ is the much larger effective mass describing the weak Josephson coupling between the planes. The formula (2.8) applies when the flux lines are dense, $n_0 \lambda_{ab}^2 \gg 1$.[5,51] In the opposite limit $n_0 \lambda_{ab}^2 \lesssim 1$, the electromagnetic coupling between CuO_2 planes is important, and one has[29]

$$\tilde{\epsilon}_1 \approx \epsilon_1/\ln \kappa \quad . \qquad (2.9)$$

Abrikosov's mean-field treatment of the transition at H_{c1} neglects disorder, and assumes that the vortices form a triangular lattice of rigid rods with density $n_0 = B/\phi_0$ parallel to the z axis, so that the last term of (2.1) vanishes. Flux lines begin to penetrate when the first term changes sign, i.e., when $H \geq H_{c1} = 4\pi \epsilon_1/\phi_0$.

A *full* statistical treatment of the partition function associated with Eq. (2.1) entails integration of $\exp(-G/k_B T)$ over all vortex trajectories $\{\mathbf{r}_i(z)\}$. The grand canonical partition function, for example, is

$$Z_{gr} = \sum_{N=0}^{\infty} \frac{1}{N!} \int \mathcal{D}\mathbf{r}_1(z) \cdots \int \mathcal{D}\mathbf{r}_N(z) e^{-G/k_B T} \quad . \qquad (2.10)$$

If we also wish to average over the quenched random impurity disorder, we should calculate

$$\overline{\ln Z} = \frac{\int \mathcal{D} V_D(\mathbf{r}_\perp, z)(\ln Z) \exp\left[-\frac{1}{2\Delta} \int d^2 r_\perp \int dz V_D^2(\mathbf{r}_\perp, z)\right]}{\int \mathcal{D} V_D(\mathbf{r}_\perp, z) \exp\left[-\frac{1}{2\Delta} \int d^2 r_\perp \int dz V_D^2(\mathbf{r}_\perp, z)\right]} \quad . \qquad (2.11)$$

2.1.1 Thermal Wandering of a Single Line

For now, we work with flux liquids at high enough densities and temperatures so that weak, oxygen vacancy disorder can be neglected. One can then estimate when the Abrikosov theory breaks down from a simple random walk argument.[4] We consider a *single* flux line $\mathbf{r}(z)$ and determine how far it wanders perpendicular to the z axis as it traverses the sample. The relevant path integral is

$$\langle |\mathbf{r}(z) - \mathbf{r}(0)|^2 \rangle = \frac{\int \mathcal{D}\mathbf{r}(s) |\mathbf{r}(z) - \mathbf{r}(0)|^2 \exp\left[-\frac{\tilde{\epsilon}_1}{2k_B T} \int_0^L \left(\frac{d\mathbf{r}}{ds}\right)^2 ds\right]}{\int \mathcal{D}\mathbf{r}(s) \exp\left[-\frac{\tilde{\epsilon}_1}{2k_B T} \int_0^L \left(\frac{d\mathbf{r}}{ds}\right)^2 ds\right]}$$

$$= \frac{2k_B T}{\tilde{\epsilon}_1} |z| \quad , \tag{2.12}$$

which shows that the vortex "diffuses" as a function of the timelike variable z,

$$\langle |\mathbf{r}(z) - \mathbf{r}(0)|^2 \rangle^{1/2} = (2Dz)^{1/2} \tag{2.13}$$

with diffusion constant

$$D = \frac{k_B T}{\tilde{\epsilon}_1} = \frac{M_\perp}{M_z} \frac{4\pi k_B T}{\phi_0 H_{c1}} \quad . \tag{2.14}$$

At $T = 77$ K, we take $H_{c1} \approx 10^2$ G and $M_z/M_\perp \approx 10^2$ and find $D = 10^{-6}$ cm, so that vortex lines wander a distance of order 1 μm while traversing a sample of thickness 0.01 cm.

These close encounters will occur quite frequently in fields of order 1 T or more, where vortices are separated by distances of order 500 Å or less. Collisions between neighboring vortices must now be taken into account. In terms of the "entanglement length" discussed in the Introduction,

$$\ell_z \equiv \frac{1}{2Dn_0} = \frac{\tilde{\epsilon}_1}{2k_B T n_0} \quad , \tag{2.15}$$

we conclude that collisions and entanglement of vortex lines will alter the Abrikosov theory whenever

$$L > \ell_z \quad , \tag{2.16}$$

i.e., for $B > B_{x1}$.

The above criterion was derived by neglecting intervortex interactions, which should reduce the asymptotic "diffusion constant" D measured at scales large compared to ℓ_z. Marchetti has studied vortex wandering in a dense flux liquid,[52] and finds that this reduction is at most about a factor of 10. Physical arguments leading to similar conclusions for quantum liquids were presented many years ago by R.P. Feynman.[53]

Correlations and Transport in Vortex Liquids

In Figure 7, we show six neighboring flux lines, projected down the z-axis, at two different values of the diffusion constant D in Eq. (2.14). These line projections can be modeled as two-dimensional random walks. The mean square projected area occupied by the lines is of order

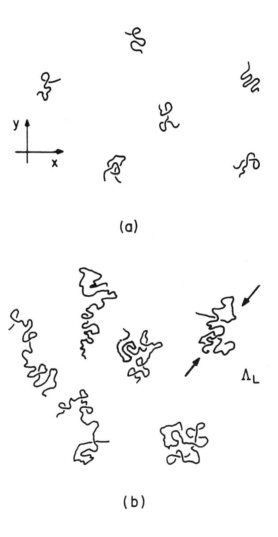

Fig. 7 Projections of flux lines onto the xy-plane in (a) the disentangled regime and (b) when the lines are on the verge of entanglement. The average projected area of one of the flux lines determines the spatial extent Λ_L of its wave function in the boson analogy.

$$\Lambda_L^2 \equiv \frac{2\pi k_B T}{\tilde{\epsilon}_1} L \quad , \tag{2.17}$$

i.e., of order the square of Eq. (2.13) evaluated with $z = L$, the sample thickness. In Fig. 7a, the vortex projections do not overlap significantly as they traverse the sample. Overlap is beginning in Fig. 7b, however, marking the onset of the new physics associated with HTC superconductors.

2.1.2 Binding Energy of a Vortex Line to Planar and Linear Pins

Consider a single flux line with trajectory $\mathbf{r}_\perp(z)$ wandering along the z-axis in a slab of thickness L near a twin boundary (see Fig. 8). This planar pinning center lies in the yz-plane, and attracts the line with a binding energy U_0 per unit length

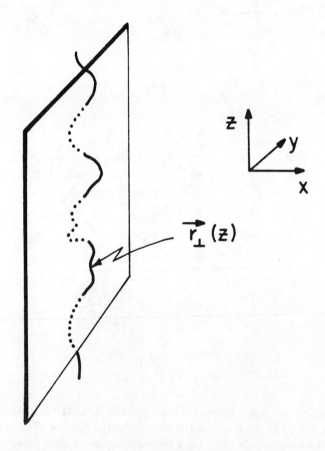

Fig. 8 Flux line trapped by a twin boundary which occupies the yz plane.

at zero temperature. The physically relevant quantity for transport experiments at liquid nitrogen temperatures, however, is the binding *free* energy $U(T) = U_0 - TS$. The binding free energy includes a contribution due to the reduced entropy S associated with localizing the line near the twin boundary. The binding free energy is given by the path integral

$$e^{U(T)L/k_BT} = \frac{\int \mathcal{D}\mathbf{r}(z) \exp\left[-\frac{\tilde{\epsilon}_1}{k_BT}\int_0^L \left(\frac{d\mathbf{r}_\perp}{dz}\right)^2 dz - \frac{1}{k_BT}\int_0^L V[\mathbf{r}_\perp(z)]dz\right]}{\int \mathcal{D}\mathbf{r}(z) \exp\left[-\frac{\tilde{\epsilon}_1}{k_BT}\int_0^L \left(\frac{d\mathbf{r}_\perp}{dz}\right)^2 dz\right]} . \quad (2.18)$$

Here, $V(x, y)$ is the binding potential due to the twin, defined to vanish as $x \to \pm\infty$. The denominator is included to subtract off the free energy of a line wandering far from the twin.

Equation (2.18) is just the Feynman path integral for a quantum mechanical particle moving in imaginary time.[54] When viewed as a problem in classical statistical mechanics, the transfer matrix along z for this partition function is the exponential of the Hamiltonian of a fictitious quantum system. As $L \to \infty$, the binding free energy $U(T)$ is then given by the negative of the lowest eigenvalue of the associated "Schroedinger equation,"[54]

$$\left[-\frac{(k_BT)^2}{2\tilde{\epsilon}_1}\nabla_\perp^2 + V(\mathbf{r}_\perp)\right]\psi(\mathbf{r}_\perp) = E\psi(\mathbf{r}_\perp) . \quad (2.19)$$

Note that k_BT plays the role of $\hbar$, and $\tilde{\epsilon}_1$ that of the mass of the fictitious particle. The density distribution of flux lines projected down the z-axis is given by the modulus squared of the ground-state wave function. We model the twin by a potential $V(\mathbf{r}_\perp)$ which is independent of y, $V(\mathbf{r}_\perp) \equiv V(x)$, and take $V(x)$ to be a square well of depth U_0 and width $2a$ in the x-direction (see Fig. 9),

$$V(x) = \begin{cases} -U_0, & -a < x < a, \\ 0, & \text{otherwise.} \end{cases} \quad (2.20)$$

Upon assuming a plane-wave state along y,

$$\psi(x, y) = e^{ik_y y}\psi(x) , \quad (2.21)$$

we recover the Schroedinger equation for a one-dimensional particle in a box,

$$\left[-\frac{(k_BT)^2}{2\tilde{\epsilon}_1}\frac{d^2}{dx^2} + V(x)\right]\psi(x) = E'\psi(x) \quad (2.22)$$

with $E = E' + (k_BT)^2 k_y^2/2\tilde{\epsilon}_1$. Assume for simplicity that the system is infinite in the y-direction, so we can set $k_y = 0$ to obtain the lowest eigenvalue, and simply replace E' by E.

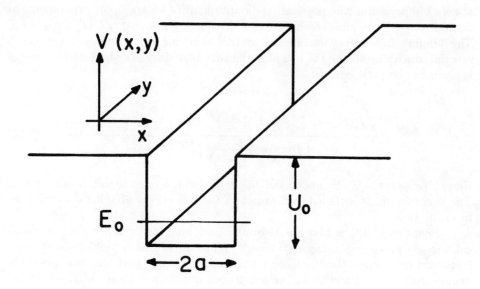

Fig. 9 Potential $V(x,y)$ presented to a vortex by the planar pin of Fig. 8.

The solution of this problem is well known from elementary quantum mechanics.[55] The ground state energy is given by the smallest positive real value of ξ which solves the transcendental equation

$$\cot \xi = \frac{\xi x}{\sqrt{1 - x^2 \xi^2}} \quad , \tag{2.23}$$

where

$$\xi = \sqrt{(E + U_0)/[(k_B T)^2/2\tilde{\epsilon}_1 a^2]} \tag{2.24a}$$

and

$$x = \sqrt{[(k_B T)^2/2\tilde{\epsilon}_1 a^2]/U_0} \quad . \tag{2.24b}$$

When $x \ll 1$, the well is very deep and we find the usual particle-in-a-box result for $U(T) \equiv -E$,

$$U(T) \approx U_0 - \frac{\pi^2 (k_B T)^2}{4 \; 2\tilde{\epsilon}_1 a^2} \quad . \tag{2.25}$$

The second term is due to quantum zero point motion, and represents for classical flux lines the loss of entropy due to confinement. For $x \gg 1$, on the other hand, the flux line is only weakly bound,

$$U(T) \approx U_0^2/[(k_B T)^2/2\tilde{\epsilon}_1 a^2] \quad . \tag{2.26}$$

Consider, for simplicity, a vortex line well below $H_{c2}^0(T)$ so that $\tilde{\epsilon}_1$, U_0 and a are approximately temperature-independent. The binding free energy renormalized by thermal wandering then takes the form

$$U(T) = U_0 f(T/T^*) \quad , \tag{2.27}$$

where, according to Eqs. (2.25) and (2.26),

$$f(x) \approx 1 - \frac{\pi^2}{4} x^2 , \qquad x \ll 1 , \tag{2.28a}$$

$$f(x) \approx 1/x^2 , \qquad x \gg 1 . \tag{2.28b}$$

The crossover temperature in Eq. (2.27) is given by

$$k_B T^* = \sqrt{2\tilde{\epsilon}_1 a^2 U_0} \quad . \tag{2.29a}$$

A schematic plot of $U(T)$ as a function of temperature is shown in Fig. 10. Upon taking $U_0 = \epsilon_1 / \ln \kappa_{ab}$ and $a = \xi_{ab}$ to model the HTC materials, we see that there is a large downward renormalization of the binding energy above a characteristic temperature,

$$k_B T^* = \sqrt{2M_\perp/M_z} \, \epsilon_1 \xi_{ab} / \ln^{1/2} \kappa_{ab} \quad . \tag{2.29b}$$

The short coherence length and large mass anisotropies typical of HTC materials conspire to make this temperature small, of order 10 K in BSCCO ($M_z/M_\perp = 3600$, $H_{c1} \approx 100$ G), for example. If melting occurs close to $H_{c2}^0(T)$, ϵ_1 and ξ_{ab} are strongly temperature-dependent in the liquid phase. One must now solve (2.29) self-consistently to determine the crossover temperature. Thermal renormalization always becomes important sufficiently close to H_{c2}^0.

A similar analysis applies to linear pins oriented along the z-axis. Pins of this kind have recently been created by bombarding initially clean YBCO samples with tin ions.[56] The statistical mechanics of one flux line is now equivalent to the quantum mechanics of a particle in a cylindrically symmetric well. Particles are more weakly bound by an attractive potential in two dimensions than in the one-dimensional example discussed above when the "zero-point energy" $\frac{(k_B T)^2}{2\tilde{\epsilon}_1 a^2}$ is large. The renormalized binding energy still takes the form (2.27). Although a result similar to (2.28a) holds for small x, Eq. (2.28b) is now replaced by[57]

$$f(x) \approx x^2 e^{-x^2} \tag{2.30}$$

when x is large.

The loss of wandering entropy when a line is pinned by planar or linear pins represents an important contribution to a strongly temperature-dependent activation barrier in transport experiments.

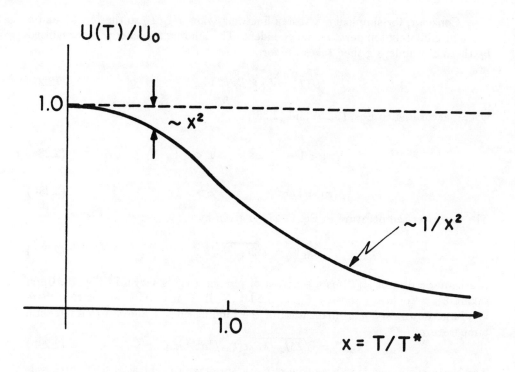

Fig. 10 Thermal renormalization of the binding free energy of a flux line to a twin boundary.

2.2 Analogy with Boson Statistical Mechanics in Two Dimensions

Even for simplified model systems, the exact calculation of partition sums like Eq. (2.10) for many *interacting* vortex lines is a formidable problem. Fortunately, we can draw on many years of experience in dealing with strongly interacting quantum fluids and exploit an analogy with the physics of boson superfluids in two dimensions.[4,5] This analogy is a natural extension of the "quantum mechanical" treatment of the binding of isolated vortex lines to planar and linear pins presented in the previous subsection.

2.2.1 Boson Hamiltonian

Consider the problem of calculating one term in Eq. (2.10), the N vortex line partition sum without disorder,

$$Z_N = \int \mathcal{D}\mathbf{r}_1(z) \cdots \mathcal{D}\mathbf{r}_N(z) e^{-G_N/k_B T} \quad , \quad . \tag{2.31}$$

where

$$G_N = \frac{1}{2}\tilde{\epsilon}_1 \sum_{j=1}^{N} \int_0^L \left(\frac{d\mathbf{r}_j(z)}{dz}\right)^2 dz + \sum_{i>j} \int_0^L V[\mathbf{r}_{ij}(z)]\, dz \quad . \tag{2.32}$$

The grand partition function is then $Z_{gr} = \sum_{N=0}^{\infty} \frac{1}{N!} e^{\mu L N} Z_N$, where $\mu = H\phi_0/4\pi - \epsilon_1$. The standard way of attacking such problems in classical statistical mechanics is to use the transfer matrix. It is not hard to show that the transfer matrix connecting neighboring constant-z slices of the partition function (2.31) is just the exponential of an N-particle Hamiltonian for quantum mechanical particles propagating in imaginary time and interacting with a pair potential $V(\mathbf{r}_{ij}(z))$.[54] All this means in practice is that, up to a multiplicative constant, Eq. (2.31) can be regarded as an integral over a quantum mechanical matrix element,

$$Z_N = \int d\mathbf{r}'_1 \cdots d\mathbf{r}'_N \int d\mathbf{r}_1 \cdots d\mathbf{r}_N \langle \mathbf{r}'_1 \cdots \mathbf{r}'_N | e^{-\mathcal{H}_N L/k_B T} | \mathbf{r}_1 \cdots \mathbf{r}_N \rangle \tag{2.33}$$

where the Hamiltonian is

$$\mathcal{H}_N = -\frac{(k_B T)^2}{2\tilde{\epsilon}_1} \sum_{j=1}^{N} \nabla_j^2 + \sum_{i>j} V[|\mathbf{r}_i - \mathbf{r}_j|] \quad . \tag{2.34}$$

The states $|r_1, \ldots, r_N\rangle$ and $\langle r'_1, \ldots, r'_N|$ describe respectively the entry and exit points for the vortices in Fig. 4.

We now insert a complete set of many-particle energy eigenstates $|m\rangle$ of $\mathcal{H}_N$ in Eq. (2.33), and find that

$$Z_N = \sum_m |\langle m | \mathbf{p}_1 = 0, \ldots, \mathbf{p}_N = 0 \rangle|^2 e^{-E_m L/k_B T} \tag{2.35}$$

where $|p_1 = 0, \ldots, p_N = 0\rangle$ is the zero momentum state,

$$|\mathbf{p}_1 = 0, \ldots \mathbf{p}_N = 0\rangle = \int d\mathbf{r}_1 \cdots \int d\mathbf{r}_N |\mathbf{r}_1, \ldots, \mathbf{r}_N\rangle \quad . \tag{2.36}$$

The eigenstates of the permutation-symmetric Hamiltonian (2.34) may be classified by their transformation properties under the permutation group. In principle, all types of states, bosonic, fermionic and those obeying "parastatistics" should enter the sum (2.35). Because the state (2.36) is itself *symmetric* under permutations, however, the matrix element in Eq. (2.35) vanishes for all but the permutation symmetric boson eigenstates. We may thus rewrite Eq. (2.35) as a restricted sum,

$$Z_N = \sum_{\substack{\text{boson} \\ \text{states } m}} |\langle m | \mathbf{p}_1 = 0, \ldots, \mathbf{p}_N = 0 \rangle|^2 e^{-E_m L/k_B T} \quad . \tag{2.37}$$

The partition sum (2.37) should be contrasted with the sum appropriate for real bosons at a finite fictitious "temperature" β,

$$Z'_N = \sum_{\substack{\text{boson} \\ \text{states } m}} e^{-\beta E_m} \quad , \qquad (2.38)$$

where $\beta = L/k_B T$. The difference between Z_N and Z'_N is in the weights involved in the projection onto the zero-momentum ground state in Eq. (2.37). As $L \to \infty$, the lowest energy bosonic state dominates in both cases, however. The partition function Z'_N is in fact the *correct* quantity even for finite L for flux lines in a special toroidal geometry, which would be especially interesting to investigate experimentally.[5,6] In this case, one expects a sharp Kosterlitz-Thouless transition from an entangled "superfluid" phase at large L to a disentangled "normal" phase at small L. This entanglement transition may become a more gradual crossover for the free boundary conditions embodied in Eq. (2.37), however.[58]

The boson analogy is summarized in Table I. The partition function (2.10) is just the grand canonical partition function for interacting bosons in two dimensions with chemical potential $\mu = H\phi_0/4\pi - \epsilon_1$. As in Section 2.2, the trajectories of vortices across the sample are isomorphic to boson world lines. The thermal energy $k_B T$ plays the role of $\hbar$, with L corresponding to the distance $\beta\hbar$ traveled in the imaginary time direction. The parameter $\tilde{\epsilon}_1$ plays the role of the boson mass. Table I shows clearly one reason why high temperature superconductors are interesting: They allow us to explore a world of exceptionally light ($\tilde{\epsilon}_1 << \epsilon_1$) bosons in which "Planck's constant" (i.e., $k_B T$) is ten times larger than in conventional materials.

Although the boson analogy is powerful, it does not extend in any simple way to *dynamic* phenomena. Real boson superfluids like ^{4}He flow without resistance through the finest capillary tubes. Entangled flux liquids, on the other hand, may be sufficiently viscous to cause a polymeric glass transition!

Table I.

Vortex lines	$\tilde{\epsilon}_1$	$k_B T$	L	$H\phi_0/4\pi - \epsilon_1$	$V[r_{ij}(z)]$
Two-dimensional Bosons	m	$\hbar$	$\beta\hbar$	μ	Boson pair Potential

Detailed correspondence of the parameters of flux line liquid with the mass, value of Planck's constant, reciprocal temperature β, and pair potential of two-dimensional bosons.

2.2.2 Coherent State Path Integral

As discussed in Refs. 5 and 43, practical calculations for flux lines in the boson representation are conveniently carried out using the coherent state path integral formulation of boson statistical mechanics in the grand canonical ensemble.[59,60] The bosons are represented by a complex field $\psi(\mathbf{r}_\perp, z)$, and the grand partition function is

$$Z_{gr} = \int \mathcal{D}\psi(\mathbf{r}_\perp, z) \int \mathcal{D}\psi^*(\mathbf{r}_\perp, z) e^{-S(\psi, \psi^*)/k_B T} \tag{2.39}$$

where the "action" is

$$S = \int_0^L dz \int d^2 r_\perp \left[k_B T \psi^* \partial_z \psi + \frac{(k_B T)^2}{2\tilde{\epsilon}_1} |\nabla_\perp^2 \psi|^2 - \mu |\psi|^2 + \frac{1}{2} v_0 |\psi|^4 + V_D(\mathbf{r}_\perp, z) |\psi|^2 \right]. \tag{2.40}$$

We have, for simplicity, used the cutoff delta function pair potential (2.3); see Ref. 43 for arbitrary in-plane pair potentials. Note that we have now retained the disorder potential.

The amplitude of the field $\psi(\mathbf{r}_\perp, z)$ is related to the fluctuating local flux line density

$$n(\mathbf{r}_\perp, z) = \sum_{j=1}^{N} \delta[\mathbf{r}_\perp - \mathbf{r}_j(z)]$$
$$= |\psi(\mathbf{r}_\perp, z)|^2. \tag{2.41}$$

"Superfluidity" of the bosons is equivalent to entanglement of the flux lines,[54] so we expect that entanglement is accompanied by coherence in the *phase* of the order parameter

$$\psi(\mathbf{r}_\perp, z) = |\psi(\mathbf{r}_\perp, z)| e^{i\theta(\mathbf{r}_\perp, z)}, \tag{2.42}$$

i.e., long-range order in

$$G(\mathbf{r}_\perp, \mathbf{r}'_\perp; z, z') = \langle \psi(\mathbf{r}_\perp, z) \psi^*(\mathbf{r}'_\perp, z') \rangle. \tag{2.43}$$

Line crystals, on the other hand, are characterized by the *absence* of long-range order in $\psi(\mathbf{r}_\perp, z)$, even though the average flux line density $|\psi(\mathbf{r}_\perp, z)|^2$ is nonzero, (see Sec. 2.2.3).

In the entangled phase, correlation functions can be calculated by expanding about the minimum of the action in Eq. (2.40),

$$\psi(\mathbf{r}_\perp, z) = \sqrt{n_0 + \pi(\mathbf{r}_\perp, z)} \, e^{i\theta(\mathbf{r}_\perp, z)}$$
$$\approx \sqrt{n_0} \left(1 + \frac{\pi}{2 n_0}\right) e^{i\theta} \tag{2.44}$$

where the average vortex density is given by

$$n_0 = \mu/v_0 \; , \tag{2.45}$$

and we keep only terms quadratic in $\theta(\mathbf{r}_\perp, z)$ and $\pi(\mathbf{r}_\perp, z)$. We can now easily calculate, for example, the structure function

$$S(q_\perp, q_z) = \overline{\langle |\hat{n}(\mathbf{q}_\perp, q_z)|^2 \rangle} \; , \tag{2.46}$$

where $\hat{n}(\mathbf{q}_\perp, q)$ is the Fourier transform of the fluxon density. The brackets represent a thermal average, while the bar means a quenched average over the Gaussian pinning potential (2.5). The structure function takes the form,[43]

$$S(q_\perp, q_z) = \frac{k_B T(n_0 q_\perp^2 / \tilde{\epsilon}_1)}{q_z^2 + [\epsilon(q_\perp)/k_B T]^2} + \Delta \left\{ \frac{(n_0 q_\perp^2 / \tilde{\epsilon}_1)}{q_z^2 + [\epsilon(q_\perp)/k_B T]^2} \right\}^2 + \mathcal{O}(\Delta^2) \tag{2.47}$$

where $\epsilon(q_\perp)/k_B T$ is the "Bogoliubov spectrum,"

$$\epsilon(q_\perp)/k_B T = \sqrt{\left(\frac{k_B T q_\perp^2}{2\tilde{\epsilon}_1}\right)^2 + \frac{n_0 V_0}{\tilde{\epsilon}_1} q_\perp^2} \; . \tag{2.48}$$

The meaning of this result (and its generalization to fluids with nonlocal interactions) will be discussed further in Section 3. For now, we simply note that Eq. (2.47) is the beginning of a systematic expansion in the disorder strength. To leading order in Δ, disorder simply introduces "Lorentzian squared" corrections to the correlations which characterize pure systems.

2.2.3 Inverse Lindemann Criterion

If the "superfluid" flux liquid loses its phase coherence even though the flux line density $|\psi(\mathbf{r}_\perp, z)|^2$ remains finite, this presumably signals a transition to new phase without significant entanglement. If the disorder is negligible, this phase is presumably a line crystal.[61]

To estimate when long-range order in $G(\mathbf{r}_\perp, \mathbf{r}'_\perp; z, z')$ disappears, we determine the parameter values such that the fluctuations in the phase exceed unity,

$$\overline{\langle \theta^2(\mathbf{r}_\perp, z) \rangle} \gtrsim 1 \tag{2.49}$$

To evaluate this "inverse Lindemann criterion," we first integrate out the amplitude fluctuation field $\pi(\mathbf{r}_\perp, z)$ in Eq. (2.40), and obtain an effective action which depends on the phase only,

$$S_{\text{eff}} = \int d^2 r_\perp \int dz \left\{ \frac{(k_B T)^2 n_0^2}{2} \left[\frac{1}{K} |\nabla_\perp \theta|^2 + \frac{1}{B} (\partial_z \theta)^2 \right] + i \frac{k_B T}{v_0} V_D(r_\perp, z) \partial_z \theta \right\} . \tag{2.50}$$

Here,
$$K = n_0\tilde{\epsilon}_1 \qquad (2.51a)$$
and
$$B = n_0^2 v_0 \qquad (2.51b)$$
are the tilt and bulk modulus, respectively, of this simple model liquid of interacting flux lines.[5]

The average in Eq. (2.49) is conveniently evaluated in Fourier space. In the absence of disorder, for example, we find

$$\langle \theta^2 \rangle = \frac{BK}{2\pi k_B T n_0} \int_0^\Lambda q_\perp dq_\perp \int_{-\infty}^\infty dq_z \frac{1}{Bq_\perp^2 + Kq_z^2} \qquad (2.52)$$

where the integral over $q_\perp$ is restricted by a cutoff Λ related to the intervortex spacing $d \approx n_0^{-1/2}$,

$$\Lambda = \sqrt{4\pi n_0} \quad . \qquad (2.53)$$

Upon approximating the Fourier integrals in this way (and including disorder), we find

$$\overline{\langle \theta^2 \rangle} \approx \frac{1}{\sqrt{\pi}} \bar{v}^{1/2} \left[1 - \frac{1}{16} \frac{\bar{\Delta}}{\bar{v}} \right] , \qquad (2.54)$$

where $\bar{v}$ and $\bar{\Delta}$ are important dimensionless coupling constants,

$$\bar{v} = v_0 \tilde{\epsilon}_1 / (k_B T)^2 \qquad (2.55a)$$

and

$$\bar{\Delta} = \Delta \tilde{\epsilon}_1 / (k_B T)^3 \quad . \qquad (2.55b)$$

It can be shown that thermal fluctuations in clean materials renormalize the coupling $\bar{v}$ in Eq. (2.54) down to very small values when the flux liquid is dilute.[5] Indeed, summing the ladder interactions between the two flux lines shown in Fig. 11 leads to the renormalized effective interaction,

$$\bar{v}_R = \frac{\bar{v}}{1 + \frac{\bar{v}}{4\pi} \ln(1/n_0 \lambda_{ab}^2)}$$

$$\approx 4\pi / \ln(1/n_0 \lambda_{ab}^2) \quad . \qquad (2.56)$$

Here, λ_{ab} represents the interaction range. Upon neglecting the term proportional to disorder in Eq. (2.54), we conclude using (2.49) that dilute line liquids in clean materials are unstable to a new phase whenever

$$\ln\left(\frac{1}{n_0 \lambda^2}\right) \lesssim \mathcal{O}(1) \quad . \qquad (2.57)$$

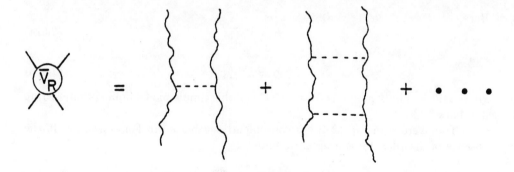

Fig. 11 Repeated interactions between two flux lines leading to the renormalized effective interaction discussed in the text.

An equivalent formula for the reentrant part of the melting line $H_m(T)$ in Fig. 1a was derived in Ref. 6. Note that inclusion of disorder in Eq. (2.54) *reduces* the phase fluctuations and hence *promotes* entanglement, as one would expect. Disorder-driven entanglement of flux lines (without thermal fluctuations) was suggested several years ago by Wordenweber and Kes and discussed by Brandt,[62] based on experimental observations of "dimensional crossover" in flux pinning experiments.

A physical argument for the loss of phase coherence, and hence long-range order, in the correlation function (2.43) for line crystals (without disorder) can be constructed as follows: $\psi(\mathbf{r}_\perp, z)$ and $\psi^*(\mathbf{r}_\perp, z)$ are, respectively, creation operators for flux line heads and tails, i.e., magnetic monopoles.[63] The composite operator in Eq. (2.43) creates an extra line at $(\mathbf{r}'_\perp, z')$, (i.e., a column of interstitials in the solid), and destroys an existing line at $(\mathbf{r}_\perp, z)$, creating a column of vacancies. The lowest energy configuration is then a line of vacancies (for $z' > z$) or interstitials (for $z' < z$) connecting the two points with an energy σs proportional to the length s of this "string." See Fig. 12. It follows that the correlation function (2.43) decays exponentially to zero (i.e., like $e^{-\sigma s/k_B T}$) for large separations in the crystalline phase. In a line *liquid*, on the other hand, it can be shown (using the hydrodynamic approach of Sec. 3.1) that the asymptotic string tension σ vanishes, implying long-range order in $G(\mathbf{r}_\perp, \mathbf{r}'_\perp; z, z')$. Note that the magnitude of the string tension in the solid will depend on the orientation of the string, and on the ordering of z and z'.

2.3 Importance of Disorder in the Dilute Limit

Results such as Eq. (2.47) are correct when the flux liquid is dense, so that the effects of disorder are screened out by flux line collisions and can be treated

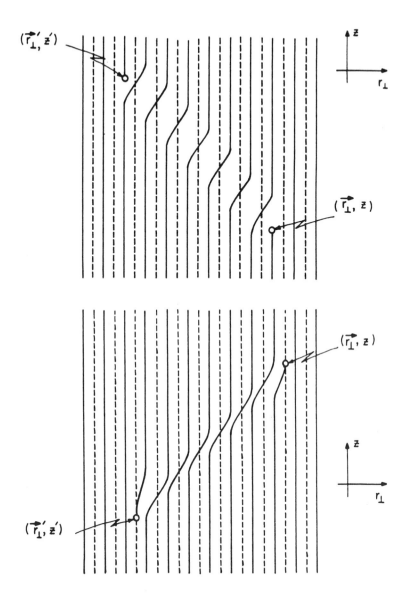

Fig. 12 Lowest energy solid phase contribution to the correlation function (2.43), which inserts a flux head and tail into a crystalline vortex array. Dashed lines represent a row of vortices slightly behind the plane of the page. In (a), a vacancy is created at "time" z, which then propagates and is destroyed at time z'. Interstitial propagation from z' to z is shown in (b). The energy of the "string" defect connecting the head to the tail increases linearly with the separation in both cases and leads to the exponential decay of $G(\mathbf{r}_\perp, \mathbf{r}'_\perp; z, z')$.

as a small perturbation. New physics arises, however, sufficiently close to H_{c1}, i.e., when the vortex lines are dilute and "screening" is no longer effective. In this limit, it can be shown that disorder must produce new physics whenever[43]

$$n_0 \lambda_{ab}^2 \lesssim e^{-4\pi(k_B T)^3/\tilde{\epsilon}_1 \Delta} . \quad (2.58)$$

Upon combining equations (2.57) and (2.58), we see that a sliver of equilibrated vortex liquid stable to both crystallization and disorder *must* exist in the density range

$$e^{-4\pi(k_B T)^3/\tilde{\epsilon}_1 \Delta} \lesssim n_0 \lambda_{ab}^2 \lesssim 1 . \quad (2.59)$$

The constitutive relation between $B = n_0 \phi_0$ and H derived for pure systems in Ref. 5,

$$B(H) = \frac{\bar{v}}{4\pi}(H - H_{c1}) \ln\left[\frac{4\pi}{\bar{v}} \frac{(\phi_0/\lambda_{ab}^2)}{H - H_{c1}}\right] , \quad (2.60)$$

can be combined with Eq. (2.58) to determine the line $H_i(T)$ in Fig. 1b, below which disorder dominates. $H_i(T)$ is given by the solution of

$$(H_i - H_{c1})\ln\left[\frac{4\pi}{\bar{v}} \frac{(\phi_0/\lambda_{ab}^2)}{H_i - H_{c1}}\right] = \frac{4\pi\phi_0}{\bar{v}\lambda_{ab}^2} e^{-4\pi(k_B T)^3/\tilde{\epsilon}_1 \Delta} . \quad (2.61)$$

Note that $\phi_0/\lambda_{ab}^2 = \mathcal{O}(H_{c1})$, so that $(H_i - H_{c1})/H_{c1} \lesssim 4\pi/\bar{v}$. Because $\bar{v}$ is so large [$\bar{v} = \mathcal{O}(10^5)$], the disorder-dominated regime is typically much *smaller* than shown in Fig. 1b. Because $\tilde{\epsilon}_1$ vanishes as $T \to T_c$, the region of Fig. 1b dominated by disorder rapidly shrinks to zero in this limit. It is not yet clear if the region below this line is a vortex glass,[29] or simply represents a crossover to new critical exponents at H_{c1}.

3. CORRELATIONS IN FLUX LIQUIDS WITH WEAK DISORDER

3.1 Hydrodynamic Treatment of Correlations

The simplified boson model treated in Section 2 neglects nonlocal effects which are known to be quantitatively important at high fields in the crystalline phase.[7,23] The long-wavelength behavior of static correlation functions can be rederived,[43] however, via a simpler (and explicitly nonlocal) approach first proposed for the *dynamics* of flux liquids.[35] Although the shear modulus of the Abrikosov flux lattice vanishes in a vortex liquid, the tilt and compressional moduli remain finite. Correlations in a dense vortex liquid with weak disorder can be described in terms of nonlocal tilt and compressional moduli which can be estimated from their crystalline phase values[7] over a wide range of fields and temperatures. In this section, we sketch this theory (which is analogous to the elastic description of a flux crystal), and discuss the physical interpretation of the correlation functions.

To describe a flux configuration such as that shown in Fig. 5, we use two basic hydrodynamic fields, a microscopic vortex density

$$n_{\text{mic}}(\mathbf{r}_\perp, z) = \sum_{j=1}^{N} \delta^{(2)}[\mathbf{r}_\perp - \mathbf{r}_j(z)] \qquad (3.1)$$

and a microscopic "tangent" field in the plane perpendicular to $\hat{z}$,

$$\mathbf{t}_{\text{mic}}(\mathbf{r}_\perp, z) = \sum_{j=1}^{N} \frac{d\mathbf{r}_j(z)}{dz} \delta^{(2)}[\mathbf{r}_\perp - \mathbf{r}_j(z)] \quad . \qquad (3.2)$$

We now coarse grain these fields over several flux line spacings to obtain smoothed density and tangent fields $n(\mathbf{r}_\perp, z)$ and $\mathbf{t}(\mathbf{r}_\perp, z)$.

The free energy of the liquid can be expanded to quadratic order in the density deviation $\delta n(\mathbf{r}_\perp, z) = n(\mathbf{r}_\perp, z) - n_0$ and in $\mathbf{t}(\mathbf{r}_\perp, z)$ as follows,

$$F = \frac{1}{2n_0^2} \int d^2 r_\perp\, dz \int d^2 r'_\perp\, dz' [K(\mathbf{r}_\perp - \mathbf{r}'_\perp, z - z')\mathbf{t}(\mathbf{r}_\perp, z)\cdot\mathbf{t}(\mathbf{r}'_\perp, z')$$
$$+ B(\mathbf{r}_\perp - \mathbf{r}'_\perp, z - z')\delta n(\mathbf{r}_\perp, z)\delta n(\mathbf{r}'_\perp, z')] + \int d^2 r_\perp\, dz V_D(\mathbf{r}_\perp, z)\delta n(\mathbf{r}_\perp, z) \quad (3.3)$$

Here, $V_D(\mathbf{r}_\perp, z)$ is the quenched random Gaussian potential discussed in Section 2.1. The parameter B is a generalized bulk modulus for areal compressions and dilations perpendicular to the z axis, while K is the modulus for tilting the lines away from the direction of the applied field. Because we are dealing with *lines*, and *not* simply oriented anisotropic particles, Eq. (3.3) must be supplemented with an "equation of continuity,"

$$\partial_z \delta n + \nabla_\perp \cdot \mathbf{t} = 0 \quad , \qquad (3.4)$$

which reflects the fact that vortex lines cannot stop or start inside the medium. Correlation functions can be calculated by assuming that the probability of a particular line configuration is proportional to $\exp(-F/k_B T)$, and imposing the constraint (3.4) on the statistical mechanics.

This condition amounts to the requirement of "no magnetic monopoles," and can be implemented using a vector potential. Alternatively, we can set

$$\delta n = -n_0 \nabla_\perp \cdot \mathbf{u} \qquad (3.5\text{a})$$

and

$$\mathbf{t} = n_0 \frac{\partial \mathbf{u}}{\partial z} \quad , \qquad (3.5\text{b})$$

where $\mathbf{u}(\mathbf{r}_\perp, z)$ is a two-component "displacement field." Equation (3.4) is now satisfied automatically, and with these substitutions, Eq. (3.3) differs from the standard elastic description of the Abrikosov flux lattice[64] only in the absence of a shear modulus. The liquid structure function is now easily found to be

$$\overline{S(\mathbf{q_\perp}, q_z)} \equiv \overline{\langle |\delta n(\mathbf{q})|^2 \rangle} = k_B T \frac{n_0^2 q_\perp^2}{\hat{K}(\mathbf{q})q_z^2 + \hat{B}(\mathbf{q})q_\perp^2} + \Delta \left(\frac{n_0^2 q_\perp^2}{\hat{K}(\mathbf{q})q_z^2 + \hat{B}(\mathbf{q})q_\perp^2} \right)^2,$$
(3.6)

where $\hat{K}(\mathbf{q})$ and $\hat{B}(\mathbf{q})$ are the Fourier transforms of the functions in Eq. (3.3). This simple hydrodynamic result is identical in form to the hydrodynamic limit of Eq. (2.47), with wave vector-dependent functions placing the constant parameters of the boson model. Weak disorder again merely leads to a small correction to the results for pure systems. To simplify the remaining discussion, we shall henceforth neglect disorder and, for the most part, simply replace $\hat{B}(\mathbf{q})$ and $\hat{K}(\mathbf{q})$ by constants B and K. We note, however, that $K(q_\perp \approx 0, q_z \approx 0)$ and $B(q_\perp \approx 0, q_z \approx 0)$ will determine the behavior near the origin of Fig. 13 while $K(q_\perp \approx \frac{\pi}{d}, q_z \approx 0)$ and $B(q_\perp \approx \frac{\pi}{d}, q_z \approx 0)$ are the controlling quantities near the peak corresponding to the first reciprocal lattice vector.

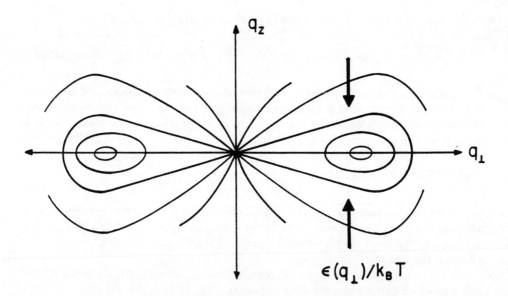

Fig. 13 Contours of constant scattering intensity for dense line liquids. The contours are linear near the origin and surround a maximum on the $q_\perp$-axis located approximately at the position of the first Bragg peak of the nearby triangular solid phase.

When $\hat{K}(\mathbf{q})$ and $\hat{B}(\mathbf{q})$ are replaced by constants the contours of constant scattering intensity are straight lines through the origin, as indicated in the small q region of Fig. 13. The conservation law embodied in Eq. (3.4) forces the structure function to vanish when $q_\perp = 0$ for nonzero q_z, and leads to scattering contours very different from those appropriate to a liquid of point particles.

The results of the boson mapping discussed in Section 2 suggest that an alternative form for Eq. (3.6), valid for a larger range of wave vectors, is

$$S(q_\perp, q_z) = \frac{n_0^2 k_B T q_\perp^2 / K}{q_z^2 + \epsilon^2(q_\perp)/(k_B T)^2} \quad . \tag{3.7}$$

The scattering function remains a Lorentzian in q_z for fixed $q_\perp$, with a width given by the function $\epsilon(q_\perp)/k_B T$ (see Fig. 13). In the Bogoliubov approximation of Section 2, this function can be expressed in terms of the elastic constants B and K as

$$\epsilon(q_\perp)/k_B T = \left[\left(\frac{k_B T n_0 q_\perp^2}{2K} \right)^2 + \frac{B}{K} q_\perp^2 \right]^{1/2} , \tag{3.8}$$

which reduces to the prediction $\epsilon(q_\perp)/k_B T = \sqrt{\frac{B}{K}} q_\perp$ for small $q_\perp$. A better approximation for $\epsilon(q_\perp)$ when the flux liquid is dense is the Feynman formula,[54]

$$\epsilon(q_\perp) = \frac{(k_B T)^2 n_0 q_\perp^2}{2 K S_2(q_\perp)} , \tag{3.9}$$

which expresses the "phonon-roton spectrum" in terms of the *two*-dimensional structure function of a constant-z cross-section, $S_2(q_\perp)$. If we insist that the three-dimensional structure function has the Lorentzian form (3.7), Eq. (3.9) is the only result consistent with the sum rule

$$\int_{-\infty}^{\infty} \frac{dq_z}{2\pi} S(q_\perp, q_z) = n_0 S_2(q_\perp) \quad . \tag{3.10}$$

The physical meaning of the function $\epsilon(q_\perp)$ for classical line liquids is illustrated in Fig. 14. Upon using (3.7) and (3.9) to determine the decay of in-plane density fluctuations $\delta\hat{n}(\mathbf{q}_\perp, z)$ along the z-axis, we find

$$\tilde{S}(\mathbf{q}_\perp, z) \equiv \langle \delta\hat{n}(\mathbf{q}_\perp, z) \delta\hat{n}^*(\mathbf{q}_\perp, 0) \rangle$$
$$= S_2(q_\perp) e^{-\epsilon(q_\perp)|z|/k_B T} , \tag{3.11}$$

which is the behavior summarized in Eq. (1.5) of the Introduction. The qualitative behavior of $\epsilon(q_\perp)$ obtained by using a realistic two-dimensional structure function in Eq. (3.9) is shown in Fig. 15. Density fluctuations with wavelength $q_\perp$ evidently decay due to line wandering along the z axis over a length scale given by

$$\ell_z(q_\perp) = k_B T / \epsilon(q_\perp)$$
$$\underset{q_\perp \to 0}{\approx} (K/B)^{1/2}/q_\perp \quad . \tag{3.12}$$

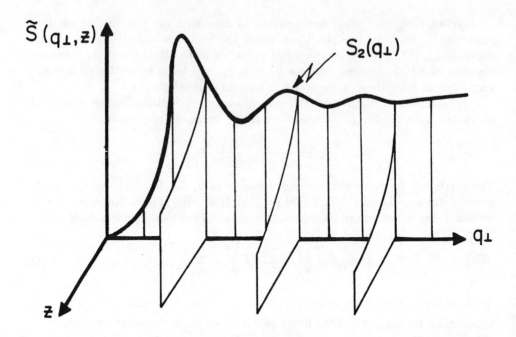

Fig. 14 Mixed structure function $\tilde{S}(q_\perp, z)$ showing the decay of the in-plane Fourier components of the density in a line liquid. For fixed $q_\perp$, the decay is exponential in $|z|$.

The two-dimensional structure function $S_2(q_\perp)$ shown in Fig. 14 is similar to that of a dense liquid of points in two dimensions. It is unusual only in that it goes to *zero* (linearly) as $q_\perp \to 0$. This incompressibility of a constant-z cross-section of a line liquid is a direct consequence of the hydrodynamic theory. Indeed, upon substituting the first term of Eq. (3.6) in Eq. (3.10), we find that

$$S_2(q_\perp) \underset{q_\perp \to 0}{\approx} \frac{k_B T n_0}{B \ell_z(q_\perp)} = \frac{k_B T n_0}{(KB)^{1/2}} q_\perp \quad , \tag{3.13}$$

where $\ell_z(q_\perp)$ is given by the second line of Eq. (3.12). Equation (3.13) resembles the usual fluctuation-dissipation theorem for point liquids, provided we identify the product $B\ell_z(q_\perp)$ with a two-dimensional bulk modulus.

Close to $H_{c2}^0(T)$, $\lambda_{ab} \gg n_0^{-1/2}$, and the flux liquid can be treated as a fluid of *logarithmically* interacting particles, as pointed out by Feigel'man.[61] The bulk modulus is approximately

$$B(q_\perp) \approx \frac{\phi_0^2}{4\pi \lambda_{ab}^2 q_\perp^2} \quad , \tag{3.14}$$

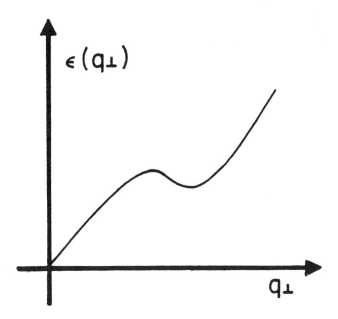

Fig. 15 Schematic of the function $\epsilon(q_\perp)$ which controls the decay of density fluctuations along the z-axis. A minimum in this "phonon-roton spectrum" appears when the two-dimensional structure function of a dense liquid is inserted in Eq. (3.9).

under these conditions. Upon inserting this result into Eq. (3.8), we see that the excitation spectrum $\epsilon(q_\perp)$ acquires a "plasmon" gap as $q_\perp \to 0$. Both $\ell_z(q_\perp)$ and the structure function $S_2(q_\perp)$ will appear to approach finite limits as $q_\perp \to 0$ in this regime. For $q_\perp \lambda_{ab} \ll 1$, however, the system should cross over to the description for vortices with short-range interactions sketched above.

Important information about vortex line liquids could be deduced from a hypothetical experiment which measures the density distribution in a constant-z cross-section.[65] The linear vanishing of the 2D structure function, in particular, shows immediately that one is dealing with a three-dimensional line liquid rather than a two-dimensional liquid of points.[66] We see from Eq. (3.13) that the slope determines the product of elastic constants KB. For cross-sections in the middle of a slab of finite thickness L, the 2D structure function will finally level out at very small $q_\perp$. The approach to a constant begins for $q_\perp \lesssim q_\perp^*$, where $\ell_z(q_\perp^*) = L/2$, i.e., for wavelengths less than

$$q_\perp^* = 2(K/B)^{1/2}/L . \qquad (3.15)$$

An experimental determination of the crossover wave vector $q_\perp^*$ would fix the *ratio* of elastic constants K/B.

3.2 Hexatic Order in Vortex Liquids

Thus far, we have said relatively little about the potentially important issue of hexatic order in vortex liquids. As pointed out by Chudnovsky,[36] the Larkin-Ovchinnikov model[3] of impurity disorder acting on an Abrikosov flux lattice destroys translational order but leaves the orientational order intact. A high temperature equilibrium hexatic liquid was proposed on the basis of a dislocation loop model at about the same time.[26] The definition of the orientational order parameter[31] is illustrated in Fig. 16, for a constant-z cross-section with seven neighboring flux lines. The local orientational order parameter associated with the central site is obtained by averaging over its six bond angles,

$$\psi_6(\mathbf{r}_\perp, z) = \frac{1}{6} \sum_{j=1}^{N} e^{6i\theta(\mathbf{r}_\perp, z)} \quad . \tag{3.16}$$

The recent observations of a low temperature hexatic vortex glass in BSCCO[11,12] suggest the possibility of an equilibrated hexatic vortex liquid at higher temperatures, and are consistent with a suggestion by Worthington et al.[16] for

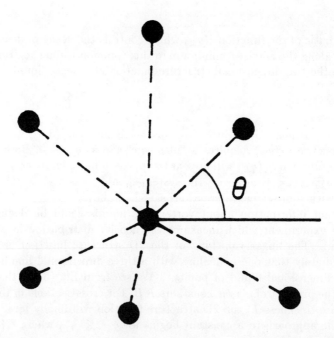

Fig. 16 Definition of the bond angles entering $\psi_6(\mathbf{r}_\perp, z)$. The bonds run between nearest neighbor flux lines in a constant-z cross section. The angles are defined with respect to an external reference axis.

YBCO. The potential for hexatic order in vortex liquids is important because a nonzero hexatic stiffness constant amplifies the intervortex shear viscosity, as discussed in Ref. 35. A broken orientational symmetry at low temperatures would, moreover, make some sort of phase transition with increasing temperature *inevitable* when this symmetry was restored.

In the absence of disorder, the hexatic order should appear just above the melting line in Fig. 1a.[26] The hexatic phase results if melting of the Abrikosov flux lattice is driven by a proliferation of thermally activated dislocations. A heavily defected vortex crystal with a finite concentration of unbound dislocation loops is one possible description of an entangled flux liquid. As pointed out by Brandt,[62] screw dislocations are an especially effective way to produce entanglement. A dislocation loop in a flux-line lattice with $\hat{\mathbf{H}}||z$ is shown in Fig. 17. All such loops are highly constrained, in the sense that they must lie in the plane defined by the z axis and their Burgers vector.[67,68] Note that these loops have a mixed edge and screw character. As the figure indicates, the average distance between the edge and screw components of this loop gas determines the translational correlation lengths parallel and perpendicular to the z-axis in the hexatic liquid.

Two basic results can be proved about this model of the flux liquid.[26] (1) Unbound dislocation loops are sufficient to melt the lattice, in the sense that the equilibrium shear modulus vanishes. Although this conclusion may seem obvious, we are unaware of an analogous result for the Larkin-Ovchinnikov model of disorder,[3] which neglects dislocations entirely. The second result is: (2) dislocations do *not* destroy the long-range order in the orientational correlation function, $G_6(\mathbf{r}_\perp) = \langle \psi_6(\mathbf{r}_\perp, z)\psi_6^*(0,0)\rangle$, i.e.,

$$\lim_{\substack{r_\perp \to \infty \\ z \to \infty}} G_6(r_\perp, z) \neq 0 \quad . \tag{3.17}$$

This result is suggested by the geometrical construction in Fig. 17: The three triangles lie in different constant-z planes, but all have the same orientation, suggesting that dislocation loops have only a minor effect on bond-orientational order. This broken rotational symmetry means that the flux liquid resists deformations in the bond angle field, as in the hexatic phase predicted by the theory of two-dimensional melting.[69] As a result, the hydrodynamic free energy in Eq. (3.3) must be augmented by a hexatic contribution,[26]

$$\delta F_H = \frac{1}{2}\int d^2r_\perp \, dz [K_\perp |\nabla_\perp \theta|^2 + K_z |\partial_z \theta|^2] \quad . \tag{3.18}$$

The Frank constants $K_\perp$ and K_z are related to the edge and screw dislocation core energies.

Note that the loop in Fig. 17 may be viewed as a virtual edge dislocation pair propagating along the z axis in the boson picture of Section 2. Loop unbinding thus provides a mechanism by which crystalline two-dimensional quantum crystals could melt into quantum hexatics, entangled in imaginary time. Little is known about the hexatic superfluid phase predicted by this approach.

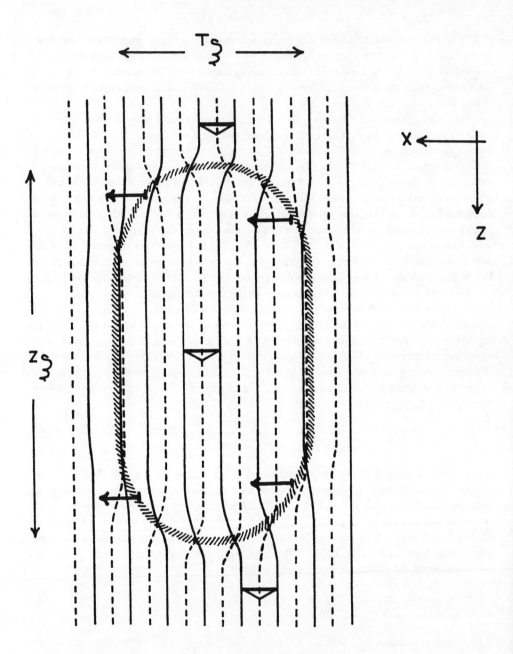

Fig. 17 Dislocation loop in a flux line solid. Dashed lines represent vortices just behind the plane of the figure. Such loops always lie in the plane spanned by their Burgers vector and the z-axis. The orientation of the three triangles is the same, showing that the loop has only a small effect on orientational order.

When disorder is included, the hexatic-to-crystal transition will be washed out, with hexatic order remaining down to $T = 0$.[36] As pointed out in Ref. 26, Chudnovsky's analysis of hexatic order within the Larkin-Ovchinnikov models does not allow for a random field which acts directly on the bond angle field. A small random field of this kind will be generated in realistic samples whenever two impurities in close proximity single out a preferred direction for the local crystallographic axes. This field will, in principle, eventually destroy long-range orientational order at sufficiently large length scales. The orientational correlation length can, however, greatly exceed the translational correlation length, in qualitative agreement with Chudnovsky's ideas. Toner has recently developed this point further,[70] and finds that a coupling to an underlying lattice with a four-fold symmetry can in fact stabilize hexatic order out to *arbitrarily* large distances, because the broken orientational symmetry then becomes Ising-like.

Hexatic order could arise at low temperatures in another way. Suppose that entanglement in an isotropic vortex liquid creates long relaxation times sufficient to prohibit the formation of *any* significant translational order upon cooling below $H_m(T)$. Hexatics may appear upon undercooling simply because this is the most ordered state compatible with entanglement.

3.3 Planar Pins, Linear Pins and Boson Localization

Thus far, we have discussed only the effects of point disorder, such as oxygen vacancies, on correlation functions. Since pinning by oxygen vacancies at liquid nitrogen temperatures is weak,[18] one can imagine regimes where the physics on large scales is dominated instead by strongly pinning *correlated* structures, such as planar twin boundaries[71] or linear defects produced by heavy ion bombardment.[56] It is straightforward to determine how the addition of correlated disorder modifies the structure function in dense flux liquids far from the irreversibility line by the methods sketched above. One can go somewhat further in understanding flux line configurations, however, if point pins are ignored entirely and one considers only disorder which is perfectly correlated along the field direction. The behavior of flux line configurations with increasing magnetic field is then closely related to the physics of boson localization.[72-75]

We have already seen how an isolated planar or linear defect interacts with one flux line in Section 2.1.2. There is always at least one bound state, although the binding weakens at high temperatures. The squared modulus of the "wave function" $\psi(\mathbf{r}_\perp)$ associated with such states determines the spatial distribution of a typical flux line configuration projected down the z-axis. The spatial extent of such wave functions will increase as the binding becomes weak.

Suppose now that *many* planar or linear defects oriented along the z-axis (and passing completely through the sample) are present, and let us imagine that the magnetic field is slowly increased above H_{c1} at a fixed temperature. The first few flux lines which enter will be trapped in the bound states discussed above. Repulsive interactions between lines will eventually limit the capacity of the defect array to absorb more flux quanta, however. At some point, the localization length $\ell_\perp$ which describes the spatial extent (perpendicular to the z-axis) of the most energetic flux lines will diverge at a two-dimensional localization transition,[75]

$$\ell_\perp \sim \frac{1}{|H_\ell(T) - H|^\nu} , \qquad (3.19)$$

where ν is a localization exponent (known to be greater than one[75]), and $H_\ell(T)$ is a critical field above which localization takes place.

Above $H_\ell(T)$, a finite fraction of the vortex lines will be part of a relatively mobile flux liquid, the dynamics of which will be discussed in Section 4. Flux flow of this viscous liquid near a twin boundary or around a cylindrical linear pin is discussed in Ref. 35. Below $H_\ell(T)$, all flux lines will be localized in a "bose glass" phase[75] with qualitatively different (and more sluggish) transport properties. More work is needed to elucidate the dynamical consequences of the transition to the bose glass, which is presumably in a *different* universality class than the vortex glass transition considered in Ref. 29. Linear defects may be particularly helpful in pinning the flux liquid above $H_\ell(T)$: Mobile flux lines will presumably entangle, not only with each other, but also with the lines which remain anchored to linear defects. This may be one reason why linear defects are so effective in raising the temperature of the irreversibility line.[57] An underlying "bose glass" transition due to twinning disorder provides an interesting alternative to "vortex glass" explanations of resistivity experiments,[30] which assume that microscopic point-like disorder dominates at large length scales. The correct physics in this case probably involves a *combination* of pointlike and correlated disorder, with an enhanced concentration of point disorder concentrated within the twin planes.[71]

4. DYNAMICS NEAR THE IRREVERSIBILITY LINE

In this Section we discuss the dynamics of flux liquids. This can be done by adapting extensive studies of the hydrodynamics of point vortices in superfluid and superconducting films carried out over a decade ago.[76,77] The hydrodynamic normal modes of both isotropic and hexatic flux liquids have now been worked out.[35] The characteristic frequencies of these modes depend on *both* the vortex friction $\gamma(T)$ and the intervortex viscosity $\eta(T)$ discussed in the Introduction. The tilt mode, in particular, may be responsible for the peaks at the irreversibility line observed in Refs. 13 and 14. We shall restrict our attention here, however, to flux flow transport experiments, focusing on four different possible explanations for the irreversibility line in Fig. 1b.

There are two broad categories of pins which can keep flux lines in place. *Pointlike microscopic disorder* includes microscopic defects such as oxygen vacancies, and can be represented by a weak potential which fluctuates on scales much shorter than the vortex spacing. Intervortex distances exceed 100 angstroms even in the most intense magnetic fields. Pointlike disorder affects vortices the same way a sheet of sandpaper influences the motion of poker chips sliding on its surface. Although oxygen vacancy disorder may be effective in pinning flux lines at low temperatures (typical pinning energies are 4 K or less), it is unlikely that this sort of pinning, by itself, could keep flux lines in place at liquid nitrogen temperatures. Even if there are many (weakly pinning) oxygen vacancies per flux line, vortices can move in response to an external current by displacing only a small line segment at a time. The energetic barrier to motion is not substantial at high temperatures, provided the vortex lines are dense enough to screen out the randomness at large length scales.[43]

Correlated or *mesoscopic disorder* appears to be necessary to pin flux lines at the high fields and temperatures envisioned for many practical applications. Examples of inhomogeneous disorder are inclusions of other phases, twin and grain boundaries, and the linear defects of Civale *et al.*[57] The pinning energies are typically much stronger than those provided by oxygen vacancies, and the spacing between these pins is usually much *larger* than the intervortex spacing. The elastic constants of the underlying Abrikosov flux lattice play an important role when a few strong pins are present because they allow flux lines to be pinned collectively. The strong pins play the role of nails which hold a carpet to a wooden floor. The entire carpet can be fixed in place with just a few nails because it has a shear modulus.

4.1 Explanations of the Irreversibility Line

Whether the irreversibility line in disordered systems (see Fig. 1b) is related to melting in pure systems is still the subject of controversy. Here, we shall simply define the irreversibility line to be the locus in the temperature-field phase diagram where the magnetization for field cooled samples disagrees with the result obtained in samples cooled initially at zero field.[48] The temperature-dependent resistivity in a magnetic field drops to zero at or near the irreversibility line. The irreversibility temperature of the bismuth based superconductors (BSCCO) at $H = 1$ T is only 30 K, even though the temperature corresponding to the upper critical field is 85 K. Understanding the physical origin of this line is of considerable practical importance, since dissipationless supercurrents will not flow above it. Four very different explanations of the irreversibility line have been proposed. Unfortunately, we do not yet have detailed knowledge of flux line configurations, except via flux decoration experiments which are restricted to low fields and temperatures. It is difficult to distinguish between the above alternatives using measurements of the resistivity alone. Here, we briefly describe each of these hypotheses, and then propose an experiment which distinguishes unambiguously between them.

4.1.1 Thermally Assisted Flux Flow

In its simplest form, the thermally assisted flux flow (TAFF) model ignores both the connectivity of individual flux lines and their mutual interactions. The system is then regarded as an ideal gas of disconnected flux bits subject to a Lorentz force (induced by externally imposed currents) and pinning. The combined effects of pinning and a uniform Lorentz force are often idealized in a tilted washboard potential. In this oversimplified form, the TAFF model is more a set of assumptions useful for parametrizing data than a real theory. The model leads to an Arrhenius temperature dependence for the flux flow resistivity,[50]

$$\rho_f = \rho_0 e^{-U_p/k_B T} \tag{4.1}$$

where U_p is interpreted as a pinning energy.

4.1.2 Melting

Because the shear modulus of the Abrikosov flux lattice drops at a melting transition, flux flow in the presence of a few strong pins should change dramatically at this point. The melting curve in the (H, T) plane determined from a simple Lindemann criterion[7] is in fact often quite close to the observed "melting" or irreversibility line determined by vibrating reed experiments.[13] Although intervortex interactions are taken into account automatically in this approach, it has the drawback that melting of vortex lines in three dimensions is probably a first-order transition. The experimental changes observed so far in the vicinity of the irreversibility line are smooth and continuous, suggesting that disorder has smeared out any underlying melting transition. A weakly first order or continuous melting transition cannot be ruled out, however.

4.1.3 The Vortex Glass Hypothesis

Disorder acting on the complex condensate order parameter of superconductors is reminiscent of disordered XY spin systems. This suggests the possibility of a finite temperature spin-glass-like transition.[29] Unlike the TAFF model, which leads to a nonzero (but possibly quite small) flux flow resistivity at all nonzero temperatures, the vortex glass hypothesis predicts that the linear resistivity vanishes below a well-defined temperature T_g. The way in which it vanishes is determined by a universal critical exponent. The onset of this transition is signalled by a diverging disorder correlation length, which measures the range over which a (disordered) condensate order parameter configuration responds to, say, a change in boundary conditions. Although the theory is almost entirely phenomenological, and relies on experimental and numerical fitting procedures to determine the relevant critical exponents, good fits to transport experiments have been obtained with this approach.[30]

4.1.4 "Polymer" Glass Hypothesis

The vortex glass hypothesis assumes that the system undergoes a glass transition due to extrinsic, impurity-induced pinning constraints. An alternative source of slow relaxation times is the *intrinsic* constraints associated with entanglement in a liquid of vortex lines.[5,45] Even in the absence of point-like (or correlated) disorder, one might expect a polymer-like glass transition with decreasing temperature, i.e., a temperature at which the intervortex viscosity becomes effectively infinite. Although the system may remain disordered upon cooling, it would behave as if it had a shear modulus because of this large viscosity. The effects of a few strong pins could then propagate, just as in an Abrikosov flux lattice with a shear modulus. As discussed in the Introduction, the validity of this picture requires a large (several times $k_B T$[35]) barrier to flux cutting.

4.2 "Viscous Electricity"

In the liquid phase, all the above ideas are consistent with the flow velocity field Eq. (1.10), which describes flux liquids on scales large compared to the intervortex spacing. Weak microscopic pinning centers are assumed to be incorporated into a renormalized Bardeen-Stephen friction constant γ. Once the vortex velocity field is known, we also know the distribution of the electric field perpendicular to $\hat{z}$ inside the superconductor,[20]

$$\mathbf{E}_\perp(\mathbf{r}) = \frac{n_v^0 \phi_0}{c} \hat{z} \times \mathbf{v}(\mathbf{r}) \quad . \tag{4.2}$$

The steady state flux flow velocity is thus linearly related to the in-plane electric field. The Lorentz force is linear in the external current $\mathbf{j}_T$, see Eq. (1.9). Using these well-known relations, we can rewrite Eq. (1.10) in the suggestive form,

$$-\delta^2 \nabla_\perp^2 \mathbf{E}_\perp + \mathbf{E}_\perp = \rho_0 \mathbf{j}_T \tag{4.3}$$

where $\rho_0 = (n_0 \phi_0/c)^2/\gamma$ is the flux flow resistivity of a sample with only homogeneous disorder and $\mathbf{j}$ is the current. We refer to this result as the equation of "viscous electricity," because it is just Ohm's law modified by a viscous term proportional to the Laplacian of the E-field and the viscous length scale

$$\delta(T, H) = \sqrt{\eta(T, H)/\gamma(T, H)} \quad . \tag{4.4}$$

The four competing theories of flux flow resistivity near the irreversibility line all give different predictions for the parameters γ and η, as summarized in Table 2.

If there is no finite temperature phase transition, as predicted by the TAFF theory, then $\gamma(T) \sim e^{U_p/k_B T}$ at low temperatures. The TAFF hypothesis makes no prediction for $\eta(T)$. If the irreversibility line is associated with freezing of the vortex liquid into a translationally ordered Abrikosov flux lattice, $\gamma(T)$ should remain *finite* at the melting temperature. If the melting transition is weakly first order,

$\eta(T)$ will increase above the transition, and then jump discontinuously to infinity. If a polymer-like glass transition is responsible for the irreversibility line, one again expects a finite $\gamma(T)$ near a glass transition T_g, and it may be possible to fit $\eta(T)$ to the standard Vogel-Fulcher form,[78]

$$\eta(T) \approx \eta_0 \exp[c/(T - T_g)] \quad . \tag{4.5}$$

In contrast, to the above three possibilities, $\gamma(T)$ is expected to *diverge* at the vortex glass transition, provided the effect of weak pins is incorporated into a renormalized friction coefficient. If we identify the length δ in Eq. (4.4) with the diverging vortex glass correlation length ξ_G of Fisher *et al.*,[29] we see that the viscosity $\eta = \delta^2 \gamma$ should also diverge. The vortex glass theory predicts that at the transition both γ and ξ_G diverge according to[29]

$$\gamma(T) \sim \frac{1}{\rho(T)} \sim \frac{1}{|T - T_{vg}|^{\nu(z-1)}}$$

$$\xi_G(T) \sim \frac{1}{|T - T_{vg}|^{\nu}} \quad , \tag{4.6}$$

where $\rho(T)$ is the resistivity, T_{vg} is the vortex glass transition temperature and the exponents were obtained by fits to the experiments of Ref. 30. Assuming $\eta \sim \xi_G^2 \gamma$, we find

$$\eta(T) \sim \frac{1}{|T - T_{vg}|^{\nu(z+1)}} \quad . \tag{4.7}$$

The exponents ν and z must, unfortunately, be fit to simulation or experiments at the moment, since an analytical theory in three dimensions does not exist. The experimental estimates[30] lie in the range $\nu \sim 1.7$-1.8 and $z \sim 4.7$-5.0.

The predictions of these different theories of the irreversibility line are summarized in Table 2. An experiment which is capable of measurng $\eta(T)$ and $\gamma(T)$, and thus distinguishing between the four competing explanations for the irreversibility line is shown in Fig. 18. The left-hand portion of an initially clean (twin free!) superconducting sample with homogeneous disorder only is irradiated by protons or neutrons to produce a high density of much stronger pins, so that the linear resistivity is effectively zero in this region. The right-hand portion remains clean. With the magnetic field normal to the plane of the figure, a current flows through the interface between the clean and dirty regions. It is desirable for this interface to be very sharp; interfacial widths of a few hundred or thousand angstroms may be possible. A series of voltage taps which bisects the interface can then be used to obtain the electric field distribution inside the sample. The electric field $E_x(x)$ determines via Eq. (4.2) the hydrodynamic flux line velocity field $v_y(x)$ so one can obtain the flux flow velocity profile inside the sample. As shown in Fig. 18 the solution of Eq. (1.10) for $\mathbf{v}(x,y)$ rises from zero to its limiting homogeneous disorder value over a distance δ. By measuring both the asympotic velocity v_∞ and the width δ (which we assume is greater than the width of the defect profile), one can determine both $\gamma(T)$ and $\eta(T)$. A discussion of experiments of this kind with a finite frequency driving term has been given by Marchetti.[79]

Table 2

Theory	$\gamma(T)$	$\eta(T)$
TAFF (Traditional)	$\gamma \sim e^{U_p/k_B T}$	$\eta \sim\ ?$
Vortex Glass	$\gamma \sim \frac{1}{|T-T_{vg}|^{6.6}}$	$\eta \sim \frac{1}{|T-T_{vg}|^{10}}$
"Polymer" Glass Transition	$\gamma \sim$ Finite	$\eta \sim e^{\frac{1}{|T-T_g|}}$
Freezing/Melting	$\gamma \sim$ Finite	η large; Grows and jumps discontinuously to infinity at T_m if transition weakly first order

Predictions of four competing explanations of the physics of the irreversibility line. The TAFF theory does not make an explicit prediction for the parameter η.

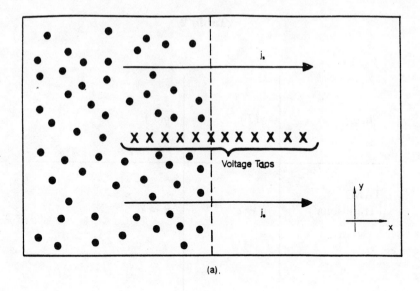

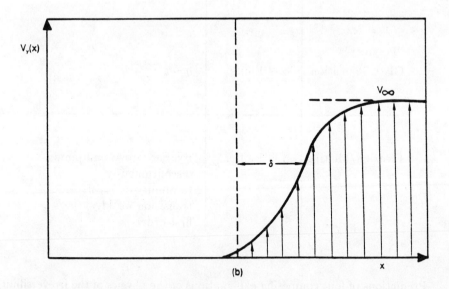

Fig. 18 Idealized experiment allowing an independent determination of the key hydrodynamic parameters γ and η.[35] The left half of the sample in (a) has high density of pinning centers. A series of voltage taps indicated by crosses determines the velocity profile shown in (b): the voltage drop between any two crosses is proportional to the *area* under the corresponding portion of the velocity profile curve.

ACKNOWLEDGMENTS

This work was supported by the National Science Foundation, through grant DMR88-17291 and through the Harvard University Materials Research Laboratory. Many of the results described here were obtained in collaboration with H.S. Seung, M.C. Marchetti and P. Le Doussal. I am grateful for these collaborations, as well as for stimulating discussions with D.J. Bishop, D.S. Fisher, P.L. Gammel, A. Kapitulnik, P.H. Kes, C.A. Murray, M. Tinkham, and T. K. Worthington. D. S. Fisher kindly provided comments on the manuscript.

REFERENCES

1. A. A. Abrikosov, Zh. Eksperim. i Teor. Fiz. **32**, 1442 (1957) [Soviet Phys.-JETP **5**, 1174 (1957)].
2. The possibility of a melting transition *very* close to H_{c2} (triggered by a vanishing shear modulus) was suggested long ago in R. Labusch, Phys. Status Solid **32**, 439 (1969).
3. A. I. Larkin, Zh. Eksp. Teor. Fiz. **58**, 1466 (1970) [Soviet Phys.-JETP **31**, 784 (1970)]; A. I. Larkin and Yu N. Ovchinnikov, J. Low Temp. Phys. **34**, 409 (1979).
4. D. R. Nelson, Phys. Rev. Lett. **60**, 1973 (1988).
5. D. R. Nelson and S. Seung, Phys. Rev. **B39**, 9158 (1989).
6. D. R. Nelson, J. Stat. Phys. **57**, 511 (1989).
7. A. Houghton, R. A. Pelcovits, and A. Sudbo, Phys. Rev. **B40**, 6763 (1989).
8. P. L. Gammel, D. J. Bishop, G. J. Dolan, J. R. Kwo, C. A. Murray, L. F. Schneemeyer, J. V. Waszczak, Phys. Rev. Lett. **59**, 2592 (1987).
9. G. T. Dolan, G. V. Chandrasekar, T. R. Dinger, C. Feild, and F. Holtzberg, Phys. Rev. Lett. **62**, 827 (1989).
10. Even disordered flux arrays can "melt," if vortex lines begin to move appreciably on experimental time scales at sufficiently high temperatures. See the striking decoration photographs in R. N. Kleiman, P. L. Gammel, L. F. Schneemeyer, J. V. Waszczak, and D. J. Bishop, Phys. Rev. Lett. **62**, 2331 (1989).
11. C. A. Murray, P. L. Gammel, D. J. Bishop, D. B. Mitzi, and A. Kapitulnik, Phys. Rev. Lett. **64**, 2312 (1990).
12. D. G. Grier, C. A. Murray, C. A. Bolle, P. L. Gammel, D. J. Bishop, D. B. Mitzi, and A. Kapitulnik, Phys. Rev. Lett. **66**, 2270 (1991).
13. P. L. Gammel, L. F. Schneemeyer, J. V. Waszczak, and D. J. Bishop, Phys. Rev. Lett. **61**, 1666 (1988).
14. D. E. Farrell, J. P. Rice, and D. M. Ginzberg, Phys. Rev. Lett. **67**, 1165 (1991).
15. E. H. Brandt, P. Esquinazi, and G. Weiss, Phys. Rev. Lett. **62**, 2330 (1989).
16. T. K. Worthington, F. H. Holtzberg, and C. A. Field, Cryogenics **30**, 417 (1990).

17. A similar interpretation of *low* field resistivity data was proposed earlier by R. B. van Dover, L. F. Schneemeyer, E. M. Gyorgy, and J. V. Waszczak, Phys. Rev. **B39**, 4800 (1989).
18. M. Tinkham, Helv. Phys. Acta **61**, 443 (1988).
19. Melting near H_{c1} has now been seen in computer simulations by L. Xing and Z. Tesanovic, Phys. Rev. Lett. **65**, 794 (1990). As predicted in Ref. 4, this transition occurs at very small reduced fields, $(H_m - H_{c1})/H_{c1} \ll 1$. There are, however, large demagnetizing factors in the usual slab-like single crystal geometries. As a result, $B = H$ to an excellent approximation, and it is the vortex density which is held fixed experimentally. The melting transition then occurs for $B \lesssim B_m = \phi_0/\lambda^2 \approx 100$ G, a region which is relatively easy to probe experimentally. For a detailed discussion of melting near H_{c1} from this point of view see Ref. 29. For an interesting simulation of a lattice model of flux lines at *high* fields, see Y.-H. Li and S. Teitel, Phys. Rev. Lett. **66**, 3301 (1991).
20. See, e.g., the articles in *Superconductivity*, edited by R. D. Parks (Dekker, New York, 1969) Vol. II.
21. I am indebted to discussions with M. Feigel'man on this point. A similar argument for reentrant melting in a real quantum mechanical system of two-dimensional fermions has been applied to electrons (and their image charges) on a thin substrate of liquid helium by P. Platzman, Phys. Rev. Lett. **50**, 2021 (1983).
22. See comment (4) on p. 9157 of Ref. 5.
23. E. H. Brandt, Phys. Rev. **B34**, 6514 (1986); Phys. Rev. Lett. **63**, 1106 (1989).
24. S. Sengupta, C. Dasgupta, H. R. Krishnamurthy, G. I. Menon, and T. V. Ramakrishnan, Bangalore preprint.
25. S.-T. Chui, Bartol Foundation preprint.
26. M. C. Marchetti and D. R. Nelson, Phys. Rev. **B42**, 9938 (1990).
27. E. Brezin, D. R. Nelson, and A. Thiaville, Phys. Rev. **B31**, 7124 (1985).
28. *The Collected Papers of L. D. Landau* (D. ter Haar, ed.), p. 193 (Gordon and Breach – Pergamon, New York, 1965).
29. M. P. A. Fisher, Phys. Rev. Lett. **62**, 1415 (1989); D. S. Fisher, M. P. A. Fisher, and D. Huse, Phys. Rev. **B43**, 130 (1991).
30. R. H. Koch, V. Foglietti, W. J. Gallagher, G. Koren, A. Gupta, and M. P. A. Fisher, Phys. Rev. Lett. **63**, 1151 (1989).
31. D. R. Nelson, in *Phase Transitions and Critical Phenomena*, Vol. 7, edited by C. Domb and J. L. Lebowitz (Academic Press, New York, 1983), and references therein. See pp. 69-71 for a formula which relates the shear viscosity to the density of free dislocations.
32. S. Doniach and B. Huberman, Phys. Rev. Lett. **17**, 1169 (1979).
33. D. S. Fisher, Phys. Rev. **B22**, 1190 (1980).
34. In this review, we reserve the word "viscosity" to apply to shear forces in a plane perpendicular to the magnetic field due to *inter*vortex interactions. The Bardeen-Stephen coupling of vortex motion to the underlying ionic lattice will be called a "friction," not a "viscosity."

35. C. M. Marchetti and D. R. Nelson, Phys. Rev. **B42**, 9938 (1990); Physica **C174**, 40 (1991).
36. E. M. Chudnovsky, Phys. Rev. **B40**, 11355 (1989); Phys. Rev. Lett. **65**, 3060 (1990); Phys. Rev. **B43**, 7831 (1991).
37. R. Seshadri and R. M. Westervelt, Phys. Rev. Lett. **66**, 2774 (1991).
38. This length has been denoted ξ_z elsewhere. We use the notation ℓ_z here to avoid confusion with the superconducting coherence length.
39. As will be discussed in Sec. 2, one should make the replacement $\frac{M_\perp}{M_z} \to 1/\ln\kappa$ in this formula when $n_0\lambda_{ab}^2 \lesssim 1$, where n_0 is the areal vortex density.
40. W. R. White, A. Kapitulnik and M. R. Beasley, Phys. Rev. Lett. **66**, 2826 (1991).
41. L. I. Glazman and A. E. Koshelev, Phys. Rev. **B43**, 2835 (1991). These authors suggest that there is a sharp phase transition at B_{x2}, instead of the gradual crossover discussed here.
42. E. M. Forgan, D. McK. Paul, H. A. Mook, P. A. Timmins, H. Keller, S. Sutton, and J. S. Abell, Nature **343**, 735 (1990).
43. D. R. Nelson and P. Le Doussal, Phys. Rev. **B42**, 10113 (1990).
44. *Contrary* to the implication in E. H. Brandt, Physica **B169**, 91 (1991), the line crossing energy in these models is *not* unphysically large, and can, in fact, be made *arbitrarily* small. See the cutoff potential used, for example, in Eq. (5.3) of Ref. 5, which leads to Eq. (2.4) in this review.
45. S. P. Obukhov and M. Rubinstein, Phys. Rev. Lett. **65**, 1279 (1990).
46. J. R. Clem, private communication 1989; E. H. Brandt and A. Sudbo, Phys. Rev. Lett. **66**, 2378 (1991), and references therein.
47. S. P. Obukhov and M. Rubinstein, Phys. Rev. Lett. **66**, 2279 (1991).
48. A. P. Malozemoff, T. K. Worthington, Y. Yeshurun, and F. Holtzberg, Phys. Rev. **B38**, 7203 (1988), and references therein.
49. This still may be an appropriate description below the melting line, however. See M. B. Feigel'man and V. M. Vinokur, Phys. Rev. **B41**, 8986 (1990); M. V. Feigel'man, V. B. Geshkenbein, A. I. Larkin and V. M. Vinokur, Phys. Rev. Lett. **63**, 2303 (1989).
50. For a nice review, see P. H. Kes, J. Aarts, J. van den Berg and J. A. Mydosh, Supercond. Sci. Technol. **1**, 241 (1989).
51. V. G. Kogan, Phys. Rev. **B24**, 1572 (1981).
52. M. C. Marchetti, Phys. Rev. **B43**, 8012 (1991).
53. R. P. Feynman, Phys. Rev. **91**, 1291 (1953).
54. R. P. Feynman and A. R. Hibbs, *Quantum Mechanics and Path Integrals* (McGraw-Hill, New York, 1965); R. P. Feynman, *Statistical Mechanics* (W. A. Benjamin, Reading, Massachusetts, 1972).
55. See, e.g., E. Merzbacher, *Quantum Mechanics*, 2nd Edition (John Wiley & Sons, New York, 1970), pp. 105–108.
56. L. Civale, A. D. Marwick, T. K. Worthington, M. A. Kirk, J. R. Thompson, L. Krusin-Elbaum, Y. Sun, J. R. Clem, and F. Holtzberg, Phys. Rev. Lett. **67**, 648 (1991).

57. L. D. Landau and E. M. Lifshitz *Quantum Mechanics* (Pergamon, New York, 1965), Sec. 45.
58. M. P. A. Fisher and D. H. Lee, Phys. Rev. **B39**, 2756 (1989).
59. N. V. Popov, *Functional Integrals and Collective Excitations* (Cambridge University Press, New York, 1981).
60. J. W. Negele and J. Orland, *Quantum Many-Particle Systems* (Addison-Wesley, New York, 1988), Chapters 1 and 2.
61. Feigel'man has made the intriguing alternative suggestion that the loss of phase coherence leads to a *disentangled* flux liquid even in the limit $L \to \infty$. Such a phase would correspond to a normal liquid at T=0 in the boson language. See M. Feigel'man, Physica **A168**, 319 (1990).
62. R. Wordenweber and P. H. Kes, Phys. Rev. **B34**, 494 (1986); E. H. Brandt, Phys. Rev. **B34**, 6514 (1986); Jap. J. Appl. Phys. **26**, 151 (1987).
63. For a related application of this idea to directed polymer melts, see P. Le Doussal and D. R. Nelson, Europhys. Lett. **15**, 161 (1991). See also J. Selinger and R. Bruinsma, to be published.
64. P. G. de Gennes and J. Matricon, Rev. Mod. Phys. **36**, 45 (1964).
65. Flux decoration measurements probing the distribution of vortex lines which enter or leave the surface of a superconducting sample are close to this idealized experiment. Here, however, one must also account for the special nature of a bounding surface and the magnetic field outside the sample. See M. C. Marchetti and D. R. Nelson, to be published, and D. Huse, AT&T Bell Laboratories preprint.
66. The only caveat is that a two-dimensional liquid of points could mimic the correlations in a cross section of a line liquid if its interactions were sufficiently long range. A $1/r$ pair potential, in particular, leads to a 2D structure function which vanishes linearly with $q_\perp$.
67. F. R. N. Nabarro and A. T. Quintanilha, in *Dislocations in Solids*, edited by F. R. N. Nabarro (North-Holland, Amsterdam, 1980), Vol. 5.
68. For important early work on this subject, see R. Labusch, Phys. Rev. Lett. **22**, 9 (1966).
69. D. R. Nelson and B. I. Halperin, Phys. Rev. **B19**, 2457 (1979).
70. J. Toner, Phys. Rev. Lett. **66**, 2523 (1991).
71. W. K. Kwok, U. Welp, G. W. Crabtree, K. G. Vandervoort, R. Hulschere and J. Z. Liu, Phys. Rev. Lett. **64**, 966 (1990).
72. J. A. Hertz, L. Fleishman and P. W. Anderson, Phys. Rev. Lett. **43**, 942 (1979).
73. M. Ma, B. I. Halperin, and P. A. Lee, Phys. Rev. **B34**, 3136 (1986).
74. T. Giamarchi and H. J. Schulz, Europhys. Lett. **3**, 1287 (1987).
75. D. S. Fisher and M. P. A. Fisher, Phys. Rev. Lett. **61**, 1847 (1988); M. P. A. Fisher, P. B. Weichman, G. Grinstein and D. S. Fisher, Phys. Rev. **B40**, 546 (1989).
76. V. Ambegaokar, B. I. Halperin, D. R. Nelson and E. D. Siggia, Phys. Rev. **B21**, 1806 (1980).
77. B. I. Halperin and D. R. Nelson, J. Low Temp. Phys. **36**, 599 (1979).

78. See, e.g., R. Zallen, *The Physics of Amorphous Solids* (Wiley, New York, 1983).
79. M. C. Marchetti, J. Appl. Phys. 69, 5185 (1991).

DISCUSSION

S. Doniach: I'd like to go back to the low-field region of your phase diagram where you, I think correctly, suggest that there should be a re-entrance in the melting line. But I feel that the kinetics will never see that, and an alternative explanation to your entanglement would be that in fact at very low densities a single vortex line would always find a strong pin, and so not move until you get to very high temperatures. And the only way such a dilute gas could move would be by cutting, essentially, the lines ... these very dilute lines would have to cut each other so that you could then transport flux across the sample. So I think there's an alternative to the entanglement picture; that's a cutting picture. And I'd like to add that the cutting ... of course, if two vortex lines are at right angles they can always cut, and the question is how much energy does it cost to make them go at right angles. And if you have a very anisotropic system where you can have lines just forming kinks between planes then the cutting energy, I think, could be a lot smaller than what you estimate.

D. R. Nelson: Okay, let me answer the various parts of that question. I've given a discussion which is pertinent to very thick samples. Of course if you make the system dilute enough there will be no entanglement, and the transport will be dominated by disconnected flux lines – as you point out – sitting in a random impurity potential. And I think one thing we certainly know – and this really comes, I think, largely from the work by Feigel'man, Vinokur, Larkin, and Geshkenbein – is that isolated flux lines, by now well-accepted results in statistical mechanics, are themselves – if sufficiently long – in a vortex glass phase. And something like that ought to be happening, although it'll be cut off by the finite sample thickness.

Flux cutting is a subtle question. It's undoubtedly very easy above the decoupling field in BSCCO, which has the weakly coupled planes. I'm more skeptical that it's easy at low fields in YBCO.

A. P. Malozemoff: In your discussion of those vacancy interstitial pairs, you indicated that there might be a point where they would proliferate, but they might be present for entropic reasons at all temperatures. Does that mean that you expect at all temperatures some linear region of the I-V characteristics?

D. R. Nelson: No, the proliferation I was alluding to – and as I say, this is work that is ongoing with Daniel – occurs only above a certain temperature. In other words, there can be regimes in the phase diagram where as you vary temperature, the vacancies never manage to proliferate before the crystal actually melts.

A. P. Malozemoff: So in spite of entropy, there are none.

D. R. Nelson: Yes. Those I would call "type-I" flux crystals. There can be regions, however – it's especially likely at high fields – where the vacancies proliferate at a finite temperature. So below that temperature there are no vacancies, essentially, in thermodynamic equilibrium. But they do exist above this temperature, in a band of crystallinity, up to the melting point. So it's very much like the distinction between type-I and type-II superconductors, where type-I superconductors are such that flux lines never enter the system before it becomes normal. But type-II allows them to enter at some finite field. And this proliferation temperature of vacancies is the analogue of H_{c1}, if you think of the vacancies as line-like excitations in a medium provided by the flux crystal.

P. L. Gammel: In principle, some of your experiments on η and γ, I suppose, have already been done by the groups who are looking at transport in crystals with a low density of twins.

D. R. Nelson: Yes, I've been in correspondence with George Crabtree, and we're hoping – I'm hoping, at least – to get together with him at this conference and see. They have some very remarkable and striking results for transport in these systems with very few twin boundaries. There are also systems he has with aligned twin boundaries so you can send the current parallel and perpendicular to the twins. The resistivity data is absolutely fascinating, and I think in principle it could be used to pull out the η and γ that I'm talking about. Also, the kinds of stuff that Peter Kes does, where he has, at least in two dimensions, fingers of strong-pinning materials with channels of weak-pinning superconductor in between. And again, that would be an excellent way to pull out these two important parameters independently.

M.P. Maley: Would you care to speculate on the possibility of moving the BSCCO irreversibility line upward by the introduction of strong pinners such as the Civale experiment in YBCO? Or is the BSCCO viscosity just too small to allow that to occur?

D. R. Nelson: My own guess – it's speculation – is that this beautiful idea of introducing linear pins ought to work fairly well, as long as you have one pin per flux line. The issue of whether they'll work at even higher magnetic fields, and hence higher flux-line densities, is related to this boson delocalization transition where you no longer have room to fit any more flux lines on the linear pins. If they're entangled and if flux cutting is hard, then you might get some response. The viscosity of the flux liquid would help you with a few linear pins, just as it would with a few planar pins. However in BSCCO is seems likely that flux cutting would be very easy at the high fields that one is usually talking about. So probably the best you could do would be to raise it up to the magnetic field intensity given by the density of linear pins.

D. S. Fisher: I think in BSCCO the linear pins may be less different from randomly-placed pins than they are in YBCO because you can pin the individual vortex points in each layer, and whether they are pinned in a line or pinned like this may not make a huge amount of difference. But presumably it would make some.

Coming back to Seb's comment about the transport at low fields. The theorist always likes to pretend that if you put a current through, the current puts a force on all the vortex lines. But of course, particularly at low fields when the vortex separations are large compared to λ, that's not really true. It only puts forces on lines near to the edge, and those lines have to move enough from the force to push the others, and so on. So in some sense, by the time one gets to a regime where you're measuring some resistance with a very small number of flux lines in the system, the flux lines will all be bunched up and the spatial inhomogeneities are going to play a very important role, and so on. I doubt anyone has really worked through an analysis of what goes on in that regime.

D. R. Nelson: The only comment was to add to Daniel's remark about the BSCCO being as effectively pinned by isolated strong pins. I think that's correct above the decoupling fields. The decoupling field for BSCCO is, I don't know – Daniel estimates three kilogauss; I got a tesla – below that field maybe linear pins would be especially helpful, but I agree that above that decoupling field, probably it wouldn't necessarily be that great.

P. H. Kes: This description in terms of interstitial vacancy pairs and the type-II behavior: would that be a kind of description you can use for the 2D-3D transition in collective pinning where you go from 2D disorder to 3D disorder, which you would call type-II?

D. R. Nelson: I think you could probably think in those terms. In the type-I flux crystal, although you have no vacancies or interstitials that go across the entire system, there will be these loops. And, there's a characteristic loop size which will be determined by thermal equilibrium considerations. And that will, I presume – as you're suggesting – determine the critical thickness where one crosses over from two- to three-dimensional behavior.

Notice that the loop is a planar excitation; it's *not* the volumetric excitation of Larkin and Ovchinnikov.

P. H. Kes: Could you make it into a kind of spiral?

D. R. Nelson: You can, and entropy will, make it into . . . you know, fluctuating out of the plane and so forth. I mean, one way to visualize it is, we probably all played with this game of sliding squares as children, where the squares are numbered and there's one vacancy that allows us to slide them around and put them in a particular permutation. You follow the evolution of those squares in time and you get the evolution of a vacancy in a square Abrikosov flux lattice.

S. Doniach: One quick question about, you know, neutrons. Can you distinguish an entanglement length from a breaking length, like in the polymer liquid?

D. R. Nelson: The way I use entanglement length it is the distance between close encounters, whether those result in a cutting, or a repulsion, or merely one line slipping through the gap between two others. So the way I'm using it doesn't distinguish. It seems to me the most direct way to distinguish things with regards to cutting is in dynamical experiments. In conventional statistical mechanics, the equilibrium things that I've calculated and discussed are remarkably insensitive to the flux-cutting barriers. Because again, you populate all regions of phase space. It doesn't matter if there's a large barrier.

One other addendum. I've spoken as if vacancies are the things that are going to proliferate in this type-II flux crystal. In Daniel Fisher's thesis, he shows that interstitials for a $1/r$ potential have a lower energy than vacancies. So it could be a proliferation of interstitials, and I think that may actually happen when you have tilted magnetic fields in these systems.

L. J. Campbell: I'd like to just make a comment. In the case that you're considering, flux liquids, the analogy to polymers may break down because the constitutive equation for a flux liquid involves a density which is necessarily intensive, because there is no conservation of flux lines as there is conservation of polymer particles. And in fact even in two dimensions one can have flux lines either up or down, and the net B field is only the difference in the up-vortices minus the down-vortices. In three dimensions there's another variable, which is the sum of the vortices in different directions. So I'm just cautioning everyone that the fully constitutive equations for a flux liquid could be terribly complicated because of this non-conservation of flux lines and the fact that they have different directions, and the density becomes a tensor.

D. R. Nelson: I agree with your point. We worried quite hard about this ten years ago when we were writing down constitutive equations for point Kosterlitz-Thouless pairs of vortices in helium films and in superconducting films. There is a plus density and a minus density. It's not hard, though, to modify what I've said and take new constitutive relations which take this into account.

When the field is strong, the bias toward up-vortices is so overwhelming that I think we're in pretty good shape in that regime.

P. L. Gammel, D. Bishop, C. Murray and D. Huse
AT&T Bell Laboratories
Murray Hill, NJ 07974

Direct Images and Unconventional Dynamics of Vortices in the High-T_c Superconductors YBCO (123) and BSCCO (2212)

Static imaging has been used to complement transport I-V measurements in inferring the phase diagram and structure of the mixed state of the high T_c superconductors YBCO(1-2-3) and BSCCO (2-2-1-2). High resolution Bitter patterns reveal the static structure of the flux lattice in the low field mixed state of BSCCO (2212). For $B \parallel \hat{c}$, the in-plane effective mass anisotropy is found to distort the lattice. Rotation of the field away from the $\hat{c}$ axis results in the appearance of a unique structure: a striking array of vortex chains. At low fields, the pattern is otherwise amorphous. Increasing the field above $\sim 20G$ results in an abrupt transition into an hexatic phase. At high fields, I-V curves with picovolt resolution imply critical behavior within the mixed state, consistent in both 1-2-3 and 2-2-1-2 with a three-dimensional vortex glass transition well below T_c. In 2-2-2, a field- and temperature-dependent crossover in the effective dimensionality of the vortex lattice is also found.

1. INTRODUCTION

The mixed state of a type-II superconductor, permeated by a dense array of magnetic vortices, has proven to be a rich and varied system. Dimensionality and disorder couple to profoundly modify the phase diagram. The basic theory of how type-II superconductors behave in a magnetic field above H_{c1} is due to Abrikosov.[1] In the Abrikosov vortex lattice phase the vortices are in the form of rigid rods, which run parallel to the magnetic field, and are arranged in a regular hexagonal crystalline array. In this simple case the phase diagram is as shown in Fig. 1(a). The figure is not to scale, as the ratio $H_{c2}/H_{c1} \sim \kappa \sim 200$ is large[2] for the cuprate superconductors.

Soon after the discovery of the cuprate superconductors it was shown[3] that the resistivity in the temperature and magnetic field regime where the vortex lattice was expected to form behaves in a qualitatively different fashion from previously studied type-II superconductors. The reason for this, we now know, is that strong thermal fluctuations cause the vortex lattice to melt[4] into a vortex fluid well below the temperature where the local superconducting order parameter is driven to zero. A new phase diagram, shown schematically in Fig. 1(b) has been proposed.[4,5,6] Note the scale in this figure is broken and highly distorted. Further, the range of stability of the states with hexatic order is uncertain. Some of the evidence supporting this phase diagram will be documented here.

For magnetic fields greater than the upper critical field H_{c2}^{MF}, estimated within mean field theory ignoring thermal fluctuations, the local superconducting order parameter ψ is driven to zero: this is the normal state. The vortices form in the vicinity of H_{c2}, and a large heat capacity signature associated with this formation is observed.[7] Due to thermal fluctuations, the vortices remain, however, in a strongly fluctuating vortex fluid state down to significantly lower temperatures before freezing. The frozen state is believed to be a vortex glass. At the vortex glass transition the thermodynamics will have an essential singularity, but it is likely to be immeasurably small. Such a vortex fluid regime has been known to occur in thin film[8] superconductors; the cuprates are the first bulk materials where its existence has become readily apparent. The arguments used here apply, quite generally, to any superconductor; however they usually only affect a miniscule region of the phase diagram.

In low fields, where the spacing between flux lines a_o is greater than λ, the tubes of flux do not strongly overlap. For $\lambda = 1400$ Å this corresponds to fields below 1000 gauss. Well into this regime, the pattern of flux emerging from the surface can be imaged directly as discussed below. The interaction between flux lines decays exponentially with decay length λ; for a perfectly clean material the vortex lattice should melt at low fields[5] when this interaction energy, which stabilizes the lattice, becomes too small to withstand the thermal fluctuations. At such low fields the vortex lattice will also be easily disordered by random pinning.[9]

The high field regime is where the flux lines are strongly overlapping: the spacing between them is significantly less than the magnetic penetration depth λ. In this regime the magnetic field in the material is fairly uniform; it remains higher at the vortex cores than between vortices, but the corrugation is much smaller than the average field. In the extreme type-II limit the corrugation is simply $\Delta B \sim \phi_o \lambda^{-2}$. Thus, in this regime, it is inappropriate to describe the system as one of flux lines.

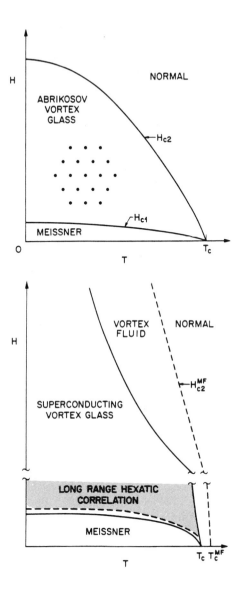

Fig. 1 (a) Schematic phase digram for a conventional superconductor without disorder or thermal fluctuations as a function of temperature T and applied magnetic field H. (b) Phase diagram incorporating disorder and thermal fluctuations. Note that the scale is broken and highly distorted. For the cuprates, H_{c1} is several orders of magnitude smaller than H_{c2}. The range of stability of the states with hexatic order is uncertain. Although observed at low temperatures, they may only represent a frozen vestige of equilibrium near T_c.

There are four (not unrelated) factors that make the vortex fluid regime large in the cuprate superconductors. These are: (i) high temperatures leading to large thermal fluctuations; (ii) small coherence lengths which allow the vortices to form at high fields since $H_{c2}^{MF} \sim \phi_o/(2\pi\xi^2)$; (iii) large penetration depths since the interactions in the high field regime are proportional to $1/\lambda^2$; (iv) strong anisotropy: these layered materials have an effective mass anisotropy $\Gamma^2 = (m_c/m_a)$ that can be at least as high as 3000. This anisotropy results in a reduced interaction between pancake vortices in different conductive layers.

For fields below H_{c2}, the primary source of resistivity in a superconductor is dissipation due to motion of vortices, driven by the Lorentz force, across the current.[10] Thus, in order to achieve zero resistivity in the mixed state, one must prevent all vortices from moving. This can happen as vortices get pinned to imperfections in the material. The microscopics of pinning can be due to the local environment affecting the order parameter, or due to surface effects on the length of the vortex lines.[11,12] The importance of the surface term can be seen by examining the line energy of a single vortex,[11] which is as large as 100K/Å for YBCO.

Larkin and Ovchinnikov[13] calculated the distortions of the vortex lattice due to randomly placed pinning centers. In all dimensions less than four, the infinitely long range positional order of a perfect lattice is destroyed even by arbitrarily weak pinning. The resultant effects on orientational order have been more vigorously debated,[14] as outlined below. An important question to ask about the resulting vortex pattern centers on the assumption of nonzero vortex mobility, leading to a nonzero resistivity. The total pinning energy for each finite region with short range vortex lattice order is finite. Thus, if each such region were assumed to be able to move without consideration of its interactions with other regions, the free energy barriers U_o that would have to be surmounted are finite. Such an assumption leads to a thermally activated resistivity $\rho \sim \exp(-U_o/kT)$ which is nonzero for all positive temperatures.

It has been recently argued that random pinning actually turns the vortex lattice phase into a vortex glass phase,[5] where the vortices are frozen into a particular random pattern that is determined by the details of the pinning in the particular sample being considered. In this vortex glass phase the vortices are not mobile, and the resistivity is strictly zero at temperatures below the transition temperature T_g. While the resistivity defined by $\rho = \lim_{I \to 0} dV/dI$ is zero in this phase, for any nonzero current, there will still be a nonzero voltage. Hence, this state is dissipationless only in a limited sense. This phase is named the vortex glass by analogy with the spin glass phase of random magnetic materials, as first introduced for random arrays of Josephson junctions by Shih, Ebner and Stroud.[15]

In both the vortex fluid and vortex glass phases, an instantaneous snapshot of the vortex pattern shows no apparent long range order. However, in the vortex fluid phase the pattern is constantly rearranging, so the time averaged correlations between the superconducting order parameter $\psi(\vec{r})$ at pairs of points in space decay with distance between the two points, vanishing at large distances. In the vortex glass phase, on the other hand, there are correlations between $\psi(r)$ and $\psi(r')$ even for pairs of points r and r' that are well separated. These correlations are not in a simple pattern, but rather in a static but random pattern that is determined by where all the vortices are located in their frozen configuration. As the vortex glass phase is approached from the vortex fluid phase, these correlations develop continuously, with the vortex-glass correlation length $\xi \sim (T-T_g)^{-\nu}$ diverging as a power of the temperature difference from the transition.

Having briefly summarized some of the theoretical ideas about the vortex patterns, it is now appropriate to consider some of the experiments in detail. These experiments are of two basic types: direct imaging of the static vortex configuration and the influence of the vortex dynamics on the transport properties.

2. STATIC STRUCTURES

Almost coincident with the discovery of the oxide superconductors was the realization of their extreme $\hat{c}$-axis anisotropy. Less dramatic, but significant, anisotropy is now known to exist in the $\hat{a}\hat{b}$ plane as well. Whether or not this provides the elusive clue to understanding the microscopics of superconductivity in these compounds, the consequences of anisotropy for the vortex lattice are readily observed. Here, direct images of the low field vortex lattice will be presented. Calculations of vortex lattice properties begin with Ginzburg-Landau (G-L) theory which has been extended to include both anisotropies and nonlocalities.[16] The simplified anisotropic London equations, valid in a wide domain of $H \ll H_{c2}$, lead to predictions of interesting distortions and orientational effects of the lattice.[17] For both the applied field normal to the planes and at other angles these are nontrivial and will be compared with experiments here. The reduced vortex lattice moduli in an anisotropic system, coupled with the other factors discussed above, have provided a rich phase diagram. Images as a function of field show a striking and prominent order-disorder transition.[18] The ordered phase is orientational only - a candidate hexatic, at intermediate fields $\sim 20G$, but develops essentially crystalline order at 100G. The field scale for the hexatic phase shows a strong dependence on the annealing history of the crystals. The low field part of the phase diagram, and its sensitive dependence on pinning, are easily explored in these images.

While only information derived from high resolution Bitter patterns will be described here, many other probes are sensitive to the static structure of the vortex lattice. Real space images of stunning beauty have been obtained with STM,[19] but only in the case of $NbSe_2$. In a sense, this provides complementary information to the magnetic field profile of the Bitter patterns, as the STM images probe the vortex core structure and location. The extreme surface sensitivity of tunneling probes may limit their range of applicability. More recently another probe of surface magnetic fields, scanning micro Hall probes,[20] have also demonstrated single vortex resolution. This, less surface specific, probe may prove to have a wider range of applicability. SEM[21] and magnetooptic probes[22] have occasionally identified isolated individual vortices, but seem far to limited to be of general use. More frequently, magneto-optic or Hall images reflect the large-scale magnetic field distribution, rather than imaging the individual vortices. This, of course, is important in determining such characteristics as current paths and densities, the effect of local sample morphology and time dependent diffusion[23] of the magnetic field, but these phenomena will not be covered here.

The structure of the vortex lattice may also be imaged through its Fourier transform, as in the neutron scattering results of Forgan et al.[24] They showed that the diffracted signal from the (10) Bragg reflection of the vortex lattice is proportional to $d_{10}\phi_o^2/\lambda^4$ where d_{10} is the (10) plane spacing. While their YBCO signals were barely observable above background, neutron scattering has, in principle, the potential for studying parallel and perpendicular order as a function of field and temperature. Similar information is, at least in principle, hidden in both NMR and μSR lineshapes and relaxation rates. In the NMR case, there is one report[25] of the asymmetric lineshape due to the vortex lattice in Tl, but generally only Gaussian lines are seen.[26] In the μSR data,[27] there is a yet no evidence of the static vortex lattice structure in the lineshape. In both resonance experiments, effects due to both the dynamics of vortices on the experimental time scale and static disorder of the vortex lattice may come into play and destroy the image of a perfect, frozen lattice. These will be mentioned along with the pertinent data here.

3. THE BITTER-PATTERN TECHNIQUE

The experiments discussed here consist of images extracted from high-resolution Bitter patterns on BSCCO (2212) crystals and YBCO (123) crystals. The technique was pioneered[28] by Trauble and Essman and Sarma to study individual vortices, and has been extensively reviewed by Huebner.[29] In the BSCCO implementation of this technique a freshly cleaved sample is cooled to 4.2K in a fixed applied field. The cleaved surface is found to be clean and smooth in SEM images. Some studies have indicated that the Bi double layer serves as the atomic cleavage location.[30] YBCO is more difficult to study. Frequently, crystals having clean, smooth facets are selected from a batch without additional surface preparation. Occasionally, the surface is etched[31] either with 1% Br/methanol or 10mM $HClO_4$/1M $NaClO_4$ which is known to result in reproducible tunneling spectra. In such cases, decoration quality is interrupted in regions with a large etch pit density. Vortices are actually present both in the etch-pits and on their sidewalls, but the details of the order are hard to establish in these cases.

The sample geometry is important, as one must take into account the demagnetization factor of the sample to determine the actual field in the bulk. Most of the samples were thin slabs of $\sim$ 1 mm extent and $5-30$ μm thickness. For $H \parallel \hat{c}$, the vortices penetrate the sample at H $\sim$ 0.5 Oe due to this demagnetizing effect. Even absent demagnetization corrections, disk shaped samples tend to trap fields due to dynamic factors. Consider an isolated vortex in the center of a disk. In order for the effects of the diameter to be important, the penetration depth must be similar to the diameter of the disk. For these ~ 1mm samples, this implies temperatures within 1 μK of T_c. Hence, effective flux expulsion from the sample center occurs only extremely close to T_c. In a quenched experiment with this geometry, essentially all flux will be trapped even for fields below H_{c1}.

The samples studied here were all field cooled. A ferromagnetic material is then evaporated at low temperatures onto the prepared sample surface. A sketch of the decoration apparatus is shown in Fig. 2. The quality and reproducibility of the images are remarkably sensitive to external parameters such as helium level and the apparatus dimensions. However, the images are quite insensitive to the magnitude of the applied field over a large range. Typically 0.2mg Fe evaporated to completion produced good pictures with a 2cm sample to boat spacing. High contrast images were also obtained with Ni. Attempts to use other ferromagnetic materials-Co and Gd failed. A background pressure (0.06 Torr) of He gas during evaporation causes the formation of ~ 50 Å thermalized ferromagnetic "smoke" particles. Particles

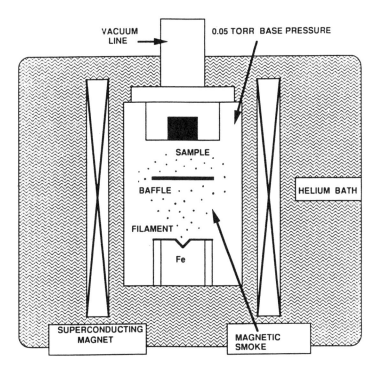

Fig. 2 Sketch of the decoration apparatus. Both nickel and iron were used to generate ferromagnetic particles in these experiments.

which drift to the surface preferentially decorate the regions of high magnetic field, thereby locating the vortices. The Bitter decoration is limited to a static snapshot of the vortex arrangement averaged over the time required to decorate, $\sim$ 1 sec in this case. Sample heating depends somewhat upon the choice of filament. Since the vortices respond to temperature gradients, excessive sample heating can cause motion of the vortices. Under optimum conditions, using a .005″ W wire as a filament, general heating was less than 0.5K. Many early experiments used .010″ wire (and 0.11 Torr base pressure) which resulted in nearly 3K heating. Following decoration, the sample is warmed to room temperature and the Fe particles, held immobile by van der Waals forces, are imaged in an SEM. Pictures of this type are shown in two of the figures discussed below.

The Bitter-pattern technique has many limitations. The most obvious is its one-shot nature: for each crystal surface, only one field and temperature can be imaged. To obtain systematics, it is necessary to resort to statistical methods. The field range accessible by the decoration technique is limited by a number of factors. In these experiments, dipole-dipole interactions tended to lead to the formation of strings of particles for lattice spacings $a_o < 0.3$ μm. This corresponds to H $\sim$ 200 Oe. More generally, the applicability of surface field map probes will break down at high fields due to the divergence of the field lines, which begins a distance λ below the surface. This means that appreciable surface field modulation exists only for $H \ll H_{c2}/\kappa \sim$ 1 kOe, still outside the range studied here. It is possible to enhance the resolution of the Bitter technique through use of a replica, in which a thin carbon film is first deposited on the same surface. After decoration, the carbon film, and the ferromagnetic particles on it, are floated off the sample and examined by TEM. The extra steps involved here impede its practicality, although it does allow fields up to 1kOe to be used.

The SEM pictures were digitized using a video frame grabber at a resolution of 512×480 pixels and a dynamic range of 256 gray levels. Two or three pictures were taken from the center of each sample from regions separated by more than 500 μm. The positions of the vortices can be determined with subpixel accuracy from the center of mass of the piles of particles in a digitized image. Regions of digitization were typically limited to 60x60 vortices due to the number of pixels available and the requirements that the vortex locations be $\sim$ 10 pixels apart. A typical digitized image contains approximately 4000 decoration clusters separated by approximately 8 pixels. Images containing obvious defects such as striations or grain boundaries were discarded. For larger scale phenomena, it is possible to piece together several adjacent images while retaining subpixel accuracy. Collages of up to 3×3 images have been usefully analyzed.

Of paramount concern is the effective temperature represented by the decoration. Consider the evolution of a quenched sample. At some time during cooling, the sample passes through T_c, and the vortices appear. Then, at a slightly lower temperature T_{irr}, the long wavelength modes (of order the sample size) of the vortex lattice fall out of equilibrium: this is the temperature where the magnetization shows the irreversibility transition. At some lower temperature T_f, the topology of the vortex lattice becomes fixed as the short wavelength modes also fall out of equilibrium. T_f, of which there is, as yet, no experimental measure, is the effective temperature imaged by the decoration. The region between T_{irr} and T_f may well be described by the type of polymer glass models advocated by Nelson.[32] It has generally be assumed that these are nearly equal, $T_f \sim T_{irr} > 0.95~T_c$ for the field range under study.[33] In this range, the irreversibility transition scales as $T_c - T_{irr} \sim H^{4/3}$. At the highest fields probed by the decorations, 100G, $T_{irr} = 0.90~T_c$. If, in fact $T_{irr} \sim T_f$, then the penetration depth where the picture is frozen in may be $\lambda_f \sim 3\lambda(0) \sim 0.9\mu$m using the two-fluid form $\lambda(T) = \lambda(0)(1-(T/T_c)^4)^{-1/2}$ for the temperature dependence of λ. This is crucial in understanding, for example, the role of nonlocal interactions in the decoration images, as λ_f calculated under this assumption is comparable to the intervortex spacing.

Processing of the digitized images begins with the construction of a defect map, the unique Delaunay triangulation.[34] Each vortex core is represented by a vertex, with bonds drawn to all nearest neighbor vortices. In all samples studied, only vortices carrying the ubiquitous superconducting flux quantum $\phi_o = hc/2e$ have been seen.[35] In addition to the BSCCO and fully oxygenated YBCO experiments detailed here, LaSrCuO (214) and oxygen deficient YBCO crystals have been examined. This places a severe limit on various candidate theories of superconductivity. Combined with Andreev reflection[36] results, this implies that the superconducting ground state must be formed from electron pairs with opposite momenta.

Detailed analysis of the images begins with the radially averaged Fourier transform. An example for an as-made BSCCO crystal in 27 Oe is shown in Fig. 3. Apart from the anisotropy in the nearest neighbor spacing described below, important information is contained in the limit $k \to 0$. An expanded view of the small k portion of the data is shown in the inset to Fig. 3. The long wavelength tails have been examined for the entire field range, and can be summarized by a power spectral density $S(k \to 0) \sim k/n^{1/2}$, where n is the vortex density. This implies the spectrum arises from the structure of the vortex lattice in the bulk of the crystal. While there is some residual debate about the theoretical estimates of the proportionality constant, the data are certainly within a factor of 10 of preliminary calculations by Nelson.[37] It had been suggested that surface interactions should dominate, since the vortices emerging from the crystal surface can be mapped to a charge $1/\alpha = 137.04$ Coulomb gas.[38] The resultant surface, an array of magnetic monopoles, would undergo a Wigner[39] crystallization at $T_M \sim 120\sqrt{B}$ with B in Oe and T in Kelvin, and would be crystallized for all fields and temperatures studied here. In this case, the long wavelength limit is $S(k) \sim k^2$, with a proportionality 10^4 different from observation. Hence, there is reason to believe the vortex pattern at the surface is representative of the structure in the bulk of the sample. At yet longer wavelengths, there should be new information in S(k). For example, in Nelson's model of an entangled vortex state, S(k) should saturate at q*, corresponding to the entanglement length along the $\hat{c}$ axis. Analysis of larger images to determine the role of entanglement are underway. Of course, one might also expect S(k) to reflect the finite sample thickness. Samples studied here were $5 \to 30~\mu$m in thickness.

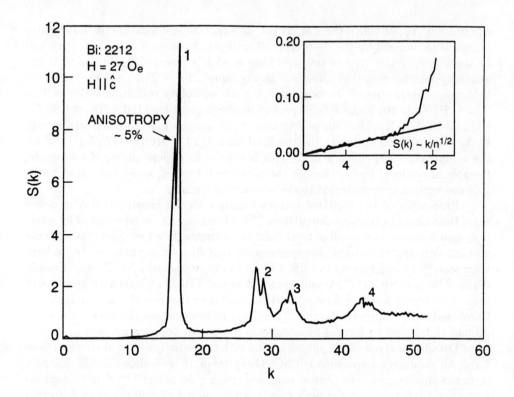

Fig. 3 Radially averaged spectral power density at 27 Oe. The first peak is split due to effective mass anisotropy in the plane. Subsequent peaks are all associated with the hexagonal lattice. In the inset, the long wavelength tail is shown. The approximate form $S(k) \sim k/n^{1/2}$ suggests these images reflect the vortex configuration in the bulk, rather than on the surface, of the sample.

In a Delaunay triangulation, triangles bordering nonsix-fold-coordinated topological defects such as disclinations and edge dislocations are shaded. The most relevant 2D defects for the destruction of long range translational order are free dislocations or isolated pairs of fivefold- and sevenfold coordinated particles, interstitials that exist as fourfold-coordinated particles with a pair of sevenfold-coordinated neighbors and vacancies, which often show up as eightfold-coordinated particles with two fivefold coordinated neighbors. These defects are responsible for the finite values of both the positional and orientational correlation lengths, as discussed below. The peaks in the spectral density of Fig. 3 can all be indexed from the hexagonal lattice. Peak shapes were more difficult to analyze quantitatively. Rather than make inferences from peak width and shape, correlation functions will be quoted here.

The orientational order is sufficient in these pictures to determine the positions of the first reciprocal lattice vectors to within 0.05%. The anisotropy in the magnitudes $|\vec{G}| = 4\pi/a_o\sqrt{3}$ ranges from 3% to 9%. This results in the noticeable splitting of the first peak in Fig. 3. Effective mass anisotropy, as determined from resistivity,[40] would have predicted a much more distorted lattice, approximately 20%. In YBCO, anisotropy determined from decoration[41] and effective mass[42] agree well. Shown in Fig. 4(a) is a two-dimensional digital Fourier transform of a YBCO decoration in an untwinned crystal at 65 Oe. The eccentricity of the ellipse f=1.15 is a measure of the penetration depth ratio $f = \Gamma = \lambda_a/\lambda_b = (m_b/m_a)^{1/2}$. The arrow in the figure is the $\hat{a}$ axis. Note that while this is parallel to the major axis of the ellipse, the reciprocal lattice vectors, as indicated by the points, are rotated slightly from this direction. In the figure, the nearest reciprocal lattice vector is 10° from the $\hat{a}$ axis.

An extreme example of this effective mass distortion occurs for decorations of the surface of thick YBCO crystals on a face perpendicular to the $\hat{a}\hat{b}$ plane. A dramatically distorted lattice, although still consisting of singly quantized flux lines was observed[41] at 4 Oe as in Fig. 4(b). The distortion is explained with an effective mass ratio $\Gamma^2 = 30$, in rough agreement with other estimates.[41] The oval structure of the individual vortices is likewise a result of the penetration depth, and hence field profile, anisotropy. The picture shown here is somewhat misleading, as the vortices are aligned by a twin boundary piercing the surface at a 45° angle at the location indicated by the arrow. More often, the rows of vortices showed substantial wandering. One interpretation of this is the response of the vortex lattice to small field gradients. Since the vortex lattice moduli are reduced by a factor $\Gamma^4 \sim 900$ in this direction, the lattice is extremely unstable. Computer simulations have resulted[43] in pictures giving qualitatively similar lattice distortions in this case.

The reason for the failure of the effective mass model in BSCCO is unclear. One possibility is that stacking faults cause $\hat{a}$ and $\hat{b}$ to alternate through the thickness of the crystal, yielding a lower averaged anisotropy. X-ray analysis of these crystals always showed some residual superlattice modulation characteristic of the $\hat{b}$ axis in the nominal $\hat{a}$ direction, consistent with this hypothesis. Resistivity would only probe the surface layer, due to the large $\hat{c}$ anisotropy. A more exciting, and fundamental, cause would result from weakly coupled two-dimensional vortices in the layers. These pancake vortices have been described by Clem.[44] The interplane interactions may be magnetic or due to Josephson coupling. If the two-dimensional vortex lattices were uncorrelated from layer to layer, then the magnetic field penetrating the surface would represent an average over many possible orientations, weighted by the distance from the surface. Hence, the net pattern would be substantially less anisotropic than the constituent patterns of the individual layers. Such an interpretation is also motivated by the μSR data.[27] In those data, a Gaussian line corresponding to $\lambda \sim 4200$ Å is seen. If the line is narrowed by a factor $(a_o/d)^{1/4}$ due to uncorrelated[45] vortex lattices with lattice constant a_o in layers a

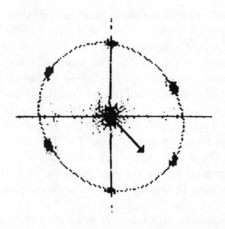

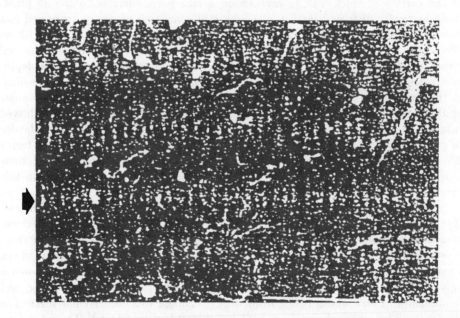

Fig. 4 (a) Digital Fourier transform of the vortex pattern in YBCO at 65 Oe. The ellipsoid superimposed has an eccentricity corresponding to the anisotropy in λ. The arrow indicates the crystal $\hat{a}$ axis, along the major axis of the ellipse and rotated $\sim 10°$ from the nearest vortex lattice reciprocal lattice vector. (b) Horizontal chains of oval vortices at a field of 4 Oe perpendicular to the $\hat{c}$ axis. The arrow indicates the position of a twin boundary which aligns a row of vortices. The bright, irregular objects are irrelevant debris. The marker is 10 μm.

distance d apart, the intrinsic linewidth corresponds to $\lambda \sim 3000$ Å. This latter value is closer to estimates from magnetic data.[46] Such an assertion would also explain the Gaussian lineshape.

Equally important is the observation that the reciprocal lattice vectors show no alignment either with the crystal axes, as determined by x-ray analysis, or with sample topography. It is important to note the distinction made clear in Fig. 4(a). The ellipse determined by the anisotropy is aligned with the crystallographic direction in the sense that the major axis is parallel to $\hat{a}$. However, the points associated with the reciprocal lattice vectors of the vortex lattice are randomly oriented on the perimeter of the ellipse. In Fig. 4(a), the nearest reciprocal lattice vector is $\sim 10°$ from the $\hat{a}$ axis. Within G-L theory, all orientations of the vortex lattice are degenerate for the case[17] of orthorhombic symmetry of the effective mass. In previous decorations, however, the orientations were found to be pinned to the crystal.[29] These early results tended to be on materials, such as Nb, where $\kappa \sim 1/\sqrt{2}$, close to the intermediate-mixed state. While, in general, it is believed that the effects of the crystal lattice become less important as κ increases, BSCCO and YBCO represent the first materials where the orientation of a vortex lattice arises spontaneously. This means that the vortex lattice is not aligned by edges, defects or the crystal, and yet has long range orientational order. In such a case, the broken symmetry associated with this off-diagonal long range order is unambiguous evidence of a phase transition, and has been identified as the boundary of the amorphous and hexatic glass phases.[18]

It should be noted that the STM images of Hess[19] show vortices aligned with the crystal axes of $NbSe_2$. However, in the case of hexagonal crystal lattice symmetry, there is a natural alignment for the hexagonal vortex lattice as well. Decoration experiments are in progress to measure the strength of this pinning to the crystallographic axes. If the field is placed at an angle with respect to the $\hat{c}$ axis, the vortex lattice will align[17] with the plane defined by $\vec{B}$ and $\hat{c}$, as discussed below. The energy associated with this alignment on tilting has been calculated numerically within the effective mass approximation. A measure of the angle at which the vortex lattice rotates away from the crystal axes to the $(\vec{B}, \hat{c})$ plane will provide a measure of the interaction with the underlying crystal orientation.

4. FIELD DEPENDENCE FOR B $\parallel \hat{c}$

Early decoration images[35] on heavily twinned and disordered YBCO crystals showed only amorphous vortex lattices: both positional and orientational order decayed exponentially with a correlation length of a few lattice constants. While most results, even on the best untwinned YBCO crystals, remain largely amorphous, recent BSCCO results are strikingly different.

Two types of BSCCO crystals were used in these experiments, both prepared by directional solidification.[47] One set of crystals was annealed at 600° C in 1 atm oxygen for 24 hours and then cooled to room temperature, while the other set was examined as made. Magnetization measurements of the as-made samples in an 0.5 Oe ambient field indicate homogeneous bulk superconductivity with onset at T_c =88.5K and a transition width of 5K. The annealed samples have an onset of 86K and a 2K width. X-ray diffraction shows that annealing shrinks the c-axis lattice constant from 30.89 Å to 30.80 Å while the $\hat{a}$ and $\hat{b}$ spacings remain unchanged at 5.413 Å and 5.411 Å respectively. High resolution XPS studies[48] of the O_{1s} core levels show that additional oxygen fills vacancies in the Bi-O plane. Despite this structural change, the room temperature resistivity anisotropy changes by less than 10% upon annealing.[49] The effective mass anisotropy should likewise remain unchanged, unless any changes were offset by compensatory shifts in the scattering time.

As the magnitude of the applied field is increased, there is a dramatic change in the density of defects observed[18] in the vortex lattice. In the annealed samples, the fraction of disclination sites in the field of view f_d varies from 0.26 at 8G to .07 at 11G to 0.00 at 69G. Examples of Delaunay triangulations at 8, 23 and 69 Oe are shown in Fig. 5. The vortices are quite disordered at the low 8 Oe field and much less so at 23 Oe. In 69 Oe no defects are seen in the field of view. These experiments could only probe fields up to 100G.

To elucidate the nature of the long range order, the bond orientational and translational correlation lengths were calculated.[50] The translational correlation function g_G (r), is calculated from the order parameter $\psi_G(r_j) = \exp(iG \cdot r_j)$ for each site j, where G is a first reciprocal lattice vector. The correlation functions are fit to a simple exponential decay $g \sim \exp(-r/\xi)$. There is some nonsystematic dependence of ξ on the reciprocal lattice vector chosen. The variation with reciprocal lattice vector has been averaged and folded into the quoted error bars. The correlation lengths increase monotonically from $\sim 1a_o$ at 5 Oe to $20a_o$ at 97 Oe for the annealed samples as a function of field as shown in Fig. 6. The as made samples have ξ_G approximately half the annealed value at each field. Our imaging system was accurate for correlation lengths up to $50a_o$. The translational order increases with magnetic field with no sign of saturation. The measured translational correlation length is in all cases quite close to the intrinsic free dislocation density average separation. Hence, the positional correlation lengths are determined by this concentration.

In the 23 Oe and 69 Oe defect maps, Fig. 5, one can easily sight down rows of vortices, despite the relatively small value of ξ_G compared to the size of the image. This is strikingly reminiscent of results[51] in 2D colloids, and is the definitive signature of an hexatic, which exhibits short range translational order and long range bond-orientational order. The sixfold bond orientational order parameter $\psi_6(r) = n_j^{-1}\Sigma\exp(i6\theta_{jk})$ is summed over the set n_j of nearest neighbors with θ the bond angle. The correlation function for this order parameter $g_6(r)$ measures the correlation of a bond angle $\theta(\vec{r})$ at $\vec{r}$ (modulo $2\pi/6$) with that at the origin. Assuming a power law dependence for decay of the correlations of the form $\langle\psi_6^*(0) \cdot \psi_6(r)\rangle \sim r^{-\eta_6}$, one can extract a correlation exponent η_6 from fits to

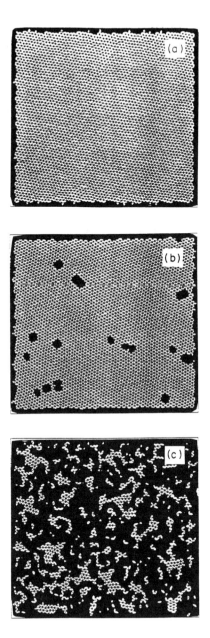

Fig. 5 Delaunay triangulations for image processed scanning electron micrographs of decorated BSCCO crystals. The freshly cleaved samples were decorated at 4.2K in fields of (a) 69 Oe (image $27 \times 25 \mu m^2$), (b) 23 Oe (image $48 \times 45 \mu m^2$ and (c) 8 Oe (image $72 \times 68 \mu m^2$). Shaded triangles join vertices which are not sixfold coordinated. In images such as (b) the density of free dislocations determines both the translational and orientational order.

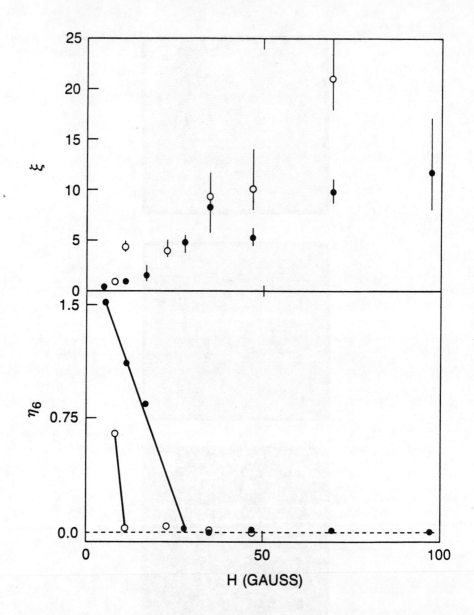

Fig. 6 The upper panel shows the translational correlation lengths ξ_G for annealed (open) and as made (closed) BSCCO samples in units of nearest neighbor spacings. The error bars represent the range of ξ_G for different G. The lower panel shows the bond-orientational correlation exponents η_6. The line at $\eta_6 \sim .06$ is our resolution limit. The hexatic boundary is at $\eta_6 = 0.25$. The two data sets may be superimposed by changing the field scale.

computed bond-orientational correlation functions.[52] These exponents are shown for the same series of decorations in Fig. 6. Immediately obvious from the figure is the abrupt change of the fitted η_6 from a relatively large value of 0.8, to the limits of our experimental resolution 0.06. This resolution limit was checked by analyzing perfect lattices with our video system. The change occurs rapidly, in a change of only a few Oe applied field, but for different fields for the as made and annealed samples, presumably reflecting the change of the concentration of intrinsic pinning sites. As in the case of the translational order, the two data sets can be matched by scaling the applied field by a factor of roughly two.

Simulations[53] in systems of 2D arrays of hard spheres with trapped random dislocations find $\eta_6 = 9c/\pi$ where c is the fraction of dislocation cores. This relation was obtained through a cumulant expansion of $g_6(r)$ within continuum elastic theory. If, in our case, we take c to be the fractional area of Voronoi polyhedra in which there is a net Burgers vector this gives values close to the observed results. Therfore, we conclude that dislocations are responsible for both the short range positional order as well as the quasi long range orientational order. Theories of the vortex lattice involving elastic properties[14] explicitly do not include any dislocations. These data suggest that such theories are irrelevant for the description of the low field vortex lattice.

After the idea of a hexatic phase was first introduced by Nelson and Halperin,[50] the issues of disorder and the decay of positional and orientational order were widely discussed. Nelson, Rubinstein and Spaepen[53] discussed the effects of disorder on correlations in their simulations on planar arrays of spheres. They were the first to note that disorder couples much more strongly to positional than to orientational correlations.

The phase with long range orientational order has been called hexatic. Nelson and Halperin's theory for 2D hexatics with thermal disorder, as opposed to those with the quenched in disorder found here, indicate that $\eta \sim 1/4$ at the hexatic to fluid phase boundary. This is consistent with observations in 2D colloidal systems with thermal disorder[51] and 2D magnetic bubble lattices.[54] For the vortex lattice, the hexatic phase boundary defined in this fashion occurs at about 20G for the as made samples. In the annealed samples, this boundary was reduced to 8G, showing a strong sensitivity to disorder. The translation and orientation correlation lengths for the as made and annealed samples are equal if scaled by the field of the phase boundary. At the hexatic phase boundary, the Fourier transform lineshape should go from Gaussian to Lorentzian.[55] Our lineshapes are not sufficient to resolve this difference, as least in part due to in plane anisotropy.

The detailed nature of the low temperature ordered phase has been discussed by Chudnovsky.[14] He suggests that in the limit of strong pinning disorder, an hexatic vortex glass is stabilized. In this case, the hexatic glass is the true equilibrium ground state produced by the competition between disorder and the repulsive interactions between vortex lines. However, that work ignored the role of dislocations. Marchetti and Nelson[14] separately considered the effects of dislocation loops of combined screw and edge character in the 3D vortex lattice. They also derive hexatic order. In this scenario, the hexatic vortex glass we observe is a frozen vestige of a high temperature hexatic vortex liquid, which is quenched in due to the low mobility of edge dislocations or vortices. Finally, Toner[14] has proposed that while the coupling of disorder to orientational correlations is weaker, it too eventually decays exponentially, but with a much longer correlation length. These data are consistent with the predictions of an isotropic vortex fluid-hexatic vortex glass[5] or hexatic vortex fluid transition,[14] or a transition between a strongly pinned disordered glassy phase to a less strongly pinned hexatic[56] near H_{c1}. Two types of data would help distinguish these: vortex mobility vs. temperature, and the microscopic irreversibility temperature T_f vs. H. Finally, it is important to note that only the simplest field history was considered here. Data on zero field cooled samples, or even more complex field histories, would help clarify the issues presented here.

5. ANGLE DEPENDENCE: THE VORTEX-CHAIN STATE

Decoration experiments have also been performed[57] with the applied field at an angle with respect to $\hat{c}$, using the geometry of the inset to Fig. 7. The samples were mounted at fixed angle and field cooled, with all decorations at 4K. In this case, the experiments probe the large anisotropy direction, so no effort was made to align the $\hat{a}$ axis. The first result of these experiments is the average density n of vortices on the cleaved face ($\perp \hat{c}$) of the crystal as a function of the angle of the applied field. For all angles examined up to 85° the density is found to be $n = (B/\Phi_o)\cos\theta$. This implies that the vortex lattice is induced by the component of B parallel to $\hat{c}$. This had been previously been inferred by Kes[58] through an analysis of torque magnetometer data.[59]

The analysis of the angle dependent structure is best divided into two regimes, $\theta < 60°$ and $\theta > 60°$. For the smaller angles, the vortex lattice is isotropic, to within the 3-9% distortions discussed previously. In addition, there remains no apparent preferred orientation of the vortex lattice, either with respect to the crystallographic axes or the tilt axis. Interestingly enough, this measured isotropy already implies extreme anisotropy in the effective mass. To clarify this point, consider the case of an isotropic system. Absent surface effects, an hexagonal vortex lattice forms perpendicular to the applied field. The decorated surface is at an angle θ to this lattice. Hence, the pattern on the surface perpendicular to $\hat{c}$ will be distorted, with a distortion $\rho = a/b = 1/\cos\theta$, simply due to projection. However, in the limit $\Gamma \to \infty$, the distortion predicted by the anisotropic London equations[17] is likewise $1/\cos\theta$. This exactly cancels the projection term, and the observed vortex lattice should be isotropic. Assuming a maximum distortion of 5% at 60°, the full expression derived from the anisotropic London equations implies that the effective mass ratio is at least $\Gamma^2 > 8$. While this is certainly consistent with other measurements, this measurement is insensitive to Γ when the anisotropy is large. Torque magnetometer data,[59] for example, give $\Gamma = 55$, a widely accepted value.

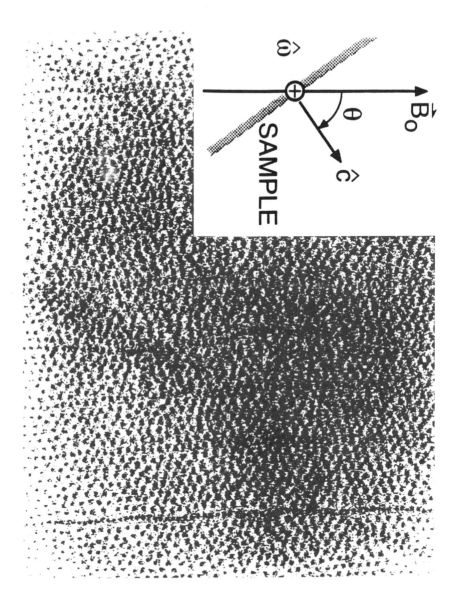

Fig. 7 Vortex chain state observed for $\theta = 70°$ and H=35 Oe. Shown in the inset is the sample geometry, which is fixed during field cooling. The chains lie in the plane defined by $\vec{B}$ and $\hat{c}$. The increased chain density with respect to the background lattice is equivalent to one flux quantum per superlattice unit cell. The Abrikosov lattice between chains is oriented with one lattice vector parallel to the chain.

For angles $\theta > 60°$, a dramatic new structure is observed as pictured in Fig. 7. This image is taken with an applied field of 35 Oe and $\theta = 70°$. The striking chains of vortices lie in the plane defined by $\vec{B}$ and $\hat{c}$. The chains have an increased density with respect to the background lattice as discussed below. The chains also orient the intervening vortex lattice so that one of the lattice vectors lies parallel to the chain axis. This is independent of the orientation of the $\hat{a}$ axis. This is the orientation of the vortex lattice predicted by anisotropic G-L theory, although without mention of the prominent chain structure seen in the experiment. Note, however, that this orientation was not defined prior to the chain formation. It has been widely reported that neutron scattering experiments actually observe a vortex lattice rotated 90° from this. This report is based on experiments on elemental Tc,[60] YBCO[24] and UPt_3.[61] In all cases, the field orientation corresponded to $\theta = 90°$, in which case all vortex lattice orientations are again degenerate within G-L theory. The observation is that one of the vortex lattice vectors lies in the basal plane in this case. The $\theta = 90°$ limit of the above observation is that one vortex lattice vector would lie parallel to the $\hat{c}$ axis. Since the distinction is only sharp for intermediate angles, this decoration experiment, therefore, is the first test of this prediction.

The chain structure appears to be independent of sample thickness. The wandering of the chains, prominent in Fig. 7, is pervasive and suggests only a weak interaction between chains. The density along a chain, however, is quite uniform, suggesting a much stronger effect in this direction. To quantify the data, in the plane of the photograph, the spacing along the chain is defined to be D and that between chains C. Although C has large variations, as can be seen from the picture, both D and C are found to scale as $B^{-1/2}$ at fixed angle. Since this is the same scaling as the vortex lattice constant, the picture at fixed angle is field independent in the sense that the field only sets the overall magnification. It is also found that for all angles and fields DC = Φ_o/B. Viewing the chains as a superlattice modulation, the increased density along the chains is equivalent to one extra flux quantum per superlattice unit cell. Hence, for the image in Fig. 7, where there are ~ 7 rows of vortices between the chains, the chains will have a vortex density $1/7 \sim 15\%$ higher than the background. Attempts to see the effect of the chains in the Fourier transform require very high resolution due both to the relatively small difference in density and their sparse occurrence. To accommodate this extra vortex, the chains acquire a necklace of dislocation pairs roughly evenly spaced a distance C apart with zero net Burgers vector, as determined from triangulation. Shown in Fig. 8 is an example of such a triangulation for 70° and 97 Oe. The chains are indicated by the heavy solid lines and the defects are shaded. A dislocation with net Burgers vector is a five-fold coordinated vortex separated by one lattice spacing from a seven-fold coordinated vortex, such as in the lower left chain. A pair of dislocations with equal and opposite Burgers vectors creates an interstitial with only local distortions, as their strain fields cancel at large distances. Such a pair is clearly seen at the bottom of the right hand chain, and is believed to be the dominant feature associated with chain formation. Note that the interstitials cannot be introduced through a simple distortion of the Abrikosov lattice, so that the dynamics of chain formation will be a complex process.

ang7097a

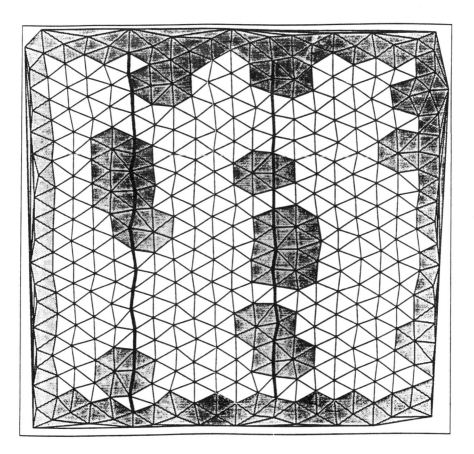

Fig. 8 To accommodate the increased density along the vortex chains, pairs of dislocation with opposite Burgers vectors are formed. In this triangulation at 70° and 97 Oe, the heavy solid lines are the chains and non six-fold coordinated regions are shaded. The necklace of dislocations along the chain is apparent. A well defined dislocation pair with zero net Burgers vector is seen on the bottom of the right hand chain.

As the angle increases, the number of vortices between chains decreases. Due to the large variation in C, the exact form of this variation is difficult to pin down. One form, generated by considering what distortions of an isotropic lattice could produce chains, and which approximately models the data, is D = $0.75[\Phi_o/B\cos(\theta)]^{-1/2}$. Of course, this distortion in inconsistent with the previous statement of the orientation of the chains and their defect structure. It should be considered only as a convenient parametrization of the data. Experimentally, the number of vortex lattice constants between chains varied from 11 at 60° to 2 at 85°.

It is clear that the chains are formed by a weakening of the repulsive interaction between vortices along the chain direction. This must be a result of the unusual current paths in a strongly anisotropic system. The chains themselves would be roughly uniformly spaced due to a repulsive interaction between them, as they are regions of increased magnetic field due to the higher vortex density.

Chain structures were anticipated theoretically, although the details do not exactly match these results.[62] Anisotropic G-L calculations near H_{c1} showed the existence of an attractive well in the vortex-vortex interaction along the chain direction. This is a remnant of the simple observation that two dipoles attract when their axes are both parallel and along the line connecting them. Parallel to the tilt axis, the interaction remains purely repulsive. Further estimates showed the state near H_{c1} to consist entirely of chains. The spacing along the chain was determined to be $\sim \Lambda = (\lambda_a \lambda_b \lambda_c)^{1/3} \sim 0.8\mu$m, and field independent. This is in obvious contradiction to these data. Further, neither calculation gave any indication of the normal vortex lattice between the chains. Recently, the finite thickness of the sample has been included.[63] The lowest energy state for intermediate sample thicknesses appropriate to this experiment was modeled with two types of vortices: chain vortices parallel to B and a hexagonal lattice parallel to $\hat{c}$. While the ratios of vortices for the lowest energy seemed to match our results, there was again predicted to be no field dependence. It may be that λ appropriate to our pictures is large, as the state is frozen in near T_c. There have also been speculations on the region of the phase diagram occupied by the vortex chain state, including arguments as to the effect of the chain formation on critical currents.[64] Further simulations exploring this and other possibilities are in progress. The role of nonlocality is also being investigated. A unique proposal by Varma[65] attributes the chain formation to a terracing of the vector potential near the tilted surface, which can qualitatively describe many of the features seen here.

The vortex lattice which evolves in the mixed state of an anisotropic superconductor embraces several unusual properties. Distortions due to effective mass were calculated on the basis of the anisotropic London equations, and generally agree with observation, although the exact magnitude in 2-2-1-2 is not fully understood. This includes distortions for the field at an angle from the $\hat{c}$ axis. Orientational effects, studied for the first time in the angle dependence of this work, also seem to follow London theory in rough outline. BSCCO and YBCO are the first examples where crystal symmetry does not seem to affect the vortex lattice orientation. For large angle tilt, the newly described vortex chain state appears. Again, the basic attractive interaction which stabilizes the structure is consistent with that derived from an anisotropic London theory.

The order, both orientational and translational, is determined by the free density of dislocations. This points to the flaws inherent in using elastic theory in describing the vortex lattice. The existence of an amorphous to hexatic transition as the vortex density is increased is probably favored by the extreme anisotropy in this system, although the details are not yet worked out.

6. VORTEX-LATTICE DYNAMICS

The first evidence for unconventional behavior of the flux lattice in the cuprates came from measurements of the decay of magnetization.[66] An irreversibility line was found and ascribed[67] to thermally activated depinning of single flux lines. High-Q mechanical oscillator data were interpreted within a more controversial interpretation, namely a melting[4] from a low temperature ordered phase into a high temperature flux liquid.

In principle, the unconventional vortex dynamics should affect any transport measurement. The vortices, which respond both to currents[10] and thermal gradients,[68] will be the major contribution to field dependent transport parameters. Naturally, an enormous variety of transport experiments have been used[69] to probe the cuprate superconductors. While the thermodynamic consequences of the proposed phase transition are likely to remain immeasurably small, the dynamics of the vortex array could be reflected in structural probes, such as neutron scattering, through effects similar to motional narrowing. Rather than attempt a review, a narrow focus will be taken here.

7. EXPERIMENTAL TECHNIQUE

The focus here will be on the results of I-V measurements with picovolt sensitivity on two select samples. The first was a thin YBCO slab of dimensions 2mm×1mm×20 μm. The crystal was heavily twinned with the thin direction parallel to the c axis. The second sample was also a thin slab, but of BSCCO. Grown[47] by directional solidification, it was of similar size and thickness to the YBCO sample, but appreciably more lamellar. Electrical contact was made using 2000 Å-thick sputtered silver pads. To optimize the measurement of ρ_{xx}, four thin parallel stripes were used as current and voltage leads. These were evaporated onto the surface soon after cleaving or otherwise cleaning the crystal. The voltage probes were $\sim$ 2mm apart, with $\sim$ 0.5mm between them and the current leads. After annealing at 400C in 1% O_2/N_2 gold or silver wires were epoxied onto the contact pads using silver epoxy. This procedure is known to produce low contact resistances.[70] Typically,

1mm² pads produced a 10-20mΩ total resistance. A large fraction of this is presumably in the epoxy connection.

Voltages were measured using a modified BTI SQUID voltmeter. The overall geometry is sketched in Fig. 9. Stray fields at the SQUID probe were less than 50 Oe, and easily screened with a superconducting shield. To adapt to the high temperatures and fields necessary for this experiment, the input circuit consisted of a NbTi copper matrix wire soldered to a .020″ silver wire and epoxied into a rigid NbTi covered copper shield at 4.2K. The copper shield was broken 3cm above the sample, over which distance the thermal gradient from 4.2K to the sample temperature was sustained. The field gradients from the magnet were designed with the known input circuit geometry to minimize excess noise in the input circuit. With the sample at 100K, the measured low frequency voltage noise was $3\mathrm{pV}/\sqrt{Hz}$. This is consistent with the measured total input circuit resistance including contacts, giving a Johnson noise equivalent resistor of 1Ω at the sample temperature of 100K. Application of the maximum 6T field increased the noise by $\sim 50\%$. Typically, an experimental bandwidth between .01Hz and .001Hz was used, resulting in an overall subpicovolt resolution. The current leads were designed similarly to the voltage leads. In addition, the current leads were broken at 4.2K with 10kHz low pass filters. Room temperature 10MHz EMI filters were used on all leads, including heaters and thermometers, passing into the cryostat.

Data were taken at fixed temperature and field as a function of current. Voltages were measured with the standard current comparator circuit, operating at fixed gain of 10^8. The voltage scale was limited on the upper end by the 10V compliance of the SQUID feedback loop, corresponding to a maximum voltage across the sample of 0.1 μV. Sample heating limited the practical current range to 5mA. Both temperature and field sweeps were of limited use due to large variations of the SQUID null with these parameters. With a SQUID gain of 10^8, the zero varied linearly by $\sim 1\mathrm{V/K}$ with temperature. The field dependence of the null, while monotonic, did not have such a simple linear form. The rough scale of the dependence was $\sim 1\mathrm{V/T}$. The temperature was measured and controlled with a homemade $SrTiO_3$ capacitance thermometer. Stability was $\sim .005\mathrm{K}$, with no measurable field dependence up to 6T.

The data consist then of a series of isothermal I-V curves for each magnetic field studied. These were taken in two ways, giving indistinguishable results. In the first, the voltage was measured at discrete current intervals. In the second, linear or logarithmic current sweeps of up to 5000 sec period were averaged to cover the entire range. In a second type of experiment, the current was square wave chopped at .001Hz-.01Hz and a digital lockin technique was used to determine the effective resistance. This value, when current independent, agreed with the value determined from fitting the linear part of the current sweep data.

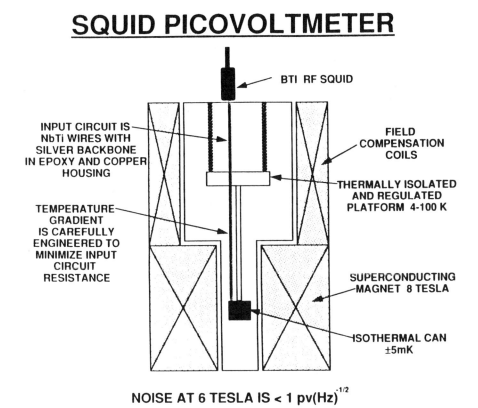

Fig. 9 Schematic of the SQUID voltmeter adapted for measurements at high temperatures and fields. Shielding and vibrations were such that the noise, $3pV/\sqrt{Hz}$, increased $\sim 50\%$ at 6 telsa applied field.

8. YBCO DATA

The YBCO crystal had a sharp zero field transition as measured in this experiment, with the resistance dropping by two orders of magnitude per 0.1K over the range $0.5 > R/R_N > 10^{-6}$. The normal state resistivity of 64 $\mu\Omega$ cm also attests to the high quality of this sample. The zero field T_c =88.033K was, however, slightly suppressed from its optimal value. While there are many fascinating predictions for the behavior of the zero field transition in terms of I-V scaling, the sharpness of the transition makes these difficult to separate from sample inhomogeneity. The focus here will be on the high field data.[71]

Data for $R/R_N > 10^{-3}$ are similar to those measured by Palstra[69] on this sample. For lower resistances, the earlier data already showed a suspiciously strong increase of the effective activation energy $U = T^2(\delta \ln R/\delta T)$ with decreasing temperature. As measured in this experiment, the early data overestimated the resistance at low temperatures, leading to a yet stronger rise in the effective U for these data and suggesting even more strongly the influence of many body effects.

Focusing for the moment on the data at 6T, consider the effect of the strong increase of U on the resistance $R \sim e^{-U/T}$. The resistance will drop sharply. With a divergent U at some temperature, the resistance will become truly zero. A good fit to this can be found in the vortex glass model,[5] in which the resistance vanishes as $R \sim (T-T_g)^{\nu(z-1)}$. The solid points in Fig. 10 show the result of such a fit. To generate the plot shown, T_g is varied to give the best straight line, whose slope is the exponent $\nu(z-1)$. This is not an especially sensitive method. A somewhat better technique will be described in connection with the BSCCO data, although the parameters derived are the same. The value $\nu(z-1) = 6.5$ is in excellent agreement

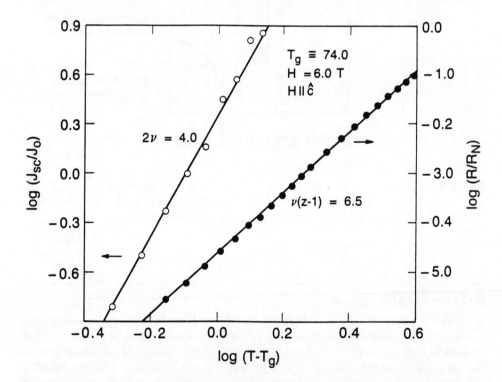

Fig. 10 Fits to the vortex glass theory for a YBCO crystal at 6 tesla. The solid symbols are the resistivity data with $R_N = 0.064\Omega$. The open symbols are the current scale defined by V/IR=2 with $J_o = 2.5 \times 10^3 A/m^2$. The values $\nu = 2$ and $z = 4.3$ are similar to thin film and BSCCO results.

with both theoretical expectation and other measurements.[5,72] Deviations from this scaling set in ~ 3K above T_g =74K. Note that no data are taken in the vortex glass phase, where the scaling of the I-V curve is expected to show qualitatively different behavior. This annoyance is a result of the limited maximum current range. Bandwidth considerations made pulsed techniques pursued by other groups impractical.

It is important to realize that this drop in resistance, corresponding to a divergent U, is unambiguous evidence for a phase transition, but not for a unique description within the vortex glass framework. Another model, based on phase transitions in structural glasses, for example, leads to an energy barrier which diverges as $R \sim e^{-U/(T-T_g)}$. This Vogel-Fulcher-type law,[73] originally used as an empirical description of structural glasses, implies time decays following the stretched exponential form $M(t) = M_o[1 + U_o/T \ln(t/t_o)]^{-1/\alpha}$. Nelson[32] has pointed out that this form is the same as would be obtained for an entangled polymer glass, homologous to his model of the entangled vortex fluid. The YBCO data can be easily fit with this expression, giving a T_g about 2-3K lower than the vortex glass fits. In addition, these fits imply values for stretched exponential decay similar to time dependent magnetization data,[74] albeit on rather different samples. In the Vogel-Fulcher law, the free energy diverges as $E \sim (T-T_g)^{-1}$. In a sense, this is a special case of the random field Ising model, in which case Fisher[75] has shown the energy to diverge as $(T-T_g)^{\nu\theta}$.

Attempts to fit these data using simple models based on thermally activated flux flow (TAFF) have been unsuccessful. The form of one model applied to these data was put forth by Griessen.[6,76] In his model, complexity is introduced in two ways. First, the sample is envisioned as a parallel summation of channels with different activation energies, distributed following a log-normal curve. In addition, each channel is shunted by a Bardeen-Stephen term. Despite these factors, simulations[77] based on this theory employing the experimental current and voltage ranges give an essentially temperature independent effective activation energy over the 3K critical regime described above, in sharp contrast to the data. In addition, the model contains a current scale, given simply by $J_{sc} \sim$ T. The difference with the data in this case is yet more remarkable, as outlined below. We have not, however, attempted simulations based on fully-3D random networks of pinning centers, perhaps a more realistic picture of the physical system within the thermal activation framework.

The width of the scaling region is strongly field dependent. Roughly, it is reduced in proportion to $T_c(H)$-T_g, where $T_c(H)$ is the mean field transition temperature. The magnitude of the scaling region $\Delta T/T_g \sim .04$ seems reasonable for a critical phenomenon. The rapid shrinking of the critical region makes it impractical to perform scaling analysis for fields below 1T, where the critical region is less than 0.5K wide. The small critical region may explain the variation of reported exponents in a variety of low field measurements.

These results, while suggestive, remain open to interpretation. More impressive are the nonlinear portions of the I-V curves. Defining a current scale I_{sc} from $V(I_{sc})/I_{sc}$ R =2 and normalizing to the value at 74.811K of $J_o = 2.5 \times 10^3$ A/m^2 gives the data shown in Fig. 10. Note that this is rather different from a critical current defined by a voltage criterion due to the factor R being included. This factor leads to the surprising decrease of the current scale with decreasing temperature shown in the plot. Also remarkable is the rate of decrease, almost a factor of 50 over a temperature interval of 3K.

Within the vortex glass scenario,[5] the current scale vanishes at T_g as $I_{sc} \sim (T-T_g)^{2\nu}$, the straight line in the figure. This can be combined with the linear response data to give $\nu = 2\pm 1$ and z=4.3 $\pm$1.5. The exponent z can also be measured from the slope of the I-V curve at T=T_g where $V \sim I^{(z+1)/2}$. Extrapolating average power law fits over different ranges gives z=3.4 $\pm$1.5 from these data. This extrapolation is difficult and fairly noisy due to the limited range of our data. A more convincing power law I-V curve at T_g was found using thick films,[72] yielding the same exponents as these single crystal results. Implicit[5] in the current data is a length scale $\xi_d = (ck_B T/\phi_o J)^{1/2}$. This is the length scale over which the current J can affect the thermal vortex distribution. For linear response, the length scale taken from the current must be longer than any intrinsic length scale. Hence, the onset of nonlinearities I_{sc} is a measure of the intrinsic correlation length. At the lowest temperature explored in the YBCO data, $\xi_d \sim 15$ μm. This implies that a cooperative region enclosing some 10^5 vortices exists here. The change in the number of cooperative vortices by a factor of $50^2 = 2500$ over 3K seems a satisfying empirical definition of a phase transition.

The two parameters R and I_{sc} can be used to collapse the I-V curves onto a universal plot. While there are different ways of expressing this, the one shown in Fig. 11 uses only parameters extracted from the data. The nonlinear resistance (V/I) is divided by the ohmic resistance R at a fixed temperature. This is plotted against the log of the current normalized to the critical current J_{sc} (T) defined above. All data sets over a range from 74.2 to 76.3K at 6T are shown. Data at other fields scaled similarly. For data sets which include V/IR > 4, R could not be measured directly, and was chosen to make the plot in Fig. 11 a smooth curve. Values chosen in this way were consistent with the current scaling shown in Fig. 10 and, by implication, the vortex glass hypothesis. Shown for comparison (open symbols) is the averaged scaling function supplied by Koch.[72] The curves were adjusted to match at V/IR =8. The simplest TAFF prediction, simply sinh (J/J_o), is also included as the solid line. As already stated, the data show significant qualitative departure from TAFF. Here, however, there is a departure from other data used to advocate a vortex glass transition. Three reasons could account for this. First, the thick film data focused on large nonlinearities. The scaling was least accurate near linear response, where this comparison is made. Alternatively, both scaling functions may be correct, with the difference attributable to the sample morphology. Of course, finally, this may indicate a failure of the vortex glass model.

A separate experiment[78] by Olsson et al. has observed critical scaling of the ac impedance. One dramatic prediction of the vortex glass model is a frequency independent phase of the impedance $\phi = (\pi/2)(2 - d + z)/z$ at the glass transition temperature T_g. The glass transition is smeared at finite frequency, but still

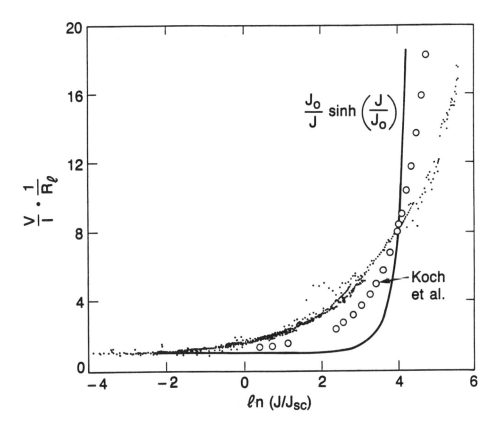

Fig. 11 Scaled I-V data over 2.5K from the 6 tesla YBCO crystal (points). For comparison the scaling function of Koch et al.[72] (open symbols) and the simplest TAFF model (line) are included. The difference from the thin film data is not completely understood.

identified through this phase. The impedance experiments found $\phi = 74°$ and thus z=5.6, somewhat higher than other reports, but generally agreeing with previous data to within quoted errors. The impedance studies were at low fields 0.55T, which seems to generally increase the measured z. This may be due either to the small width of the scaling region or to the crossover to the scaling of the zero field transition at this point.

The field dependent data can be used to construct a phase diagram as shown in Fig. 12. The H_{c2} slope is taken from magnetization data,[79] shifted to agree with the resistive zero field T_c, sharp on this scale. The solid symbols are the T_g values extracted from vortex glass fits. The general shape is similar to a wide variety of other experimental data. The solid line is a 4/3 power law, arising from the zero

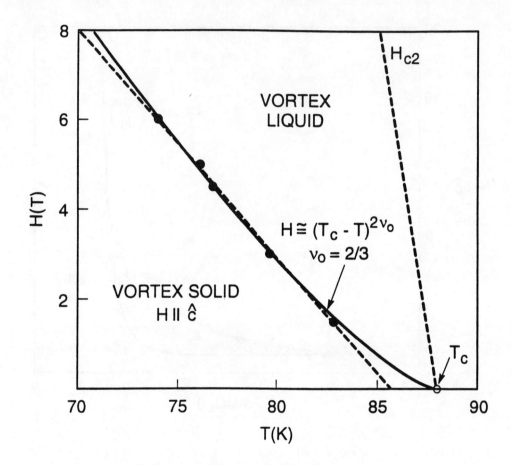

Fig. 12 YBCO vortex phase diagram. The solid symbols are the data. The H_{c2} slope is taken from Welp et al.[79] with T_c adjusted to the zero field resistance transition, which is sharp on this scale. The solid line is a 4/3 power law fit, suggested by the zero field temperature dependence of the coherence length in the fluctuation regime. The dashed line is a linear interpolation of the data.

field temperature dependence of the coherence length in the fluctuation regime. Why this should continue to hold in the high fields probed here is not known. Mean field theory would give a linear slope, as shown by the dashed line, although this doesn't match onto the zero field transition. Predictions regarding the shape are generally extremely sensitive to the low field part of the phase boundary, not probed here.

While promising, the YBCO data are subject to a variety of caveats. First, $T_g/T_c > 0.85$. The relative closeness to T_c means that the underlying superconducting properties are still evolving, affecting pinning strength and interactions. Clear interpretation is muddled by these dependencies. Further, the effects of twin boundaries, known to be important pinning centers,[80] has not been fully characterized. Recent data on nearly twin free crystals has shown remarkable sensitivity on the alignment of the current, field and twin boundary direction.[81] Discussions[82] of the dynamics in twin free YBCO have pointed to interesting possibilities, such as a two-stage melting transition with an intervening hexatic. Finally, the characteristic pinning energy in YBCO is always quite large. Studies on a twin free, cleaner system with $T_g/T_c \ll 1$ are less subject to interpretation. A first effort in this direction is the use of BSCCO.[83]

9. BSCCO DATA

The BSCCO crystal used here was not annealed. Similar crystals extracted from the melt demonstrate a Meissner transition <2K wide. Subsequent annealing, while initially sharpening the transition, tends to lower the value of T_c. Magnetic decoration, discussed earlier, of these crystals reveals extremely well ordered lattices which is evidence of large scale homogeneity. Further evidence for high crystal quality is found in the normal state resistivity of $150\mu\Omega$-cm and the sharp transition in nominally zero field. The resistance was studied over the range $10^{-8}\Omega$ to $10^{-3}\Omega$, well below the normal state resistance of $.070\Omega$. Typical maximum currents were 5mA, corresponding to current densities near the high end of 10^5 A/m^2. The assumption that the current passes uniformly through the sample will be discussed more below.

In high fields above 2T, the I-V curves were linear over the range accessible to this experiment. While the inability to identify the critical current places severe limitations on analysis of the data, several exciting points are evident. Consider the temperature dependent activation energy defined by $U = T^2 \delta \ln R/\delta T$ shown in Fig. 13 for several applied fields. The general behavior is rather different from the YBCO data. At the lowest field of 1T, the data are consistent with a simple thermal activation model down to the resolution of the SQUID.

At a higher field of 3T, however, the data show marked deviations at both low and high temperatures. The high temperature deviations have been analyzed by a variety of authors.[69,76] Generally, the temperature dependence of such quantities as the superconducting gap lead to T/T_c corrections. These have been inserted into phenomenological models to fit the resistivity in the high temperature range. Plots showing the resistance on linear or Arrhenius axes are in qualitative agreement with these corrections. Actually, these fits seem rather poor on a plot of the form

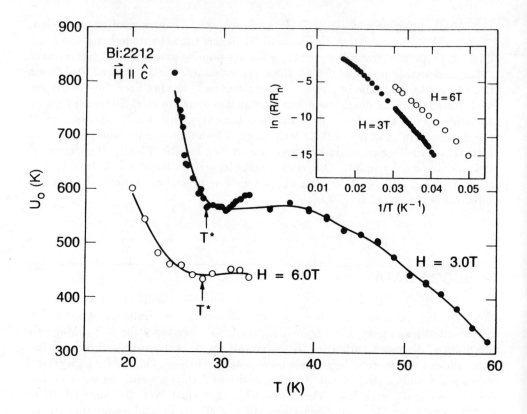

Fig. 13 Temperature dependence of the effective activation energy U_o for 3 tesla and 6 tesla magnetic fields applied parallel to the $\hat{c}$ axis in BSCCO. $U_o(T)$ is defined from the local slope of the Arrhenius plots, shown in the inset for the same magnetic fields. T^* denotes the temperature at which deviations from a constant U_o set in. The solid lines are guides to the eye.

shown here. For example, the fit of Inui et al.[76] at 3T gives a U different by a factor of 2 in this range with no noticeable temperature dependence on this scale. The focus here, however, is on the dramatic and sudden increase at low temperatures. A phenomenological crossover temperature T^* has been indicated to denote the onset of this divergence. Clearly this is the signature of new processes dominating the physics of the vortices. In contrast to YBCO, this occurs in a temperature range where the underlying superconducting parameters such as the penetration depth are completely temperature independent. Hence, the argument for a phase transition in the vortex lattice seems unavoidable.

Shown in Fig. 14 are the low temperature data at 3T plotted to emphasize the vortex glass critical behavior. The increase in the apparent activation energy is easily fit within the vortex glass framework.[5,71,72] In this case the resistance, which vanishes as $(T-T_g)^{\nu(z-1)}$ transforms into a straight line plot of $dT/d\ln\rho$ vs. T with intercept T_g and slope $1/\nu\,(z-1)$. These data give $T_g = 20.2$K. The resultant exponent $(\nu(z-1))^{-1} \sim 0.15$ is both field independent as shown in the inset and similar to the value obtained for YBCO. The possible deviation at 2T is interesting. As in the YBCO case, the larger exponent may be an artifact of the shrinking of the critical region. In BSCCO, a field dependent dimensional crossover is expected,[6] as discussed below. It may be that the exponents are changing in response to this crossover.

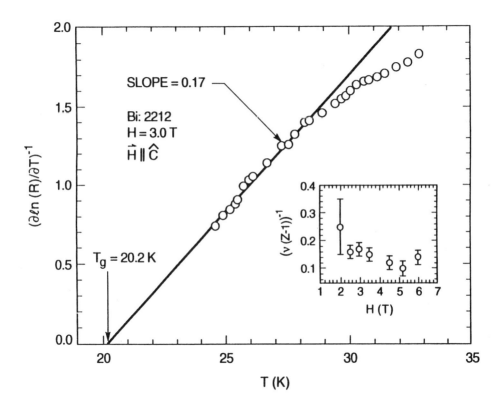

Fig. 14 Plot of the inverse logarithmic derivative of the resistance for the 3 tesla BSCCO data. The solid line represents a fit to the vortex glass model, which is linear in this plot. The intercept is $T_g = 20.2$ K and the slope is $[\nu(z-1)]^{-1} = 0.17$. Deviations from the vortex glass model are observed above 28K, indicating the limits of the critical regime (similar to T^*). Inset: the critical exponents for different fields. The value and error bars for the 2 tesla data are strongly affected by the shrinking of the critical region at low fields.

The data of Fig. 14 may be understood in terms of a temperature dependent dimensional crossover. At high temperatures, while the vortex lattice may be well ordered in the planes, there is little correlation between planes. This type of argument has separately been used to interpret[27,45] μSR data. However, as the temperature is lowered, the two-dimensional correlations grow. Despite these increasing correlations, the dynamics will remain dominated by thermal activation, as the effective barriers produced by two-dimensional correlations are likely to be small in comparison to pinning energies. The high temperature resistance will be dominated by thermal activation over the pinning barriers. The system will eventually cross over to three-dimensional behavior as the three-dimensional vortex glass correlation length ξ_{3D} grows. This should occur when $(\xi_{3D}/a_o)^2 = (\lambda/d)$ where a_o is the vortex lattice spacing and d is the interplane spacing.[5] These data are insufficient to allow numerical estimates for ξ_{3D}. Nonetheless, as a crude estimate this crossover will occur near $T^* \sim 2T_g$. This is because estimates for T^* and T_g arise from the same arguments, with only numerical differences. We postulate that this T^* is the same as the phenomenological crossover temperature T^* mentioned above. Below T^* the dynamics will contain an important contribution from critical fluctuations associated with the incipient three-dimensional vortex glass transition which ultimately occurs at T_g.

A SQUID magnetometer was used to measure[84] the field dependent magnetization of a BSCCO sample from the same batch up to 8kG, focusing on the behavior below 30K. Unfortunately, the data were contaminated by a large H/T term, which began to dominate the superconducting signal at temperatures comparable to 30K. Despite this, unusual time dependence, involving large, discrete, jumps in the magnetization, began to evolve in the vicinity of T_g. The relationship to the resistivity data is tantalizing, perhaps signaling the onset of significant cooperative phenomena.

Because of the limited current range available, the nonlinear response was not observed in the high field data. While nonlinear I-V curves have been reported on thin films,[85] these should be viewed in light of the huge anisotropy in this system, coupled with the persistent low film quality. The absence of nonlinearities, while making direct comparison to the vortex glass theory more difficult, provides a clue to the crossover from thermal activation to critical behavior. The three dimensional vortex glass correlation length $\xi_{3D} = 81.6\mu m\ (T/J)^{1/2}$ defines when the measuring current affects the thermal vortex distribution. For linear response,[5] the effective length defined by the measurement current must be longer than the important correlation lengths in the system. As the correlation length diverges, the current scale will decrease correspondingly to zero. In an anisotropic system, the correlations are, naturally, anisotropic. Generally, these can be written as $\xi_{ab}\xi_c = (\xi_{3D})^2$, with $\xi_{ab}/\xi_c = \Gamma$. In the extreme two-dimensional case this can be reduced to $\xi_{2D}d = (\xi_{3D})^2$. In the high temperature data, it is this two-dimensional length which is appropriate to the description of these data.

Due to the extreme anisotropy in this material, the current is likely to flow in only a few conducting planes. Indeed, the anisotropy ρ_c/ρ_a has been shown to increase dramatically[86] with decreasing temperature in the mixed state. At T=50K and with a maximum current of 5mA, $\xi_{2D} < 1\mu m$, assuming the current is confined to a single layer. A more uniformly distributed current only serves to raise this upper bound. At 6T, this upper bound on ξ_{2D} is more than $50a_o$, consistent with a well ordered lattice in the planes as seen in the decoration images described here. These long range two-dimensional correlations support the hypothesis that T^* is the point at which three-dimensional correlations start to become important. Unfortunately, the high field data do not extend far enough into the critical regime to be sensitive to the nonlinearities induced by the slowly diverging ξ_{3D}.

Shown in Fig. 15 is the phase diagram for T^* and T_g extracted from these data. Contrary to YBCO, the scaling region $(T^* - T_g)$ becomes extremely broad in high fields. This width, approximately linear in field, is much greater than would be expected for a critical region, with $\Delta T/T_g \sim 1$ at 6T. The YBCO data, with $\Delta T/T_g \sim .05$ are much closer to a traditional range of critical phenomena. The reason underlying this wide scaling region in BSCCO is unknown.

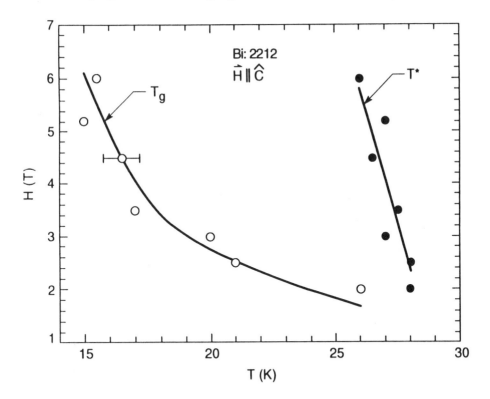

Fig. 15 Magnetic phase diagram. T_g is the vortex glass transition and T^* marks the crossover to a low temperature regime dominated by three dimensional cooperative effects. The solid lines are guides to the eye. The apparent collapse of the critical region near 2T may be related to the 2D-to-3D crossover field.

The extreme anisotropy of BSCCO leads to a vortex lattice which should become increasingly two-dimensional at high fields[6] and temperatures. In recent theoretical and experimental work, the layered structure is modeled as coupled superconducting sheets. The competition between interplane and intraplane effects can lead to various phase diagrams for the vortex lattice. The scale $B_{2D} = 0.3 \to 1$ T for this dimensional crossover depends on the details of the model. As an example, magnetic coupling between vortices in adjacent layers leads to two-dimensional behavior for fields greater than $B_{2D} = \phi_o/\Gamma d^2 \sim 0.5T$. In two dimensions, the vortex glass[5] transition is at T=0. The critical behavior in this case leads to a resistance which vanishes as $R \sim e^{-(T_o/T)^x}$ with x>1. This dramatically different critical behavior leads to the conclusion that the observed low temperature divergence is a three-dimensional effect. Indeed, due to the finite interplane coupling, the vortex lattice must become effectively three-dimensional at sufficiently low temperatures for all applied fields.

The critical region seems to collapse at finite field $\sim 1T$. In addition, the I-V curves below this field develop some, as yet not completely analyzed, nonlinear component. It would be tempting to associate this with B_{2D}. These data, however, can only be distinctly analyzed in terms of 3D vortex glass behavior at higher fields and low temperatures. Higher current data is necessary to understand in detail the dynamics of the high temperature, high field regime. Several authors have already analyzed data[87] in terms of 2D vortices. Further analysis is required of the data at yet lower fields to understand whether the character of the transition changes.

10. FINAL CONCLUSIONS

High resolution I-V curves in both YBCO and BSCCO show evidence of new collective phenomena in the mixed state of the high temperature superconductors. The YBCO data is clear evidence for the formation of a vortex glass phase. The BSCCO data has been preliminarily analyzed in the same form, but shows new complexities. Dimensionality seems to be playing an important role in this case.

The mixed state of extreme type-II superconductors has provided a rare opportunity to explore, in a well controlled way, the details of a melting transition in the presense of disorder. The Abrikosov crystal, stabilized by long range interactions has a lower critical dimension d=4. Experiments such as those presented here, in the physical dimension d=3, close to the critical value, show extreme sensitivity to both disorder and dimensionality. Disorder arises from the random pinning of the flux lines; dimensionality enters in layered materials where weak interplane coupling can lead to significant regions of quasi two-dimensional behavior. In both layered[88] materials, and high temperature superconductors, the role of thermal fluctuations is enhanced, expanding the range where they can affect the vortex dynamics. Finally, it is important to understand the type of order probed in a given experiment as there are suggestions that the melting transition in clean samples can proceed in two stages, as first the positional and then the orientational order is destroyed with increasing temperature. There is clearly much work still to be done.

The author would like to thank David Nelson, Daniel Fisher, Matthew Fisher, Peter Littlewood, Chandra Varma, Gerald Dolan, Tom Palstra and Betram Batlogg for many insights. The fruitful collaborations with Christian Bolle, Gerald Dolan, David Grier, Aharon Kapitulnik, Raynien Kwo, David Mitzi and Lynn Schneemeyer were an essential part of this work.

REFERENCES

1. A. A. Abrikosov, Zh. Eksp. Teor. Fiz. 32, 1442 (1957) [Sov. Phys. JETP 5, 1174 (1957)]. For a general review, see A. L. Fetter and P. C. Hohenberg in "Superconductivity," edited by R. D. Parks (Dekker, New York, 1969) vol. 2.
2. For a general review of experiments in the cuprates, see B. Batlogg, Physica B 169, 7 (1991); Proc. of Los Alamos Symposium-1989, K. S. Bedell, D. Coffey, D. E. Meltzer, D. Pines and J. R. Scrieffer, editors (Addison-Wesley, Redwood City, 1990) and references therein.
3. See, for example B. Oh, K. Char, A. D. Kent, M. Naito, M. R. Beasley, T. H. Geballe, R. H. Hammond, A. Kapitulnik and J. M. Gaybeal, Phys. Rev. B36, 7861 (1988); T. R. Chien, T. W. Jing, N. P. Ong and Z. Z. Wang, Phys. Rev. Lett. 66, 3075 (1991).
4. P. L. Gammel, L. F. Schneemeyer, J. V. Waszczak and D. J. Bishop, Phys. Rev. Lett. 61, 1666 (1988). D. R. Nelson and H. S. Seung, Phys. Rev. B39, 9153 (1989); A. Houghton, R. A. Pelcovits and A. Sudbo, Phys. Rev. Phys. Rev. B40, 6763 (1989); M. A. Moore, Phys. Rev. B39, 136 (1989).
5. M. P. A. Fisher, Phys. Rev. Lett. 62, 1415 (1989). D. S. Fisher, M. P. A. Fisher and D. A. Huse, Phys. Rev. B43, 130 (1991); D. S. Fisher, in this volume.
6. M. V. Feigelman, V. B. Geshkenbein, A. I. Larkin and V. M. Vinokur, Phys. Rev. Lett. 62, 3093 (1989); L. I. Glazman and A. E. Koshelev, Phys. Rev. B43, 2835 (1991); M. V. Feigelman, V. B. Geshkenbein and A. I. Larkin, Physica C167, 177 (1987).
7. For an introduction to fluctuation effects on the heat capacity signature of YBCO, see M. B. Salamon, S. E. Inderhees, J. P. Rice, B. G. Pazol, D. M. Ginsberg, Phys. Rev. B38, 885 (1968).
8. B. A. Huberman and S. Doniach, Phys. Rev. Lett. 43, 950 (1979). D. S. Fisher, Phys. Rev. B22, 1190 (1980). P. L. Gammel, A. F. Hebard and D. J. Bishop, Phys. Rev. Lett. 60, 144 (1988).
9. G. J. Dolan, G. V. Chandrashekhar, T. R. Dinger, C. Field and F. Hotzberg, Phys. Rev. Lett. 62, 827 (1989).
10. Y. B. Kim and M. J. Stephen in "Superconductivity," edited by R. D. Parks (Dekker, New York, 1969) vol. 2.

11. This, and many basic properties of type-II superconductors are reviewed in, "Introduction to Superconductivity" M. Tinkham (Krieger, Malabar, FL, 1985).
12. For a detailed analysis of approaches to flux pinning, see M. Tinkham and C. J. Lobb in "Solid State Physics," edited by H. Ehrenreich and D. Turnbull (Academic, New York, 1990) vol. 42; C. J. Lobb, Phys. Rev. **B36**, 3930 (1987); M. R. Beasley, R. Labusch and W. W. Webb, Phys. Rev. **181**, 682 (1969).
13. A. I. Larkin and Yu. N. Ovchinnikov, J. Low Temp. Phys. **34**, 409 (1979).
14. For a theoretical introduction, see J. Toner, Phys. Rev. Lett. **66**, 2523 (1991); E. M. Chudnovsky, Phys. Rev. **B40**, 11355 (1989); E.M. Chudnovsky, Phys. Rev. **B43**, 7831 (1991); M. C. Marchetti and D. R. Nelson, Phys. Rev. **B41**, 1910 (1990).
15. W. Y. Shih, C. Ebner and D. Stroud, Phys. Rev. **B30**, 134 (1984).
16. V. G. Kogan, Phys. Rev. **B24**, 1572 (1981); A. Houghton, R. A. Pelcovits and A. Sudbo, Phys. Rev. **B42**, 906 (1990); A. Sudbo and E. H. Brandt, Phys. Rev. **B43**, 10482 (1991).
17. L. J. Campbell, M. M. Doria and V. G. Kogan, Phys. Rev. **B38**, 2439 (1988); V. G. Kogan and L. J. Campbell, Phys. Rev. Lett. **62**, 1552 (1989).
18. D. G. Grier, C. A. Murray, C. A. Bolle, P. L. Gammel and D. J. Bishop, Phys. Rev. Lett. **66**, 2270 (1991); C. A. Murray, P. L. Gammel and D. J. Bishop, Phys. Rev. Lett. **64**, 2312 (1990).
19. H. F. Hess, R. B. Robinson, R. C. Dynes, J. M. Valles and J. V. Waszczak, Phys. Rev. Lett. **62**, 214 (1989). H. F. Hess, R. B. Robinson and J. V. Waszczak, Phys. Rev. Lett. **64**, 2711 (1990). R. Berthe, U. Hartmann and C. Heiden, Appl. Phys. Lett. **57**, 2351 (1990).
20. H. F. Hess, private communication. Also, H. F. Hess, proceedings M^2 SHTSC-III, Kanazawa, Japan, July 22-26, 1991.
21. J. Mannhart and R. P. Heubner, Jap. J. Appl. Phys. **26-3**, 1565 (1987).
22. N. Moser, M. R. Koblischka, H. Kronmuller, B. Gegenheimer and H. Theuss, Physica C **159**, 117 (1989).
23. This has been studied in the intermediate state of type-I superconductors and is reviewed by J. D. Livingston and W. DeSorbo in "Superconductivity," edited by R. D. Parks (Dekker, New York, 1969) vol. 2; for YBCO films, see Y. Yokoyama, Y. Hasumi, H. Obara, Y. Suzuki, T. Katayama, S. Gotoh and N. Koshizuka, Jap. J. Appl. Phys. **30**, L714 (1991).
24. E. M. Forgan, D. McK. Paul, H. A. Mook, P. A. Timmins, H. Keller, S. Sutton and J. S. Abell, Nature **343**, 735 (1990). For earlier studies, see D. Christen, K. Kerchner and S. T. Sekula, Phys. Rev. **B21**, 102 (1980).
25. M. Mehring in "Earlier and Recent Aspects of Superconductivity," Springer Series in Solid State Science vol. 90, edited by G. Bednorz and K. A. Müller; J. T. Moonen, D. Reefman, J. C. Jol, H. B. Brom, T. Zetterer, D. Hahn, H. H. Otto and K. F. Renk, Proc. M^2 SHTSC-III, July 22-26, Kanazawa, Japan.
26. F. Borsa, A. Rigamonti, M. Corti, J. Ziolo, O.-B. Hyun and D. Torgison, submitted to PRL; C. Slichter, proc. TIS^2, July 15-21, Kyoto, Japan; M. Takigawa,

P. C. Hammel, R. H. Heffner, Z. Fisk, J. D. Thompson and M. Maley, Physica C **162-164**, 175 (1989).
27. D. R. Harshman, R. N. Kleiman, M. Inui, G. P. Espinoza, D. B. Mitzi, A. Kapitulnik, T. Pfiz and D. Ll. Williams, preprint; D. R. Harshman, A. T. Fiory and R. J. Cava, Phys. Rev. Lett. **66**, 3313 (1991); E. H. Brandt, Phys. Rev. Lett. **66**, 3213 (1991); see also ref. 45.
28. H. Trauble and U. Essman, J. Appl. Phys. **25**, 273 (1968). N. V. Sarma, Phil. Mag. **17**, 1233 (1968).
29. R. P. Huebner, "Magnetic Flux Structures in Superconductors" (Springer-Verlag, Berlin, 1979).
30. M. D. Kirk, J. Nogami, A. A. Baski, D. B. Mitzi, A. Kapitulnik, T. H. Geballe and C. F. Quate, Science 242, 1673 (1988).
31. M. Gurvitch, J. M. Valles, A. M. Cucolo, R. C. Dynes, J. P. Garno, L. F. Schneemeyer and J. V. Waszczak, Phys. Rev. Lett. **63**, 1008 (1989).
32. D. R. Nelson in "Current Problems in Statistical Mechanics," Cornell U., Sept 23&24, 1991. See also ref. 3.
33. L. Lombardo, M. Ng, J. S. Urbach, D. B. Mitzi and A. Kapitulnik, preprint.
34. F. F. Preparata and M. L. Shamos in "Computational Geometry, an Introduction" (Springer-Verlag, New York, 1985); S. Fortune, Algorithmica 2, 153 (1987).
35. P. L. Gammel, D. J. Bishop, D. J. Dolan, J. R. Kwo, C. A. Murray, L. F. Schneemeyer and J. V. Waszczak, Phys. Rev. Lett. **59**, 2592 (1987).
36. N. P. Ong, T. W. Jing, Z. Z. Wang, J. Clayhold, S. J. Hagen and T. R. Chien, in "Strong Correlations and Superconductivity," edited by H. Fukuyama, S. Maekawa and A. P. Malozemoff, Springer Series in Solid State Sciences vol. 89 (Springer- Verlag, Heidelberg, 1989); K. E. Gray, Mod. Phys. Lett. B**2**, 1125 (1988); P. J. M. van Bentum, H. F. C. Hoevers, H. van Kempen, L. E. C. van de Leemput, M. J. M. F. de Nivelle, L. W. M. Schreurs, R. J. M. Smokers and P. A. A. Teunissen, Physica C **153-155**, 1718 (1988); Phys. Rev. B**36**, 843 (1987).
37. D. R. Nelson, in this volume.
38. D. A. Huse, preprint.
39. C. C. Grimes and G. Adams, Phys. Rev. Lett. **42**, 795 (1979).
40. S. Martin, A. T. Fiory, R. M. Fleming, L. F. Schneemeyer and J. V. Waszczak, Phys. Rev. Lett. **60**, 2194 (1989).
41. G. J. Dolan, F. Holtzberg, C. Field and T. R. Dinger, Phys. Rev. Lett. **62**, 2184 (1989); L. Ya. Yinnikov, I. V. Grigoriera, L. A. Gurevich and Yu. A. Ossipyan in "High Temperature Superconductivity from Russia" (World Scientific, London, 1989) edited by A. I. Larkin and N. V. Zavaritsky.
42. For c-axis, see D. E. Farrel, J. P. Rice, D. M. Ginsberg and J. Z. Liu, Phys. Rev. Lett. **64**, 1573 (1990). For ab plane, see D. M. Ginzberg, J. P. Rice, T. A. Friedman and J. M. Mochel, proc. second ISSP International Symposium on the Physics and Chemistry of Oxide Superconductors, Jan. 16-18, 1991, Tokyo, Japan; A. Umezawa, G. W. Crabtree, U. Welp, W. K. Kwok, K. G. Vandervoort and J. Z. Liu, Phys. Rev. B**42**, 8744 (1990).

43. L. J. Campbell and B. I. Ivlev, unpublished. In the initial simulations, the period of the buckling could not be defined uniquely. A more recent theory may determine this period as well as the other features.
44. J. R. Clem, Phys. Rev. B**43**, 7837 (1991); J. R. Clem, Proc. ICTPS'90, Rio de Janeiro, Brazil, April 29-May 5, 1990 (World Scientific, Singapore, 1990).
45. This particular limit was noted by D. S. Fisher, private communication. See also M. Inui and D.R. Harshman, proc. M^2 SHTSC-III, July 22-26, 1991 Kanazawa, Japan.
46. S. Mitra, J. H. Cho, W. C. Lee, D. C. Johnston and V. G. Kogan, Phys. Rev. B**40**, 2674 (1989).
47. D. B. Mitzi, L. W. Lombardo, A. Kapitulnik, S. S. Laderman and R. D. Jacowitz, Phys. Rev. B**41**, 6564 (1990).
48. F. Parmigiani, Z. X. Shen, D. B. Mitzi, I. Lindau, W. E. Spicer and A. Kapitulnik, Phys. Rev. B**43**, 3085 (1991).
49. S. Martin, private communication. Measurements on [2-2-0-1] materials also suggest that Γ should change by less than a factor of three: S. Martin, A. T. Fiory, R. M. Fleming, L. F. Schneemeyer and J. V. Waszczak, Phys. Rev. B**41**, 846 (1990).
50. B. I. Halperin and D. R. Nelson, Phys. Rev. Lett. **41**, 121 (1978). D. R. Nelson and B. I. Halperin, Phys. Rev. B**19**, 2457 (1979).
51. For colloids, see C. A. Murray, W. O. Sprenger and R. A. Wenk, Phys. Rev. B**42**, 688 (1990) and references therein; the general 2D case is discussed in K.J. Strandburg, Rev. Mod. Phys. **60**, 161 (1988).
52. For a review, see D. R. Nelson in "Phase Transitions and Critical Phenomena" vol. 7, edited by C. Domb and J. L. Lebowitz (Academic, London, 1985).
53. D. R. Nelson, M. Rubinstein and F. Spaepen, Phil. Mag. A**46**, 105 (1982).
54. R. Seshadri and R. M. Westervelt, Phys. Rev. Lett. **66**, 2774 (1991).
55. R. K. Kalia and P. Vashishta, J. Phys. C **14**, L643 (1981); V. M. Bedanov, G. V. Gadiyak and Yu. E. Lozovik, Phys. Lett. **92A**, 400 (1982).
56. D. R. Nelson and P. Ledousal, Phys. Rev. B**42**, 10113 (1990).
57. C. A. Bolle, P. L. Gammel, D. G. Grier, C. A. Murray, D. J. Bishop, D. B. Mitzi and A. Kapitulnik, Phys. Rev. Lett. **60**, 112 (1991).
58. P. H. Kes, J. Aarts, V. M. Vinokur and W. van der Beek, Phys. Rev. Lett. **64**, 1063 (1990).
59. D. E. Ferell, S. Bonham, J. Foster, Y. C. Chang, P. Z. Siang, K. G. Vandervoort, D. J. Lam and V. E. Kogan, Phys. Rev. Lett. **63**, 782 (1989).
60. J. Schelten, G. Lippmann and H. Ullmaier, J. Low Temp. Phys. **14**, 231 (1974).
61. R. N. Kleiman, C. Broholm, G. Aeppli, E. Bucher, N. Stuchelli, D. J. Bishop, K. N. Clausen, B. Howard, K. Mortensen and J. S. Pedersen, preprint.
62. A. I. Buzdin and A. Yu. Simonov, JETP Lett. **51**, 191 (1990). A. M. Grishin, A. Yu. Martynovich and S. V. Yampolskii, Sov. Phys. JETP **70**, 1089 (1990). V. G. Kogan, N. Nakagawa and S. L. Theiman, Phys. Rev. B**42**, 2631 (1990).
63. L. Daemon and L. J. Campbell, Bull. Am. Phys. Soc. **30**, 1064 (1991).
64. B. I. Ivlev, N. B. Kopnin and M. M. Salomaa, Phys. Rev. B**43**, 2631 (1991); B. I. Ivlev and N. B. Kopnin, Phys. Rev. B**44**, 2747 (1991).

65. C. M. Varma, private communication.
66. For a review, see A. P. Malozemoff in "Strong Correlations and Superconductivity," edited by H. Fukuyama, S. Maekawa and A. P. Malozemoff, Springer Series in Solid State Sciences vol. 89 (Springer-Verlag, Heidelberg, 1989).
67. A. P. Malozemoff, L. Krusin-Elbaum, D. C. Cronemeyer, Y. Yeshurun and F. Holtzberg, Phys. Rev. B**38**, 6490 (1988). Y. Yeshurun and A. P. Malozemoff, Phys. Rev. Lett. **60**, 2202 (1988).
68. For cuprates, see T. T. M. Palstra, B. Batlogg, L. F. Schneemeyer and J. V. Waszczak, Phys. Rev. Lett. **64**, 3090 (1990); M. Zeh, H.-C. Ri, F. Kober, R. P. Huebner, A. V. Ustinov, J. Mannhart, R. Gross and A. Gupta, Phys. Rev. Lett. **64**, 3195 (1990).
69. For a review, see T. T. M. Palstra, B. Batlogg, L. F. Schneemeyer and J. V. Waszczak, Phys. Rev. Lett. **61**, 1662 (1988); Appl. Phys. Lett. **54**, 764 (1989); Phys. Rev. B**41**, 6621 (1990).
70. R. B. van Dover, private communication
71. P. L. Gammel, L. F. Schneemeyer and D. J. Bishop, Phys. Rev. Lett. **66**, 953 (1991).
72. For films, see R. H. Koch, V. Foglietti, W. J. Gallagher, G. Koren, A. Gupta and M. P. A. Fisher, Phys. Rev. Lett. **63**, 1511 (1989); R. H. Koch, V. Foglietti and M. P. A. Fisher, Phys. Rev. Lett. **64**, 2586 (1990); L. Civale, T. K. Worthington and A. Gupta, Phys. Rev. B**43**, 5425 (1991). For ceramics, see T. K. Worthinton, E. Olsson, C. S. Nichols, T. M. Shaw and D. R. Clarke, Phys. Rev. B**43**, 10538.
73. J. P. Sethna, Europhys. Lett. **6**, 529 (1988); S. Brawer, "Relaxation in Viscous Liquids and Glasses" (American Ceramic Society, Ohio, 1985).
74. See A. P. Malozemoff and M. P. A. Fisher, Phys. Rev. B**42**, 6784 (1990) for a discussion of magnetization relaxation within the vortex glass framework.
75. D. S. Fisher, Phys. Rev. Lett. **56**, 416 (1986).
76. Various formulations can be found in R. Griessen, Phys. Rev. Lett. **64**, 1674 (1990); S. N. Coppersmith, M. Inui and P. B. Littlewood, Phys. Rev. Lett. **64**, 2585 (1990); M. Tinkham, Phys. Rev. Lett. **61**, 1658 (1988); M. Inui, P. B. Littlewood and S. N. Coppersmith, Phys. Rev. Lett. **63**, 2421 (1989); C. W. Hagen and R. Griessen, Phys. Rev. Lett. **62**, 2857 (1989).
77. For comparison, values B=6.0T, $U_o^a st = 85 meV$, $\gamma = 1.4$, $A = 3 \times 10^{-10}$, $S_o = 4 \times 10^{-7}$ and $\rho_n = 6.4 \times 10^{-7} \Omega - m$ were used in the simulations.
78. H. K. Olsson, R. H. Koch, W. Eidelloth and R. P. Robertazzi, Phys. Rev. Lett. **66**, 2261 (1991).
79. U. Welp, W. K. Kwok, G. W. Crabtree, K. G. Vandervoort and J. Z. Liu, Phys. Rev. Lett. **62**, 1908 (1989)
80. L. J. Swartzendruber, A. Roitburd, D. L. Kaiser, F. W. Gayle and L. H. Bennett, Phys. Rev. Lett. **64**, 483 (1990); W. K. Kwok, U. Welp, G. W. Crabtree, K. G. Vandervoort, R. Hulscher and J. Z. Liu, Phys. Rev. Lett. **64**, 966 (1990); E. M. Gyorgy, R. B. van Dover, L. F. Schneemeyer, A. E. White, H. M. O'Bryan, R. J. Felder, J. V. Waszczak, W. W. Rhobes and F. Hellman, Appl. Phys. Lett. **56**, 2465 (1990).

81. G. W. Crabtree, in this volume.
82. T. K. Worthington, F. Holtzberg and C. A. Field, Cryogenics **30**, 417 (1990); D. E. Farell, J. P. Rice and D. M. Ginzberg, to be published in Phys. Rev. Lett. (1991).
83. H. Safar, P. L. Gammel, D. J. Bishop, D. B. Mitzi and A. Kapitulnik, preprint.
84. B. Batlogg, private communication. Also compare ref. 27.
85. E. Zeldov, N. M. Amer, G. Koren and A. Gupta, Appl. Phys. Lett. **56**, 1700 (1990).
86. G. Briceno, M. F. Crommie and A. Zettl, Phys. Rev. Lett. **66**, 2164 (1991).
87. C. J. van der Beek and P. H. Kes, Phys. Rev. **B43**, 13032 (1991).
88. For a recent comparison of layered materials and the cuprates, see W. R. White, A. Kapitulnik and M. R. Beasley, Phys. Rev. Lett. **66**, 2826 (1991).

DISCUSSION

A. P. Malozemoff: Could you tell me how you defined your crossover?

P. L. Gammel: λ is the penetration depth.

As to the crossover, I wish I could define it in any sensible way.

Here is what Daniel Fisher told me when I asked him, "What would you guess about where the dimensional crossover occurs?" His claim was that, in the vortex glass picture, the type of scaling arguments you get for the dimensional crossover are really the same type of arguments you're using for the vortex glass transition itself. So one might expect that the dimensional crossover occurred at some constant times T_g, which was the guess I showed for dimensional crossover. The three-dimensional vortex glass correlation length in terms of lattice constants was going to have to be λ/d so this would be some fairly large number of 3D lattice spacings. That's what I know about the crossover.

S. Doniach: Just a general question: Are you saying the irreversibility line is the same as T_g, or do you have any opinion about that?

P. L. Gammel: I'm not really an expert in magnetization, but I suppose the irreversibility line, measured in the limit of zero field-step and zero frequency would be the same as T_g. You're doing the same experiment. After all, this is all of transport, right? So if you're measuring in the limit of zero current, and you're doing a zero-frequency experiment, you should find a transition at T_g.

A. Kapitulnik: On the same crystals, or similar crystals from the same crucible, we can put the irreversibility line on the same plot.

P. L. Gammel: Yes. I have done it too. The two data sets are quite similar. Then there's the question of nonlinearity – because generally, when you measure the irreversibility line, I think the current scale is a bit higher, although the voltage scales are very low.

D. S. Fisher: The effective time scales are longer.

P. L. Gammel: The effective time scale is longer. Well, it's not that much longer because we're looking at transport in millihertz range.

A. Kapitulnik: It's 0.05 hertz for the irreversibility line.

P. L. Gammel: Yes. So this transport data we're taking at one millihertz is comparable to magnetization.

D. S. Fisher: There's an effective frequency which you can get out from your z and ν, and you can ask how does that frequency compare with the time scale of the irreversibility-line data.

M. P. Maley: In the BSCCO data you and Tom Palstra both see this rapid decrease of the activation energy measured from the resistance curves at a low temperature. Now there are two mechanisms that have been proposed for this. One is the 3D-to-2D crossover and the other one, as we heard from David Nelson's talk, is a thermal smearing model. Now, how are these two things related, and how do you distinguish between, say, a rapid decrease with temperature of activation energies from those two mechanisms?

P. L. Gammel: I'm not really sure. Quite honestly, I haven't thought about the high-temperature range. I think Masahiko would be the person to talk to. Well, you at one point tried to fit the high-temperature part.

M. Inui: Right.

P. L. Gammel: I'd say plotted in this way, the high temperature fits are not all that successful.

D.S. Fisher: The nice thing about some of the BSCCO data is all of these temperatures are sufficiently far below T_c that all the bare parameters like temperature, penetration length, and other things are only changing very slightly, so it's hard to see how you would get the rather sharp temperature changes from the thermal smearing of pinning. But you'd certainly get some.

P. L. Gammel: Yes. I think Marty was asking about the part of the data I didn't discuss.

M. P. Maley: I was actually asking about the low-end temperature . . .

P. L. Gammel: This is the part of the data which I know has been tried to fit with these various models that you were talking about, and as to the fit . . . I'm not really all that sure about, but, you know, you can talk to people who have done some modeling, here. I really haven't done much modeling in this high-temperature range, and this low-temperature range is what I tried to explain.

D. R. Nelson: Just a comment on Daniel's remark on the constancy of these parameters in BSCCO. If you have planar objects that are pinning, or linear objects, then there's still a lot of thermal action in the effective binding energy. There's no phase transition in this little model, but there's still a lot of thermal changes.

A. P. Malozemoff: I'd like to comment on what Marty Maley just mentioned about the thermal renormalization of core pinning, also going back to David's talk. I think there's always a question of to what extent it plays a role in these kinds of effects. My understanding, going back to Vinokur's earlier work with Feigel'man [Feigel'man and Vinokur, PRB **41** 8986 (1990)], is that, in fact, the thermal renormalization effect has a boundary which crosses the usual irreversibility line. In other words, roughly, B goes as T^2. I want to mention some very interesting evidence for this phenomenon that Krusin-Elbaum has obtained at IBM. Studying the irreversibility line, the anomaly in the lossy part of the susceptibility, she observed that there was a break in the curve and that at low fields it dropped down. And this was in a field range in yttrium barium copper oxide crystals which was well above H_{c1} and actually fitted quite nicely the kind of predictions that Vinokur has made for the thermal renormalization phenomenon. But because it's a phenomenon which crosses the irreversibility line, I think it's not plausible to imagine that the phenomena which you see at very different fields are related to that particular effect.

D. R. Nelson: If the vortex glass phase is there, does it matter whether it's orientationally ordered or not? Is that just a spectator, the orientational order, or would there be different exponents or something?

D. S. Fisher: In the limit of very weak pinning, there should be a long positional correlation length – and then all vortex glass stuff only takes place on longer distances than that – and then there'll be a much longer orientational correlation length which could get to be enormously long. So in that limit, where we're ignoring disclination, we're assuming the orientational correlation length is infinite. It's quite plausible that the vortex glass transition which would happen on top of that – in some sense, from a hexatic fluid to a hexatic vortex glass – could well have different critical behavior than a normal one.

Daniel S. Fisher
Harvard University
Cambridge, MA 02138

Phase Transitions and Transport in Anisotropic Superconductors with Large Thermal Fluctuations

The current understanding of transport and phase transitions in the cuprate superconductors—particularly $YBCO$ and $BSCCO$—is reviewed. New results are presented on the two-dimensional regimes and 2D–3D crossover in the strongly anisotropic case of $BSCCO$. The emphasis is on pinning and vortex glass behavior.

1. INTRODUCTION

Fluctuation effects in conventional superconductors such as broadening of phase transitions and flux creep tend to be very small primarily because of the large coherence lengths. Thus mean field theory, with only small fluctuation corrections, usually provides an adequate description of these systems. Regimes in which fluctuation effects cause qualitatively different physics are very difficult to study as they typically occur in very small regions of the phase diagram or, in transport, require measuring extremely small voltages.

In striking contrast, in the high temperature cuprate superconductors a combination of factors—short coherence lengths, anisotropy and higher temperatures—make fluctuation effects many orders of magnitude larger.[1] For very anisotropic materials such as $Bi_2Sr_2CaCu_2O_{8+\delta}$ (BSCCO) the parameter which controls the strength of thermal fluctuations becomes of order 10^{-1} or larger!

Questions of the existence of true superconductivity—i.e. vanishing linear resistivity—which are of primarily academic interest in conventional superconductors, thus become crucial for understanding electrical transport in BSCCO.[1] Indeed, the contrast between the behavior of BSCCO and conventional superconductors in this regard is arguably the most dramatic difference between cuprate and conventional superconductivity. Although in the Meissner phase, fluctuations in the form of vortex loops have long been known to cause *non-linear* dissipation while leaving the linear conductivity infinite,[2] in the mixed state, the interplay between thermal fluctuations and pinning of vortex lines complicates matters sufficiently that the existence of true superconductivity has remained an open—although usually unstated—question.

The conventional Anderson-Kim[3] picture of vortex creep is that thermal fluctuations cause vortex "bundles" to fluctuate over the barriers caused by the pinning potential, thus giving rise to a *finite* linear resistivity of the form

$$\rho \sim \rho_f e^{-U(T)/T} \tag{1.1}$$

with ρ_f the flux-flow resistivity in the absence of pinning and $U(T)$ a typical barrier height. For conventional materials, U/T is so large (except extremely close to T_c) that the linear resistivity becomes immeasurably small. In the cuprates, however, one would expect measurable resistivity considerably below the mean field transition.

Experiments on BSCCO[4] have indeed found non-zero resistivity in magnetic fields of a few tesla down to temperatures of order 20K, a factor of four below the nominal "transition" temperature $T_{c2}(H)$ in these relatively small magnetic fields! In contrast, in $YBa_2Cu_3O_7$ ($YBCO$) the linear resistivity appears to *vanish* at a field dependent transition only ten or so degrees below the upper critical field line $T_{c2}(H)$.[5,6] In this paper our current understanding of fluctuation effects and transport in the cuprate superconductors is summarized and some of the important open issues are discussed.

The fundamental phenomenon which underlies the behavior of the mixed state of superconductors, in the presence of random impurities or other small-scale inhomogeneities that act to pin vortices, is the existence of an underlying *equilibrium phase transition* from a vortex liquid (which has properties qualitatively like that of a "normal," albeit somewhat peculiar, metal) to a *vortex glass phase*[1,7] with vanishing linear resistivity and a form of off-diagonal long-range order analogous to that in spin glasses. At this point, the detailed properties of this vortex glass phase have neither been studied in detail experimentally nor been fully analyzed theoretically; the essential roles of nonequilibrium effects and hysteresis make such studies especially difficult (although some progress has been made on analogous problems in spin glasses).[8] Thus in this paper we will give only a brief review of properties of the vortex glass phase to set the stage for the later discussion of the approach to this phase from above and phase transitions into it.

In $YBCO$, the experimental and theoretical picture is becoming quite clear:[1] as we shall see there is rather compelling evidence, from a series of transport measurements, for a vortex glass phase transition,[5,6] although much theoretical and experimental work remains to be done. After discussing experiments on the weakly anisotropic case of $YBCO$, we will focus most of our attention on the much more strongly anisotropic $BSCCO$ in which two-dimensional effects play an essential role. [Note that the thallium-based compounds also appear to be strongly anisotropic,[9] although they are currently not as well characterized as $BSCCO$; the situation for the lower-T_c lanthanum cuprates is less clear at this point.] It will be argued that the experimental evidence for both $YBCO$ and $BSCCO$ suggests relatively strong pinning for fields in the tesla range. Thus the phase transitions and much of the transport will be dominated by the physics of the vortex-glass phase transition and the approach to it.

1.1 Outline

In the remainder of the Introduction, we review Ginzburg-Landau theory for an anisotropic superconductor, introduce the appropriate parameters, and then discuss the qualitative phase diagram. In Section II, the conjectured properties of the vortex glass phase are reviewed, while in Section III the transition from vortex fluid to vortex glass is discussed. Section IV contains a quantitative discussion of phase transitions and interlayer couplings in *clean* anisotropic layered superconductors. In Section V, the effects of pinning in strongly anisotropic materials is discussed and the properties of an incipient two-dimensional vortex glass regime introduced. The crossover to three-dimensional vortex glass behavior at lower temperatures is also analyzed. Finally, Section VI contains brief conclusions and issues for future experiments. Some of the elastic properties of vortex lattices used here and previously[1] are relegated to an Appendix.

Most of Sections I, II, and III review the results of reference 1. Section IV and Appendix A are expansions of results used in ref. 1, while most of Section V is new.

1.2 Ginzburg–Landau Theory Parameters

In order to specify notation and to discuss the scale of the relevant parameters, we briefly review Ginzburg–Landau theory for an anisotropic, uniaxial superconductor. We denote the unique axis perpendicular to the layers by z and in-plane directions either by $\perp$ or without subscripts. Currents in the plane decay with a penetration length $\lambda \equiv \lambda_\perp$ while those normal to the planes decay with a much longer penetration length λ_z. The ratio

$$\gamma \equiv \frac{\lambda_\perp}{\lambda_z} \qquad (1.2)$$

is the square root of the mass anisotropy $m_\perp/m_z$ in an effective mass approximation.

For $YBCO$, $\gamma \approx 0.18$ [10] and at low temperatures $\lambda_\perp \approx 1400$ Å with $\lambda_\perp(T)$ apparently having a roughly BCS-like temperature dependence.[11] $BSCCO$ is much more strongly anisotropic with $\gamma \lesssim 1/50$ [12] although it is probably true that there is *no* experimental *lower bound* on γ. The low temperature penetration length in $BSCCO$ is reported to be $\lambda_\perp > 2500$ Å from muon spin resonance,[13] but this value is perhaps an overestimate as we discuss below. The in-plane coherence length is somewhat ill-defined in the presence of fluctuations but, from fits to the initial drop of the resistivity in a field, a $\xi \equiv \xi_\perp$ of order 15Å at low temperatures can be inferred, yielding for $YBCO$

$$\kappa \equiv \frac{\lambda_\perp}{\xi_\perp} \approx 100 , \qquad (1.3)$$

which is possibly an overestimate.[14] Nevertheless, it is extremely type II. We will thus restrict consideration to the strongly type-II limit.

From Ginzburg–Landau theory, one expects the coherence length in the z-direction to be

$$\xi_z = \gamma \xi_\perp \qquad (1.4)$$

yielding, even for $YBCO$, ξ_z less than the interlayer spacing d at low temperatures, so that it is even more poorly defined than $\xi_\perp$. Nevertheless, the combination $\gamma \xi_\perp$ will enter in various quantities such as energies at longer length scales, so it is still an important parameter.

As we shall see, H_{c2} is not well-defined in the presence of strong thermal fluctuations. However the characteristic field

$$H_X = \frac{\Phi_0}{2\pi \xi_\perp^2} \qquad (1.5)$$

will set the scale of the field dependence.[1] For simplicity, we only consider magnetic fields in the z-direction throughout this paper.

In a strongly type-II material, the fluctuations in the magnetic field are usually less important than fluctuations in the phase of the order parameter. The dimensionless quantity which controls the strength of the phase fluctuations is[1]

$$\delta(T) \equiv \frac{16\pi^2 T \lambda_\perp^2}{\Phi_0^2 \gamma \xi_\perp} = \frac{\kappa \lambda_\perp / \gamma}{2 \times 10^6 \text{Å}} (T/100K) \qquad (1.6)$$

so that planar anisotropy ($\gamma < 1$), large κ and high temperatures all increase δ, making it orders of magnitude larger than in conventional materials. For $YBCO$ with a zero field superconducting transition of $T_{c0} \approx 90K$,

$$\delta \approx 0.3(T/T_c)(\kappa/100) \qquad (1.7)$$

for $T \ll T_c$ and

$$\delta \approx \left(\frac{T_c - T}{T_c}\right)^{-\frac{1}{2}} \left(\frac{\kappa}{500}\right) \qquad (1.8)$$

in the Ginzburg-Landau regime relatively near T_c.

The standard Ginzburg criterion for the importance of critical fluctuations in zero field is that $\delta \sim 4$. However a better criterion, at least in the superconducting phase, can be obtained by considering the behavior of δ in the XY critical region in which order parameter (but not magnetic field) fluctuations are important. In this region δ goes to *a universal* constant[1]

$$\delta \approx \delta_c \approx 0.8 \; ; \tag{1.9}$$

from Eq (1.8) this yields a width of the critical region of order 10% for $\kappa = 100$ in $YBCO$.[1,15] In this XY critical regime,

$$\xi \sim (T_c - T)^{-\nu} \tag{1.10}$$

with $\nu \approx 2/3$ and

$$\lambda_\perp \sim (T_c - T)^{-\nu/2} \tag{1.11}$$

so that the system becomes less type II.[1,16]

In the strongly anisotropic case of $BSCCO$, the two-dimensional nature of the fluctuations becomes important. In this case, a related 2D fluctuation parameter can be defined by

$$\delta_2 \equiv \frac{16\pi^2 T \lambda_\perp^2}{\Phi_0^2 d} = \left(\frac{T}{100K}\right)\left(\frac{\lambda_\perp}{1500\text{Å}}\right)^2 \left(\frac{15\text{Å}}{d}\right) 0.08 \tag{1.12}$$

which is just proportional to the ratio of the temperature to the 2D superfluid density, $\rho_{s2} = d\rho_{s\perp}$ yielding

$$\delta_2 = \frac{T}{\pi \rho_{s2}} = \frac{T}{\Theta_2} \tag{1.13}$$

where we have defined a characteristic temperature

$$\Theta_2(T) \equiv \frac{\Phi_0^2 d}{16\pi^2 \lambda_\perp^2}. \tag{1.14}$$

In the layered limit, the $\frac{1}{2}(\nabla_z \varphi - A_z)^2$ term in the Ginzburg-Landau free energy must be replaced by

$$\left[1 - \cos\left[\varphi_{\zeta+1} - \varphi_\zeta - \int_{\zeta d}^{(\zeta+1)d} A_z(z) dz\right]\right]/d^2 \tag{1.15}$$

with φ_ζ the phase in layer ζ. The energy of a point (or "pancake") vortex in an isolated layer is

$$E_v \approx \Theta_2 \ln\left(\frac{\lambda^2}{d\xi}\right) \tag{1.16}$$

where the long length scale cutoff is the transverse penetration length $\Lambda_\perp = 2\lambda^2/d$. We will return to the consequences of the strong anisotropy in later sections.

1.3 Phase Diagram

For strongly type-II materials the lower critical field H_{c1} is much smaller than the characteristic field H_X (= H_{c2} in mean field theory). For the cuprates, $H_{c1}^z \approx 300G$. Although there are many interesting effects for fields near H_{c1}, we will primarily consider fields $H \gg H_{c1}$. The mean field phase boundary at $H_{c2} = H_X = \Phi_0/2\pi\xi^2$ becomes only a crossover line in the presence of fluctuations near which the resistivity starts to decrease and the system starts to become diamagnetic. Below H_{c2}, the system can usefully be described in terms of vortices. The typical spacing between the vortices in the mixed state is

$$a_v = \left(\frac{\Phi_0}{B}\right)^{\frac{1}{2}} \quad (1.17)$$

with $B \approx H$ for $H \gg H_{c1}$ or in a film geometry.

With fluctuations, the phase boundary at H_{c2} gets replaced by either the melting transition of the Abrikosov lattice at $T_M(H)$ [or $H_M(T)$] in a perfectly clean material,[17] or the vortex glass phase transition at $T_G(H)$ [or $H_G(T)$] for a dirty material.[1] Between H_G (or H_M) and H_X, the system is a vortex fluid [possibly with a hexatic bond orientationally ordered fluid phase existing, in a clean system, between the isotropic fluid and Abrikosov lattice phases].[17,18] A schematic phase diagram is shown in Fig. 1 for a material such as $YBCO$.

In the next sections we discuss the vortex glass phase which replaces the lattice in the presence of random pinning and then return in Section IV to a more quantitative analysis.

2. VORTEX-GLASS PHASE

In this section we summarize the theoretical picture of a three-dimensional vortex glass phase.[1,7,19] For simplicity, we will consider only the extreme type II limit with penetration length λ much longer than any other characteristic length; the effects of current inhomogeneities on longer scales and sample size dependent effects have not yet been worked out. In addition, for this qualitative discussion, we will ignore the effects of crystalline anisotropy.

Since the pioneering work of Larkin and Ovchinnikov,[20] it has been known that random impurities destroy Abrikosov vortex lattices on length scales longer than a characteristic length scale L_p which depends on the elastic moduli and strength and correlations of the pinning potential felt by the vortices. Yet what happens on scales larger than L_p has remained an open question, although the Anderson-Kim[3] theory implicitly presumes that the system is fluid-like with only finite, although large, energy barriers for motion of bundles of vortices of size $\sim L_p$.[21] Because the vortices are extended objects, however, there is another more interesting possibility:

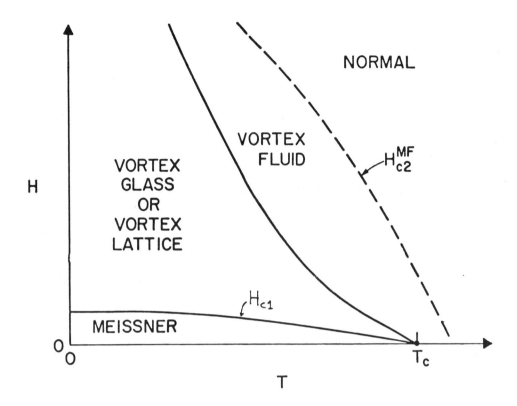

Fig. 1 Schematic phase diagram of a three–dimensional type–II superconductor with strong thermal fluctuations. The crossover from the normal regime to the vortex–fluid regime is not a sharp phase transition and occurs near the location of the mean–field transition H_{c2}^{MF} or, more generally at small fields $H_x = \frac{I_o}{2\pi\xi^2}$, with ξ the *real* correlation length. Without random pinning, a vortex–lattice phase is present and the vortex–fluid phase also intrudes at H_{c1} between the Meissner and vortex–lattice phases. This intrusion is not shown here since, for parameters appropriate to the cuprate superconductors, it occurs over too narrow an interval of H to be seen in this figure. With random pinning we expect the vortex–lattice phase to be replaced with a vortex–glass phase.

on length scales larger than the lattice correlation length L_p (essentially the size of well-ordered microcrystallites of vortices), the system is in a vortex glass phase in which the vortices fluctuate about a set of preferred positions determined— for a given sample—by the competition between the intervortex repulsion and the random potential. Associated with this vortex glass phase is off-diagonal long-range order in the pair order parameter $\langle\psi(r)\rangle^{1,7}$ in a fixed gauge $\nabla \cdot A = 0$. The *phase of* the pair order parameter will vary from one point in space to another as determined by the preferred positions of the vortices, although its magnitude will be non-zero almost everywhere; this order is thus analogous to the magnetic Edwards-Anderson order of spins in an X–Y spin glass.[22]

Note that the physically observable *static* correlations of the vortex positions in a vortex glass phase will exhibit only rather subtle differences from those of a vortex fluid phase (e.g. in both phases the thermal averaged vortex density will be inhomogeneous due to the random potential), however the *dynamic correlations* will be markedly different.

2.1 Nonlinear Current-Voltage Response

The dynamics of the vortices are reflected in the non-linear response of the vortex glass to an applied transport current density J. The current will exert a force on all the vortices and the system will try to lower its free energy by moving vortices across the current. In a vortex glass phase, this can only be accomplished by collective excitations which are either virtual vortex loops (see Fig. 2) or some kind of vortex bundle,[1] which have free energy typically growing as a power of their linear length scale L:

$$F_L \sim \Upsilon L^\theta. \tag{2.1}$$

Stability of the vortex glass phase requires $\theta > 0$. We will return to the possibility of $\theta < 0$ in Section VI. In D dimensions, a rigorous upper bound on θ for a closely related model is $\theta \leq (D-2)/2$ yielding $D \leq 1/2$ in 3D.[23]

The current yields a negative contribution to the free energy of an excitation of the form

$$F_J \sim -JL^\kappa \tag{2.2}$$

where L^κ is the effective projected area of the excitation normal to the applied current, which for vortex loops is just L^2; thus $\kappa \geq 2$.[21] For sufficiently large scale excitations, with

$$L > L_J \sim J^{-1/(\kappa-\theta)}, \tag{2.3}$$

$F_L - F_J$ becomes negative and the vortices can move to lower the free energy. This type of motion has to surmount barriers of typical size

$$B_L \sim L^\psi. \tag{2.4}$$

A vortex excitation of size $\sim L_J$ going over such a barrier is the basic unit of dissipation: On scales larger than L_J, sections of size L_J can move approximately independently causing macroscopic dissipation associated with the resulting vortex current perpendicular to J. The non-linear electric field will thus be of the form

$$E \sim \rho_f J \, e^{-(J_T/J)^\mu} \tag{2.5}$$

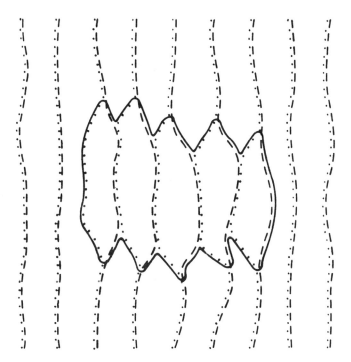

Fig. 2 Schematic of a vortex–loop–like excitation in a vortex glass phase. The dashed lines are the initial configuration, the dotted lines are the final configuration, the solid line is the relative vortex loop, and the current is normal to the paper.

where we have written the factor associated with activation over the barriers $\exp[-U(J)/T]$ in the form

$$\frac{U(J)}{T} = \left(\frac{J_T}{J}\right)^\mu \qquad (2.6)$$

so that the characteristic current scale of non-linear dissipation is

$$J_T \sim \left(\frac{1}{T}\right)^{\frac{1}{\mu}} \qquad (2.7)$$

with the exponent

$$\mu = \psi/(\kappa - \theta) \leq 1. \qquad (2.8)$$

The expression Eq (2.5) with $\mu = 1$ has the same form as that due to vortex loop nucleation in the Meissner phase and the upper bound on μ in the vortex glass arises from consideration of similar vortex-loop like excitations.[2] We thus see that the vortex glass phase is, like the Meissner phase, a *linear superconductor*

($E/J \to 0$ as $J \to 0$) but does *not have a non-zero critical current*. As the current is increased, however, there will become a point at which the forces from the pinning can no longer resist the force from the current; above this current J_F, the vortices will break free and flow without barriers and the differential resistivity will rapidly increase towards ρ_f. [Note however, that the geometry of the vortex flow in this regime has not been investigated in detail; it may well be very inhomogeneous. There are many interesting issues associated with the initiation of this strongly non-linear flow which we will not delve into here, but see e.g. Ref. 24.] The current J_F is determined by the pinning *forces* on the vortices, while J_T is related to pinning *energies*.

In the mixed state of conventional superconductors, $J_F \ll J_T$ so that even just below J_F, the thermal creep of vortices is very slow. In this case, J_F appears as an almost sharp critical current density. For the cuprate superconductors, on the other hand, this limit only obtains at very low temperatures. In most of the mixed state (as well as very near $H_{c2}(T)$ in conventional materials), J_F is not too much smaller than J_T so that the I-V curves are much smoother and there is no well-defined critical current. These two different behaviors are sketched in Figure 3. In the Meissner phase, the analogous J_F/J_T is just proportional to the dimensionless thermal fluctuation parameter $\delta(T)$, Eq. (1.6), in the presence of pinning. In the vortex-glass phase $T/U = (J_F/J_T)^\mu$ is again a dimensionless measure of the strength of thermal fluctuations.

A direct consequence of the two current scales is an altered form of the non-equilibrium decay of supercurrents, and concomitantly, decay of magnetization.[1] A useful approximate form for such current decay which correctly reproduces both the conventional and the strong fluctuation regime is

$$J(t) \approx \frac{J_F}{[1 + (T/U) \ln(t/t_0)]^{\frac{1}{\mu}}} \qquad (2.9)$$

preceded by rapid decay at times $t < t_0 \sim 10^{-9}$sec to the "critical current" J_F, followed by much slower creep over the larger and larger barriers as J decreases further. If $U/T \gg 1$, the current will never decay much below J_F, and the long time limit of Eq (2.9), which comes from Eq (2.5), never obtains; i.e. it is largely irrelevant whether or not the linear resistivity really vanishes. This is the conventional Anderson-Kim flux creep regime.[3] For $U/T \lesssim 20$, on the other hand, J will decay well below J_F on laboratory time scales so that the form of the low-current I-V curve becomes relevant. Magnetization decay experiments on $YBCO$[25] show a temperature dependence of $dJ/d\ln t$ with a *maximum* far below T_c consistent with the form Eq. (2.9) with roughly temperature independent J_F and U, although it is unclear whether this is the correct interpretation of these data. More detailed work on this problem—including careful analysis of size-dependent effects—is clearly needed.

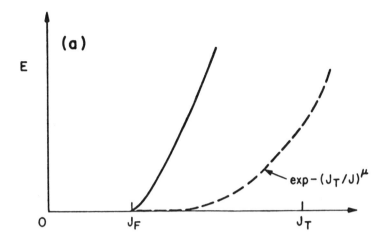
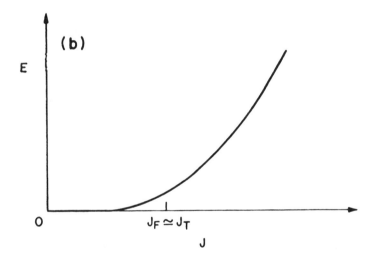

Fig. 3 Current-voltage characteristics in a superconducting (vortex–glass or Meissner) phase with J the current density and E the electric field in the sample. (a) In the conventional weak fluctuation regime there is a fairly sharp onset of nonactivated vortex flow and dissipation at current density J_F. The dissipation for $J \ll J_F$ is due to activated nucleation and growth of vortex-loop or bundle excitations and vanishes as $\exp[-(J_T/J)^\mu]$ for $J \to 0$. In this regime $J_T \gg J_F$, so the dissipation is still very small just below J_F. (b) In the thermal-fluctuation–dominated regime, J_F becomes of order J_T so the onset of dissipation is smoother.

2.2 Linear AC Response

The linear ac response of a vortex glass is rather subtle due to the intrinsic inhomogeneities. Much of the linear response is dominated by regions with anomalously low energy, and thus thermally active, vortex excitations. The vortex glass phase does *not* have a superfluid density as it is conventionally defined in terms of the dc response to a finite twist in the phase of the order parameter across a large sample. It nevertheless *does* have a superfluid density in the zero-frequency limit of an ac linear response measurement. Indeed at low frequencies, the ac conductivity has the form

$$\sigma(\omega) \approx \frac{\rho_s}{-i\omega + 0^+} + O\left[\frac{1}{|\ln(\omega t_o)|^{\theta/\psi}}\right], \qquad (2.10)$$

the (weakly) dominant first term having the same form as in a Meissner phase but with large corrections due to the inhomogeneously distributed active excitations.

The penetration of a very weak additional ac magnetic field decays exponentially with a penetration length

$$\lambda_G(\omega) \sim |\omega\sigma(\omega)|^{-\frac{1}{2}}. \qquad (2.11)$$

In the limit of zero frequency $\lambda_G(0) \sim \rho_s^{-1/2}$ is *finite*. Thus an *extra infinitesimal field will penetrate only a finite distance* even in the limit of zero frequency (as in the Meissner phase) even though the thermodynamic dB/dH is non-zero in this mixed state! This apparent paradox is resolved by considering the amplitude dependence of the ac field penetration: in the limit of zero frequency, the amplitude must be made smaller and smaller to stay in the linear regime; any fixed amplitude field *will* penetrate in the limit of zero frequency.

There are various subtle effects with interesting experimental consequences caused by the inhomogeneity of the penetrating field,[1] but we will not discuss these here.

2.3 Nonequilibrium Effects

In the vortex glass phase, there are relaxation processes at arbitrarily low frequencies. These give rise to hysteresis and pronounced history dependence, even at macroscopic time scales (for example, the oft-observed dependence on whether a sample is cooled before or after a magnetic field is turned on). These will eventually be important for a proper understanding of the vortex glass phase, but have not yet been investigated in any detail. At this point, we can only refer the interested reader to papers on analogous effects in spin glasses.[8]

Because of these and other subtleties (such as the effects of transport current inhomogeneities caused by finite penetration length effects) it is more fruitful to first understand the evidence for the *existence* of a vortex glass phase obtained from measurements near the phase transition before returning in the future to properties of the vortex glass phase itself. The remainder of this paper is thus primarily concerned with this and related issues.

3. VORTEX–GLASS PHASE TRANSITION

In this section we briefly review the predicted behavior near to a putative phase transition at a temperature $T_G(H)$ from a vortex fluid phase with finite linear dc conductivity to a superconducting vortex glass phase.[1,5] We will assume that this transition is second order; in any case, there does not seem to be evidence for discontinuities experimentally, although one might well expect them in the limit of very weak pinning; we return to this question later.

By analogy with conventional continuous phase transitions, we expect a characteristic length ξ_G diverging near T_G as

$$\xi_G \sim |T - T_G|^{-\nu} \tag{3.1}$$

and a characteristic rate for relaxation of critical fluctuations

$$\Omega_G \sim \xi_G^{-z} \tag{3.2}$$

with the dynamic exponent $z > 1$. From scaling of the free energy and the vector potential, we expect that the characteristic current scale near the transition, J_X, will scale as

$$J_X \sim \frac{cT}{\Phi_0} \xi_G^{-2} \tag{3.3}$$

with c the speed of light (in general dimensions, D, $J_X \sim \xi_G^{1-D}$).

3.1 Nonlinear I-V Curves

The nonlinear I-V curves should exhibit a universal scaling form

$$\frac{E}{J} \sim |T - T_G|^{(z-1)\nu} \Theta_\pm \left(\frac{J}{J_X}\right). \tag{3.4}$$

Above T_G, the scaling function $\Theta_+(j) \to$ constant as $j \to 0$ yielding a linear resistivity vanishing as

$$\rho \sim |T - T_G|^{(z-1)\nu} \tag{3.5}$$

Close to the transition, the electric field will become appreciably nonlinear for currents above $J_X(T)$ and at larger currents will be similar to that at T_G. At T_G,

$$E \sim J^{(z+1)\nu} \tag{3.6}$$

for a wide range of currents limited at high currents by corrections to scaling. *Below* T_G, the electric field will be exponentially small for $J < J_X$ and

$$\Theta_-(j) \sim e^{-c/j^\mu} \tag{3.7}$$

as $j \to 0$. Thus as $T \to T_G$ from below, the two current scales in the vortex glass phase, J_T and J_F, will both scale as J_X so that the I-V curves will not show a well-defined critical current, as shown in Fig. 3b.

The first evidence for scaling of I-V curves (and hence for a vortex glass phase transition) was found by Koch *et al.*[5] in $YBCO$ films which are quite strongly disordered, probably including very small-scale microtwins. Good scaling was found for fields from 10 to 40 kG. The *upper* limit of the power law I-V curve found at the apparent transition gives, via Eq (3.3), a length scale above which pinning-dominated vortex glass behavior occurs. Putting in anisotropy factors,[26] this yields a length scale of order the inter-vortex spacing—suggesting that the vortex lattice order probably only extends out to a few vortex spacings. The existence of a wide critical regime with $|T - T_G(H)| \sim T_{c0} - T_G(H)$, also suggests that the pinning is strong and lattice correlations short-range; very weak pinning will tend to lead to a sharp drop in the the resistivity.

The exponents (and scaling functions) found by Koch *et al.*[5] are roughly independent of field: $\nu = 1.8$ and $z = 5$ corresponding to linear conductivity diverging as $|T - T_G|^{-7}$. Note that $z \geq 4$ is expected from analysis of the transition in high dimensions,[27] and a relatively large ν is expected from the proximity of the lower critical dimension to three. Numerically, an exponent $\nu = 1.2 \pm 0.5$ has been found for the related "gauge glass" model.[28] This is somewhat smaller than, but not inconsistent with, the experimental value of ν.

Experiments on good single crystals of $YBCO$ by Worthington *et al.*[29] found behavior, over a similar range of currents, strikingly different from Koch *et al.*[5] data on films. This is presumably due to the weaker pinning in single crystals. However, Gammel *et al.*[6] have recently measured $YBCO$ crystals down to very low current densities, and again found scaling behavior—particularly linear resistivity vanishing as a power of $(T - T_G)$ over $4\frac{1}{2}$ decades in ρ with an exponent $(z - 1)\nu = 6.5 \pm 1.5$ consistent with Koch *et al.*[5] The width of the critical region is of order a few K with $T_c - T_G(60kG) \sim 15K$.[6] The scaling function Θ_+ has been measured[6] but appears to be somewhat different in shape from the Koch *et al.*[5] scaling function, although how significant these differences are is not clear. Unfortunately, due to technical differences, it has not been possible so far to measure I-V curves for $YBCO$ films over the full range from Koch *et al.*[5] data to that probed by Gammel *et al.*[6] This would certainly be very useful to see if indeed scaling (perhaps with some corrections) works over the whole range. If so, it would provide very compelling evidence for the existence of a true vortex glass transition. However, even the present situation with similar critical behavior found in very different current ranges for two rather different systems—$YBCO$ crystals and much dirtier $YBCO$ films—already strongly suggests that a true phase transition underlies the onset of superconductivity in the mixed state.

Several comments are in order at this stage. Firstly, one should note that the regime very near to T_G in which *magnetic* fluctuations become important (which occurs for $T < T_G$ when $\lambda_G(0) \sim \xi_G$)[16] has only barely been reached in the experiments so far. Thus one might argue that a vortex glass phase transition only strictly exists in the extreme type-II limit $\lambda \to \infty$. This would, by itself, already be very interesting and certainly different from the conventional view. At this point, all that can be usefully said is that there does not seem to be any compelling reason to believe that the existence of a three-dimensional vortex glass phase should be sensitive to the range of the intervortex interactions, even though this might in fact be the case.

Secondly, it is useful, though perhaps premature considering the state of the data and sample preparation, to explore possible reasons for apparent differences in the scaling functions between $YBCO$ crystals and films. The simplest possible reasons are experimental: differences in the regimes of sensitivity to the scaling function between the experiments, as discussed in Ref. 6, and different fitting procedures. Two alternative possibilities involve the differences in the pinning: in crystals, the dominant pinning is presumably oxygen vacancies, while in the films microtwins probably dominate. The latter are spatially extended in the $\hat{c}$ direction, which might cause a reduction in the asymptotic critical region[30] with an intermediate regime controlled by a Bose-glass-like transition.[31] A study of this possibility is underway.[30]

Alternatively, the very weakness of the pinning in crystals may yield two characteristic lengths: a short scale L_p above which vortex lattice translational order is destroyed, and a much longer length L_h up to which bond-orientational (hexatic) order persists.[18] If the scales probed by the current are less than L_h, the orientational order may well affect the I-V curves. In neither of these scenarios, however, is it clear why the exponents should remain similar, although the size of the errors are large enough that the measured exponents are quite uncertain. Clearly, this issue requires further investigation.

Finally, we comment on so-called "thermally assisted flux flow" (TAFF) theories.[32] These essentially amount to parametrizing the I-V data with a temperature- and current-dependent barrier $U(J,T)$, and then assuming a simple form for U without much justification. As Gammel *et al.*[6] convincingly show, however, such forms *cannot fit* their data without a current scale which is very strongly temperature dependent near to the apparent T_G. *The existence of this rapidly decreasing current scale is strong evidence for a diverging length scale and hence collective effects.* Nevertheless, more direct measurements of a diverging correlation length would certainly be useful.

3.2 Frequency-Dependent Conductivity

Another useful probe of the behavior near to a putative vortex glass transition is the *ac* linear conductivity. This will have the scaling form

$$\sigma(\omega) \sim |T - T_G|^{-(z-1)\nu} S_{\pm}\left(\frac{\omega}{\Omega_G}\right) \qquad (3.8)$$

with $\Omega_G \sim |T - T_G|^{z\nu}$. *Above* T_G, at low frequencies

$$\sigma(\omega) \sim (T - T_G)^{-(z-1)\nu} \left(1 + C_+ i\omega/\Omega_G + ...\right) \qquad (3.9)$$

so that the phase angle

$$\phi_\sigma = \tan^{-1}\frac{Im\sigma}{Re\sigma} \to 0,$$

while *below* T_G,

$$\sigma(\omega) \sim \frac{(T_G - T)^\nu}{-i\omega + 0^+} \qquad (3.10)$$

at low frequencies so that[33]

$$\phi_\sigma \approx \frac{\pi}{2} - \frac{C}{(\ln \Omega_G/\omega)^x}.$$

At T_G, the phase angle will be frequency independent at low frequencies with

$$\phi_\sigma^c \approx \frac{\pi}{2}\left(\frac{z-1}{z}\right) \qquad (3.11)$$

so that with large z it is quite close to $\pi/2$. At T_G,

$$\sigma(\omega) \sim e^{i\phi_\sigma^c} \omega^{-(z-1)/z}. \qquad (3.12)$$

Plotting $\phi_\sigma(T,\omega)$ as a function of T for various ω should yield a crossing point at T_G. Measurements over a wide range of frequency have not yet been performed, but should be done to test the scaling theory. The exponents z and ν and T_G itself can be extracted *independently* from both *I-V* curves and $\sigma(\omega)$ and the scaling laws thereby checked.

3.3 Thermodynamics

Near to a second order vortex glass transition, which we have assumed, there will, in principle, be singularities in thermodynamic quantities, in particular the specific heat C_v and magnetic susceptibility $\chi = dM/dH$. Both of these should have $|T_G - T|^{-\alpha}$ singularities, but for two reasons these are probably unobservably small. Firstly, if the exponent $\nu > 1$, hyperscaling implies $\alpha = 2 - 3\nu < -1$, thus only the derivative of C_v or χ will exhibit a cusp. Higher derivatives (for example $\frac{d\chi}{dH}$ or $\frac{d^2\chi}{dH^2}$) might be measured by low frequency harmonic generation, but the overall amplitude of the singularities will be small because the entropy associated with the vortex glass ordering is of order a fraction of a boltzmann per "unit cell" of size γa_v^3. Nevertheless, high enough sensitivity measurements might be possible, but it is not clear how useful the results would be except to convince those sceptics who doubt the role of collective effects in the onset of superconductivity in the mixed state of a dirty material.

4. PHASE TRANSITIONS IN CLEAN SUPERCONDUCTORS

In order to make quantitative comparisons between theory and experiment—in particular to understand phase diagrams—it is necessary to first consider phase transitions in ideal clean systems in various regimes. These will roughly set the scale of the phase diagrams in the presence of pinning, especially if it is not too strong. For strongly anisotropic layered superconductors such as $BSCCO$, much of the physics is controlled by two-dimensional effects in individual layers. We thus first consider phase transitions in individual layers and then turn to the effects of interlayer couplings.

4.1 Individual Layers in Zero Magnetic Field

In the absence of an applied field, an *isolated* layer of thickness d only has a true phase transition in the approximation that the transverse penetration length $\Lambda_\perp \equiv 2\lambda^2/d$ becomes infinite. This Kosterlitz-Thouless transition occurs at a temperature T_{KT} determined by[34]

$$\Lambda_\perp \left(T_{KT} \right) = \frac{\Phi_0^2}{16\pi^2 T_{KT}} \tag{4.1}$$

which, even for temperatures of order 100K, corresponds to $\Lambda_\perp \approx 200\mu m$. The criterion Eq (4.1) is equivalent to the fluctuation parameter $\delta_2 = 1/2$. For $d = 15 \text{Å}$ and $T_{KT} \approx 85K$ appropriate to $BSCCO$, the Kosterlitz-Thouless criterion corresponds to a penetration length $\lambda(T_{KT}) \approx 4000\text{Å}$. Even without taking into account the fluctuation renormalizations which will increase λ from its Ginzburg-Landau value,[34] one would expect that with a reported low temperature λ of $> 2500\text{Å}$,[13] the transition temperature would thus be strongly suppressed from its mean field value, T_{c0}. The experimental temperature dependence of the resistivity has been fitted to a 2D form[35] but with a Ginzburg-Landau amplitude $\lambda_0 = \lambda(T)\left(\frac{T_{c0}-T}{T_{c0}}\right)^{1/2}$ of 600Å, more consistent with a smaller low temperature penetration length. We return to this point later.

Vortex pairs in an isolated layer can be created with finite energy cost at separations larger than $\Lambda_\perp$ due to screening by magnetic fields out of the film. But the activation energy for an isolated vortex just below the "transition"

$$E_v \approx 2T_{KT} \ln \Lambda_\perp/\xi \tag{4.2}$$

is large and yields a very small Boltzmann factor of

$$e^{-E_v/T} \approx (\Lambda_\perp/\xi)^{-2} \sim O(10^{-8} \sim 10^{-10}) \tag{4.3}$$

depending on the Ginzburg-Landau coherence length at the transition. (See below.) As temperature is decreased, this factor becomes still smaller and soon immeasurable; thus for practical purposes, the transition at T_{KT} in an isolated layer is almost sharp, and the resistivity below T_{KT} still linear but exponentially small.

In an *infinite stack* of uncoupled layers, on the other hand, the magnetic fields of a vortex in one layer are screened by the other layers. This causes the interaction between vortices in one layer to remain logarithmic at all distances but reduces the strength of the interaction (from the short-distance form for an isolated layer) by a factor of $(1 - \frac{d}{2\lambda})$ at distances larger than λ.[36,37] This results in a *sharp phase transition* but with a small reduction of T_{KT}. Below the transition, the energy to separate single-layer vortex pairs in such a stack grows logarithmically at all distances[37] and this will result in a power law I-V curve of the form[34]

$$V \sim I^{1-\frac{1}{\zeta(T)}} \tag{4.4}$$

where $\zeta(T) = \frac{T}{2\pi\rho_s(T)d} = \frac{1}{2}\delta_2(T)$ which increases from 0 at $T = 0$ to $1/4$ at T_{KT} yielding $V \sim I^3$ at the transition. [For an isolated film, at very low currents this crosses over to linear resistivity.] Note that the dominant processes will be *vortex pairs* in *individual layers*; the extra in-plane energy needed to make vortex loops makes such loops too costly.

4.2 Vortex-Lattice Melting in Two Dimensions

In the presence of a penetrating magnetic field, the vortices in a single layer repel logarithmically out to distances of order $\Lambda_\perp$ (or to infinite distance in a stack). Thus as long as the intervortex spacing $a_v = (\Phi/B)^{1/2} \ll \Lambda_\perp$, the near neighbors will interact logarithmically. For temperatures much lower than T_{KT}, the effects of thermally induced vortices are negligible and the magnetically induced extra vortices (all of the same sign) form a vortex fluid. At still lower temperatures these vortices will form a 2D triangular lattice with a non-zero shear modulus but only quasi–long–range order. The melting temperature of this lattice is given by[38]

$$T_{2M} \approx (1 \sim 2) \times 10^{-2} \, \Theta_2 \tag{4.5}$$

corresponding to

$$T_{2M} \approx \left(\frac{1500 \text{Å}}{\lambda_\perp}\right)^2 \left(\frac{d}{15 \text{Å}}\right)(15 \sim 30)K. \tag{4.6}$$

Well below T_{2M}, the shear modulus of the lattice is given by[38]

$$\mu_2 = \mu d = \frac{\Phi_0 B d}{64\pi^3 \lambda_\perp^2} = \frac{\Theta_2}{4\pi a_v^2}, \tag{4.7}$$

but this will be decreased somewhat as T_M is approached. [A reduction factor of $0.4 - 0.8$ is included in Eq (4.5).][38] Note that the 2D vortex lattice melts when the mean square fluctuations of the nearest-neighbor relative displacements are of order $0.03 a_v^2$. For weakly coupled layers, the 2D melting temperature sets the scale of the collective interactions between vortices. Note the appearance of the small dimensionless parameter T_{2M}/Θ_2.

4.3 Interlayer Vortex Interactions

We next consider the magnitudes of the different interlayer couplings, and then discuss 2D-3D crossovers in various regimes.

We first ignore Josephson coupling between the layers. As discussed earlier, the vortex interactions *within* a layer are just

$$V(\rho) = -2\Theta_2 \ln(\rho/\xi) \tag{4.8}$$

where we use ρ to indicate positions within layers. If we consider the displacement of one vortex by a small amount $\vec{u}$, then the energy associated with its two neighbors in the direction of $\vec{u}$ at distance a_v is

$$E_{nn} \approx \frac{1}{2}u^2 \frac{4\Theta_2}{a_v^2} . \tag{4.9}$$

The magnetic interaction with a vortex in another layer a distance ζd directly above it, (with $\zeta = 0, \pm 1, \pm 2...$ labelling the layers) costs an energy[37]

$$E_{M\zeta} = \frac{1}{2}u^2 \frac{\Theta_2 e^{-\zeta d/\lambda}}{2\zeta \lambda^2} \tag{4.10}$$

and that with all of a vortex line above and below it an energy

$$E_M = \frac{1}{2}u^2 \frac{\ln(\lambda/d)\Theta_2}{\lambda^2} . \tag{4.11}$$

Thus we see that for large magnetic fields, the interactions within the layers become much stronger than the interlayer magnetic couplings:

$$\frac{E_M}{E_{nn}} \approx \frac{\pi H_{c1}}{B} \tag{4.12}$$

since ξ and d are comparable.

With weak Josephson coupling between the layers, characterized by a superfluid density

$$\rho_{sz} = \gamma^2 \rho_{s\perp} \tag{4.13}$$

and a corresponding $\lambda_z = \lambda/\gamma \gg \lambda$, the energy cost of a single layer vortex displacement due to the Josephson coupling to the vortex immediately above or below is

$$E_J = \frac{1}{2}u^2 \ln(\lambda_z/\xi)\gamma^2 \Theta_2/d^2. \tag{4.14}$$

[Note that the long length scale cutoff in the logarithm depends somewhat on how the various energies are grouped, but does not much affect the results.] This energy is smaller than E_M if the system is so anisotropic that

$$\gamma\lambda < d . \tag{4.15}$$

This corresponds to $\gamma < 10^{-2}$ at low temperatures, which for $BSCCO$ is probably not ruled out by experiments. For $\gamma \ll 10^{-2}$, the interlayer Josephson couplings hardly affect the vortex correlations; their main role is to make the system three-dimensionally superconducting. With γ not so small, there will still be a regime in high magnetic fields where the interlayer interctions are small: this occurs for

$$B > B_A \equiv \Phi_0 \gamma^2 / d^2 \; ; \quad (4.16)$$

the (criterion–dependent) crossover occurs when B is several times B_A.[39] For $BSCCO$, $B_A \approx 3kG$ with $\gamma = 1/50$.

These considerations are rather crude, and more detailed calculations are needed to estimate coefficients; nevertheless, the field and anisotropy dependences of interlayer couplings are correctly given by the balance between E_{nn}, E_M, and E_J.

We now consider 2D-3D crossovers associated with the various phase transitions.

4.4 Zero-Field Superconducting Transition: 2D–3D Crossover

In a stack of closely-spaced superconducting layers with no Josephson coupling, the effects of the interlayer magnetic interactions sharpen the Kosterlitz-Thouless transition to a true phase transition and lowers its temperature by a small amount. (They may also give rise to interesting interactions between nonlinear currents in different layers, but we will not investigate these here.) In the presence of a weak Josephson coupling, however, the system will cross over, just above T_{KT}, to a three-dimensionally correlated system with strongly anisotropic correlations. Crudely, this crossover will occur when the in-plane correlation length ξ_{KT} becomes sufficiently large that the Josephson coupling energy over a correlation area ξ_{KT}^2 becomes of order T. A better treatment yields an extra suppression factor of order $(\xi_\perp/\xi_{KT})^{\frac{1}{4}}$ where $\xi_\perp(T)$ is the temperature-dependent in-plane Ginzburg-Landau coherence length.[40] Thus the 2D-3D crossover will occur when

$$\gamma^2 \left(\frac{\xi_{KT}}{d}\right)^2 \left(\frac{\xi_\perp}{\xi_{KT}}\right)^{\frac{1}{4}} \sim 1 \quad (4.17)$$

so that $\xi_{KT} > d/\gamma$ at the crossover. Since with uncoupled layers

$$\xi_{KT} \sim \xi_\perp(T_{KT}) \exp\left[b\left(\frac{T_{c0} - T_{KT}}{T - T_{KT}}\right)^{1/2}\right] \quad (4.18)$$

with b of order unity, Eq. (4.17) yields a shift in the 3D transition temperature from T_{KT}, relative to the shift from the mean-field T_{c0}, of order[34]

$$\frac{T_c - T_{KT}}{T_{c0} - T_{KT}} \sim \left[\ln \frac{d}{\gamma \xi_\perp(T_{KT})}\right]^{-2} \quad (4.19)$$

which is not very small. In 2D, the resistivity vanishes as ξ_{KT}^{-2}, thus it will drop by a factor of order $(\frac{\gamma \xi_\perp(T_{KT})}{d})^2$ due to 2D critical fluctuations before the 3D coupling takes over.

Resistivity data on $BSCCO$ has been fitted by Martin et al.[35] to a Kosterlitz-Thouless form with $T_{c0} - T_{KT} \sim 2K$, consistent with a Ginzburg-Landau temperature dependent

$$\lambda_{GL} \approx 600\text{Å}(\frac{T_{c0} - T}{T_{c0}})^{-1/2} \, , \quad (4.20)$$

roughly consistent with a low temperature $\lambda \approx 1500\text{Å}$ and a dirty-limit temperature dependence. This suggests $\xi_\perp(T_{KT}) \sim 50 - 100\text{Å}$ yielding, with γ of 50, a drop in the resistivity of order 10^{-2} before the crossover to 3D behavior. If γ is considerably smaller, a wider 2D region might be observed.

Concurrent measurements of the resistivities ρ_{ab} and ρ_c on very good quality single crystals should be able to directly distinguish a 2D from a 3D regime: In a 2D critical regime, ρ_{ab} should plummet but ρ_c should not be strongly affected; only in the 3D critical regime will ρ_c drop, eventually vanishing with ρ_{ab} at the bulk T_c.

In the 3D critical region, the ratio ρ_{ab}/ρ_c will approach γ_c^2 where γ_c is the value of the penetration length anisotropy just below T_c. This will be somewhat decreased (i.e. be *more* anisotropic) from its value well below T_c by the 2D critical fluctuations. [c.f. the factor $(\xi_\perp/\xi_{KT})^{1/4}$ in Eq. (4.17)[41]] Shorts can of course cause problems; these are likely to dominate for very small γ.[9]

4.5 Vortex-Lattice Melting

We now consider the Abrikosov-lattice melting transition in a clean system. If the Josephson coupling between the layers is sufficiently strong ($\gamma > 10^{-2}$) or the lattice spacing not much less than λ, the melting in intermediate fields will be three-dimensional. Estimates of the melting temperature have been made using the Lindemann criterion that the rms displacement of a vortex line from its lattice position at the melting temperature T_M is

$$\langle u^2 \rangle_{T_M} \approx 0.02 \, a_v^2 \, . \quad (4.21)$$

For $H_X \gg B \gg \max(B_A, H_{c1})$, [with $B_A \equiv \Phi_0 \gamma^2/d^2$] one finds[1,42]

$$T_M \approx 0.1 \, \gamma a_v \frac{\Phi_0^2}{16\pi^2 \lambda^2(T_M)} \approx \frac{0.1 \gamma a_v}{d} \Theta_2 \, . \quad (4.22)$$

The prefactor in Eq. (4.22) is quite uncertain given the crudeness of the Lindemann criterion. However, the choice of coefficient 0.02 in Eq. (4.21) is consistent with the mean square nearest-neighbor displacement at the 2D KT transition.[1] Comparing Eq.(4.22) with the 2D melting temperature Eq.(4.5), one finds a crossover for $a_v > d/5\gamma$ for γ not too small, corresponding to a crossover field[39]

$$B_{2\to 3} \geq 30 B_A \approx 30 kG \, (100\gamma)^2, \quad (4.23)$$

which corresponds to a field of order $10^6 G$ for $YBCO$, by which point the approximation $B \ll H_X (\approx H_{c2})$ breaks down. Vortex-lattice melting in $YBCO$ is thus virtually always 3D.

For T_M relatively close to T_c, the penetration length $\lambda(T) \sim (T_c - T)^{-\frac{1}{3}}$ due to critical fluctuations (see Section I). Thus, from Eq.(4.22)

$$T_c - T_M \sim B^{\frac{3}{4}} . \tag{4.24}$$

In this regime, the melting field

$$B_M \approx 0.1 H_X = 0.1 \frac{\Phi_0}{2\pi \xi^2}, \tag{4.25}$$

so that the assumption $B \ll H_X$ is valid, and the vortex lines are well formed near B_M.

For fields well above $B_{2\to 3}$, the wavelength in the z-direction of the dominant fluctuations is of order the lattice spacing, d, while for lower fields, it is of order $5\gamma a_v$. [Of course, at fields close to H_{c1}, vortex spacings becomes larger than λ and the system is described in terms of weakly coupled vortex lines. See below.] In the extreme anisotropy limit, $\gamma \lambda \ll d$, where the interlayer interactions are primarily magnetic, the melting becomes quasi-two-dimensional for fields

$$B > 30 H_{c1} \approx 10^4 G. \tag{4.26}$$

In the *high field limit*, (but with $B \ll H_{c2}(0)$) the melting temperature generally approaches that of a single layer: T_{2M}. If the 2D melting transition is of Kosterlitz-Thouless type, then the shift in T_M due to interlayer couplings will be of order

$$T_M \approx T_{2M} \left[1 + \frac{C}{(\ln B/B_{2\to 3})^2} . \right] \tag{4.27}$$

If on the other hand, the 2D transition is first order, the shift will only be of order $1/B$. Note that the 3D melting transition is likely to be first order,[43] but for large fields the discontinuities will be reduced by factors of order $\frac{T_M - T_{2M}}{T_M}$ if melting in 2D is second order.

For *very low fields*, $B \lesssim H_{c1}$, the intervortex interactions become weak, since $a_v > \lambda$, and a fluid phase can intervene between the Meissner and vortex lattice phases.[17] This will be very narrow, however, and may be hard to observe experimentally.[1]

5. PINNING AND VORTEX–GLASS BEHAVIOR IN ANISOTROPIC SUPERCONDUCTORS

In this section, we consider, quantitatively, the effects of random pinning in anisotropic layered superconductors and the resulting phase diagrams. We will assume that the pinning is primarily due to *microscopic* scale inhomogeneities (oxygen vacancies, microtwin boundaries, radiation damaged regions, etc.); there is of course a lot of interesting physics associated with macroscopic defects, but we will not go into this here.

5.1 Weakly Anisotropic Materials

In the presence of pinning, the vortex lattice melting transition is generally replaced by a vortex glass transition. In the limit of *extremely weak pinning*, the vortex glass transition is likely to be close to the clean system melting transition $T_M(B)$ except for very close to H_{c1} or for very large fields. If the melting transition is first order, the resistivity should jump to zero also at the resulting vortex glass transition.[44] In general, if the pinning is very weak, the system will *not* have a large decrease in the resistivity until very close to T_M where the pinning takes over. Thus experimentally, the resistivity data in $YBCO$ crystals and films, as well as $BSCCO$ crystals, suggests relatively strong pinning with only rather short-range lattice correlations. [The apparently contrary evidence from vortex lattice imaging at very low fields will be discussed in the conclusion.]

Even for relatively strong pinning, we expect that the vortex glass phase boundary, in regimes where the interlayer couplings make the system fully three-dimensional, will have similar field dependence to the clean-system melting transition. Thus for fields in the range $H_{c1} \ll B \ll B_A$, (Eq. 4.16), $T_G(B)$ will have a form similar to Eq.(4.22)[1]:

$$T_G \approx C_G \, \gamma a_v \, \frac{\Phi_0^2}{16\pi^2 \lambda^2(T_G)} \sim T_c - CB^{\frac{3}{4}} \qquad (5.1)$$

with the coefficient C_G, which may be somewhat field-dependent, determined by the form and strength of the random pinning. For intermediate strength pinning, we guess $C_G \sim 10^{-1}$, but it may increase for very strong pinning.[45] For $YBCO$, B_A is very large so that the vortex glass transition will be three-dimensional except for extremely large fields.

[In very low fields, $B \leq H_{c1}$, the intervortex interactions become weak and the phase boundary may change. However for all but very weak pinning, much of the low field fluid phase which exists in a perfectly clean system[17] will probably be destroyed and replaced by the vortex glass phase[1], which for H very near H_{c1} will have quasi one-dimensional characteristics associated with the almost–individually–pinned vortex lines. Nevertheless, a sliver of vortex fluid may persist.][46]

We now turn to the interesting case of strongly anisotropic materials such as *BSCCO*; for fields $B > B_A$, these will exhibit regimes of quasi-two-dimensional behavior.

5.2 Incipient Vortex–Glass Behavior in Two Dimensions

We now discuss the behavior of an isolated 2D layer in a magnetic field in the presence of small scale random pinning. For simplicity, we ignore the finiteness of $\Lambda_\perp$ so that the vortices interact logarithmically at all distances. We first present a qualitative picture and then attempt to make it more quantitative.

As in 3D, any random pinning destroys the vortex lattice in 2D (and also the hexatic phase,[18] although only at longer scales). However, unlike in 3D, we do *not* expect a finite temperature vortex glass transition in 2D;[47] The rigorous bound $\theta \leq \frac{D-2}{2}$ for the closely related gauge glass[23] strongly suggests that $\theta \leq 0$ in 2D. Nevertheless, at *zero temperature*, the vortices will be frozen into a 2D vortex glass phase with a characteristic energy scale for deformations at scale L behaving as

$$F_L \sim L^{\theta_2} \tag{5.2}$$

with

$$\theta_2 < 0 . \tag{5.3}$$

Thus the system will be unstable to large scale deformations at low temperatures for $L > \xi_{2G}(T)$ with the 2D vortex glass correlation length diverging as

$$\xi_{2G}(T) \sim T^{-\nu_2} \tag{5.4}$$

with

$$\nu_2 = -\frac{1}{\theta_2} . \tag{5.5}$$

Numerical estimates give $\nu_2 = 2.1 \pm 0.1$.[48] On scales smaller than ξ_{2G}, there will be barriers to vortex motion growing with length scale as

$$B_L \sim L^{\psi_2} . \tag{5.6}$$

On scales larger than ξ_{2G}, regions of size ξ_{2G}^2 behave roughly independently so that the characteristic rate of vortex motion, Ω_{2G}, is given by

$$|\ln \Omega_{2G} t_0| \sim \frac{\xi_{2G}^{\psi_2}}{T} \sim T^{-(1+\psi_2\nu_2)} \tag{5.7}$$

(with t_0 a microscopic time of order 10^{-9} sec). The linear resistivity will vanish at low temperatures as

$$\rho \sim \rho_f \, e^{-K/T^{(1+\psi_2\nu_2)}} . \tag{5.8}$$

If, as is quite possible, ψ_2 is zero, the $T^{-\psi_2 \nu_2}$ factor is probably replaced by a (roughly) $\ln(\Theta_2/T)$ factor, yielding only a slightly faster than Arhennius temperature dependence,

$$\rho \sim \rho_f \, e^{-\tilde{K}\ln(\Theta_2/T)/T} \tag{5.9}$$

Because of a distribution of barrier heights, however, there will be a broad distribution of relaxation times, thus strictly speaking it is only $\ln \Omega_{2G}$ that should scale simply as in Eq. (5.7). Different measures of characteristic frequencies, Ω_{2G}, will thus yield different frequencies

$$\Omega_{2G} \sim e^{-K/T^{1+\psi_2 \nu_2}} \tag{5.10}$$

with *different coefficients* K for different measures. These effects also make scaling of nonlinear I-V curves subtle.

Nonlinear Conductivity

In 2D, the characteristic scale L_{2J} probed by a nonlinear current density J is

$$L_{2J} \sim \frac{cT}{J\Phi_0 d} \, . \tag{5.11}$$

Thus one would expect a crossover from linear to nonlinear conduction when $L_{2J} \sim \xi_{2G}$, yielding a characteristic current scale

$$J_{2X} \sim \frac{cT}{d\Phi_0 \xi_{2G}} \sim T^{1+\nu_2} \, . \tag{5.12}$$

[In the usual collective flux creep picture, one has, instead, $J_X \sim T$, associated with the *absence* of a diverging length scale.]

A natural guess for the form of the nonlinear I-V curves is that

$$\ln\left(\frac{E}{\rho_f J}\right) \sim \frac{1}{T^{1+\psi_2 \nu_2}} \mathcal{L}\left(\frac{J}{J_{2X}}\right) \tag{5.13}$$

with $\mathcal{L}(j) \to$ const for small j and $\mathcal{L}(j) \sim j^{-\psi}$ for large j yielding, for $J \gg J_{2X}$,

$$\ln\left(\frac{E}{\rho_f J}\right) \sim \frac{J^{-\psi_2}}{T^{1-\psi_2}} \, . \tag{5.14}$$

Changes in barrier heights due to currents could perhaps change these scaling forms, but in any case, the nonlinear effects will be hard to measure because of the factor $1/d$ in Eq. (5.12) which makes the characteristic current scale large for a 3D stack of layers. If $J_{2X}(T)$ can be measured, however, it would provide direct evidence for a divergent length scale at low temperatures and hence, for correlation effects.

We now make some comments on relevant experimental parameters.

5.3 Single-Vortex Pinning

Pinning of point vortices in a single layer of $BSCCO$ could be by oxygen vacancies or other impurities, by the somewhat large scale defects associated with orthorhombicity, or by larger scale inhomogeneities, cracks, etc. Small scale defects will primarily pin vortex cores. The maximum such pinning energy, $\hat{U}_1$, is of order

$$\hat{U}_1 \approx \frac{\Theta_2}{4} \tag{5.15}$$

caused by punching a vortex-core sized hole in the layer. At temperatures of order the 2D melting temperature, such pinning could give rise to Arrhenius factors in the resistivity of order 10^{-5}-10^{-10}. For $d = 15\text{Å}$,

$$\hat{U}_1(T) \approx 300K \left(\frac{1500\text{Å}}{\lambda(T)}\right)^2. \tag{5.16}$$

Experiments on $BSCCO$ have followed the resistivity down to less than 10^{-6} of the normal state resistivity in fields between $20kG$ and $60kG$.[4] If we account for some of this decrease from the B/H_{c2} factor in the Bardeen-Stephen theory (i.e. just the fraction of the volume occupied by vortex cores) then at the lowest temperatures, a fit to

$$\rho \approx \rho_N \left(\frac{B}{H_X}\right) e^{-U(T)/T} \tag{5.17}$$

yields a U of $280K$ at $30kG$ and $240K$ at $60kG$.[4] This can just barely be accounted for by individual layer vortex pinning if $\lambda \approx 1500\text{Å}$. More careful thought and examination of the data both suggest that this is an implausible interpretation: Firstly, one would expect the linear resistivity to be dominated by the more weakly pinned vortices, and it seems unlikely that very strong pinning sites would occupy the of order 5% of the volume necessary to pin each vortex at $60kG$. Even if they did, the intervortex interactions within each layer would play a substantial role since they are comparable to $\hat{U}_1$.

As shown by Safar *et al.*,[4] however, analysis of the data directly indicates more collective effects. A plot of

$$\tilde{U}(T) \equiv \frac{d \ln \rho}{d(1/T)} \tag{5.18}$$

shows, at $30kG$, a slowly increasing regime with $\tilde{U}$ reaching $\sim 550K$ as T decreases to $30K$ and then a rapid rise of $\tilde{U}$ over a $5K$ temperature range to $800K$ (whereupon the data stop). At $60kG$, a less pronounced increase in $\tilde{U}$ is seen from $25K$ to $20K$. This strongly suggests the appearance of collective effects. Although the three-dimensional interlayer couplings will eventually play a crucial role in driving the resistivity to zero, they cannot become effective unless the stronger intralayer interactions have first increased the correlations within each layer. How much of the increase in $\tilde{U}$ is due to the intralayer correlations is unclear. At $30kG$, the temperature dependence of $\tilde{U}(T)$ down to $T \sim 30K$ is roughly consistent with a BCS-like temperature dependence of $\hat{U}_1 \sim \lambda^{-2}(T)$.[4] Thus this regime may well be predominantly single vortex pinning, although the overall scale of $\tilde{U}$ seems somewhat too large. There is not, however, a good enough theory of weakly correlated vortex motion in a strong random potential on which to base a quantitative analysis. Note also that a temperature dependence of ρ in a 2D vortex glass regime as given by Eq (5.9) with $\psi_2 = 0$ is not easy to distinguish from a simple Arrhenius dependence. Quantitative estimates of ξ_{2G} with simple models of the pinning would be very useful.

If few- (or better, one-) layer thick $BSCCO$ crystals can be made, the role of intralayer correlations might be clarified. Nonlinear I-V measurements, if they are able to measure $J_{2X}(T)$, would of course also be very useful, as they could distinguish between sub-vortex-spacing scale pinning and collective pinning of correlated regions. Experiments in higher fields, where interlayer interactions play even less role and strong pinning sites may be saturated, may be the best hope for experimentally addressing this issue.

This brings us naturally to a more detailed analysis of the interlayer interactions and the two-dimensional to three-dimensional crossover.

5.4 2D–3D Vortex–Glass Crossover

We next consider the crossover from incipient 2D vortex–glass behavior to fully three-dimensional vortex glass ordering. The above discussion suggests that for $BSCCO$, the pinning in tesla fields is quite strong (indeed in the next subsection we will make a better case for this on the basis of muon spin resonance data[13]). We thus consider the behavior of a strongly anisotropic material such as $BSCCO$ in the presence of *relatively strong pinning*.

As the 2D vortex glass correlations grow, the coupling between the layers becomes stronger. A crossover to 3D behavior and a 3D vortex glass transition will occur when the interlayer coupling between regions of area ξ_{2G}^2 becomes of order T.

We first consider magnetic interactions. The magnetic interlayer couplings extend out to layers separated by distances, z, of order λ, but the long–range interactions within each layer make the long wavelength components of the vortex density small. The effects of the short wavelength components, however, fall off rapidly for $qz > 1$ so that these are also ineffective. The interactions between vortex glass correlation areas will thus decay rapidly for $z \geq a_v$. The interlayer magnetic interactions are effectively random in sign due to the randomness in the vortex positions as determined by the random potential in each layer. The root mean square interlayer magnetic interactions are of order

$$E_{GM} \approx \frac{\xi_G}{a_v} \frac{1}{3} \left(\frac{d}{a_v} \right)^{\frac{1}{2}} \frac{a_v^2}{\lambda^2} \Theta_2 \qquad (5.19)$$

with the first factor arising from the incoherent sum of $(\xi_{2G}/a_v)^2$ terms, and the factor $(d/a_v)^{1/2}$ arising from the incoherent sum over $\sim (a_v/d)$ layers.

The Josephson coupling between layers is also the sum of effectively random terms. In a 2D vortex glass coherence area, the phase is correlated in time although it varies approximately randomly in space over distances of order a_v. The resulting rms coupling between nearest neighbor layers is of order

$$E_{GJ} \sim \frac{\gamma^2 a_v^2}{d^2} \frac{\xi_G}{a_v} \Theta_2 \qquad (5.20)$$

with considerable uncertainty in the coefficient. Unless γ is so small that $\frac{\gamma \lambda}{d} < (\frac{d}{a_v})^{1/4}$, the Josephson interlayer interactions will dominate over the magnetic interactions.

The Josephson-dominated crossover temperature $T_{G2 \to 3}$, from two- to three-dimensional behavior is given by $E_{GJ}(T_{G2 \to 3}) \sim T_{G2 \to 3}$ and the resulting 3D vortex glass transition temperature will be some fraction of $T_{G2 \to 3}$. With

$$(\xi_{2G}/a_v) \sim (\Theta_2/T)^{\nu_2} \qquad (5.21)$$

this yields

$$T_G \sim \Theta_2 \left(\frac{\gamma a_v}{d} \right)^{\frac{2}{1+\nu_2}} \sim B^{\frac{-1}{1+\nu_2}}. \qquad (5.22)$$

[Note that the width of the 3D critical region will also be some fraction of T_G] With the numerical estimate $\nu_2 \approx 1/2$, we have $T_G \sim B^{-1/3}$ for small B, assuming a field independent effective pinning strength. It would clearly be useful to obtain an estimate of the coefficient in Eq.(5.22), to compare with experiment. This, of course, would require some realistic model of the pinning potential. [Note that it is quite possible that T_G can be greater than T_{2M} even for weakly coupled layers.]

Near the 2D-3D crossover temperature, $T_{G2 \to 3}$, the vortex glass correlation length

$$\xi_{2G}(T_{G2 \to 3})/a_v \sim \left(\frac{\gamma a_v}{d} \right)^{\frac{-2\nu_2}{1+\nu_2}} \sim B^{\frac{2}{3}}. \qquad (5.23)$$

Only for sufficiently large B will this ratio be much larger than one. Thus for smaller B or weaker anisotropy (e.g. in $YBCO$) there will *not* be a well-defined 2D vortex glass regime. A schematic phase diagram for $BSCCO$ is shown in Fig. 4.

The data on $BSCCO$[4] in fields of tens of kG suggest a quite sudden crossover indicated by the sharp rise in the effective barrier $\tilde{U}(T)$, consistent with a 2D-3D crossover. The $\sim 20\%$ decrease in the crossover temperature as B increases from $30kG$ to $60kG$[4] is not inconsistent with a $B^{-1/3}$ dependence, but the effective pinning could be field dependent, and these fields may not be in the asymptotic high field regime. Thus the agreement may well be fortuitous.

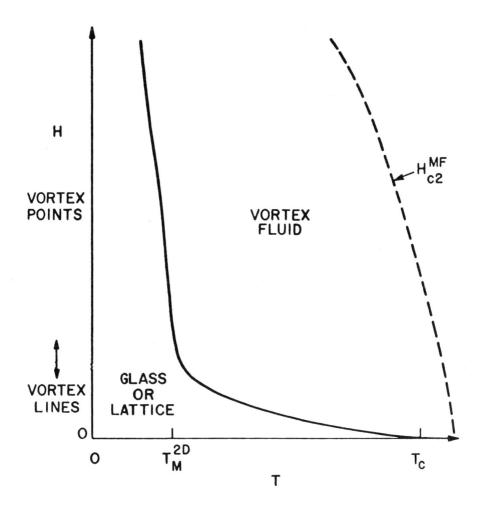

Fig. 4 Schematic phase diagram for a highly layered strongly type–II material like $BiSrCaCuO$. The crossover from the high–field two–dimensional regime of single vortices (vortex points) in each layer with weak interlayer interactions to the three–dimensional vortex–line regime at low fields is discussed in Section VI. For a perfectly clean system, the melting line at high fields approaches that of an individual layer T_{2M}. The field is oriented normal to the layers.

An irreversibility line has been measured by Lombardo et al.[45] for $BSCCO$ crystals. This should roughly coincide with $T_G(B)$. For fields in the range $1-50kG$, and T_G decreasing from 40 to 14K, their data can be reasonably well fit by $T_G \sim B^{-1/4}$, but again it is not clear how much significance should be attached to this.

Note, however, that after irradiating their samples with protons, the irreversibility temperature is found to increase by an amount of order 30%,[45] but $T_G(B)$ still has a roughly $B^{-1/4}$ field dependence—perhaps lending some support to the interpretation in terms of 2D–3D crossover.

In an ideal system, the resistivity perpendicular to the layers (the $\hat{z}$ direction) will not start to drop appreciably until $T_{G2\to 3}$ when the order parameter correlations between the layers start to develop. It will then drop to zero between $T_{G2\to 3}$ and T_G.[41] In practice, this behavior will be obscured by shorts between the layers and thus be hard to observe due to the extreme resistance anisotropy in this regime. Nevertheless, this behavior has been seen qualitatively by Kapitulnik *et al.*[49] in $BSCCO$ crystals: in tesla–range fields, ρ_c continues to *increase* as T is lowered and then plummets only when ρ_{ab} has become very small.

Before continuing the discussion of another experiment which provides strong evidence for short–range vortex positional correlations, we raise an interesting theoretical question which might be relevant for artificially layered materials for which one can obtain $\gamma\lambda \ll d$ so that the magnetic interactions between vortices on different layers dominate over the Josephson coupling.

In the absence of any interlayer Josephson coupling, a layered system obviously cannot be a superconductor in the z–direction. But can the magnetic interactions cause a type of vortex–glass transition into a phase with long–range (or quasi-long–range) superconducting correlations and vanishing linear resistivity in each layer? This novel phase would be quite different from a proper 3D vortex glass. For example, non–linear dissipation would presumably proceed by vortex–pair–like rather than vortex–loop–like excitations. The phase transition into such a phase would presumably also be strongly anisotropic. The primary issue is whether the logarithmic attraction between vortices on neighboring layers is long enough range to cause a transition. We leave this for future consideration.

5.5 Muon Spin Resonance and Distribution of Local Magnetic Fields

Distributions of local magnetic fields in the mixed state of superconductors can be obtained by several methods, including muon spin resonance. For an average field $\bar{B} \gg H_{c1}$ in the z–direction, the distribution of field variations $\delta B_z = B_z - \bar{B}$ can be measured. For $YBCO$ crystals data[11] at low temperatures yield behavior consistent with a conventional vortex lattice with variance

$$\overline{\delta B_z^2} \approx \frac{\Phi_0^2}{(4\pi\lambda^2)^2}(0.57) \qquad (5.24)$$

with $\lambda \approx 1400\text{Å}$ consistent with other measurements, and a characteristic asymmetric line shape, associated with large magnetic fields near vortex cores and minima in B_z between them.

For $BSSCO$ crystals, the behavior is very different: in a field of $15kG$, the distribution of B_z obtained from the resonance line is much more symmetric and far narrower with a close-to-gaussian shape.[13] This suggests that the magnetic field at a typical muon site is the sum of a number of random contributions from different vortices. We thus consider the effects of different distributions of vortex positions in a strongly anisotropic material. Long wavelength variations in vortex density will increase the variance of B_z, while thermal fluctuations on the time scales probed ($\sim 10^{-7}$sec) can round out the high field tail somewhat, but, at low temperatures, cannot significantly decrease the linewidth or make it close to gaussian. Static short-wave length displacements of vortex lines further than a few vortex lattice constants from the muon also do not have much effect.

The only apparent way to decrease the variance as well as making the distribution more gaussian is to invoke strong disorder between layers.[50] Since the effects of intralayer disorder (with the constraints caused by the interactions) are not large, we consider for simplicity a set of perfect lattices in each layer with *uniform* displacements $\vec{u}_\zeta$ within each layer, ζ. With no correlations in the $\vec{u}_\zeta$, corresponding to very weakly coupled layers, this yields

$$\overline{\delta B_z^2} \approx \frac{\Phi_0^2}{(4\pi\lambda^2)^2} 1.3 \frac{d}{a_v}. \tag{5.25}$$

Roughly, the a_v/d layers which contributed coherently to the variance for a 3D lattice, now contribute incoherently—yielding a closer-to-gaussian distribution. Any positive correlations between the vortices in each layer will *increase* $\overline{\delta B_z^2}$, and make the distribution less gaussian. However, correlations beyond distances $\sim a_v$ do not appreciably affect the result. For a field of $15kG$, $\overline{\delta B_z^2}$ is suppressed from Eq.(5.24) to Eq.(5.25) by about a factor of 10, so that the apparent λ inferred from a fit to Eq.(5.24) (lattice) is too large by a factor of 1.8. The quoted[13] λ of 3300Å thus becomes of order 1800Å. This λ, which is almost surely a lower bound, is comparable to that of $YBCO$. A better result should be obtainable by fitting the variances and line shapes for a range of fields to a calculation with a non-zero field dependent correlation length $L_{pz}(B)$ in the z-direction but the *same* λ for each field. The existing data[13] show broader, more asymmetric lines for lower fields, consistent with the increasing effects of interlayer couplings as B decreases.

Several important points are in order at this stage: firstly, a λ of $\geq 1800\text{Å}$ is more consistent with the sharp zero field transition than $\lambda \sim 3000\text{Å}$, as noted in Section IV. Secondly, the gaussian nature of the δB_z distribution at $15kG$[13] implies *strong interlayer disorder in BSCCO*. Since the Josephson coupling between 2D lattice-like regions of size L_p^2 is of order

$$E_{LJ} \sim \Theta_2 \frac{\gamma^2 a_v^2}{d^2} \left(\frac{L_p}{a_v}\right)^2, \tag{5.26}$$

the pinning energy in a region of size L_p^2, which is of order Θ_2 for $L_p > a_v$ must be at least as large as E_{LJ}. This implies that

$$\frac{L_p}{a_v} < C \frac{d}{\gamma a_v} \tag{5.27}$$

with coefficient C probably of order one. Thus strong interlayer disorder at fields of order $10kG$ in $BSCCO$ implies *very short range lattice correlations* within the layers. It is hard to see how to get around this conclusion, unless γ is much less than the measured value of 1/50.

6. CONCLUSIONS

We have seen that strong thermal fluctuations can dramatically change the macroscopic properties of superconductors. Our main focus has been the mixed state for fields in the range $H_{c2}(T) \gg B \gg H_{c1}(T)$. In a perfectly clean material, the primary effect of fluctuations is to lower the melting line of the Abrikosov vortex lattice to below the mean field H_{c2} which becomes merely a crossover line below which the system can be described as a vortex fluid. [The properties of the vortex fluid have been analyzed in detail by Nelson and collaborators.[17]] In the presence of random pinning, the vortex lattice is destroyed and replaced by a new equilibrium phase: a vortex glass. The vortex glass phase has vanishing linear resistivity, exponentially small resistance at small currents with a characteristic scale J_T, and rapid increase in the dissipation above a current scale J_F. In conventional materials (and in general at low temperatures), $J_T \gg J_F$ so that J_F becomes an almost sharp critical current (Fig. 3a). With strong fluctuations, or in general near to the vortex glass transition T_G, $J_T \sim J_F$ so that the I-V curves are smooth with no well-defined critical current (Fig. 3b). In this regime, the logarithmic decay of currents (and magnetizations) can proceed to well below J_F at macroscopic times.

As the temperature is lowered in a magnetic field, there is a phase transition to the vortex glass characterized by diverging correlation length and times and a vanishing current scale which characterizes the nonlinear response. Both I-V curves and $\sigma(\omega)$ exhibit scaling behavior as T_G is approached. For strongly anisotropic layered materials, various two-dimensional regimes can exist, particularly a range of 2D incipient vortex glass behavior with in-plane resistivity decreasing faster than Arrhenius and growing vortex correlations within each layer.

We now briefly summarize the current state of our understanding of the two most widely-studied cuprate superconductors, $YBCO$ and $BSCCO$, as well as suggesting a few needed experiments and possible applications to artificially layered materials.

6.1 YBCO

The anisotropy in $YBCO$, $\gamma \approx 0.2$, is weak enough that the system behaves three-dimensionally in all but perhaps extremely large magnetic fields. For fields from $10 - 50kG$ good evidence for a vortex glass transition has been found in both films and single crystals with power-law-vanishing resistivity and scaling of nonlinear I-V curves.[5,6] The films are much more strongly disordered and the range of current density over which scaling is observed suggests vortex glass behavior down to distances of a few intervortex spacings.[5] The data on crystals[6] is taken at much lower currents and scaling is only found nearer to the transition, suggesting intermediate range positional correlations—perhaps out to tens of lattice spacings. At this point, the exponents and scaling functions obtained from the two sets of data[5,6] are relatively consistent, although apparent differences do exist and the errors bars are large. Non–linear I-V curves for films in the range down to Gammel et al.,[6] sensitivity could markedly strengthen the evidence for scaling or show up problems. Note that at this point, none of the data get into the regime where the finite screening length and magnetic fluctuations become important.[16] It may be that this can be reached in some range of fields.

AC conductivity measurements over a wide range of frequencies would also be useful: T_G, ν and z could all be extracted independently and checked with the exponents obtained from nonlinear I-V curves. Preliminary data are consistent with the scaling laws.[51]

Over the narrow range of B analyzed in detail,[5,6] $T_c - T_G(B)$ is consistent with the $B^{3/4}$ behavior expected in the low field critical regime. The *coefficient* is also in reasonable agreement with simple theoretical estimates.[1] Over a wider field range, the "irreversibility line,"[52] which roughly coincides with $T_G(B)$, also scales in this way. These suggest a reasonably wide zero field critical region at least for the penetration length $\lambda(T)$, consistent with expectations. Analysis[1] of the *zero field critical behavior* $\sigma(\omega)$, scaling of I-V curves, etc. may be possible in very uniform crystals, although primarily in a narrower critical regime. Quantitative comparison with theory for the magnitude of fluctuations in various regimes as well as the zero field critical behavior itself could then be attempted.

6.2 BSCCO

The measured anisotropy in $BSCCO$ is $\gamma \approx 0.02$,[12] but this may well be just an upper bound, with the measured anisotropy dominated by shorts (screw dislocations, mosaic spread of the layers, etc.). In any case, the measured γ is small enough that two–dimensional effects dominate over a wide region of the phase diagram. In low fields the system will be three–dimensional, but when $\gamma\sqrt{\Phi_0/B}$ becomes considerably less than the layer spacing d, the layers become only weakly coupled.

At higher fields, there should be a wide range of temperatures with a 2D vortex glass correlation length growing as $T^{-\nu_2}$ with faster than Arrhenius growth (but perhaps only $\exp[\tilde{K}(\ln 1/T)/T])$ of the conductivity. At this point, it is not clear experimentally[4] how collective the transport is in this two–dimensional regime, although we expect that the vortex glass correlation length, ξ_{2G}, can become quite large. Eventually, the system crosses over to 3D behavior, with power–law diverging conductivity at temperatures of order $25K$ in fields of tens of kG.[4] Non–linear I-V data could directly probe the vortex glass correlation length, but the necessary current scales are much larger than for $YBCO$ near T_G, since the vortex "loops" have "area" of only $d\xi_{2G}$ (rather than ξ_G^2). AC conductivity measurements would also be useful to check the tracking of characteristic frequency with resistivity and the *broadening* on a *logarithmic frequency scale* of the frequency dependence of $\sigma(\omega)$, in the 2D incipient vortex glass regime.

The crossover at low temperatures to 3D vortex glass is probably dominated by Josephson coupling and occurs, (with a field independent effective pinning potential) at $T_{G2\to 3} \sim B^{-1/(1+\nu_2)}$. The 3D vortex glass transition T_G is also expected to scale as $B^{-1/(1+\nu_2)}$. The irreversibility lines in single crystals of $BSCCO$, both with and without proton irradiation, are found to scale roughly as $B^{-1/4}$, (but with a temperature scale increased by 30% after irradiation), perhaps suggesting that this regime is being obtained.[45]

Muon spin resonance in good $BSCCO$ crystals[13] shows a very narrow, roughly gaussian distribution of local magnetic fields in an applied field of $15kG$ at low temperatures. This strongly suggests very short-range correlations of vortex positions between layers and, unless γ is much smaller than the measured value, only very short-range lattice correlations within each layer. Further analysis of μSR data, both in clean crystals and perhaps in radiation damaged ones, is clearly called for. Spin-echo NMR measurements might also yield useful information on intermediate-time-scale vortex motion and field distributions.

Note that the conclusion that the pinning in $BSCCO$ is strong is not inconsistent with comparable pinning strength in single crystals of $YBCO$: near the measured 3D vortex glass transition in $YBCO$, thermal fluctuations of the vortex lines smooth out the pinning, thereby increasing the vortex lattice correlations.

Flux Imaging

For both $YBCO$[53] and $BSCCO$,[54] flux imaging in very low fields of order H_{c1}, show well-developed lattice correlations, disrupted by dislocations, and very long-range hexatic orientational order at temperatures of $4K$. There are several reasons, however, why these observations do *not* imply long-range correlations at the much higher fields where transport measurements have been made.

Firstly, the vortices at $4K$ probably exhibit frozen-in correlations from the temperature—very close to T_c—where the lattice formed on cooling. At that temperature, the thermal fluctuations can average out the pinning potential quite effectively and the orientational correlations will then persist to very long length scales.[55] A second effect, as pointed out by Huse,[56] is the extra strong ($\sim \Phi_0^2/r$) interactions between the ends of vortex lines at a surface; this may help account for the long-range correlations. More theoretical and experimental work—particularly on the nonequilibrium history-dependent aspects—is definitely needed.

6.3 Artificially Layered Superconductors

Materials more strongly anisotropic than $BSCCO$ can be made by growing regular stacks of alternating superconducting and insulating layers, with period d. In terms of the measured zero field penetration lengths of the composite, λ and

λ_z for currents in and between planes, respectively, one can again define a (T-dependent) anisotropy $\gamma \equiv \lambda/\lambda_z$. In a penetrating magnetic field, two-dimensional effects will be important when $\gamma\sqrt{\Phi_0/B} \leq d$. However for large spacing between the superconducting layers (but still smaller than λ), one can easily reach the limit $\gamma\lambda \ll d$ in which *magnetic* interlayer interactions between vortices dominate. The most intriguing question, for $\gamma \to 0$, is whether such a system can form a three-dimensionally ordered "stacked vortex glass" phase with vanishing resistivity within the layers but no superconductivity in the z-direction.

Because of the wide range of parameters that can be achieved in such systems, artificially layered superconductors are a very fruitful avenue for future research, one that is already being pursued by the Stanford group.[57]

ACKNOWLEDGMENTS

Much of this work reviewed here has been done in collaboration with David Huse and Matthew Fisher to whom I am indebted. I have also benefitted greatly from discussions with David Nelson and with numerous experimental colleagues, particularly, Peter Gammel, Aharon Kapitulnik, Tony Fiory, Peter Kes and David Bishop; David Nelson and David Bishop also provided useful comments on the manuscript.

This work was supported in part by the NSF under grant DMR 9096267, and through Harvard University's Materials Research Laboratory, as well as by the AP Sloan Foundation. I would also like to thank AT&T Bell Labs and the Aspen Physics Center for hospitality during the period when some of this work was carried out.

APPENDIX: ELASTIC MODULI OF VORTEX LATTICES

There is a large body of recent literature on the elastic properties of vortex lattices in anisotropic superconductors. The detailed wavevector dependences of the elastic constants are needed to calculate thermal fluctuations,[1,42] the perturbative effects of weak pinning, etc. Unfortunately, there are serious flaws in several of the calculations—particularly concerning the tilt modulus of the vortex lines.[58] Since this behavior is important for much of the quantitative analysis in the present paper (and also in Ref. 1), we quote results in various regimes, restricting consideration to magnetic fields $\ll H_{c2}$ normal to the planes; and also to layer spacing d much smaller than both λ and a_v.

Isolated Flux–lines

Vortices are coupled via both magnetic and Josephson coupling. In terms of the Fourier transform $\vec{u}(q_z)$ of the vortex displacements, the tilt elastic energy of a single line can be written as

$$\mathcal{H}_1 = \frac{1}{2} L \int_{q_z} K_1(q_z) \, |\vec{u}(q_z)|^2 \qquad (A1)$$

with L the sample thickness and

$$K_1(q_z) \approx \frac{\Phi_0^2}{16\pi^2\lambda^2} \left\{ \frac{1}{2\lambda^2} \ln(1+\lambda^2 q_z^2) + \gamma^2 q_z^2 \left[\ln(\lambda/\xi\gamma) - \frac{1}{2}\ln(1+\lambda^2 q_z^2) - \frac{1}{2}\right] \right\} \qquad (A2)$$

with λ the in-plane penetration length and

$$\gamma = \lambda/\lambda_z < 1 \qquad (A3)$$

the anisotropy. The fourth term in Eq.(A2) and the cutoff ξ inside the logarithm of the second term depend on the short scale cutoff (i.e. vortex core size): the choice used reproduces the isotropic limit. Note that either an anisotropic cutoff with $\xi_z = \gamma\xi$ or a discrete set of layers is needed to get the result [Eq. (A2)]. The first term survives in the limit $\gamma \to 0$ of *no* Josephson coupling. It has been derived by Clem[37] in this limit. It is the *dominant* term[1] for long wavelength distortions, even for $\gamma \approx 0.2$ appropriate for $YBCO$, but has been missed by other authors.[58] For large $q_z \sim 1/d$, the q_z^2 in front of all but the first term should be replaced by $\frac{1}{d^2}[2 - 2\cos q_z d]$ and $|q_z|$ restricted to $< \pi/d$.

Vortex Lattice

In a vortex lattice, the general form of the elastic energy can be written as

$$\mathcal{H}_{el} = \frac{1}{2} \sum_{\vec{q}} K^{\alpha\beta}(\vec{q}) u^\alpha(\vec{q}) u^\beta(\vec{q}) \qquad (A4)$$

with $\alpha, \beta = x, y, z$ and

$$K^{\alpha\beta}(\vec{q}) = q_z^2 \, \delta^{\alpha\beta} C_{44}(\vec{q}) + q_\perp^2 \, \delta^{\alpha\beta} \, C_{66}(\vec{q}) \qquad (A5)$$

$$+ q_\perp^\alpha q_\perp^\beta \, C_L(\vec{q}),$$

with $\vec{q}_\perp$ the in-plane component of the wavevector.

Intermediate Fields

For $H_{c2} \gg B \gg H_{c1}$, the long wavelength shear modulus

$$C_{66}(0) \approx \frac{1}{4} \frac{\Phi_0 B}{16\pi^2 \lambda^2}, \tag{A6}$$

which is just the 2D result.[59] The q dependent corrections to C_{66} are relatively small even up to the zone boundary.

The bulk modulus is much larger due to the long-range interactions:[59]

$$C_L(\vec{q}) \approx \frac{B^2}{4\pi} \left[\frac{1 + q^2 \lambda_z^2}{(1 + q^2 \lambda^2)(1 + q_z^2 \lambda^2 + q_\perp^2 \lambda_z^2)} \right] \tag{A7}$$

for not too large q. Near the zone boundary $q_\perp \sim \pi/a_v$, C_L is of order C_{66}. Compressional deformations are thus strongly suppressed by the large C_L, except for wavelengths of order a_v. Over the dominant range of wavevectors, the tilt modulus is given by

$$C_{44}(\vec{q}) \approx \frac{B^2}{4\pi} \frac{1}{1 + q_z^2 \lambda^2 + q_\perp^2 \lambda_z^2} \tag{A8}$$

$$+ \frac{\Phi_0 B}{16\pi^2 \lambda^2} \gamma^2 \ln\left[1/(\gamma \xi \sqrt{\lambda^{-2} + q^2})\right]$$

$$+ \frac{\Phi_0 B}{32\pi^2 \lambda^4 q_z^2} \ln\left(1 + \frac{q_z^2 \Phi_0}{25 B}\right)$$

The first term in Eq.(A8) is the long-range magnetic energy, but it is strongly suppressed, for $q_\perp > 1/\lambda_z$, by the $q_\perp^2 \lambda_z^2$ in the denominator, and thus does not play a major role. The third term is due to the magnetic interlayer interactions and persists without Josephson coupling; the form of the logarithmic dependence is an interpolation between the high and low q_z behavior. Note that the scale a_v has replaced, in this term, the λ appearing in the magnetic contribution to the single line tilt modulus, Eq.(A2). Again the q_z^2 outside of logarithms in $q_z^2 C_{66}(q)$ should be replaced by $(2 - 2\cos q_z d)/d^2$ to include the effects of discrete layers.

The mean square thermal displacements $\langle u^2 \rangle$ are dominated by the shear modulus C_{66} with $q_\perp$ near the zone boundary, combined with the second term of C_{44} with $\pi/q_z \sim \max(d, a_v/5\gamma)$ for the Josephson dominated regime with $\gamma \lambda > d$. For the extremely anisotropic magnetically dominated case $\gamma \lambda < d$, the third term of C_{44} dominates instead.

Small Fields

If the field is decreased so that $a_v \gg \lambda_z$, the interactions between the vortices become exponentially small. However, in the intermediate regime with $\lambda \ll a_v \ll \lambda_z$, the interactions are relatively weak and the tilt modulus is somewhat modified from its large a_v limit,

$$q_z^2 \, C_{44}(\vec{q}\,) \approx \frac{K_1}{\Phi_0}(\vec{q}\,) + \frac{B^2}{4\pi} \, q_z^2 \left[\frac{1}{1 + \lambda^2 q_z^2 + \lambda_z^2 q_\perp^2} - \frac{1}{2} \frac{1}{1 + \lambda^2 q_z^2} \right]. \quad (A9)$$

The collective second term does not play a major role except for very small $\vec{q}$. Thus for $a_v \gg \lambda$, the *interactions* between different vortices are essentially the same as in an isotropic system.

REFERENCES

1. D. S. Fisher, M. P. A. Fisher, and D. A. Huse, Phys. Rev. B**43**, 130 (1991). Much of the present paper is a review of results in this paper.
2. See e.g., J. S. Langer and M. E. Fisher, Phys. Rev. Lett. **19**, 560 (1967).
3. P. W. Anderson and Y. B. Kim, Rev. Mod. Phys. **36**, 39 (1964).
4. H. Safar, P. L. Gammel, D. J. Bishop, D. B. Mitzi, and A. Kapitulnik, preprint.
5. R. H. Koch, V. Foglietti, W. J. Gallagher, G. Koren, A. Gupta, and M. P. A. Fisher, Phys. Rev. Lett. **63**, 1511 (1989).
6. P. L. Gammel, L. F. Schneemeyer, and D. J. Bishop, Phys. Rev. Lett. **66**, 953 (1991).
7. M. P. A. Fisher, Phys. Rev. Lett. **62**, 1415 (1989).
8. D. S. Fisher and D. A. Huse, Phys. Rev. B**38**, 386 (1988); **38**, 373 (1988).
9. Strongly anisotropic materials are hard to measure due to "shorts" between the layers.
10. G. J. Dolan, F. Holtzberg, C. Field, and T. R. Dinger, Phys. Rev. Lett. **62**, 2184 (1989).
11. See, L. Krusin-Elbaum, R. L. Greene, F. Holtzberg, A. P. Malozemoff, and Y. Yeshurun, Phys. Rev. Lett. **62**, 217 (1989), and references therein.
12. D. E. Farrell, S. Bonham, J. Foster, V. C. Chang, P. Z. Jiang, K. G. Vandervoort, D. J. Lam, and V. G. Kogan, Phys. Rev. Lett. **63**, 782 (1989).
13. D. R. Harshman, E. H. Brandt, A. T. Fiory, M. Inui, D. M. Mitzi, L. F. Schneemeyer, and J. V. Waszczak, preprints.
14. Because H_{c2} is not well-defined (for $T \leq T_c$), ξ and hence κ is hard to measure; see, however, Ref. 11. A value for $\xi(0)$ of approximately 15 Å has also been inferred from a Little-Parks experiment; P. L. Gammel, D. A. Polakov, C. E. Rice, L. R. Harriott and D. J. Bishop, Phys. Rev. B**41**, 2593 (1990).
15. We assume a (roughly) *BCS* temperature dependence for $\lambda(T)$ outside of the critical region; this is consistent with penetration length measurements

as $YBCO$ (ref. 10), although the measured $\lambda(T)$ may be slightly flatter than the BCS form.

16. Very close to the zero field transition, λ/ξ becomes of order one and magnetic fluctuations become important. Similar behavior occurs near the vortex glass transition when $\lambda_G/\xi_G = O(1)$. We will not investigate these regimes, in which the critical behavior will change; but see Ref. 1.
17. D. R. Nelson and H. S. Seung, Phys. Rev. B**39**, 9153 (1989). For a recent review, see D. R. Nelson in proceedings of this Los Alamos Superconductivity conference.
18. M. C. Marchetti and D. R. Nelson, Phys. Rev. B**41**, 1910 (1990).
19. Several other authors have developed theories of a vortex-glass-like phase in terms of an *elastic* medium of vortices in a random potential. Since these ignore the effects of dislocations, which will occur on scales $\geq L_p$, these theories cannot correctly describe the long length (and time) scale physics. They may however, in some cases, have a regime of validity at intermediate length scales. (See also ref. 21.) This work includes, for example: M.V. Feigel'man, V. B. Geshkenbein, A.I. Larkin, and V.M. Vinokur, Phys. Rev. Lett. **63**, 2303 (1989); M.V. Feigel'man and V.M. Vinokur, Phys. Rev. B**41**, 8986 (1990); and T. Nattermann, *Phys. Rev. Lett.* **64**, 2454 (1990).
20. A. I. Larkin and Yu. N. Ovchinnikov, J. Low Temp. Phys. **34**, 409 (1979).
21. In general, L_p will be anisotropic with $L_{pz}/L_{p\perp}$ depending on the vortex elasticity and the pinning potential. For weak pinning, there is likely to be a scale much larger than L_p out to which hexatic orientational order exists (Refs. 18 and 55). We will not consider the effects of this here. In addition, for weak pinning with small scale correlations at very low temperatures, there may be an intermediate range of length scales for which pinning is important but dislocations can be ignored, as in Ref. 19. In this paper we also ignore the possible effects of anisotropy of excitations.
22. For a review of spin glasses, see K. Binder and A. P. Young, Rev. Mod. Phys. **58**, 801 (1986).
23. M. Aizenman and D. S. Fisher, unpublished. The rigorous bound is for the closely related gauge-glass model (see Ref. 28) and may thus not be correct for the 3D vortex glass.
24. O. Narayan and D. S. Fisher, preprint .
25. Y. Yeshurun, A. P. Malozemoff, and F. Holtzberg, J. Appl. Phys. **64**, 5797 (1988) and references therein; see also A. P. Malozemoff and M. P. A. Fisher, Phys. Rev. B**42** 6784 (1990).
26. The vortex glass correlation length will be anisotropic. A crude estimate yields $\xi_{Gz}/\xi_{G\perp} \sim \gamma$. The current scale will then be $J_X \sim (\xi_{G\perp}\xi_{Gz})^{-1}$.
27. Expansions in $D = 6 - \epsilon$ have been performed for the vortex glass critical behavior, M. A. Moore and S. Murphy, Phys. Rev. B**42**, 2587 (1990).
28. J. D. Reger, T. A. Tokuyasu, A. P. Young, and M. P. A. Fisher, preprint.
29. T. K. Worthington, F. H. Holtzberg, and C. A. Field, Cryogenics **30**, 417 (1990).
30. See, D. R. Nelson in Ref. 17, and private communication.

31. M. P. A. Fisher, P. B. Weichmann, G. Grinstein, and D. S. Fisher, Phys. Rev. B**40**, 546 (1989).
32. See e.g., M. Inui, P. B. Littlewood, and S. N. Coppersmith, Phys. Rev. Lett. **63**, 2421 (1989); C. W. Hagan and R. Griessen, Phys. Rev. Lett. **62**, 2857 (1989).
33. A. T. Dorsey, Phys. Rev. B**43**, 7575 (1991).
34. B. I. Halperin and D. R. Nelson, J. Low Temp. Phys. **36**, 599 (1979).
35. S. Martin, A. T. Fiory, R. M. Fleming, G.P. Espinosa, and A. S. Cooper, Phys. Rev. Lett. **62**, 677 (1989).
36. The layer spacing, d, is defined as the distance between the centers of the layers. For $BSCCO$, the layers are a double layer of $Cu - O$ planes. See Ref. 37 for a more detailed analysis. Note that the in-plane penetration length $\lambda \equiv \lambda_\perp$ is defined in terms of the properties of the *stack* of layers rather than an individual layer.
37. J. R. Clem, Phys. Rev. B**43**, 7837 (1991), has analyzed in detail the magnetic interactions between vortices in a stack of Josephson-uncoupled layers.
38. D. S. Fisher, Phys. Rev. B**22**, 1190 (1980).
39. The author would like to thank Mikhail Feigel'man for pointing out that the estimate for the crossover field from 2D-3D melting in Ref. 1 is too small by a numerical factor. See also discussion in the Appendix.
40. This, and the other 2D-3D crossovers are simple applications of the general theory of dimensional crossover.
41. Very near to the 3D zero field superconducting transition, we expect that the resistivity ratio ρ_c/ρ_{ab} will approach a constant, $1/\gamma_c^2$, which is the inverse superfluid density anisotropy in the critical region just below T_c. For a weakly coupled stack of layers, γ_c may be considerably smaller than the anisotropy, γ, well below T_c. The correlation length anisotropy in the critical region will generally be $\xi_z/\xi_\perp \approx \gamma_c$. Related behavior can occur near a 3D vortex glass transition.
42. A. Houghton, R. A. Pelcovits, and S. Sudbo, Phys. Rev. B**40**, 6763 (1989); E. H. Brandt, Phys. Rev. Lett. B**63**, 1106 (1989).
43. E. Brezin, D. R. Nelson, and A. Thiaville, Phys. Rev. B**31**, 7124 (1985).
44. For very weak pinning, a first order melting transition (if it occurs in the clean system) will persist as a discontinuity in the entropy, resistivity, etc. Naively, the vortex glass transition will coincide with this discontinuity. It is possible, however, that the vortex glass transition lies below (or perhaps even above) this first order transition. This would occur, for example, in a 2D system with a first order melting transition, since in this case, $T_G = 0$.
45. L. Lombardo, A. Kapitulnik, A. Leone and D. Dammann, preprint.
46. T. Nattermann, M. Feigel'man, and I. Lyuksyntov (preprint) have analyzed the $B(H)$ curve near H_{c1} in the presence of pinning. I believe that their results are incorrect, for subtle reasons, which will be discussed elsewhere, and the earlier result [T. Natterman and R. Lipowsky, Phys. Rev. Lett. **61**, 2508 (1988)] should obtain.

47. D. S. Fisher and D. A. Huse, Phys. Rev. B36, 8937 (1987) study the analogous problem of 2D spin glasses and 2D–3D crossover.
48. M. P. A. Fisher, T. A. Tokuyasu, and A. P. Young, Phys. Rev. Lett. 66, 2931 (1991).
49. A. Kapitulnik, private communication.
50. E. H. Brandt, Phys. Rev. Lett. 66, 3213 (1991) has performed a somewhat related analysis of the distribution of local fields, but in a less physical approximation.
51. H. K. Olsson, R. H. Koch, W. Eidelloth, and R. P. Robertazzi, Phys. Rev. Lett. 66, 2661 (1991).
52. Oh., K. Car, A. D. Kent, M. Naito, M. R. Beasley, T. H. Geballe, R. H. Hammond, A. Kapitulnik, and J. M. Graybeal, Phys. Rev. B37, 7861 (1988); T. K. Worthington, W. J. Gallagher, D. L. Kaiser, F. H. Holtzberg, and T. R. Dinger, Physica C B153–155, 32 (1987).
53. P. L. Gammel, D. J. Bishop, G. J. Dolan, J. R. Kwo, C. A. Murray, L. F. Schneemeyer, and J. V. Waszczak, Phys. Rev. Lett. 59, 2592 (1987); G. J. Dolan, G. V. Chandrashekhar, T. R. Dinger, C. Field, and F. Holtzberg, *ibid.* 62, 829 (1989).
54. C. A. Murray, P. L. Gammel, and D. J. Bishop, Phys. Rev. Lett. 64, 2312 (1990); D. G. Grier, C. A. Murray, C. A. Bolle, P. L. Gammel, D. J. Bishop, D. B. Mitzi, and A. Kapitulnik, Phys. Rev. Lett. 66, 2270 (1991).
55. E. M. Chudnovsky, Phys. Rev. B40, 11355 (1989).
56. D. A. Huse, preprint.
57. W. R. White, A. Kapitulnik, and M. R. Beasley, Phys. Rev. Lett. 66, 2826 (1991).
58. A. Sudbo and E. H. Brandt, Phys. Rev. Lett. 66, 1781 (1991); and preprint.
59. See Refs. 38 and 42 and references therein.

DISCUSSION

A. P. Malozemoff: I'd like to ask about the size of the critical region for the vortex glass behavior for the case of yttrium barium copper oxide. Peter gave some evidence that above T_G there might be a three-degree-wide region, but the flux creep data that you referred to is pretty good evidence that one gets this kind of behavior (where U goes as $1/J$ to some power) all the way down to 20 K. Is it plausible for this range to go from 20 K up to 88 K, or whatever?

D. S. Fisher: There are two issues which one must be careful to separate. The first issue is how wide is the critical regime, which is an issue of the phase transition from the fluid to the vortex glass. That seems to be relatively wide in the YBCO films, much narrower in the crystals. It's exactly that which I was using to get out this estimate that there is of order tens of vortex spacings of lattice-like correlations in the crystals.

There is another issue, which is over what regime do I expect to see vortex glass behavior. As soon as you get below the vortex glass transition temperature, if you go to low enough currents you should always see this kind of behavior. At intermediate currents you may not see it just below the vortex glass transition, in the regime where you can still measure I-V curves. So one would expect that that kind of behavior with the effective barrier growing as an inverse power of J would work all the way from just below the vortex glass transition all the way down to zero temperature. Well, it's of course harder to measure different things in different regimes. So those are I think perfectly consistent, but one is the issue of what is the regime of the vortex glass behavior, the other is the question of what is the regime of the critical behavior.

I should just mention that even in a system which didn't have a vortex glass transition, it is probably still true that at low temperatures some of the behavior of creep and so on and these kind of crossovers, as well as the apparent inverse-power-of-J-variation of the barrier, will still persist. They will however be cut off eventually at very low currents. And that would happen if you had say an isolated layer, or thin system with only a finite number of layers, at low temperatures.

V. M. Vinokur: I'd like to make some small comments on the theoretical basis of this. First of all, what is important and what is the difference between the Anderson-Kim model and the more modern theories of vortex glass? In my personal opinion, it's not the elastic interaction between vortex bundles but the elasticity of a vortex bundle itself. This is quite obvious, because already in David Nelson's paper we have in hand a system, an isolated vortex line, and the static behavior of this system in a random field is known quite well. So we could treat its dynamics equally well. We can find that an isolated vortex line is also in the state of a vortex glass, and this is the consequence of its elasticity. We have no other vortex to interact with for an isolated vortex line. In contrast, the Anderson-Kim model treated vortex bundles like particles, and that was their basic assumption which produced their result. As soon as we take into account flexibility of one isolated vortex bundle we immediately get these diverging barriers.

Another comment concerning this issue of dislocations. This is probably a kind of confusion. The point is that Larkin and Ovchinnikov introduced two different pinning lengths. Unfortunately they used the same notation, and probably this is the origin of misunderstanding. One of them, the shortest correlation length, is just the pinning length which is relevant for all this vortex glass behavior. The other one which is, in the most important cases, exponentially larger than the first, is just the length at which a dislocation should arise. But what is more important, probably, is that you can be easily convinced that individual dislocations cannot affect the existence of vortex glass in the three-dimensional case. Because, individual dislocations as well as dislocational loops cannot carry magnetic flux and it's well known from elasticity theory that the motion of the individual dislocations is massless. So the motion of dislocations does not contribute to the flux motion.

D. R. Nelson: But vacancy interstitial pairs can.

V. M. Vinokur: Yes, just pairs, but that is just what kills the vortex glass in two dimensions because you need finite energy to construct these interstitial vacancy pairs and you need infinite energy in the three-dimensional case to construct these infinite pairs of interstitial-vacancies or dislocations.

D. R. Nelson: Unless it's a type-II solid and entropy allows you to have them in 3D.

V. M. Vinokur: Yes.

D. S. Fisher: Regarding the first point, if the vortex bundles don't interact with each other, no matter how much elasticity is within each bundle the system will still have linear resistivity. I agree with you that sometimes one's best understanding comes from the understanding of single vortex lines, which I should have mentioned.

On the second point, I think I disagree rather strongly. In all the temperature regimes where the data associated with the question of whether it's a vortex glass regime or not, and also all the regimes where you can actually see enough flux creep that you can get down into this regime of currents that are much less than the critical current ... In all those regimes, there is enough smearing of the pinning by thermal fluctuations that there is no separation of length scales between these different lengths which Larkin and Ovchinnikov introduced. There may be slightly longer length scales at which the dislocations come in than where the lattice distortions start becoming of order one, but it's a numerical factor there.

There is however an important question of principle, which is whether or not it is clear that, if I take a phase which has elasticity in it – and I assume that the elasticity always persists – then I obviously have a phase which is different than the fluid, okay. And by ignoring dislocations and other defects one is basically doing that. So unless one allows oneself to go into a regime where the dislocations come in, one is not getting at all to the question which I'm trying to address here, which is whether or not there is really a real superconducting phase.

Nevertheless, if you stay at relatively low temperatures, in a system with not such strong pinning, there are certain intermediate regimes where quite a good description of the system is in terms of an elastic system without dislocations. So quantitatively it can be very useful, but it does not really get at the qualitative issue which I was trying to focus on.

D. R. Nelson: I was interested in your observation that the pictures that show up in these decorations, which sometimes show a lot of orientational and modest translational order, are a nonequilibrium vestige of what's going on at high temperatures. That seems sensible especially because as you know disorder is blurred out by thermal fluctuations at those higher temperatures. Could it not be therefore that the vortex glass behavior that you want to focus on – the underlying vortex glass transition – is really not close to melting, and is in fact *lower* than the melting temperature.

D. S. Fisher: We don't really know where the vortex glass transition is. If one has a first-order or even second-order three-dimensional vortex lattice melting transition, and one has very weak pinning – meaning that the lattice correlations can be very long on both sides of the transition – then I would expect that the vortex glass transition will coincide with the melting transition.

If on the other hand one has a film, or a single layer, then the melting transition in the absence of pinning is just some temperature which is field-independent. But if one has pinning we know that there's no phase transition at all, so then in some sense the melting transition doesn't have much to do with the pinning transition but it may be a point where the resistivity will go down a lot if the pinning is very weak.

For the realistic cases, where the intermediate strengths of pinning and so on, we really don't have any idea of how to estimate the vortex glass transition beyond doing things like the Lindemann criteria where you just balance various energies and ask when they balance. Those are things which are dangerous to at least factors of ten, and maybe dangerous to factors of a hundred. Qualitatively, it seems that the vortex glass transition in YBCO is roughly where one would have expected the Lindemann estimates of the three-dimensional melting transition to be, but we know if we put stronger pinning defects in it gets pushed up. So I think it depends a lot on the regime, and it's only in the limit of very weak pinning that one can really describe at least some of the scales where things happen in terms of melting.

V. G. Kogan: Looking at your phase diagram with the crossover at so-called H_{c2}, and then a phase transition at T_G, where do you expect to have a specific heat anomaly?

D. S. Fisher: That's a very good question. Let me just put up the phase diagram again to answer that.

Let's look at this one. If one estimates how much specific heat anomaly there should be at the vortex glass transition, from very general grounds one believes that the specific heat exponent should be negative, so the specific heat shouldn't diverge. From estimating of the correlation length exponent, we believe that it's actually even weaker than that: the specific heat shouldn't even have a cusp. A large amount of entropy of the system could come out in this regime here [between H_{c2} and T_G].

In BSCCO it's probably even stronger. Probably by the time you get down to here [$\frac{1}{2} T_c$], almost all the entropy's gone and there's very little left in this whole low temperature regime. So the amount of entropy associated with the [vortex glass ordering] is probably going to be quite small. The amount of entropy associated with this transition is down by a factor of order the two-dimensional vortex glass correlation length squared, at the point where you cross over to three-dimensional behavior.

A. Kapitulnik: I gave you a bound.

D. S. Fisher: Right. And it is very small. I think the bound is actually potentially very useful because I think it may tell one about how correlated the motion is.

D. R. Nelson: But to drive your point home, the peak is close to the mean-field H_{c2} where the vortices themselves are beginning to form.

D. S. Fisher: I would guess there would be a specific heat peak there, maybe in YBCO there will still be some signatures of it through the fluid regime, but that's something which one could perhaps do a better estimate of. But I think Aharon's data is very useful as far as setting bounds on what's going on.

If you said this whole regime was single-vortex behavior in the individual layers, then I think it would be hard to have so little entropy associated with this. I think that data suggests that things have already got pretty collective in the layers before you get down here [3D crossover]. Or already collective between the layers, but with these anisotropies that's less likely.

J. Clarke: I want to comment about noise.

As I mentioned briefly yesterday, we've done some measurements in which we bring a high-T_c superconductor very close to a SQUID, and measure the noise that's produced by the motion of vortices pinned in the material. In the low-field limit, this gives you information about the effects of uncorrelated vortices. What I wanted to ask was, if you go into the high-field regime, which would be a much more difficult experiment, would there be interesting signatures of any of these theories?

D. S. Fisher: There's actually some very interesting possibilities if one could make a very small pick-up coil. Particularly interesting. And one could look for space-, history- and time-dependent effects.

J. Clarke: How small?

D. S. Fisher: Well, supposing we're working in a tesla field so we've got 400 Å spacing between the vortices, then we're talking about a micron-sized pick-up coil. Well, that's pretty big by today's standards!

But there's also things to do with bigger coils. If one put a big coil in, one would expect to see almost perfect $1/f$ noise in the magnetization with just logarithmic corrections to it, and also in various other quantities. However, if you had a small probe, say a micron-sized probe that you could move around, you could find the regions which were contributing to that noise. But then what you could do is you could change the temperature and you would find that there were very different regions that would contribute. And that's definitely one of the predictions of the theory.

Of course, some of the noise is just directly related — if it's the linear regime — to $\sigma(\omega)$, but there's a lot of interesting history-dependent behavior that you could also get. If you cool the system and then start measuring noise right away, then you would see that it would probably be bigger than $1/f$ at low frequencies and then as you waited, it would come down toward $1/f$ at low frequencies. And, the crossover frequency would be, say, three decades away from how long you had waited for. That kind of behavior has been seen in spin glasses, and we actually understand it semi-quantitatively. And at least, at our level of understanding, analogous behavior would be expected here.

J. Clarke: Just one comment on that. Is it obvious that the noise you would measure in this kind of experiment is related to $\sigma(\omega)$? $\sigma(\omega)$ relates to a transport property of a current flowing through the film, albeit a high-frequency current, and the magnetic noise measurements measure the motion of the flux.

D. S. Fisher: There is this penetration length which I call λ_G, which is the zero-frequency limit of the penetration length of ac noise. If you're working in a sample which is of order that size, then they will definitely be related to each other. If you're working in a sample which is much bigger than that, then they're not directly related. In fact, then, to get the things I said about $\sigma(\omega)$ one has to go through and do more carefully because of course the currents are all inhomogenous and one has to go through all the normal things that determine these currents . . .

Once you get into the nonlinear regime, the inhomogeneities actually play less of a role, though the relations [between nonlinear $\sigma(\omega)$ and noise] are much more subtle. But they're essentially the same relations that they would be in a Meissner phase, at least naively.

J. Clarke: OK.

4. Vortex Pinning and Creep

Chair — Marty Maley

G. Aeppli
AT&T Bell Laboratories
Murray Hill, NJ 07974

Muon and Neutron Investigations of Vortex Correlations in High-T_c Superconductors

The principles of muon spin relaxation (μSR) and neutron scattering as applied to the study of vortex correlations in superconductors are reviewed. We also describe corresponding experiments involving the determination of magnetic penetration depths, history-dependent phenomena, phase purity, vortex dynamics and flux lattice structure.

1. INTRODUCTION

Most of the experiments described in this book probe the bulk response of flux lattices rather than their microscopic properties. We review here two techniques more directly sensitive to microscopic field distributions and their dynamics. In particular, the neutron scattering cross-section is directly proportional to the Fourier transform of the internal field autocorrelation function,

$$g(x, y, t, t') = \langle h(x,t) h(y,t') \rangle , \qquad (1)$$

while the muon spin relaxation (μSR) function is derivable from the local field autocorrelation function $g(x, x, t, t')$. The present chapter consists of a description of the two techniques and their applications to the study of flux lattices in exotic superconductors.

2. MUON SPIN RELAXATION (μSR)

2.1 What the technique measures[1]

In time-differential μSR experiments, muons with known initial spin polarization are stopped one at a time in a sample, where they decay via a parity-violating process emitting positrons preferentially along their final polarization. The relevant processes in the generation and subsequent decay of the muons (generally positive) are as follows:

$$p + p \rightarrow \begin{cases} \pi^+ + d \\ \pi^+ + n + p \\ \pi^+ + \pi^- + p + p \end{cases} \qquad (2)$$

$$\pi^+ \rightarrow \mu^+ + \nu_\mu \qquad (3)$$

$$\mu^+ \rightarrow e^+ + \nu_e + \bar{\nu}_\mu \qquad (4)$$

The proton-proton collision of Eq. (2) is produced by the acceleration (by e.g., a cyclotron) of protons into a fixed target. The pion (π^+) and muon (μ^+) rest masses and lifetimes are 139.8 MeV, 105.7 MeV, 2.6×10^{-8} s, and 2.2×10^{-6} s, respectively. At the surface muon channels of the Tri-University Meson Facility (TRIUMF) in Vancouver, Canada, where most of the experiments to be described below were performed, the incident muon energies are of order 4 MeV. Data are collected as a function of time t after arrival of the individual muons in the sample by counting the numbers $N_+(t)$ and $N_-(t)$ of positrons emitted in the directions parallel and antiparallel to the incident muon spin. Typically, more than 10^6 muon decays are timed to yield suitable $N_+(t)$ and $N_-(t)$ histograms. The resulting ratio, $[N_+(t) - N_-(t)]/[N_+(t) + N_-(t)]$, is proportional to the time-dependent polarization of the muon ensemble. A magnetic field can be applied either parallel (longitudinal) or perpendicular (transverse) to the initial muon polarization, yielding, respectively, a T_1-type relaxation function $G_{zz}(t)$ or a T_2-like relaxation envelope $G_{xx}(t)$ modulating a precessing muon decay asymmetry, just as for the free induction decay in NMR.

More specifically,

$$G_{ij}(t) = \langle S_\mu^i(t) S_\mu^j(t' = 0) \rangle \qquad (5)$$

where $\vec{S}_\mu(t)$ denotes the muon spin operator at time t and the indices i and j refer to different spatial components. The angular brackets represent the average over all muon sites and possible muon spin histories, obtainable from integration of Heisenberg's equation of motion for $\vec{S}_\mu(t)$,

$$\frac{\hbar}{i}\frac{\partial}{\partial t}\vec{S}_\mu(t) = [\mathcal{H},\vec{S}_\mu(t)] \tag{6}$$

subject to the boundary condition that $\vec{S}_\mu(t=0)$ is as fixed at the time of muon implantation. The interaction Hamiltonian,

$$\mathcal{H} = \vec{h}(t)\cdot\vec{S}_\mu \tag{7}$$

contains the potentially time-dependent internal and external fields $\vec{h}(t)$ to which the muon is subject. When $\vec{h}(t) = H\hat{z}$ is time-dependent and spatially homogeneous, Eqs. (6) and (7) can be solved to show that the muon spins simply undergo precession, with

$$\begin{cases} G_{xx}(t) = \cos(\gamma_\mu Ht + \phi) \\ G_{yy}(t) = \sin(\gamma_\mu Ht + \phi) \end{cases} \tag{8}$$

where $\gamma_\mu/2\pi = 135.5$ MHz/tesla is the gyromagnetic ratio of the muon and the (fixed) phase angle ϕ depends on the experimental configuration. The simple behavior (8) actually can obtain in real experiments: Fig. 1(a) shows precession in the plane perpendicular to an external field of 80 gauss for muons implanted at time t=0 in $La_{1.85}Sr_{0.15}CuO_4$ at room temperature.

If there are a variety of possible muon sites characterized by different static fields along $\hat{z}$ Eq. (8) is readily generalized to

$$\begin{cases} G_{xx}(t) = \int \cos(\gamma_\mu Ht + \phi)\rho(H)\,dH \\ G_{yy}(t) = \int \sin(\gamma_\mu Ht + \phi)\rho(H)\,dH \end{cases} \tag{9}$$

where $\rho(H)$ is the (normalized) distribution of field strengths. Thus, in the static limit, μSR asymmetry spectra are simply proportional to the Fourier transforms of the internal field distribution $\rho(H)$. It is this property that has made μSR a technique for studying flux lattices in superconductors. Specifically, flux lattices at achievable fields (~ 1T) have lattice constants of order 100 Å, which is much larger than the spacing between candidate muon sites in most solids, so that $\rho(H)$ as sampled by the muons is virtually indistinguishable from that averaged throughout the solid.

While the shape of $\rho(H)$ depends on many details characterizing the flux lattice in question, most experiments have concentrated on establishing the second moment $\langle|\Delta H|^2\rangle$ of $\rho(H)$.[2] This has occurred because $\langle|\Delta H|^2\rangle$ is easily extracted from raw data and also because at equilibrium and in the London limit where

$$\xi \ll d \ll \lambda, \tag{10}$$

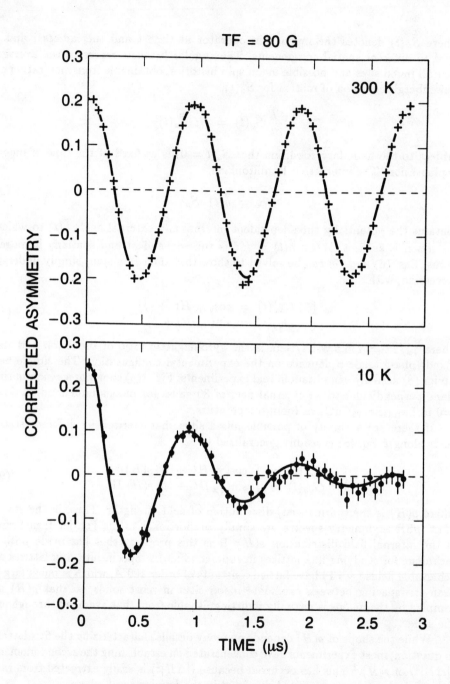

Fig. 1 Muon spin relaxation signal for $La_{1.85}Sr_{0.15}CuO_4$ in a transverse field of 80 G for T above and below $T_c = 37K$. The 10K data were obtained after cooling in a 0.4T field. From Ref. 5.

$$\langle|\Delta H|^2\rangle = 3.706 \times 10^{-3} \phi_o^2 \lambda^{-4} . \quad (11)$$

In Eqs. (10) and (11) the symbols have the usual meanings: ξ is the pair coherence length, d the lattice constant for the (triangular) flux lattice, λ the magnetic penetration depth, and $\phi_o = 2.068 \times 10^{-7}$ G·cm^2, the magnetic flux quantum. Condition (10) has generally been satisfied in investigations of high-T_c superconductors, but not always in studies of materials, such as heavy fermion systems, with smaller upper critical fields H_{c2}. Indeed, because condition (10) is occasionally forgotten, published literature[3] contains inappropriate applications of Eq. (11).

If h(t) is time-dependent as it will be for a vortex liquid the interpretation of $G_{ij}(t)$ becomes considerably more difficult. However, when certain simplifying assumptions are made, it is possible to extract interesting dynamical information from $G_{ij}(t)$. Consider, for example, a random local field $h_z(t)$ with autocorrelation function

$$\langle h_z(t) h_z(t') \rangle = \sigma^2 \exp(-|t-t'|/\tau) . \quad (12)$$

The corresponding power spectrum is

$$S(\omega) = \langle|h(\omega)|^2\rangle = \frac{\sigma^2}{2\pi} \frac{\tau}{(\omega\tau)^2 + 1} = |\sigma_\omega|^2 \quad (13)$$

$S(\omega)$ is related to the local susceptibility $\chi_o(\omega)$ by the fluctuation-dissipation theorem,

$$S(\omega) = \chi_o''(\omega)(n(\omega) + 1) \quad (14)$$

which implies that comparison of bulk (wavevector q=0) ac susceptibility data and μSR measurements can yield information about q-dependent effects in a vortex lattice. We sketch here a derivation of $G_{ij}(t)$ from Eq. (12). First, note that for a particular field history h(t), integration of the Bloch-Heisenberg Eq. (6) gives

$$S_\mu(t) = \exp\left[i \int_o^t h_z(t')\,dt'\right]$$

$$= \exp\left[i \int_{-\infty}^\infty d\omega \frac{h(\omega)}{\omega}(\exp i\omega t - 1)\right] \quad (15)$$

where we consider $S_\mu(t)$ to be a complex number representing the components of $\vec{S}_\mu$ perpendicular to $\hat{z}$, the direction of $h_z(t)$, and $S(t=0) = 1$. Also, for simplicity, we are using units where $\gamma_\mu = 1$. The correlation functions G (t) are then proportional to

$$\langle S_\mu(t) S_\mu(0) \rangle = \langle S_\mu(t) \rangle$$
$$= \iint [\mathcal{D}h] \exp\left[i \int_{-\infty}^{\infty} d\omega\, h(\omega)(\exp i\omega t - 1)\right] \exp\left[-\int \frac{1}{2} \left|\frac{h(\omega)}{\sigma(\omega)}\right|^2 d\omega\right] \quad (16)$$

where $\iint [\mathcal{D}h]$ represents the functional integration over all possible field histories h(t). Eq. (16) is readily evaluated by quadrature, with the result that

$$\langle S_\mu(t) S_\mu(0) \rangle = \exp\left\{i \int_{-\infty}^{\infty} d\omega\, |\sigma(\omega)|^2 \frac{1}{\omega^2} (\exp i\omega t - 1)^2\right\}$$
$$= \exp\left\{-[\sigma^2 \tau^2 (1 - \exp(-t/\tau)^2)]\right\} \quad (17)$$

Eq. (17) is known as the Abragam form. In the static limit, $\tau \to \infty$ and Eq. (17) reduces to the Gaussian form $\exp(-\sigma^2 t^2)$ associated with inhomogeneous broadening. On the other hand, if $\tau \to 0$, $\langle S_\mu(t) S_\mu(0)\rangle = 1$ which represents extreme motional narrowing. For a vortex liquid in the London limit, $\sigma \sim \lambda^{-2}$, as given by Eq. (11), so that Eq. (17) implies that if $\lambda(T)$ is known, the internal field fluctuation rate τ could be readily extracted from μSR experiments.

Below the melting transition of a particular vortex configuration, the internal fields have both static and dynamic components,

$$h(t) = h_o + \Delta h(t), \quad (18)$$

where $\langle \Delta h(t) \rangle = 0$. The Fourier product rule then implies that if h_o and $\Delta h(t)$ are uncorrelated $\langle S_\mu(t) S_\mu(0) \rangle$ is simply the product of an expression like Eq. (17), accounting for $\Delta h(t)$, and Eq. (9), accounting for the distribution of static fields. For a vortex lattice, the Debye-Waller effect determines the relative weights of the static and dynamic contributions to $\langle S_\mu(t) S_\mu(0) \rangle$.

2.2 Measurements of the magnetic penetration depth

By far the most popular[4-9] application of μSR to superconductors has been to the determination of the magnetic penetration depths. The technique has many advantages over more traditional methods for obtaining $\lambda(T)$, most notable that it is a bulk rather than a surface probe. As long as the crystals constituting the sample are substantially larger than the separation (100-1000 Å in typical experiments) between field-induced vortices, grain boundary and surface effects are usually negligible. Thus, μSR does not require special purpose samples, such as thin films or spheres of controlled dimensions. It also yields absolute values for λ which are relatively insensitive to the details of the experiment and analysis: one can see from

Eq. (11) that even an error of a factor of two in the determination of $\langle|\Delta H|^2\rangle$ leads to an error of less than 20% in λ. Fig. 1(b) shows the first microscopic evidence[5] obtained for the formation of a vortex lattice-like state in a high-T_c compound, $La_{.85}Sr_{0.15}CuO_4$. The decaying envelope of the precession signal corresponds to nonzero $\langle|\Delta H|^2\rangle$, in this case measured for a field close to H_{c1} at a temperature T (10K) well below $T_c = 37$ K. Fig. 2 shows the temperature dependence of the decay constant $\Lambda \sim \sqrt{\langle|\Delta H|^2\rangle}$ for the precession signal measured at a field $H = 0.4T \gg H_{c1}$.

To interpret $\lambda(T)$, it is useful to recall the London formula,

$$\lambda^2 = \frac{m^*/m_e}{4\pi n_s r_e} \qquad (19)$$

where $r_e = c^2/m_e{}^2 = 2.82 \times 10^{-5}\text{Å}$ is the classical radius of the electron, m^* the effective carrier mass, and n_s the superfluid density. From Eq. (11), it is apparent that $\Lambda(T) \sim n_s(T)$, which implies that $n_s(T)$ is reduced from its T=0 value

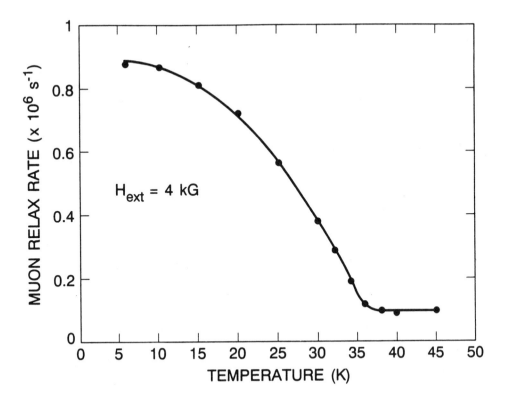

Fig. 2 Temperature dependence of the Gaussian relaxation rate Λ obtained for $La_{1.85}Sr_{0.15}CuO_4$ in a transverse field $H_{ext} = 0.4T$. From Ref. 5.

because of thermally induced excitations across the gap. Thus, $\Lambda(T)$ can be used to determine the nature of the gap function. For $La_{1.85}Sr_{0.15}CuO_4$, conventional weak coupling (BCS) and two-fluid (where $n_s \sim 1 - (T/T_c)^4$) behaviors are not obtained. Instead, as $T \to 0, n_s(0) - n_s(T) \sim T^\alpha$ with $2 < \alpha < 3$, suggesting gapless behavior potentially due to the combined effects of disorder and low-energy magnetic fluctuations visible in similar material by inelastic neutron scattering.[10]

Matters are very different for the stoichiometric compound $YBa_2Cu_3O_7$, and its substoichiometric relative, $YBa_2Cu_3O_{6.7}$, which, when properly prepared, has a somewhat ordered oxygen vacancy array.[11] As described in the article by Batlogg[12] in a previous volume in this series, the consensus among experimentalists is that as found in the original[7] measurements, $\lambda(T)$ is consistent with a conventional gap function in these two compounds. Not only μSR measurements[6-8] performed for $H > H_{c1}$ in the vortex state, but also bulk data[13,14] obtained for $H < H_{c1}$ lead to this conclusion. For example, Fig. 3 shows the agreement between the in-plane $\lambda(T)$ obtained from screening[14] in a 500 Å film, single crystal μSR data,[8] and the BCS theory. There is also agreement on an absolute value of 1400 Å for $\lambda_{ab}(T=0)$ in $YBa_2Cu_3O_7$. For $YBa_2Cu_3O_{6.7}$ ($T_c = 65K$), $\lambda(T)$ is also consistent with conventional pairing, and λ_{ab} (T=0) = 2550 Å,[8] considerably longer than one might guess using the popular hypothesis[3] that $\lambda^{-2} \sim n_s \sim T_c$ and the values of λ and T_c for the stoichiometric compound.

Having described how unconventional behavior obtains in a disordered material ($La_{1.05}Sr_{0.15}CuO_4$) and conventional behavior occurs in a clean compound ($YBa_2Cu_3O_7$), we close this section by showing that unconventional behavior can arise in a clean compound. Fig. 4 displays the squared inverse magnetic penetration depths measured (by μSR) parallel and perpendicular to the $\hat{c}$ axis of the hexagonal heavy fermion compound UPt_3. In this experiment,[9] the unique feature of the μSR technique, that λ can be measured in all directions for a single physical configuration of the sample and apparatus, has been exploited. Specifically, orthogonal Helmholtz coils around the sample are used to prepare flux lattices in either the ab or ac planes of the hexagonal close packed material, while an electromagnetic spin rotator in front of the sample is set so that $\vec{S}_\mu(0)$ is perpendicular to the applied field. The resulting data show that both $\lambda_\parallel^{-2}$ and $\lambda_\perp^{-2}$ vary more rapidly with T at low T than conventional theory predicts. In addition, they have different T-dependencies with $\lambda_\parallel^{-2}(0) - \lambda_\parallel^{-2}(T) \sim T^2$ and $\lambda_\perp^{-2}(0) - \lambda_\perp^{-2}(T) \sim T$ as $T \to 0$. From these data, which together with earlier transverse ultrasound measurements[15] constitute the most direct evidence for anisotropic superconductivity in UPt_3, it is possible to conclude that the gap function has a line of nodes parallel to the planes and pointed nodes along $\hat{c}$ on an (assumed) spherical Fermi surface. Indeed, the solid lines in Fig. 4 are the outcome of a weak-coupling calculation[9,16] based on such a gap function.

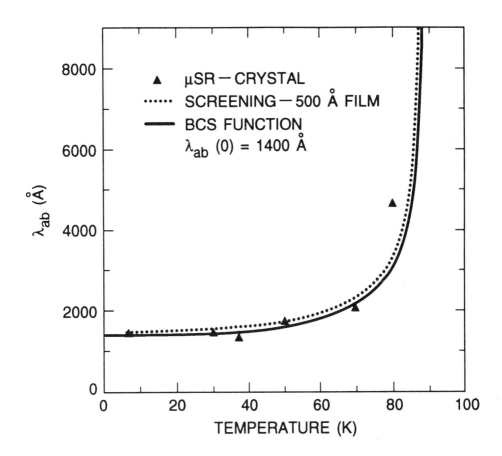

Fig. 3 In-plane magnetic penetration depth in $YBa_2Cu_3O_{7-\delta}$ measured by μSR on single crystals (from Ref. 8) and screening in a 500 Å thick film (from Ref. 14).

2.3 Two-phase coexistence at "non-optimal" compositions

An early and well-known result on high-T_c superconductors was the strong composition dependence of Meissner fractions in series of materials such as $La_{2-x}Sr_xCuO_4$.[17] This result is common to samples prepared by different groups,[17-19] as shown in Fig. 5(a) where the diamagnetic fraction is plotted as a function of x for $La_{2-x}Sr_xCuO_4$. The issue which arises is whether the reduced Meissner fractions for x far from 0.15 are due to anomalous flux pinning effects or

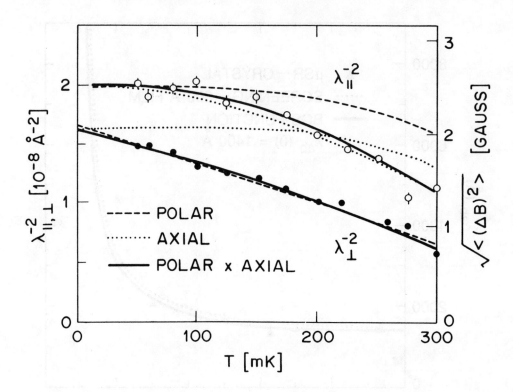

Fig. 4 Temperature dependence of magnetic penetration depth measured parallel ($\lambda_\parallel$) and perpendicular ($\lambda_\perp$) to the c axis in the heavy-fermion superconductor UPt$_3$. From Ref. 9.

simply the fact that the samples in question are not homogeneous bulk superconductors. Fig. 5(b),[19,20] which shows the T-dependence of the Meissner fraction for several samples, provides an argument in favor of the latter hypothesis: the onset temperature for nonzero Meissner fraction does not seem to change with x $\geq$ 0.15. This is also a feature of magnetic susceptibility curves which have been used[18] to derive, via a mid-point criterion, superconducting "phase diagrams" for La$_{2-x}$Sr$_x$CuO$_4$ with x $\geq$ 0.15. The μSR data of Fig. 6 provide a more definite indication of multiphase behavior, not only for x = 0.25, but also for x = 0.075. The samples are the same ceramic samples as those whose Meissner fractions are shown in Fig. 5(b).

The applied field is 0.3 T $\gg H_{c1}$, and the asymmetry functions are plotted in the reference frame rotating at the best fit precession frequency. The solid lines represent the best descriptions of the data using a Gaussian envelope modulating a single precession signal, a form which gives an excellent description of the μSR

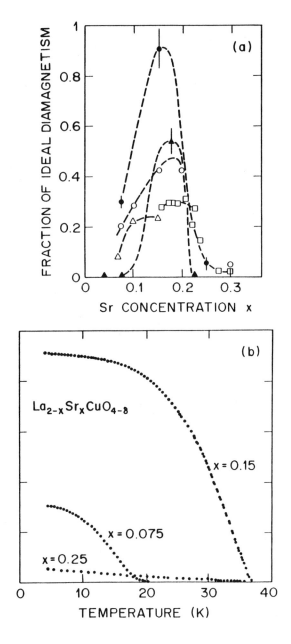

Fig. 5 The (a) diamagnetic fraction for various $La_{2-x}Sr_xCuO_4$ samples as a function of x. The filled circles correspond to samples for which μSR data are shown in Fig. 6. The other symbols correspond to samples made by various groups Ref. (17, 18) at other times. Frame (b) shows T-dependent Meissner fraction for our three samples. From Ref. 19 and Bertram Batlogg, private communication.

signal for flux lattices in optimized CuO-based superconducting ceramics such as YBa$_2$Cu$_3$O$_7$ and La$_{1.85}$Sr$_{0.15}$CuO$_4$.[5-8] From the residuals (shown in bottom half of figure) yielded by these, it is clear that while the model of a single precession frequency is a good description for x = 0.15, it fails for x = 0.075 and 0.25. The presence of a second precession signal, as indicated by the repeated zero crossings of the residuals for x $\neq$ 0.15, is not expected for homogeneous type-II superconductors where, as is the case here, λ and the flux lattice constant are much larger than the distance between possible muon stopping sites, and there is generally (as in 6 (b)) a precession signal characterized by a frequency incorporating a single magnetic shift. Thus, the additional oscillating components for x = 0.075 (where the envelope of the residuals does not even seem to decay as rapidly as A (t) itself) and 0.25 in the vortex regime where the external field $\gg H_{c1}$, are clear indications that the simplest possible explanation of the maximum in the low-field diamagnetism, namely that samples with x far from 0.15 contain substantial non-superconducting inclusions,

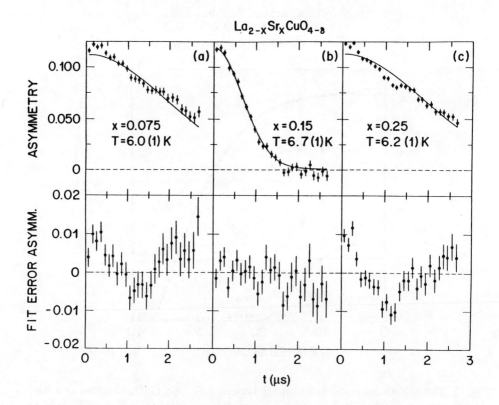

Fig. 6 The muon asymmetry (top frames) and associated residual spectra for La$_{2-x}$Sr$_x$CuO$_{4-\delta}$ with (a) x = 0.075; (b) 0.15; and 0.25 (c). From Ref. 19.

is also the most likely. This result has far reaching implications. Most notable are (i) that "systematic" studies[3,18] of how superconducting properties change with material parameters may simply be studies of inhomogeneous mixtures of various superconducting and non-superconducting materials and (ii) that superconductivity may exist only for certain special carrier concentrations.

2.4 Nonequilbrium behavior of vortex state

Given the ease with which $\langle|\Delta H|^2\rangle$ is measured by μSR, it was natural even at the outset of studies of the CuO-based superconductors to establish the history dependence of $\langle|\Delta H|^2\rangle$. Figs. 7 and 8 show such data for ceramic $La_{1.85}Sr_{0.15}CuO_4$[5] and $YBa_2Cu_3O_7$.[7] Of special interest is the reversibility of $\langle|\Delta H|^2\rangle$ for temperatures well below the respective critical temperatures of the two compounds. This is indicative of the relatively weak pinning forces characteristic of layered CuO materials. The fact that such effects are observed using a *microscopic* probe of the flux lattice demonstrates that the phenomenon, probed in much greater detail by the traditional techniques reviewed elsewhere in this book, of irreversibility lines occurring well below T_c is not simply a surface or grain boundary effect.

2.5 Vortex dynamics

Even though (see (2.1)) μSR has clear promise as a technique for measuring vortex dynamics, we are unaware of attempts to extract directly the parameters describing the power spectrum $S(\omega)$ for local field fluctuations in a vortex lattice. However, Harshman and collaborators[21] have collected some very intriguing data on single crystals of $Bi_{2.1}Sr_{1.94}Ca_{0.88}Cu_{2.07}O_{8+\delta}$. Fig. 9 shows the dependence on T of various moments of the field distribution $\bar{\rho}(H)$ deduced as if all internal fields H were static. The mean square inhomogeneity $\langle|\Delta H|^2\rangle$ and diamagnetic shift $\langle H \rangle - H_o$ by no means follow the trends found even in relatively poor samples of $La_{2-x}Sr_xCuO_4$ and $YBa_2Cu_3O_7$, and which one might expect based on the T-dependent low-field ($H \ll H_{c1}$) diamagnetic signal observed for samples of the type characterized in Fig. 9. Instead of two-fluid or BCS behavior, with a sharp onset for nonzero $\langle|\Delta B|^2\rangle$ at T_c, there is a smooth growth below T_c, with the sharpest rise occurring as $T \to 0$. More startling yet is the nonmonotonic behavior of $\langle B \rangle - B_o$ with a tendency towards reduced diamagnetism and (possibly) paramagnetism setting in near 25 K = T', not far from where many bulk anomalies have been found for the same compound. Finally, the third moment, $\langle|\Delta H|^3\rangle$, which measures the asymmetry of the effective field distribution $\bar{\rho}(B)$, acquires nonzero values below T^1. Nonzero $\langle|\Delta H|^3\rangle$ generally characterizes well-formed flux lattices such as those parallel to the planes of $YBa_2Cu_3O_7$.[9] Thus, it is very tempting to ascribe the effects near T' to collective vortex freezing, above which motional narrowing greatly reduces $\langle|\Delta H|^2\rangle$ and eliminates the asymmetry in $\bar{\rho}(B)$. Below T', Debye-Waller

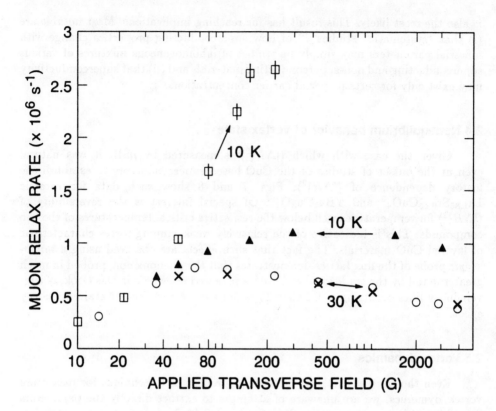

Fig. 7 Dependence of muon spin depolarization rate in $La_{1.85}Sr_{0.15}CuO_4$ on transverse field strength for various temperatures and field histories. From Ref. 5.

effects would account for the T-dependence of $\langle|\Delta H|^2\rangle$ and $\langle|\Delta H|^3\rangle$. However, before accepting a collective freezing scenario, one should be aware that like all other techniques, μSR has a finite frequency window defined by the maximum length of time (between 10 and 50 μsec, depending to a large extent on whether a continuous source such as TRIUMF or a pulsed source such as ISIS in the U.K. is used) over which muon decays can be sensibly accumulated to construct G(t). Therefore one might well attribute the sharp cross-overs seen in Fig. 9 to a thermal activation model monitored at 100 kHz and involving only individual vortices.

Because μSR measures the spectrum $S(\omega)$ of local field fluctuations and ac susceptibility measures the spectrum of long wavelength fluctuations, it is interesting to compare the fluctuation rates obtained from the two techniques. At 35 K and H=0.4 T, assuming the Abragam form and letting $\sigma(T)$ be defined by the

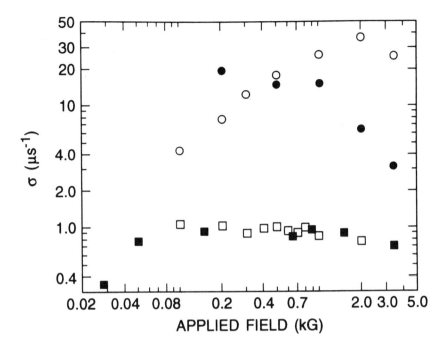

Fig. 8 Dependence on field of Gaussian linewidth σ (measured for sintered YBa$_2$Cu$_3$O$_7$) at 6K and 72K for various field-temperature histories. Open symbols represent data taken in increasing steps from 100 G, after zero-field cooling to 6K (circles) and 72 (squares), and the filled circles represent data taken as a function of decreasing field after field cooling (with H = 0.34T) to 6K (circles) and 72K (squares). From Ref. 7.

two-fluid form indicated by the dashed line in Fig. 9(a), we estimate crudely a local field fluctuation time of $\tau \cong 4 \times 10^{-5}$ sec. The ac susceptibility data[22] on similar but not identical material yield a fluctuation time τ_{ac} of order 2×10^{-5} sec for similar field (H=1.1 T) and temperature. We conclude that the vortex dynamics here are not especially wavelength-dependent. Matters change significantly on lowering T to 25 K, where the ac susceptibility indicates a characteristic time $\tau_{ac} > 0.1$ sec, while from the μSR we estimate that τ is still of order 10^{-5} sec, as one might guess based on the large separation persisting between the dashed curve (corresponding to the two-fluid model) and the data in Fig. 9(a). Thus interpreting the ac susceptibility data as representative of the bulk, cooling from 35 to 25 K has made the vortex dynamics extraordinarily wavelength dependent, again lending support for collective freezing near 25 K. We look forward to μSR and susceptibility experiments, as well as analysis of both data sets to extract the local and long wavelength fluctuation spectra $S(\omega)$, to further test this interpretation.

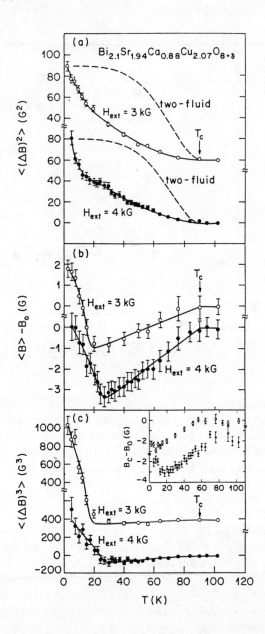

Fig. 9 T-dependence of various moments of distribution of effective fields seen by muons in $Bi_{2.1}Sr_{1.94}Ca_{0.88}Cu_{2.07}O_{8+\delta}$. From Ref. 21, with kind permission of D. Harshman.

3. NEUTRON SCATTERING

Magnetic neutron scattering measures the Fourier transform in time and space of the internal field correlation functions. Thus, the scattering cross-section

$$\partial^2 \sigma / \partial \omega \partial \Omega \sim \frac{k_f}{k_i} S(Q,\omega) , \quad (20)$$

where

$$S(Q,\omega) = \int dx\, dy\, dt\, dt'\, \langle h(x,t) h(y,t') \rangle \exp i\omega(t-t') \exp iQ(x-y) \quad (21)$$

In Eqs. (20) and (21), k_f and k_i are the incoming and outgoing neutron wavevectors, $\hbar\omega$ represents the difference between ingoing and outgoing neutron energies, and the momentum transfer $Q = k_i - k_f$. The local correlation function $S(\omega)$ (see Eq. (13)), in principle measurable by μSR, is simply

$$S(\omega) = \int S(Q,\omega) d^d Q$$

while the ω-dependent bulk susceptibility is related via the fluctuation-dissipation theorem to $S(Q=0,\omega)$. It should be obvious that in principle, complete neutron scattering experiments on the vortex states of BiSrCaCuO could settle immediately most controversies aired at this meeting. Unfortunately, sample requirements are much more stringent for neutron scattering than for μSR and so the present section of this chapter is much shorter than section 2. Indeed, only elastic ($\omega = 0$) measurements[23] have been performed on *any* superconductor and thus far, we are aware of only one group[24] which has observed flux lattices in high-T_c compounds by neutron scattering. Fig. 10 shows the lowest order Bragg peaks obtained by this group[25] from YBa$_2$Cu$_3$O$_7$ in a 0.6 T field applied parallel to $\hat{c}$. The flux lattice is clearly square, with the strong peaks along the [110] and [$\bar{1}$10] directions of the underlying crystal. Equally interesting is Fig. 11, which shows a scan through the [100] reflection of the flux lattice for T=10K and B=2T. This corresponds to a scan in momentum space along $c^* \parallel \hat{c} \parallel \vec{B}$. The peak is resolution limited, which by Fourier's theorem implies that the vortex lines are rigid rods on length scales $\gtrsim 30$ μm. In other words, there is no vortex entanglement here.

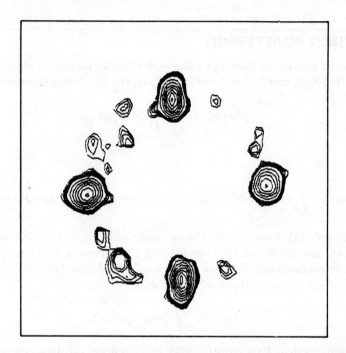

Fig. 10 Neutron diffraction pattern for flux lattice in $YBa_2Cu_3O_{7-\delta}$. Kindly provided by E. M. Forgan (see also Ref. 24).

4. SUMMARY AND CONCLUSIONS

The muon spin relaxation (μSR) and neutron scattering techniques have made the following contributions to our understanding of copper oxide superconductors.
  (i) Determination in absolute units of the magnetic length $\lambda(T)$.
 (ii) Observation in $YBa_2Cu_3O_7$ of well ordered flux lattices, composed of straight flux lines, at low T and intermediate fields.
(iii) Demonstration that sample multiphase coexistence is the most likely cause of reduced Meissner fractions found for $La_{2-x}Sr_xCuO_4$ ceramics with non-"optimal" x (i.e. x $\neq$ 0.15)
 (iv) Finding that microscopic irreversibility sets in at surprisingly low T.
  (v) Observation of microscopic vortex motion in BiSrCaCuO single crystals.

While studies of type (i) have been extensively performed, there have been relatively few experiments of the other four varieties. Therefore, we conclude that both μSR and neutron diffraction still have the potential to provide new information about the magnetic correlations in high-T_c materials.

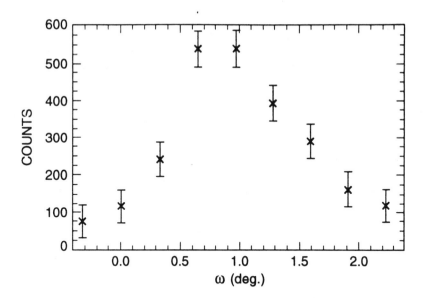

Fig. 11 Rocking curve (scan along c* direction in momentum space) of Bragg peak from flux lattice in $YBa_2Cu_3O_{7-\delta}$. Kindly provided by E. M. Forgan (see also ref. 24).

ACKNOWLEDGMENTS

The author is especially grateful to D. Harshman and E. Forgan for making unpublished figures available for reproduction in this article. He also thanks B. Batlogg, C. Broholm, D. Bishop, R. Cava, D. Harshman, R. Kleiman, D. Williams and other collaborators too numerous to mention for their role in the flux lattice experiments reviewed here as well as their considerable influence on the author's perspective on this field. Work at TRIUMF was supported by the Natural Sciences and Engineering Research Council of Canada and the Canadian National Research Council.

REFERENCES

1. A complete introduction to this field is A. Schenck, "Muon Spin Rotation Spectroscopy Principles and Applications in Solid State Physics," (Adam Hilger, Bristol, 1985).
2. E. H. Brandt, Phys. Rev. B **37**, 2349 (1988); W. Barford and J. M. F. Gunn, Physica **156C**, 515 (1988).

3. See, e.g. Y. J. Uemura *et al.*, Phys. Rev. Lett. **66**, 2665 (1991).
4. A. T. Fiory *et al.*, Phys. Rev. Lett. **33**, 969 (1974); Y. J. Uemura *et al.*, Hyperfine Int. **31**, 413 (1986).
5. G. Aeppli *et al.*, Phys. Rev. B **35**, 7129 (1987).
6. W. J. Kossler *et al.*, Phys. Rev. B **35**, 7133 (1987); F. M. Gygax *et al.*, Europhys. Lett. **4**, 473 (1987); R. Wappling *et al.*, Phys. Rev. Lett. **A122**, 209 (1987); F. W. Cooke *et al.*, Phys. Rev. B **37**, 9401 (1988); B. Pumpin *et al.*, Physica C **162-164** (1989).
7. D. R. Harshman *et al.*, Phys. Rev. B**36**, 2386 (1987).
8. D. R. Harshman *et al.*, Phys. Rev. B**39**, 851 (1989).
9. C. Broholm *et al.*, Phys. Rev. Lett. **65**, 2062 (1990).
10. See, e.g. S.-W. Cheong *et al.*, Phys. Rev. Lett. **67** 1791 (1991).
11. D. J. Werder *et al.*, Phys. Rev. B **37**, 2317 (1988); R. M. Fleming *et al.*, Phys. Rev. B **37**, 7920 (1988).
12. B. Batlogg in *High Temperature Superconductivity*, The Los Alamos Symposium 1989, edited by K. S. Bedell, D. Coffey, D. E. Meltzer, D. Pines and J. R. Schrieffer (Addison-Wesley, Redwood City, 1990).
13. A. T. Fiory *et al.*, Phys. Rev. Lett. **61** 1419 (1988).
14. L. Krusin-Elbaum *et al.*, Phys. Rev. Lett. **62**, 217 (1989).
15. B. S. Shivaram *et al.*, Phys. Rev. Lett. **56**, 1078 (1987).
16. See also F. Gross *et al.*, Z. Phys. B **64**, 175 (1987).
17. R. B. van Dover *et al.*, Phys. Rev. B**35**, 5337 (1987); R. M. Fleming *et al.*, Phys. Rev. B**35**, 7191 (1987); J. Orenstein *et al.*, Phys. Rev. B **36**, 8892 (1987).
18. J. B. Torrance *et al.*, Phys. Rev. Lett. **61**, 1127 (1988).
19. D. R. Harshman *et al.*, Phys. Rev. Lett. **63**. 1187 (1989).
20. B. Batlogg, private communication.
21. D. R. Harshman *et al.*, Phys. Rev. Lett. **67**, 3152 (1991).
22. See, e.g. review by P. Gammel, J. Appl. Phys. **67**, 4676 (1990).
23. See, R. P. Huebener, *Magnetic Flux Structures in Superconductors*, (Springer, Heidelberg, N.Y. 1979) for a review of work on "conventional" superconductors.
24. E. M. Forgan *et al.*, Nature (London) **343**, 735 (1990).
25. E. M. Forgan, private communication.

DISCUSSION

D. R. Nelson: Could you explain again why the lattice structures seen by Forgan, et al. in 1-2-3 was square? Why did we not see six spots?

G. Aeppli: Because I see only four spots. And the four spots are arranged on a square. Is that good enough?

P. L. Gammel: The vortices are heavily pinned by twin boundaries.

G. Aeppli: I agree. I'm just saying that the best bet is that it's a square lattice, just from looking at the data.

D. R. Nelson: What did that mean?

G. Aeppli: But remember, the twin boundaries . . .

P. L. Gammel: I think it's likely, and I think Forgan would agree, that you have very well oriented vortices which are pinned by the twin boundaries, and then you can have the rest of the vortices which are less well correlated in the remainder of the sample.

D. S. Fisher: The scattering is dominated by the Bragg peaks?

P.L. Gammel: . . . And the Bragg peaks are dominated by the vortices which are along the twin boundaries. That doesn't say anything about most of the vortices, which are the ones somewhere else.

G. Aeppli: But there's a total-moment sum rule here, Peter. The point is that these things more or less exhaust the sum rule and give you the muon λ. I mean that there's a relationship between the μSR results and these results, since μSR measures $(\delta H)^2$. The sum over our data is also $(\delta H)^2$. Not having done this [the neutron] experiment, my feeling is that they would have never seen these results if they hadn't accounted for a large part of $(\delta H)^2$ in the sample.

P. L. Gammel: I didn't realize that.

G. Aeppli: You may be completely correct in that the orientation of the thing is fixed by the twin boundaries which happen to be along these directions, but I think also that these peaks come from the bulk of the grains. I don't think . . .

D. R. Nelson: Well why are there not six spots?

G. Aeppli: There are not six spots. There are four spots, and then there's some stuff in between which accounts for some other part of $(\delta H)^2$. But I don't believe, having tried measurements ourselves, that these signals account for anything less than 30 percent of the bulk of the 1-2-3 crystals. But again . . .

B. R. Coles: Can you get more detailed information by using polarized neutrons?

G. Aeppli: No, not really, because you have a field at the sample when you do the experiment. So, you're always in this limit where it would be very difficult to arrange a polarized beam experiment near the big magnet that you have here. You wouldn't learn much anyway, because from the point of view of scattering it's very much like a ferromagnet. Things get more interesting, in fact, if you can do inelastic scattering. Because then, just as in the ferromagnet, you might actually conceive of seeing some [neutron spin] flips.

D. K. Finnemore: I'm a little surprised you didn't take more data points in the λ versus T curves in the range between 70 K and 85 K. I realize that λ gets long so it's hard to measure, but . . .

G. Aeppli: We have other experiments where we take more. My view is that the μSR works best where the other techniques work worst, and vice versa. And so it's clear that careful magnetometry near where λ is big, you could get many more data points and much higher accuracy and make, also, potentially more interesting conclusions especially since you are, after all, applying a rather large field to get λ. You should be somewhat leery of making too many conclusions based on μSR determinations of λ, especially in the 2-2-1-2 but of course even in the 1-2-3 once λ gets to be of order 10 000 Å or above. You're better off then doing a real bulk experiment.

T. M. Rice: Can't you decide this issue – whether or not you are missing a lot of vortices – by just comparing the lattice parameter of your lattice with the total flux through the sample?

G. Aeppli: No, the lattice parameter is right. Roughly right. You have to look at the absolute intensities, Maurice.

P. L. Gammel: The lattice parameter is going to be roughly the same in the planes, or in between them.

D. R. Nelson: It's the average intervortex spacing in both cases.

P. L. Gammel: The average intervortex spacing is pretty much fixed by the applied field.

D. R. Nelson: I see, Peter's explanation is you just have one-dimensional crystals, of lines . . .

G. Aeppli: No, one-dimensional crystals will not give you spots like that. One-dimensional crystals will give you streaks on this plane. That's not correct.

D. R. Nelson: Okay, then why are there four spots Gabe? Why are there four spots there?

G. Aeppli: Look, I don't know why there are four spots. I can only say that if there are four spots I can apply Fourier's theorem and get a square lattice.

D. R. Nelson: So can I.

G. Aeppli: No, but I mean the data are the data. Why it's square, I don't know. By the way, most of you have probably not seen these data. The earlier data is much less clear than these data, so if you've been used to looking at the old data – these are very new data and much, much cleaner than anything they've shown in the past.

D. S. Fisher: The width of these Bragg peaks gives you how long a correlation length?

G. Aeppli: This you can actually see, Daniel, because you can look at these widths and compare them with the distance from the origin and you can then decide how much you want to trust them to extract correlation lengths after having made resolution corrections. And you can just see by inspection here that you're dealing with something of order ten, fifteen lattice spacings in the plane.

S. A. Trugman: Are there other spots further off? I mean, there should be a whole lattice of these things, right?

G. Aeppli: You bet.

S. A. Trugman: So where are the other ones?

G. Aeppli: The other ones are very very weak because the intensity goes down like $1/q^4$. And, I'm also not sure whether they rotated the cryostat far enough to satisfy the Bragg condition for those.

S. A. Trugman: I guess this could be something trying to be a triangular lattice slightly distorted, or something.

G. Aeppli: Look, all I could do is apply Fourier's theorem to the data, and I think it's then further . . .

D. R. Nelson: You can apply it, but you need to know the phases, of course, to uniquely determine this in real space. And this could be a superposition of two orthogonal hexagonal lattices, something like that . . . [Laughter]

G. Aeppli: No, the superposition of hexagonal lattices would give you, I think, something quite different.

S. Doniach: Maybe it's telling us something about the physics in the order parameter.

B. R. Coles: How many different temperatures have they done?

G. Aeppli: This was just given to me. I don't know how many different temperatures they did this optimized experiment at. In the *Nature* article there are of order ten temperatures. That's the only published thing that I know of. There may be something from the Japanese meetings.

D. R. Nelson: There is a paper with this picture in it, actually.

G. Aeppli: Okay, I haven't seen that paper. That must be for the Japanese meeting that was just held a few weeks ago.

P. L. Gammel: I also wanted to point out something about the μSR. Can you put up your slide of the 2-2-1-2 μSR data again. I'm interested in the data on the first moment, which is similar to magnetization.

G. Aeppli: Yes.

P. L. Gammel: Bertram Batlogg and I have been taking some high field magnetization data, and we see a large H/T contamination of the high-field magnetization. This gives you similar magnitude to the upturn in the μSR just from a free-spin term.

G. Aeppli: That's why I didn't put it into my description of the physics.

P. L. Gammel: The second question is, why not do the neutron scattering in 2-1-4, where large crystals are now available.

G. Aeppli: That's actually being done, up to 30 K, which is unfortunately just where we start seeing irreversibility in the muons. What Herb Mook and his collaborators have found is that there's a very clean flux lattice, again which is resolution-limited in all directions including the z-direction. So that's being done.

M. P. Maley: When you do the irreversibility line measurements and you ramp the field from zero, and you compute the gradient of the flux across the sample, does that agree with the magnetic measurements?

G. Aeppli: This is a different thing, actually. These gradients are not across the entire sample. They're microscopic gradients that we measure, not a macroscopic gradient. And so what you can say is that the microscopic gradients at equilibrium correspond nowadays to what everybody believes the λ value is.

B. I. Ivlev*
University of Southern California
Los Angeles, CA 90089-0484

Resistance, Flux Motion and Pinning in High Temperature Superconductors

1. INTRODUCTION

Properties of vortex systems in high-temperature superconductors are now under intense investigation. As observed by Müller et al.[1] and Yeshurun and Malozemoff,[2] the relaxation of magnetization supposes that flux creep plays an important role in magnetic measurements. The experiments by Palstra et al.[3] clearly demonstrate the thermally activated nature of resistivity in a wide region below T_c. The activation energy depends on the magnetic field, the temperature, and the transport current. The current dependence of the activation energy is a very remarkable feature of a vortex state. If in the limit of small current the activation energy tends to infinity (flux creep), it can be considered a true superconducting state with zero ohmic resistivity. If the small current limit of the activation energy is finite, then ohmic resistivity exists and this region is called thermally assisted flux flow (TAFF). (See paper by Kes et al.[4]) The transition between ohmic and nonohmic resistivity was observed by Koch et al.[5]

---

*Permanent Address: Landau Institute for Theoretical Physics, USSR Academy of Sciences, Kosygin St. 2, 117334 Moscow

Reviews of flux motion in high-temperature superconductors have been given by Doniach,[6] Tinkham,[7] and Brandt.[8] Flux creep and TAFF correspond to different states of a vortex system. Flux creep occurs in a solid vortex state, which can be realized either as the Abrikosov lattice or as a "glass state" considered by Fisher.[9] TAFF in a bulk superconductor should correspond to a vortex liquid state. (See the papers by Doniach[6] and Chakravarty et al.[10]) The melting of the vortex-line lattice and the formation of a vortex liquid was studied by Nelson and Seung,[11] Brandt,[12] and Houghton et al.[13]

Experiments by Yeshurun and Malozemoff,[2] Palstra et al.,[3] Kim et al.,[14] and Brunner et al.[15] give different magnetic field dependence of the activation energy. The theoretical study of the activation processes in a liquid state and the estimation of the activation energy as a function of the magnetic field and temperature are not a simple problem. Some approaches for the liquid state are given below.

In the case of high current, thermally activated and quantum creep are considered.

2. LAYERED SUPERCONDUCTORS

The phenomenological theory of layered superconductors is the Lawrence-Doniach model.[16] This modified Ginzburg-Landau theory follows from a tight-binding approximation. The free energy for a layered superconductor is

$$F = \int d^2 r \, d \sum_n \left[-\alpha |\psi_n|^2 + \frac{\beta}{2} |\psi_n|^4 + \frac{1}{2m} \left| \left(\nabla - \frac{2ie}{c} \vec{A} \right) \psi_n \right|^2 \right.$$
$$\left. + \frac{1}{2Md^2} \left| \psi_{n+1} \exp\left(-\frac{2ie}{c} \int_{nd}^{(n+1)d} dz A_z \right) - \psi_n \right|^2 + \frac{\vec{B}^2}{8\pi} \right]. \quad (1)$$

The phenomenological values are determined as

$$\alpha = \frac{1}{2m\xi_{ab}^2} \quad \beta = \frac{8\pi^3}{m^2 \phi_o^2} \frac{\lambda_{ab}^2}{\xi_{ab}^2}.$$

The length ξ_{ab} is the coherence length in the direction parallel to the planes and λ_{ab} is the London penetration depth of currents which are parallel to the ab-planes. The large value of the effective mass ratio $M/m = \lambda_c^2/\lambda_{ab}^2 = \xi_{ab}^2/\xi_c^2$ determines the coherence length ξ_c in the c-axis direction and the London penetration depth λ_c for the current in the c-direction.

The important physical parameter is the ratio of the interlayer distance d to the coherence length ξ_c. If the value d/ξ_c is small, a superconductor can be considered as three-dimensional and anisotropic. The typical example of this type are the yttrium compounds with the anisotropy ratio $\lambda_c^2/\lambda_{ab}^2 \sim 30$. The opposite limiting case $d/\xi_c \gg 1$ corresponds to rather a two-dimensional layered superconductor with a weak Josephson coupling between the layers. Examples of this type are the bismuth and thallium compounds with a large anisotropy ratio $\lambda_c^2/\lambda_{ab}^2 \sim 3000$.

Using expression (1) for the free energy one can describe a system of vortices in any configuration. The vortices along the c-direction can be either rectilinear or curved, and they can also be placed arbitrarily within the ab-planes.

3. RESISTIVE PROPERTIES OF THE VORTEX STATE

A pure superconducting sample free of defects cannot carry nondissipative current in the mixed state, because even a small Lorentz force acting on a vortex leads to its motion. The motion of the vortex produces a voltage across the sample. But some inhomogeneities inside a sample can effectively pin the vortex. If the Lorentz force is smaller than the pinning force, the vortex structure is at rest and there is no voltage. It means that some critical current density j_c exists which is determined by the pinning properties of the sample. The critical current density j_c is called the depinning current. In microscopically homogeneous samples (isotropic or anisotropic) some structural defects can lead to the pinning of vortices. (See the review paper by Larkin and Ovchinnikov.[17]) In layered high-temperature superconductors there is an additional intrinsic pinning mechanism due to order parameter modulation.[18,19,20] This interlayer pinning is important if the magnetic field direction is parallel to the layers. In this case the vortex core is located between the superconducting planes. A shift of the core into a layer destroys the superconductivity in that layer, which costs the condensation energy. Hence, it is very important how the magnetic field is aligned with respect to the ab-planes.

If the transport current is smaller than the depinning value j_c, the vortices are pinned in the sample, but this state is metastable, because vortices can displace from one position to another. In accordance with Anderson's idea a vortex bundle jumps through a finite distance.[21] This jump can occur by means of a thermal fluctuation. In this case the resulting voltage in the pinned vortex system has a thermally activated behavior, and it gives rise to the activated flux creep.[2,3] As the temperature is lowered, the thermally activated vortex motion can change to a quantum regime.[22-26]

The flux flow regime for $j > j_c$ is characterized by ohmic resistivity, where the electric field is $E = \rho_{FF}\, j$. (See the review paper. Ref. 17.) In the flux creep region at $j < j_c$ the voltage has a more complicated dependence on the current, and one can write in this case

$$E = B \exp(-A_o) . \qquad (2)$$

The A_o and B depend on the current, magnetic field, and temperature. For the thermally activated creep the exponent is $A_o = U_o/T$, where U_o is the activation energy.

Below, we will focus on consideration of the flux creep region, and mainly discuss the exponent A_o.

4. FLUX CREEP IN THE CASE OF LARGE CURRENT

In this section we will consider the case of a large current which can be comparable to or less than the critical value, but it is essential that the current has to be finite.

4.1 Magnetic field perpendicular to the ab-planes

In this case only defects of structure can be responsible for pinning. We consider here the collective pinning mechanism produced by weak disorder.[17] In high-temperature materials, it can be due to randomly distributed defects, like oxygen vacancies. The theory of collective creep under these conditions has been developed by Feigelman and co-authors.[27] This consideration starts with an elastic Abrikosov vortex lattice with a pinning potential. The elastic and pinning energy can be written in the form given in Larkin's paper[28]

$$F_{lat} = \int \frac{d^3k}{(2\pi)^3} \left[(c_{11} - c_{66}) \frac{(\vec{k}\vec{U}_k)^2}{2} + \frac{c_{66}}{2} (k_\perp \vec{U})^2 + \frac{c_{44}}{2} k_z^2 \vec{U}^2 + \vec{U}_k \vec{f}_k \right] \quad (3)$$

where the two-dimensional vector $\vec{U}$ describes a local displacement of the flux line lattice; c_{11}, c_{44}, and c_{66} are, respectively, the bulk, tilt, and shear elasticity moduli. The random pinning force f satisfies the condition

$$\langle \vec{f}_k \vec{f}_{k1} \rangle = (2\pi)^3 W \delta(\vec{k} - \vec{k}_1) . \quad (4)$$

The magnetic field is directed along the z-axis which coincides with the c-axis.

If the transport current has the same order of magnitude as the depinning one, $j \lesssim j_c$, the hopping distance is of the order of the coherence length ξ_{ab}. Expressions for the sizes of the vortex bundle in the c-direction, L, in the direction of hopping $R_\parallel$, and in the transverse direction $R_\perp$ can be obtained from the relations (3) and (4)[27]

$$R_\perp \sim c_{44}^{1/2} c_{66}^{3/2} \xi_{ab}^2 / W$$

$$L \sim R_\parallel \sim R_\perp (c_{11}/c_{66})^{1/2} . \quad (5)$$

In the same way, the expressions for the activation energy U_o and the depinning current can be obtained,

$$U_o \sim c_{66}^{3/2} c_{11}^{3/2} \xi_{ab}^4 / W \qquad (6)$$

$$j_c \sim \frac{1}{B} \frac{W^2}{\xi_{ab}^3 c_{44} c_{66}^2} . \qquad (7)$$

A quite different situation occurs in the case of a small transport current j. In order to have a gain in the Lorentz energy, the vortex bundle has to hop on a larger distance and its transversal size scales as $R_\perp \sim j^{-5/9}$. In the limit of small transport current the scaling law for the activation energy is

$$U_o \sim j^{-7/9} . \qquad (8)$$

In accordance with formula (2), this means the absence of ohmic resistivity for a weakly pinned Abrikosov lattice.

4.2 Intrinsic interlayer pinning of vortices parallel to the ab-plane

Interlayer pinning was considered by Tachiki and Takahashi,[18] Ivlev and Kopnin,[19] and Barone et al.[20] If the vortex structure fits exactly the interlayer structure, the depinning current j_c can be estimated from following simple considerations. The interlayer pinning force acting on a vortex of length L is $F_{cond} \xi_{ab} L$, where $F_{cond} \sim \phi_o^2/\lambda_{ab}^2 \xi_{ab}^2$ is the condensation energy density (the vortex destroys the superconductivity over the area of order $\xi_{ab} L$ within the ab-plane). The Lorentz force acting on a vortex of length L is $j\phi_o L/c$. Equating this value to the interlayer pinning force, we get an estimate for the depinning current, $j_c \sim j_{GL}$, where the Ginzburg-Landau depairing current is

$$j_{GL} = \frac{\phi_o c}{12\pi^2 \sqrt{3} \lambda_{ab}^2 \xi_{ab}} . \qquad (9)$$

If the vortex system forms the Abrikosov lattice, one has to compare the elastic vortex lattice energy and the interlayer pinning energy in order to determine if the vortex lattice is commensurate with the interlayer structure or not. (Pokrovsky and Talapov[29] and Ivlev and Kopnin.[30]) Near the curve $H_{c2}(T)$ in the region $T \simeq T_c$ the elastic energy of the vortex lattice dominates and the period of the vortex structure along the c-axis is determined by the magnetic field. The depinning current in this case is an oscillating function of the magnetic field. It reaches maxima when the magnetic field is such that the c-axis period of the Abrikosov lattice is commensurate with the interlayer crystal structure. It can be true only in the case $\xi_c > d$. For Y-based compounds, it is a relatively large region below T_c. The maximum value of the critical current for the commensurate situation is[19]

$$j_c \sim j_{GL} \left(1 - \frac{B}{H_{c2}}\right) \left(\frac{\xi_c}{d}\right)^2 \exp\left(-\frac{8\xi_c^2}{d^2}\right). \qquad (9a)$$

In the opposite limit when B is near the lower critical field H_{c1}, the critical current is determined by the single vortex interlayer pinning. For $\xi_c \gg d$ it has the form[20]

$$j_c \sim j_{GL} \left(\frac{\xi_c}{d}\right)^{7/2} \exp\left(-15.77 \frac{\xi_c}{d}\right). \qquad (9b)$$

As one follows from formulas (9a,b) the large size vortices are weakly pinned by the interlayer periodic structure.

4.3 Magnetic field parallel to the ab-plane

In this case, the equilibrium positions of the vortex cores are between the superconducting planes of a strongly layered superconductor such as Bi-Sr-Ca-Cu-O. Suppose there are no other types of pinning except for intrinsic interlayer pinning. For the vortex line to move to the neighboring interlayer space, the "kinks" in the vortex line, as indicated in Fig. 1, must be nucleated. One vortex deviates from its equilibrium position into the neighboring interlayer space. The energy of this critical nucleus is the activation energy U_o.

We now consider the expression for the change in energy, U, due to the activation of a segment of the length R to the neighboring interlayer spacing. U has the form

$$U = \delta F_{cond} + V_{int}(R) - (j - j_1) \frac{\phi_o d}{c} R. \qquad (10)$$

Following the paper by Chakravarty et al.,[31] $\delta F_{cond} = (\phi_o^2 d/8\pi^2 \lambda_{ab}^2) \ln(d/\xi_c)$ is the loss of the condensation energy due to the destruction of superconductivity at those two points on the layer threaded by the vortex, separated by a distance R. $V_{int}(R)$ is the interaction energy of the two vortex kinks. The term proportional to the transport current j is due to the Lorentz force. The term proportional to j_1 is the energy that arises due to the displacement of the vortex line and is clearly proportional to the length R. The origin of this tension energy can be understood from Fig. 1. The equilibrium Abrikosov vortex lattice is represented on a cross sectional plane S (Fig. 1a) as a system of dots forming a compressed hexagonal lattice. A small shift of one vortex from the equilibrium position A to the point B in Fig. 1b produces an elastic energy in the crystal vortex lattice.

The current j_1 can be found by estimating the change in an elastic energy of the system in Fig. 1b. This gives[31]

$$j_1 \simeq \frac{cBd}{16\pi \lambda_{ab}^2}. \qquad (11)$$

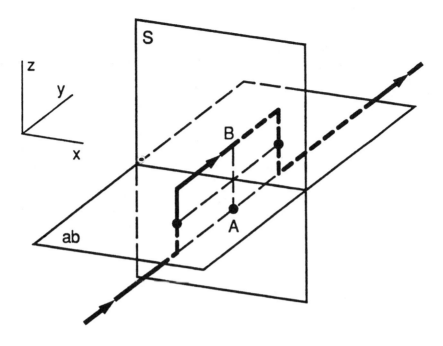

Fig. 1a Single-vortex activation when $\vec{B} \parallel ab$. The length of a shifted vortex segment is R.

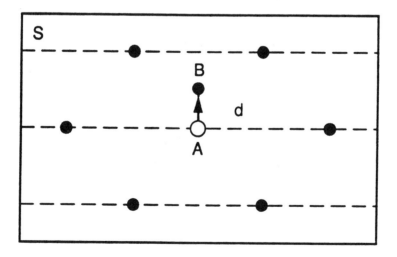

Fig. 1b The illustration of the origin of a linear tension energy when the vortex is shifted a small distance.

The activation energy is determined by the relation $U_o = U(R_c)$, where R_c is the critical distance corresponding to the extremum of the expression (10). This is a result of competition between the tension energy (the last term in Eq. 10) and the two-kink interaction energy, which in the case $d\lambda_c/\lambda_{ab} \ll R \ll \lambda_c$ can be written in the form[31]

$$V_{int}(R) = - \frac{\phi_o^2 d^2}{32\pi^2 \lambda_{ab}^3} \frac{\lambda_c}{R} . \tag{12}$$

In the same limiting case, the critical distance is $R_c \sim \xi_{ab}^2 \, d \, j_{GL}/\xi_c \, (j-j_1)$ and the activation energy is mainly $U_o = \delta F_{cond}$. If the transport current j is very close to, but larger than, the value j_1 determined by vortex lattice elasticity, then the critical size of the vortex loop R_c tends to infinity. It follows from the expression (10), at $j < j_1$, that one cannot obtain an energy gain in order to define the value of R_c. For a current j smaller than j_1, the approximation of a single vortex is not valid. In this case the critical nucleus consists of a bundle of vortices. As was shown by Chakravarty et al.,[31] the critical size of this bundle at $j < j_1$ along a field direction is $R_c \sim c\phi_o/\lambda_{ab}^2 \, j$, and the activation energy is

$$U_o = \delta F_{cond} \left(\frac{j_1}{j}\right)^2 . \tag{13}$$

Comparing the results (8) and (13) for the field perpendicular and parallel to the ab-planes one can draw an important conclusion: regardless of the particular nature of the pinning mechanism and the magnetic field direction, the thermally activated flux creep in the Abrikosov vortex lattice becomes less probable when the transport current j decreases. The same statement holds for a disordered but solid vortex system with a finite shear modulus. In accordance with the general formula (2) this corresponds to the absence of ohmic conductivity in the limit of small current in a solid vortex systems.

5. RESISTIVITY AT SMALL TRANSPORT CURRENT

There is experimental evidence that, in the limit of low current density, high-temperature superconductors manifest ohmic resistivity. Current-voltage characteristics were investigated in Refs. 3, 5, 14, 30-32, and by other authors. At small current density, the resistivity does not depend on the current, but has a thermally activated character. That is, in formula (2), coefficient B is linearly proportional to j, and the exponent is $A = U_o/T$ where U_o does not depend on j. This regime is called the thermally assisted flux flow (TAFF), proposed by Kes et al.[4]

We have seen above that it is impossible to construct a finite energy critical nucleus corresponding to the TAFF regime in the limit of small current, in the case of a 3D solid (crystal lattice or glass) vortex system. In a stack of 2D superconducting layers when the Josephson interaction between layers can be omitted, the magnetic field perpendicular to the ab-planes forms a 2D vortex lattice in each layer. The motion of a point defect through this lattice is accompanied by a plastic deformation of the lattice. The plastic energy can be evaluated in accordance with the London theory as $\phi_o d\,(\delta B)/8\pi$, where $\delta B = \phi_o/2\pi\lambda_{ab}^2$ is the magnetic field modulation in the vortex lattice. The plastic energy plays the role of the activation energy $U_o^{(2D)}$ (see the paper by Vinokur et al.[40]),

$$U_o^{(2D)} \simeq \frac{\phi_o^2 d}{16\pi^2 \lambda_{ab}^2}. \tag{14}$$

This expression for the activation energy can be generalized to account for the inhomogeneities.[39,40] Together with the formula (14), one thus obtains a good approximation for solid vortex systems in 2D layers (films) or for a stack of layers with negligible Josephson coupling.

Koch et al.[5] have reported, from current-voltage studies on YBa$_2$Cu$_3$O$_7$, the evidence for a solid-to-liquid transition at a "melting temperature" $T_g(B)$. This transition was interpreted as a transition to the "vortex-glass" phase predicted by Fisher.[9] At small current, the strongly nonlinear current-voltage characteristic in the glass state $(T < T_g)$ turns to an ohmic characteristic in the liquid state $(T > T_g)$. The "melting curve" (Gammel et al.[41]) and the "depinning curve" (Esquinazi et al.[42]) obtained from vibration experiments deviate slightly from the curve $T_g(B)$. (See also the paper by Brandt.[8]) Yeshurun and Malozemoff[2] and Tinkham[7] have discussed the irreversibility line of the type $(1 - T/T_c) \sim B^{2/3}$.

In the paper by Worthington et al.[35] a new phase boundary has been proposed which occurs at $T_k(B) > T_g(B)$. Chien et al.[37] gave detailed studies of the ohmic resistivity in a liquid vortex phase. They found the transition from an activated regime at $T < T_k$ (TAFF) to a diffusive regime at $T > T_k$. The phase diagram of the vortex system in YBa$_2$Cu$_3$O$_7$ is shown in Fig. 2.

The TAFF regime has also been seen by experimental results of Palstra et al.[3] for the magnetic field directions both parallel and perpendicular to the ab-plane.

The temperature dependence of the resistivity of Tl$_2$Ba$_2$CaCu$_2$O$_x$ reported by Kim et al.[14] is shown in Fig. 3. The authors of this paper have not found an angular dependence of resistivity as a function of the angle between the magnetic field and current when both of them are parallel to the ab-plane. The same result was obtained earlier by Iye et al.[34] A macroscopic Lorentz force independent resistivity of the yttrium compounds has also been reported by Budhani et al.[38] In the paper by Kwok et al.,[36] in addition to the Lorentz force dependent part, there was an angular independent "background" resistivity.

We will try to offer a partial explanation of the phenomena listed above by focusing on the TAFF regime when the magnetic field is either parallel or perpendicular to the ab-planes. The TAFF regime corresponds to a liquid state of the vortex system.

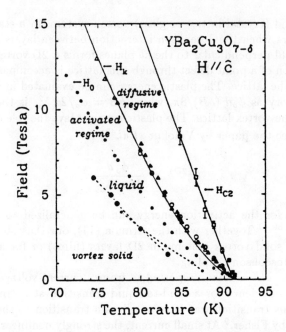

Fig. 2 Phase diagram of the vortex system in YBa$_2$Cu$_3$O$_7$ (Chien et al.[37]). The "onset" field H_o is determined from the condition $\rho = 0.01\rho_N$. The melting line H_g is located by data from Koch et al.[5] (asterisks) and Gammel et al.[41] (asterisks in circles).

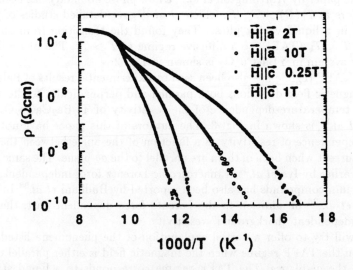

Fig. 3 The resistive tails of $Tl_2 Ba_2 CaCu_2 O_x$ film for the applied field $\vec{H} \perp \vec{a}$ and $\vec{H} \parallel \vec{a}$ at measuring current density of 10 A/cm^2 (Kim et al.[14]).

6. THERMAL ACTIVATION IN A LIQUID STATE AT SMALL CURRENTS

The thermal activation probability of a vortex motion in a liquid state is also determined by the formation of a critical nucleus, the energy for which is the activation energy. As we have seen above from formula (10), in a solid vortex state, one vortex loop activation process cannot give a finite value of a nucleation energy at small current, due to the elastic properties of a solid vortex state. This can clearly be seen in Fig. 1b where the shifted vortex B produces an elastic energy, which leads to the linear tension energy of the vortex loop in Fig. 1a. For a small current the Lorentz force cannot exceed this tension force.

In the paper by Chakravarty et al.[10] it was proposed that the tension energy can be overcome in a liquid state if the vortex B in Fig. 1b is shifted sufficiently far, over distance l_c, from its initial position A. If the length l_c is some density-density correlation length then there is no restoring force between positions B and A, and both defects A and B relax independently by diffusion processes in a vortex liquid. The creation of such a vortex loop costs the energy of the vortex segments connecting points A and B, as shown in Fig. 4 by a solid arrow.

The length of segment AB on Fig. 4 should exceed some correlation length l_c, which can be estimated for a vortex liquid state as a mean intervortex distance. Thus, the activation energy in a vortex liquid state is equal to the energy needed to create the single vortex loop shown in Fig. 5. For a small current the distance A_1A_2 or B_1B_2 is long, and the activation energy corresponds to the energy of two segments A_1B_1 and A_2B_2. Their lengths have the order of magnitude of the mean intervortex distance. This statement is applicable regardless of the relative orientation of the magnetic field and the ab-plane.

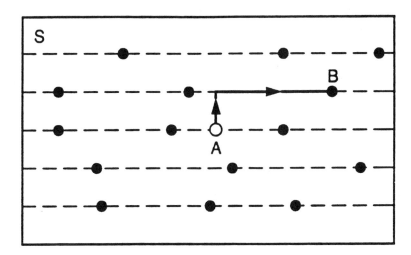

Fig. 4 Shift of a vortex position in a vortex liquid system.

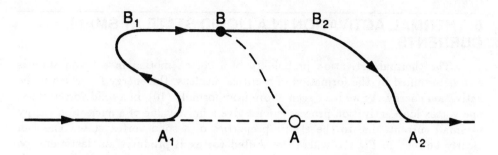

Fig. 5 Vortex configuration giving rise to the activation energy.

The process of thermal activation in a vortex liquid state is similar to that discussed by Caldeira and Leggett[43] for the activated motion of a particle with friction. Other fluctuating vortices play the role of a thermal environment. If the magnetic field is parallel to the ab-plane then the vortex activation reminds one of the motion of an overdamped elastic string through a washboard potential.

For a magnetic field perpendicular to the ab-plane the activated regime is possible, as has been noted by Vinokur et al.,[44] if the pinning time is shorter than the nucleus formation time. Otherwise pinning can be omitted, and the TAFF mechanism is replaced by a diffusive vortex motion, when the diffusion coefficient does not have an activated temperature dependence.

Let us consider now some particular cases.

6.1 Magnetic field perpendicular to ab-planes

The melting of the vortex system when $\vec{B} \parallel \vec{c}$ was studied in the papers by Brandt,[8,12] by Nelson and Seung,[11] by Houghton et al.,[13] by Feigelman et al.,[45] by Glazman and Koshelev.[46]

The lines A_1A_2 and B_1B_2 in Fig. 5 are parallel to the c-axis which is parallel to the z-axis. The length $l_\perp$ of segments A_1B_1 or A_2B_2 is

$$l_\perp \sim \left(\frac{\phi_o}{B}\right)^{1/2}. \qquad (15)$$

The configuration energy in Fig. 5 strongly depends on whether the parameter l is smaller or larger than $d\lambda_c/\lambda_{ab}$.

If $l \ll d\lambda_c/\lambda_{ab}$ one can use the Josephson approximation, which follows from the expression (1) for a free energy where the modulus of the order parameter can be considered as

$$|\psi|^2 = \frac{m\phi_o^2}{16\pi^3 \lambda_{ab}^2}. \qquad (16)$$

The activation energy, U_o, is the sum of the energies of the two segments $A_1 B_1$ and $A_2 B_2$ in Fig. 5,

$$U_o = \frac{2|\psi|^2}{Md} \int (1 - \cos \chi) dx\, dy \qquad (17)$$

where χ is a phase difference between layers defined as

$$\chi = \tan^{-1}\left(\frac{y}{x}\right) - \tan^{-1}\left(\frac{y}{x-l}\right). \qquad (18)$$

Now from Eqs. (16) - (18) one can obtain

$$U_o = \frac{\phi_o^2 l_\perp^2}{16\pi^2 d \lambda_c^2} \ln\left(\frac{d\lambda_c}{l_\perp \lambda_{ab}}\right). \qquad (19)$$

(Compare with Ref. 45.) Vortices in a Josephson-coupled layered system were studied by Bulaevsky.[47] Recently Clem and Coffey[48] have obtained current distributions near the vortex.

If the length $l_\perp$ is large enough so that $l_\perp \gg d\lambda_c/\lambda_{ab}$, one can use the London approximation for a vortex description in an anisotropic three dimensional superconductor. (See the papers by Kogan[49] and Campbell et al.[50]) In this case the activation energy is

$$U_o = \frac{\phi_o^2 l_\perp}{8\pi^2 \lambda_{ab} \lambda_c} \ln\left(\frac{\lambda_{ab} l_\perp}{d\lambda_c}\right). \qquad (20)$$

Using the definition of $l_\perp$ we can estimate the value of the crossover magnetic field B_o when the Josephson regime (18) makes a transition to the London region (19),[39,45]

$$B_o \simeq \frac{\phi_o \lambda_{ab}^2}{4d^2 \lambda_c^2}. \qquad (21)$$

An estimate of the activation energy for a vortex liquid state with a magnetic field perpendicular to the ab-plane is

$$U_o \simeq \frac{\phi_o^2 d}{32\pi^2 \lambda_{ab}^2} \begin{cases} \frac{B_o}{B} \ln \frac{B}{B_o} & B_o \lesssim B \qquad (22a) \\ \sqrt{\frac{B_o}{B}} \ln \frac{B_o}{B} & B_o \gtrsim B. \qquad (22b) \end{cases}$$

The result corresponding to the second limiting case (22b) has been obtained in the paper.[44] Two regimes (22) correspond to different mechanisms of the flux lattice melting. At the low field limit the 3D melting occurs. At the higher field limit melting is of 2D type.

The temperature T_m of 3D melting that leads to the vanishing of the shear modulus in a system of vortex lines is determined from the Lindemann criterion.[13,45,46] This can be written as $T_m \sim c_L^2 U_o$, where U_o is determined by

Eq. (22b). The Lindeman nnumber is $c_L \simeq 0.1$-0.2, and for this reason in the liquid state $(T > T_m)$ there exists a region where the parameter $U_o/T > 1$.

Note that the common factor in the formula (22) is of the order of the plastic activation energy (14) for a two dimensional vortex lattice.[40] In the paper by White, Kapitulnik, and Beasley,[51] the activation energy was measured in a multilayer system of Mo–Ge. The ten layer system considered by the authors was an intermediate case between pure 2D system (14) and the Josephson coupled stack of layers (22). In their paper, the Josephson coupling was shown to increase the activation energy.

For highly anisotropic high-T_c materials like Bi-based compounds, one can take $d = 15\text{Å}$, $\lambda_c/\lambda_{ab} \simeq 50$. It gives $B_o \simeq 0.1$ tesla. (The authors of Ref. 45 estimate this value as 3 tesla.) The activation energy for these anisotropic materials can be written as

$$U_o \simeq 10^3 \left(1 - \frac{T}{T_c}\right) (K) \begin{cases} \frac{0.1}{B} \ln \frac{B}{0.1} & 0.1 \text{ T} \lesssim B \\ \sqrt{\frac{0.1}{B}} \ln \frac{0.1}{B} & B \lesssim 0.1 \text{ T}. \end{cases} \quad (23)$$

For less anisotropic Y-based superconductors one can chose $d = 12\text{Å}$ and $\lambda_c/\lambda_{ab} \simeq 5$. In this case $B_o \simeq 14$ tesla and the activation energy is

$$U_o \simeq 10^3 \left(1 - \frac{T}{T_c}\right) (K) \begin{cases} \frac{14}{B} \ln \frac{B}{14} & 14 \text{ T} \lesssim B \\ \sqrt{\frac{14}{B}} \ln \frac{14}{B} & B \lesssim 14 \text{ T}. \end{cases} \quad (24)$$

Experimental data obtained by Kim et al.[14] give $U_o \sim (1 - T/T_c)^{1.5} B^{-0.71}$ for Tl-based superconductors in a wide interval of magnetic fields.

The formulas (23) do not fit the experimental data by Palstra et al.[3] for $Bi_2Sr_2CaCu_2O_8$, who obtained a power law field dependence of the activation energy that differed from the theoretical prediction. Brunner et al.[15] obtained a purely logarithmic dependence for a Y-based compound. Kes and van der Beek[39] obtained $U_o \sim B^{-1/4}$ for $Bi_2Sr_2CaCu_2O_8$, which matches the data[3] only for fields larger than 3 tesla.

Chien et al.[37] reported data for $YBa_2Cu_3O_7$, which are fitted well by the phenomenological formula

$$U_o = \frac{H_c^2}{8\pi} \frac{\phi_o \xi_c}{B} \simeq \frac{2.34 \times 10^5}{B} \left(1 - \frac{T}{T_c}\right)^{3/2} (K). \quad (25)$$

Previously Yeshurun and Malozemoff[2] and Tinkham[7] pointed out a formula of this type. Formula (25) for a vortex liquid can be interpreted either as the condensation energy of the volume $\phi_o \xi_c/B$ or as the energy of the ab-parallel vortex of length $\phi_o/B\, \xi_{ab}$. We see that one can ascertain only qualitative agreement between the experimental data and theoretical considerations.

6.2 Magnetic field parallel to the ab-plane

This case was studied in some experimental papers[3,14,38] and the TAFF regime was also observed. A typical activated resistivity is shown in Fig. 3. Kim et al.[14] have specially investigated the role of an unavoidable magnetic field misalignment which, according to Kes et al.,[52] is the source of the resistivity when $\vec{B}$ is generally parallel to the ab-plane. Kim et al.[14] have found the misalignment between the external magnetic field and the ab-planes to be irrelevant.

Suppose that the magnetic field is turned on or changed at sufficiently low temperature when a sample already has a pronounced layered structure. Vortices start to fill the different interlayer spacing. The system will have a minimum energy when the vortices form the Abrikosov lattice thermodynamically compressed by the ratio λ_{ab}/λ_c in the c-axis direction. (See the paper by Campbell et al.[50]) In order to do that vortices should enter only some particular periodically spaced interlayer gaps. But there is another possible way to enter such an interlayer spacing. In this case the vortex lattice period along the c-axis is forced as d and the ratio of the unit cell sizes is far from the equilibrium one. At first sight, the volume energy density of that lattice is much bigger than for equilibrium one. But as it has been shown[53] the vortex lattice rearranges itself by sliding vortices along planes and the energy density of this state has the absolute minimum value in the limit $B \gg H_{c1}$. The mean distance along the ab-plane between vortices in this rearranged state is

$$l_\parallel \simeq \left(\frac{\phi_o}{B} \frac{\lambda_c}{\lambda_{ab}}\right)^{1/2}. \qquad (26)$$

The thermodynamically compressed Abrikosov lattice could be obtained instead by cooling a sample in a fixed magnetic field and the typical intervortex distance has the same value (26). For a magnetic field $B > H_d = (\phi_o/2d^2)(\lambda_{ab}/3\lambda_c)^{1/2}$ the size of the unit cell along the c-axis equals the interlayer distance d. For Bi-based superconductors $H_d \simeq 5$ tesla and formula (26) is valid if $B \lesssim 5$ tesla.

The experimentally observed TAFF regime points out that a vortex state should be a liquid. For the magnetic field parallel to the ab-plane the vortex solid-vortex liquid phase transition is presumably of first order. According to arguments by Mikheev and Kolomeisky,[54] for that field orientation, a second order phase transition to the liquid state is impossible.

Both segments A_1B_1 and A_2B_2 of the vortex loop in Fig. 5 consist of two parts: the short segment of length d aligned perpendicular to the ab-plane and the segment of length $l_\parallel$ lying parallel to the ab-plane. The activation energy is the total energy of the two parts A_1B_1 and A_2B_2 and is given by the expression (See Chakravarty et al.[10])

$$U_o = \frac{\phi_o^2 d}{8\pi^2 \lambda_{ab}^2} \ln \frac{d}{\xi_c} + \frac{\phi_o^2}{8\pi^2} \frac{l_\parallel}{\lambda_{ab}\lambda_c} \ln\left(\frac{l_\parallel \lambda_{ab}}{d\lambda_c}\right). \qquad (27)$$

Substituting the value (26) into the formula (27) one can obtain

$$U_o = \frac{\phi_o^2 d}{8\pi^2 \lambda_{ab}^2} \left[\ln \frac{d}{\xi_c} + \frac{1}{2} \left(\frac{\phi_o \lambda_{ab}}{B \lambda_c d^2} \right)^{1/2} \ln \left(\frac{\phi_o \lambda_{ab}}{B \lambda_c d^2} \right) \right]. \qquad (28)$$

Using the values for highly layered Bi- or Tl-based compounds in the previous section gives the following estimate for the activation energy:

$$U_o = 10^3 \left(1 - \frac{T}{T_c}\right)^q \left(\alpha + \frac{\beta}{\sqrt{B}} \ln \frac{5}{B} \right) (K) \qquad (29)$$

$$q = 1 \qquad B \lesssim 5T$$

with B in Tesla. Numerical constants α and β are of order unity.

Experimental data by Palstra et al.[3] give the same magnetic field dependence of the activation energy for $B \lesssim 1$ tesla. Kim et al.[14] reported the B^{-1} type dependence of the activation energy for $\vec{B} \perp \vec{c}$. The exponent q could not be exactly defined from these experimental data, but the minimal value of q was estimated as 1.5.

Another interesting feature of layered superconductors is the macroscopic Lorentz force independence of the resistivity when both the external magnetic field and the current are parallel to the ab-plane.[14,34,38] It means that the resistivity is independent of the angle between the field and the current, even when they are parallel to each other.

If the magnetic field is parallel to the ab-plane, there is a finite concentration of thermally excited kinks (c-axis parallel segments of vortices of length d) which is proportional to exp $(-2\varepsilon_o/T)$. The value of $2\varepsilon_o$ is the first term of the right hand side of Eq. (28) which is equal to twice the energy of one kink. The total flux in the c-direction is zero, and the vortex system forms a 3D liquid plasma of kinks where there is no linear tension between any two kinks. Two kinks are shown in Fig. 6. If the current is perpendicular to the magnetic field the activation energy is given by expression (28). For any arbitrary direction of the current the motion of a kink is accompanied by vortex reconnection processes, as considered by Blatter et al.[55] The reconnection energy has the same order of magnitude as the second term of the right hand side of Eq. (28). For this reason, when the macroscopic magnetic field is parallel to the layers and also parallel to the transport current, the microscopic Lorentz force is not zero due to the finite thermally activated kink density. Kwok et al.[36] have reported the partial angular dependence of resistivity when some "background" resistivity remains even if magnetic field and current are parallel.

In conclusion, one has to point out that the pre-exponential factor B in the formula (2) can play an important role and, probably, can compete with the main effect given by the exponent A_o.

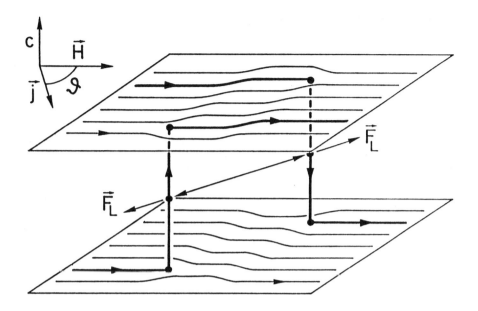

Fig. 6 Motion of kinks for the case $\vec{B} \perp \vec{c}$. The Lorentz force $\vec{F}_L$ is perpendicular to the current direction.

7. QUANTUM FLUX CREEP

The activation energy is determined by the classical saddle point solution for the vortex system. When the temperature decreases, the probability of thermal activation processes goes down. From the consideration of microscopic quantum tunneling in Josephson junctions, (see the paper by Devoret et al.[56]) it is known that, at sufficiently low temperatures, thermal activation is replaced by quantum tunneling. In the Josephson junction, the problem is similar to a one-particle problem (Caldeira and Leggett[43]).

The possibility of quantum creep can also be considered for a vortex system. Mitin[57] has reported experimental data on vortex creep in molybdenium sulfides which can be interpreted in terms of quantum flux creep. The experiments by Mota et al.[58] on high-T_c, heavy fermion, and organic systems also comment on that possibility. Quantum creep in granular systems was considered by Larkin et al.[22] In papers by Glazman and Fogel[23] and by Fisher et al.,[24] data on creep in thin films are reported.

Recently, quantum flux creep was considered by Blatter et al.[25] and by Ivlev et al.[26] In order to consider quantum tunneling one should know the dynamics of the system in an imaginary time $\tau = -it$. Instead of a static energy optimization, one now has to optimize the action A, which gives the exponent A_o in Eq. (2). The action has the form

$$A = \int_0^{1/T} d\tau \int_{-\infty}^{\infty} dy \left[\frac{\varepsilon}{2} \left(\frac{\partial u}{\partial y}\right)^2 + V_{pin}(u) - \frac{\phi_o j}{c} u + E_D \left\{\frac{\partial u}{\partial \tau}\right\} \right] \quad (30)$$

where T is a temperature. Vortices are supposed to align along the y-axis and the displacement u is transverse with respect to the y-axis. The pinning energy, V_{pin}, depends on the physical situation. The less trivial term is the dynamical one $E_D\{\partial u/\partial \tau\}$. In the case of a pure hamiltonian system this term is simply proportional to $(\partial u/\partial \tau)^2$. But the vortex system is an example of a system with dissipation. The dynamics of a vortex system in real time can be described by the equation, (see the review paper by Larkin and Ovchinnikov[17])

$$\eta \frac{\partial u}{\partial t} - \varepsilon \frac{\partial^2 u}{\partial y^2} + V'_{pin}(u) = \frac{\phi_o j}{c} \quad (31)$$

where η is a friction coefficient and ε is an elastic modulus.

Doniach and Inui[59] have argued that in a strongly layered superconductor, Coulomb effects (layer capacitance) add terms involving the second derivative in time to Eq. (31).

There are many realizations of quantum systems in imaginary time which reduce to Eq. (31) in real time. One of the possible realizations is given by Caldeira and Leggett[43] when the dynamical term has the form

$$E_D \left\{\frac{\partial u}{\partial \tau}\right\} = -\frac{\eta}{2\pi} \frac{\partial u}{\partial \tau} \int_0^{1/T} d\tau_1 \frac{\partial u}{\partial \tau_1} \ln |\sin[\pi t(\tau - \tau_1)]| . \quad (32)$$

This approximation can be justified considering the Matsubara representation of the equation of motion, which follows from minimizing the action (30). It reads

$$-\varepsilon \frac{\partial^2 u_\omega}{\partial y^2} + \eta |\omega| u_\omega + V'_{pin}(u)_\omega = \frac{\phi_o j}{c} \delta_{\omega,o} \quad (33)$$

where $\omega = 2\pi nT$ is the Matsubara frequency. Equation (33) can be obtained from the time-dependent Ginzburg-Landau theory which holds for superconductors with short inelastic scattering time. (See papers by Gorkov and Kopnin[60] and by Hu and Thompson[61])

The friction coefficient η is expressed through the vortex flow resistivity ρ_f as

$$\eta = \frac{\phi_o B}{c^2 \rho_f} .$$

The vortex flow resistivity is determined by the relation $E = \rho_f j$. Taking the usual approximation for the flow resistivity $\rho_f = \rho_N B/H_{c2}$ one can get

$$\eta = \frac{\phi_o H_{c2}}{c^2 \rho_N}. \tag{34}$$

The transition to the quantum creep region occurs at some crossover temperature T_o. For $T > T_o$ one has the activation regime, at the point $T = T_o$ the exponent in the formula (2) is $A_o = U_o/T_o$, and for $T < T_o$ this exponent remains finite down to zero temperature.

In order to find the crossover temperature T_o one has to solve the equation of vortex motion in imaginary time (33). An estimate of T_o can be found from comparison of the first two terms of the right hand side of Eq. (33), where $\omega \sim T$. It gives the relation

$$T_o \sim \frac{\varepsilon}{\eta l^2} \tag{35}$$

where l is a length scale along the vortex direction. The temperature T_o can be too small in order to make the quantum regime observable. For this reason the length scale l should be small. The regime with small length scale corresponds to large currents, when only one or few vortices are involved in the process. It correlates with a statement that a vortex bundle is a rather classical object for which the tunneling motion is unprobable.

Suppose that the magnetic field is aligned along the ab-planes. In this case the elastic modulus is[26]

$$\varepsilon = \frac{\phi_o^2}{8\pi} \frac{\lambda_c}{\lambda_{ab}^2} \ln \frac{\lambda_c}{\xi_{ab}}. \tag{36}$$

If we substitute the value of η, Eq. (34), the value of ε, Eq. (36) and the value of the upper critical field $H_{c2} = \phi_o/2\pi\xi_{ab}\xi_c$ into formula (35) we get

$$T_o \sim \frac{c^2 \rho_N}{\lambda_{ab}^2} \frac{\xi_{ab}^2}{l^2}. \tag{37}$$

Expression (37) can be considered as a rough estimate of the crossover temperature. For example, for a vortex aligned along the c-axis the result (37) is also true because the ratio ε/H_{c2} is the same order of magnitude for both orientations. The estimate (37) also follows from results (25) for $\vec{B} \parallel \vec{c}$, if $l \sim L_c$, where L_c is the pinning length in the c-direction.

In expression (37) only the length scale l depends on the particular nature of the pinning.

7.1 Magnetic field parallel to the ab-planes

Suppose that the coherence length ξ_c in the c-direction is not small compared to the interlayer distance. This condition holds for materials like YBa$_2$Cu$_3$O$_7$ for $T_c - T \leq 30K$. Under these conditions of the interlayer pinning potential, as was shown in Ref. 26 at $j \sim j_c$ (j_c is an interlayer depinning current), the crossover temperature is

$$T_o = \frac{5}{2} \frac{c j_c}{H_{c2} d} \rho_N . \qquad (38)$$

It corresponds to the length scale

$$l \simeq \xi_{ab} \left(\frac{d}{\xi_c}\right)^{1/2} \left(\frac{j_{GL}}{j_c}\right)^{1/2} \qquad (39)$$

where j_{GL} is the Ginzburg-Landau depairing current.[9] The activation energy at $T = T_o$ is

$$U_o = \frac{48\sqrt{2} d}{5\pi} \left(\frac{\varepsilon d \phi_o j_c}{2\pi c}\right)^{1/2} . \qquad (40)$$

The essential feature of formula (38) is that it expresses T_o through physically measured quantities. For YBa$_2$Cu$_3$O$_7$ compounds, $d = 12$Å, $H_{c2} = 200$ tesla, $\rho_N \simeq 10^{-4}$ Ω·cm. In the case of the interlayer pinning, the depinning current j_c has the same order of magnitude as the depairing current $j_{GL} \sim (10^8 - 10^9)$A/cm^2. Then the formulas (38) and (40) give estimates $T_o \simeq 50$ K and $U_o \simeq 500$ K. The important condition here is the large value of the depinning current j_c which determines T_o (38).

Summarizing the results one can write estimates for the crossover temperature T_o and for the exponent $A_o = U_o/T_o$ (2),

$$T_o \sim \frac{c^2 \rho_N}{\lambda_{ab}^2} \left(\frac{\xi_c}{d}\right) \left(\frac{j_c}{j_{GL}}\right) \; ; \; A_o \sim \frac{d}{e^2 \rho_N} \left(\frac{d}{\xi_c}\right)^{3/2} \left(\frac{j_{GL}}{j_c}\right)^{1/2} \qquad (41)$$

the relation $d < \xi_c$ or $d \sim \xi_c$ is supposed to hold.

7.2 Magnetic field perpendicular to the ab-planes

This case was investigated by Blatter *et al.* for the case of collective creep. In this case the length scale is $l \sim L_c \simeq \xi_{ab}(j_{GL}/j_c)^{1/2}$, where j_c is the depinning current. At a current $j \sim j_c$ and for a magnetic field $B \sim H_{c1}$, the following estimate results:

$$T_o \sim \frac{c^2 \rho_N}{\lambda_{ab}^2} \left(\frac{j_c}{j_{GL}}\right) \; ; \; A_o \sim \frac{\xi_{ab}}{e^2 \rho_N} \left(\frac{j_{GL}}{j_c}\right)^{1/2} . \qquad (42)$$

At higher magnetic fields the tunneling motion of a vortex bundle is less probable. The depinning critical current for collective pinning is supposed to satisfy the condition $j_c/j_{GL} < 10^{-2}$.

Comparing the possibility of quantum flux creep for different field orientations (41,42), one can see that the essential condition is the relatively large value of the depinning current, which has to be comparable to the depairing one.

The alternative explanation of the finite creep probability at T=0 was discussed by Doniach[6] on the basis of the model suggested by Hagen and Griessen.[62] This model supposes the pinning energy to be distributed over a range of values down to zero.

8. THE VORTEX CHAIN STATES AND CRITICAL CURRENTS

At a sufficiently low temperature, a vortex lattice is not melted, and its elastic properties can influence the critical depinning current. In the case of collective pinning the depinning current strongly depends on the elastic moduli (7). Formula (7) confirms the statement that a soft vortex lattice is more strongly pinned (See the paper by Brandt[8]). It means that some elastic-moduli-dependent rearrangement of a vortex lattice should change the critical depinning current.

Suppose that the magnetic field is tilted with respect to the ab-plane at an angle θ as shown in Fig. 7a. In this situation, Bolle et al.[63] reported the observation of a novel flux-lattice structure. For some range of tilt angles, some of the flux lines form an array of flux chains uniformly spaced along the sample.

The vortex lattice can be considered in the plane $x'z'$ which is perpendicular to the vortices. At a sufficiently large magnetic field the vortex lattice forms the perfect Abrikosov hexagonal structure in the "natural" coordinate system $x''z''$ which are

$$x'' = x' \left(\frac{\lambda(\theta)}{\lambda_c}\right)^{1/2}$$

$$z'' = z' \left(\frac{\lambda_c}{\lambda(\theta)}\right)^{1/2} . \quad (43)$$

Here the angle-dependent penetration depth is determined by the equation

$$\lambda^2(\theta) = \lambda_{ab}^2 \cos^2\theta + \lambda_c^2 \sin^2\theta . \quad (44)$$

This type of vortex lattice corresponds to the Campbell-Doria-Kogan approximation.[50] The proper unit cell is denoted in Fig. 7b by a 30° triangle.

As shown by Grishin et al.,[64] Buzdin and Simonov,[65] and in Refs. 66 and 67, for some range of tilt angles and the magnetic field, vortices form a chain structure.

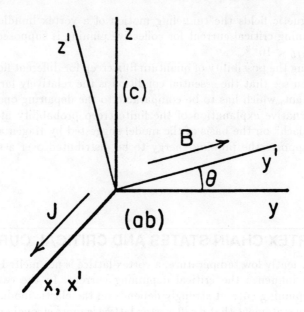

Fig. 7a Vortex cores are aligned along the y' axis.

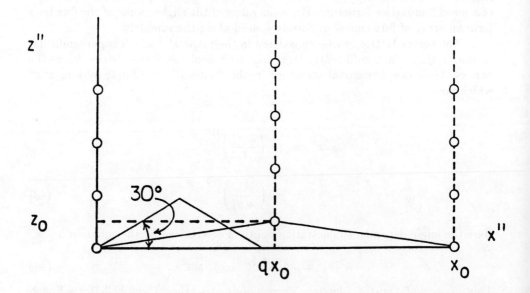

Fig. 7b A picture of the "natural" $x''z''$ plane. Vortex cores are denoted by circles ($q=\frac{1}{2}$). The normal lattice corresponds to the 30° triangle.

structure is characterized by a large interchain distance. In "natural" coordinates (43) the vortex lattice unit cell is essentially compressed in the z-direction as shown in Fig. 7b and the ratio x_o/z_o essentially exceeds its "natural" value of $2\sqrt{3}$.

The region of the vortex chain state for not extremely anisotropic material such as $YBa_2Cu_3O_7$ is shown in Fig. 8.[67] Here the upper boundary of the vortex chain state is

$$H^* = K \frac{\phi_o \lambda_c}{\lambda_{ab}^2 \lambda(\theta)} \qquad (45)$$

where K is a number of order unity. The lower critical field is $H_{c1} \sim \phi_o \lambda(\theta)/\lambda_c \lambda_{ab}^2$.

One has to say that the vortex chain state considered here differs from the experiments,[63] where the chains were observed with the totally "natural" vortex lattice. Probably, the consideration of correctly curved vortex lines should be added

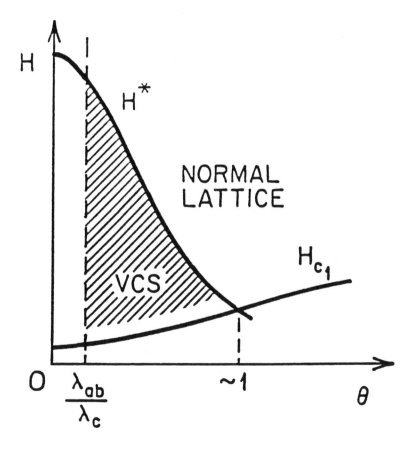

Fig. 8 The magnetic field-tilt angle diagram. There is no sharp boundary between the vortex chain state and the normal lattice, this transition is smooth.

according to Sudbø and Brandt,[68] and the surface effect can also be taken into account.

In the vortex chain state the lattice should be significantly softened. The chains of vortices are located in the planes which are parallel to $\vec{c}$ and $\vec{B}$. If the concentration of vortices in each chain increases, then the shear displacement of one chain relative to another costs little energy because the magnetic field along a chain is almost homogeneous.[53] In this softened state, the effective pinning is stronger because an isolated vortex chain can move easily and be kept fixed by pinning centers more strongly than the big region of a collectively coupled lattice. This causes an increase of the critical depinning current in the vortex chain state.

This has to lead to a peculiar angular dependence of the critical current when a magnetic field is tilted with respect to the ab-plane. In some region of width $\Delta\theta$ near $\theta = 0$ when the magnetic field is almost parallel to the layers the critical current has to increase. This $\Delta\theta$ region corresponds to the dashed area in Fig. 8. The width $\Delta\theta$ decreases with increasing magnetic field and temperature. The same type of angular dependence of the critical current near $\theta = 0$ has been observed in experiments by Roas et al.[69] in $YBa_2Cu_3O_7$. The magnetic field and temperature dependence of the experimentally measured width $\Delta\theta$ can be fitted using the formula (45), if $T_c = 82$ K, $\lambda_{ab}^2 = \lambda_{ab}^2(0)(1 - T/T_c)^{-1}$, $\lambda_{ab}/\lambda_c = 0.15$, and $K\phi_o/\lambda_{ab}^2(0) = 2$ tesla. Definite conclusions about the equivalence of theory and experiment[69] await more study.

9. CONCLUDING REMARKS

The origin of the nonlinear resistivity in a high-T_c superconductor at sufficiently large current can be understood in terms of the thermally activated flux creep phenomena. If the current is smaller than the depinning one, this regime corresponds to the activation of a vortex bundle, which leads to an increase of the activation energy. Study of the flux creep regime can be done on the basis of collective or interlayer pinning, depending on the magnetic field direction.

In the limit of a small current the TAFF regime is experimentally observed where the current-voltage characteristic is linear. In a bulk system of Josephson-coupled layers, the TAFF regime can be explained in terms of a vortex-liquid system. This situation is much more complicated to analyze though the theory can qualitatively describe the magnetic field and temperature dependences of an activation energy. A simplified theory considering critical nucleation of a loop of the mean intervortex size cannot quantitatively describe all the experimental situations. It means that the theory of the vortex-liquid state must be developed further.

The mechanism of resistivity is not completely clear in the case $\vec{B} \parallel \vec{J} \parallel \vec{c}$ (Briceno et al.,[70] Kapitulnik[71]), but one can suppose the activated vortex kinks to be responsible for resistivity.

Some experiments show the quantum flux creep regime can be reached. This field is far from being completely investigated.

Some unusual states of vortices (e.g. chain states) or of layered systems ("spiral forest"[72]) can also influence the resistivity of a superconductor.

ACKNOWLEDGMENTS

I am grateful for stimulating discussions with A. Barone, M. R. Beasley, G. Blatter, E. H. Brandt, L. J. Campbell, S. Chakravarty, S. Doniach, G. M. Eliashberg, T. H. Geballe, A. Kapitulnik, N. B. Kopnin, A. I. Larkin, K. Maki, K. Marchetti, A. Mota, Yu. N. Ovchinnikov, V. L. Pokrovsky, M. M. Salomaa, G. Schön, A. Schmid, and R. S. Thompson.

REFERENCES

1. K. A. Müller, M. Takashige and J. G. Bednorz, Phys. Rev. Lett. **58**, 1143 (1987).
2. Y. Yeshurun and A. P. Malozemoff, Phys. Rev. Lett. **60**, 2202 (1988).
3. T. T. M. Palstra, B. Batlogg, R. B. van Dover, L. F. Schneemeyer and J. V. Waszczak, Phys. Rev. B **41**, 6621 (1990).
4. P. H. Kes, J. Aarts, J. van den Berg, C. J. van der Beek and J. A. Mydosh, Supercond. Sci. Technol. **1**, 242 (1989).
5. R. H. Koch, V. Foglietti, W. J. Gallagher, G. Koren, A. Gupta and M. P. A. Fisher, Phys. Rev. Lett. **63**, 1511 (1989).
6. S. Doniach, in High Temperature Superconductivity, The Los Alamos Symposium - 1989, edited by K. S. Bedell, D. Coffey, D. E. Meltzer, D. Pines and J. R. Schrieffer (Addison-Wesley, Redwood City, 1990).
7. M. Tinkham, in Proc. 19th Internatl. Conf. on Low Temperature Physics, Brighton, Aug. 1990.
8. E. H. Brandt, Intern. Journal of Mod. Phys. B **5**, 751 (1991).
9. M. P. A. Fisher, Phys. Rev. Lett. **62**, 1415 (1989).
10. S. Chakravarty, B. I. Ivlev and Yu. N. Ovchinnikov, Phys. Rev. Lett. **64**, 3187 (1990).
11. D. R. Nelson and S. Seung, Phys. Rev. B **39**, 9153 (1989).
12. E. H. Brandt, Phys. Rev. Lett. **63**, 1106 (1989).
13. A. Houghton, R. A. Pelcovits and S. Sudbø, Phys. Rev. B **40**, 6763 (1989).
14. D. H. Kim, K. E. Gray, R. T. Kampwirth and D. M. McKay, Phys, Rev. B **42**, 6249 (1990).
15. O. Brunner, L. Antognazza, J.-M. Tricone, L. Miéville, and Ø. Fisher, preprint, 1990.
16. W. E. Lawrence and S. Doniach, in proc. 12th Internatl. Conf. on Low Temperature Physics, E. Kanda, ed. (Academic Press of Japan, Kyoto, 1971).

17. A. I. Larkin and Yu. N. Ovchinnikov, in Nonequilibrium Superconductivity, ed. by A. I. Larkin and D. N. Langenberg (Elsevier Science Publishers B.V. 1986).
18. M. Tachiki and S. Takahashi, Solid State Commun. **70**, 291 (1989).
19. B. I. Ivlev and N. B. Kopnin, ZhETF Pisma **49**, 678 (1989); Journ. Low Temp. Phys. **77**, 419 (1989).
20. A. Barone, A. I. Larkin and Yu. N. Ovchinnikov, Journ. of Superconductivity **3**, 155 (1990).
21. P. W. Anderson, Phys. Rev. Lett. **9**, 309 (1962).
22. A. I. Larkin, Yu. N. Ovchinnikov and A. Schmid, Physica B **152**, 266 (1988).
23. L. I. Glazman and N. Ya. Fogel, Fiz. Nizk. Temp. **10**, 95 (1984) [Sov. J. Low Temp. Phys. **10**, 51 (1984)].
24. M. P. A. Fisher, T. A. Tokuyasu and A. P. Young, Phys. Rev. Lett. **66**, 2931 (1991).
25. G. Blatter, V. B. Geshkenbein and V. M. Vinokur, Phys. Rev. Lett. **66**, 3297 (1991).
26. B. I. Ivlev, Yu. N. Ovchinnikov and R. S. Thompson, Phys. Rev. B **44**, (1991).
27. M. V. Feigelman, V. B. Geshkenbein, A. I. Larkin and V. M. Vinokur, Phys. Rev. Lett. **63**, 2303 (1989).
28. A. I. Larkin, ZhETF **58**, 1466 (1970) [Sov. Phys. JETP **31**, 784 (1970)].
29. V. L. Pokrovsky and A. L. Talapov, ZhETF **75**, 1151 (1978).
30. B. I. Ivlev and N. B. Kopnin, Journ. Low Temp. Phys. **80**, 161 (1990).
31. S. Chakravarty, B. I. Ivlev and Yu. N. Ovchinnikov, Phys. Rev. B **42**, 2143 (1990).
32. E. Zeldov, N. M. Amer, G. Koren, A. Gupta, R. J. Galambino and M. W. McElfresh, Phys. Rev. Lett. **62**, 3093 (1989).
33. J. D. Hettinger, A. G. Swanson, W. J. Skocpol, J. S. Brooks, J. M. Graybeal, P. M. Markiewich, R. E. Howard, B. L. Straughn and E. G. Burkhardt, Phys. Rev. Lett. **62**, 2044 (1989).
34. Y. Iye, S. Nakamura and T. Tamegai, Physica C **159**, 433 (1990).
35. T. K. Worthington, F. H. Holtzberg and C. A. Feild, Cryogenics **30**, 417 (1990).
36. W. K. Kwok, U. Welp, G. W. Crabtree, K. G. Vandervoort, R. Hulscher and J. Z. Liu, Phys. Rev. Lett. **64**, 966 (1990).
37. T. R. Chien, T. W. Jing, N. P. Ong and Z. Z. Wang, Phys. Rev. Lett. **66**, 3075 (1991).
38. R. C. Budhani, D. O. Welch, M. Suenaga and R. L. Sabatini, Phys. Rev. Lett. **64**, 1666 (1990).
39. P. H. Kes and C. J. van der Beek, in Proc. 19th Internatl. Conf. on Low Temp. Physics (Brighton, 1990).
40. V. M. Vinokur, P. H. Kes and A. E. Koshelev, Physica C **168**, 29 (1990).
41. P. L. Gammel, L. F. Schneemeyer, J. V. Waszczak and D. J. Bishop, Phys. Rev. Lett. **61**, 1666 (1988).
42. P. Esquinazi, A. Gupta and H. F. Braun, Physica B **165** & **166**, 1151 (1990).
43. A. O. Caldeira and A. J. Leggett, Ann. Phys. **149**, 374 (1983).
44. V. M. Vinokur, M. V. Feigelman, V. B. Geshkenbein and A. I. Larkin, Phys. Rev. Lett. **65**, 259 (1990).

45. M. V. Feigelman, V. B. Geshkenbein and A. I. Larkin, Physica C **167**, 177 (1990).
46. L. I. Glazman and A. E. Koshelev, Phys. Rev. B **43**, 2835 (1991).
47. L. N. Bulaevsky, Int. Journ. Modern. Phys. B **4**, 1849 (1990).
48. J. R. Clem and M. W. Coffey, Phys. Rev. B **42**, 6209 (1990).
49. V. G. Kogan, Phys. Rev. B **38**, 7049 (1988).
50. L. J. Campbell, M. M. Doria and V. G. Kogan, Phys. Rev. B **38**, 2439 (1988).
51. W. R. White, A. Kapitulnik and M. R. Beasley, Phys. Rev. Lett. **66**, 2826 (1991).
52. P. H. Kes, J. Aarts, V. M. Vinokur and C. J. van der Beek, Phys. Rev. Lett. **64**, 1063 (1990).
53. B. I. Ivlev, N. B. Kopnin and V. L. Pokrovsky, Journ. Low Temp. Phys. **80**, 187 (1990).
54. L. V. Mikheev and E. B. Kolomeisky, Phys. Rev. B **43**, 10431 (1991).
55. G. Blatter, B. I. Ivlev and J. Rhyner, Phys. Rev. Lett. **66** 2392 (1991).
56. N. Devoret, J. M. Martinis and J. Clarke, Phys. Rev. Lett. **55**, 1908 (1985).
57. A. V. Mitin, ZhETF **93**, 590 (1987) [Sov. Phys. JETP **66**, 335 (1987)].
58. A. C. Mota, A. Pollini, G. Juri, P. Visani and B. Hilti, Physica A **168**, 298 (1990).
59. S. Doniach and M. Inui, Phys. Rev. B **41**, 6668 (1990).
60. L. P. Gorkov and N. B. Kopnin, Uspekhi Fiz. Nauk **116**, 413 (1975) [Sov. Phys. Uspekhi **18**, 496 (1975)].
61. C. R. Hu and R. S. Thompson, Phys. Rev. B **6**, 110 (1972).
62. C. W. Hagen and R. Griessen, Phys. Rev. Lett. **62**, 2857 (1989).
63. C. A. Bolle, P. L. Gammel, D. G. Grier, C. A. Murray, D. J. Bishop, D. B. Mitzi and A. Kapitulnik, Phys. Rev. Lett. **66**, 112 (1991).
64. A. M. Grishin, A. Yu. Martynovich and S. V. Yampolskii, Physica B **165 & 166**, 1103 (1990).
65. A. I. Buzdin and A. Yu. Simonov, Pis'ma ZhETF **51**, 168 (1990) [JETP Lett. **51**, 191 (1990)].
66. B. I. Ivlev, N. B. Kopnin and M. M. Salomaa, Phys. Rev. B **43**, 2896 (1991).
67. B. I. Ivlev and N. B. Kopnin, Phys. Rev. B **44**, 2747 (1991).
68. A. Sudbø and E. H. Brandt, preprint (1991).
69. B. Roas, L. Schultz and G. Saemann-Ischenko, Phys. Rev. Lett. **64**, 779 (1990).
70. G. Briceno, M. F. Crommie and A. Zettl, Phys. Rev. Lett. **66**, 2164 (1991).
71. A. Kapitulnik, private communication.
72. M. Hawley, I. D. Raistrick, J. G. Beery and R. J. Houlton, Science **251**, 1587 (1991) [See also ibid. p. 1564].

DISCUSSION

K. E. Gray: Just let me make a comment about the Lorentz-force dependence. We've now studied this in a series of materials, and it seems that the amount of this

$\sin^2$ component depends on the anisotropy in the material. For example in yttrium, as you point out, it's fairly large compared to the independent background, and if we go to the one-layer thallium material, which is intermediate in anisotropy, we see a smaller component. And then in either the BSCCO or the thallium there seems to be no component. So it seems to be related to the coupling between the layers, the amount of this Lorentz-force component, when the field is parallel to the *ab* planes.

D. S. Fisher: Does it depend on the amount of mosaic spread?

K. E. Gray: We've done this in epitaxial films, and so I don't think that there's very much mosaic spread. I'm not sure what you mean by the mosaic spread now.

G. Aeppli: If you take an x-ray diffraction spectrum for the films, if you look at a [001]-type reflection, do you see a sharp peak?

K. E. Gray: Oh you mean the width of the rocking curve. They're all about 0.3 degrees or less, in these films. Probably less, in the epitaxial films.

P. H. Kes: This happens when you rotate the current and the field in the same plane?

K. E. Gray: In the same plane. Yes.

A. P. Malozemoff: And you have good *ab*-orientation also in those bismuth and thallium . . .

K. E. Gray: No, the YBCO and the thallium. The bismuth is not our work, but I think there's single crystal work in the literature that shows there's no dependence.

B. I. Ivlev: As I remember you didn't have any angular dependence . . .

K. E. Gray: Not in the thallium-2212. That's the higher anisotropy.

B. I. Ivlev: [Unintelligible] For yttrium, there is only a tendency for observation of a very small dependence for yttrium. But Kwok, Crabtree and others, they obtain it seems to me, also for yttrium.

K. E. Gray: I think in the epitaxial films we've looked at, it's similar to what Kwok has seen in the single crystal. I won't say precisely the same, but similar amounts.

P. H. Kes: If you rotate the field with respect to the *c* direction, and you have the current in the *ab* planes, you don't get this $(\sin\delta)^2$ behavior, but this $\cos\delta$.

B. I. Ivlev: Yes. Oh, no no, it's more complicated. No, it's more complicated, it's much more complicated. It oscillates, of course, but it's very . . . because it goes to the chain state; the chain state, immediately, with small angles between magnetic field and *ab* plane.

P. H. Kes: I need this explanation as an introduction for my lecture.

S. Doniach: In your quantum tunneling, you assume that it's an overdamped system, and at very low temperatures one could argue that the damping has frozen; there's no more damping.

B. I. Ivlev: Yes . . .

S. Doniach: . . . How do you treat this?

B. I. Ivlev: Yes, this is a good question. Yes, I agree. At this small temperature, when we have some T-crossover temperature from activation to the quantum limit, if we start to decrease temperature further, yes, I agree with you that we should obtain some dependence on temperature. Yes. Yes.

D. S. Fisher: Just coming back to this angular dependence . . . This is for the magnetic field and the currents both in the *ab* planes?

B. I. Ivlev: Yes.

D. S. Fisher: Now how much has been analyzed, and how much is known about the current paths? I mean, one assumes the current's going straight. How does one know, with twin boundaries and so on, that it isn't doing some complicated path?

B. I. Ivlev: Of course it's possible, if the sample is very disordered of course. No, no, no, Dan. More or less in the classic case, without . . . Even in the classic, in the parallel . . . idealized case, one can explain it qualitatively.

M. P. Maley: Of course the BSCCO doesn't have any twin planes, so the integrals . . .

G. W. Crabtree: Yes, just a further comment on that. There are a lot of explanations. All you need is either the current path or the vortex line to be nonlinear, and then you'll get a Lorentz-force component. But my personal feeling is that those explanations can't explain the magnitude of the effect. If you look in the 1-2-3 single crystal, for example, at higher temperatures where resistivity is about half the normal-state value, you find that there's broadening and this background effect is much larger than the Lorentz-force effect. And it seems to me that if you have small current-path deviations or vortex-line deviations you can account for a small effect but not for a large one. In my opinion this background term is still unexplained and is a major question.

D. C. Larbalestier: Has the experiment been done in twin-free and twinned crystals?

G. W. Crabtree: Yes, and we get essentially the same result for the $\sin^2$ part.

Can I add one more thing. We also did an experiment in several 1-2-3 single crystals that were of poorer quality (you can tell that by the width of the transition, and so on) and there you don't find such a large $\sin^2$ component, but you find about the same size of the background component. So the higher the quality of the crystal the more Lorentz-force component there is, but the background component always seems to be there, about the same magnitude, no matter what the quality.

M. P. Maley: I think it's also true in polycrystalline BSCCO tapes, where you know that the current path is kind of convoluted, contorted, that you've also got this same independence.

B. I. Ivlev: You also can analyze this angular dependence in your case?

M. P. Maley: I think there isn't any. At least in a lot of the Sumitomo data that I've seen, and also our own data, there isn't much of a dependence.

D. C. Larbalestier: I have a question. Intrinsic pinning came up a lot, but I noticed in George Crabtree's poster you actually measured this angular dependence, and found that it was extremely sharp, a degree or so. Now, the implication of this therefore is that intrinsic pinning is not practically important in most cases, because you will never be so well aligned. So I wonder if there's any comment from anybody here, therefore, about reconciling the experimental result there with the general idea that really intrinsic pinning is a source of the high critical current density when H is nominally parallel to ab.

K. E. Gray: Let me take a try at that. Certainly you could try to talk about the effects you get in the field that's parallel as a component of the field that's not parallel, which is what you're suggesting here, and I think Peter Kes was the first to point out the importance of that. However, if you take the examples of measurements we've made with the field both in the ab direction and the c direction it would require about a four-degree misorientation over the entire sample in order to explain that, and that's just not within experimental limits. So I don't think the misorientation can be the cause of this component we're seeing when the field is parallel.

And just another point in terms of the qualities of the material. We saw no essential difference between the polycrystalline samples and the epitaxial samples in this effect. So at least at some level, it doesn't seem to be dependent on current paths that aren't going as we expect them to.

B. I. Ivlev: The misalignment in your experiments were four degrees . . . ?

K. E. Gray: We would require four degrees in order to explain it as a misalignment, and it's not possible for the experiments to be that far off.

P. H. Kes
Kamerlingh Onnes Laboratory, Leiden University
2300 RA Leiden, The Netherlands

Vortex Pinning and Creep Experiments

A brief review of basic flux-pinning and flux-creep ingredients and a selection of experimental results on high-temperature-superconductivity compounds is presented. Emphasis is put on recent results and on those properties which are central to the emerging understanding of the flux-pinning and flux-creep mechanisms of these fascinating materials.

1. INTRODUCTION

The discovery of the high-temperature superconductors (HTS) has led to many speculations about superconducting applications at liquid-nitrogen temperature. However, it is now generally realized that the large critical temperature T_c and strong upper critical field $H_{c2}(0)$ are only necessary conditions. In addition, a large critical current density J_c is required; e.g. for superconducting magnets a $J_c > 10^9 \text{A m}^{-2}$ at fields B above 20 T is needed to be competitive with conventional superconductors like $(NbTi)_3Sn$. Although impressive progress has been made in the raising of J_c,[1] it also became clear that some of the characteristic properties of the HTS, like the quasi two-dimensional nature related to their layered structure or the short coherence length, lead to serious limitations of the applicability of these materials at high temperatures. From a practical point of view it doesn't seem relevant whether the low J_c at high T_c is connected to a melting transition of the vortex lattice (VL),[2-4] or a transition from a liquid to a vortex glass,[5,6] or that it is caused by thermally activated depinning.[7-9] This is the more so, as these phenomena are intimately related.[10-13] On the other hand it is very important to know if there is any reason to believe that J_c can be improved or that the irreversibility line can be moved to higher T, by some treatment of the material. Such knowledge cannot be attained without a deep insight into the properties of a VL in an anisotropic superconductor, nor without fundamental investigations of flux pinning and flux creep in such materials. This paper is intended to give a brief overview of the progress made in the last few years, especially regarding the experimental evidence.

The study of fundamental backgrounds requires materials of high quality and systematics of the experiments. In view of the large anisotropy of the HTS it is also important to clearly define the configuration of field, current and crystal orientation. With this in mind I'd like to restrict this paper mainly to investigations on single crystals and highly textured thin films. Work on $YBa_2Cu_3O_7$ (Y:123) and $Bi_2Sr_2CaCu_2O_8$ (Bi:2212) will be most presented as being representative for a very and extremely anisotropic superconductor with, respectively, proximity and Josephson coupling between the superconducting CuO_2 double layers. In Section 2 some of the basic ingredients for pinning and creep will be dealt with as an introduction to the section about the experimental situation (Section 3), which is divided into many subsections in order to present results and discussions next to each other.

2. BASIC INGREDIENTS

2.1 Properties of an Anisotropic Vortex Lattice

The coupling between the superconducting CuO_2 double or triple planes of the high-T_c cuprates can be expressed in terms of an anisotropy parameter Γ being the ratio of the effective quasi-particle masses in the c and in the ab direction (we will ignore the anisotropy in the ab plane). The anisotropy is reflected in the anisotropy of upper and lower critical fields and the Ginzburg-Landau (GL) coherence lengths ξ_{ab}, ξ_c, and the penetration depths λ_{ab} and λ_c. In the latter case the subscript denotes the direction of the screening current. The following simple relations hold: $\Gamma = (\xi_{ab}/\xi_c)^2 = (\lambda_c/\lambda_{ab})^2$.

As first pointed out by Lawrence and Doniach[14] a quasi two-dimensional (2D) situation arises when $\xi_c(T)$ becomes much less than the interlayer spacing s. In the Bi and Tl:2212 compounds very large Γ values have been reported: ≥ 3000[15] and $\geq 10^5$,[16] respectively. The crossover from 3D to 2D behavior occurs a few mK below T_c. In Y:123 with $\Gamma \geq 29$[17] the 3D regime is more extended, roughly

10 K.[18] Therefore, it seems reasonable to treat the VL properties of Y:123 in the framework of an anisotropic GL theory, whereas for the Bi and Tl compounds a quasi-2D approach is required.[19] In the first case the vortices are still considered as flux lines (tubes), but with anisotropic lattice configurations and elastic constants.

With the field along the c axis a triangular Abrikosov lattice occurs with a lattice parameter $a_0 = 1.075(\Phi_0/B)^{\frac{1}{2}}$, the usual shear modulus c_{66}, and almost the same compression modulus c_{11}, but with a tilt modulus $c_{44}(k)$ which is reduced roughly by a factor Γ.[4] Here k denotes the wave vector of the deformation field. It should be noted that the reduction by Γ is fully obtained only in the limit of nonlocal elasticity,[20] i.e. when $k^{-1} < \lambda/(1-b)^{\frac{1}{2}}$; for $k = 0$ there is no reduction. Both c_{66} and c_{44} are important parameters in describing the response to forces on the VL exerted by pinning centers [21-23]. Pinning by randomly distributed defects leads to a disordered VL which can be visualized most directly by decoration experiments.[24] The reduction of c_{44} causes a noticeable decrease of the longitudinal correlation length L_c of the VL.[25]

With the field along the ab plane the VL adopts an isosceles structure with lattice parameters $\approx a_0\Gamma^{\frac{1}{4}}$ and $a_0\Gamma^{-\frac{1}{4}}$, and very anisotropic c_{66} [26] and c_{44} [20,27] expressing the fact that deformations parallel to ab cost less energy. For arbitrary field orientations the screening currents mainly run in the ab planes leading to a torque whereby Γ can be determined.[28] New vortex configurations have been predicted for low fields and angles ϑ (between H and c axis) above ≈ 65 degrees.[29] It is not yet clear wether the recently observed vortex chains[30] in Bi:2212 single crystals are in full support of these predictions. Pinning effects for arbitrary orientations are hard to analyze.

In the quasi-2D case the screening currents are confined to the CuO_2 layers leading to a segmentation of the vortices into 2D pancake vortices.[31] The coupling of the 2D vortices in adjacent layers determines the Josephson length $R_J \equiv \Gamma^{\frac{1}{2}}s$. Phenomena on a length scale smaller than R_j will appear as if the layers are totally decoupled, whereas for collective phenomena of larger size 3D behavior is expected.[32] At low temperatures a field $B_{2D} \approx \Phi_0/\Gamma s^2$ marks the crossover from 3D (low) to 2D (high) fields. When the field is within an angle of order Γ^{-1} parallel to the ab planes a locking transition to a Josephson VL takes place.[33] The properties of such a lattice have first been described by Bulaevski[19] and are presently intensively studied.[34-38] An important difference with an Abrikosov vortex is the absence of a normal core, for the order parameter is uniform in the ab direction and only varies in the c direction with period s being large in the CuO_2 planes and almost zero between them. The screening currents between the planes are very weak so that the material is almost transparent for fields along ab.[39] When the field makes a larger angle with the planes than Γ^{-1} the relevant vortex properties are determined by the pancake vortices which can be treated as if a field $H \cos\vartheta$ is applied in the c direction.[39-41]

Both for an anisotropic 3D and quasi-2D VL in the 3D regime the reduction of c_{44} leads to a dramatic increase of the squared-average displacement of the vortices $<u^2>_T$ by thermal fluctuations.[4] When $<u^2>_T \approx (a_0/10)^2$ the 3D VL melts by the development of dislocation loops.[42,43] The above condition determines a melting line in the $B - T$ plane.[4,6] In the 2D regime thermal fluctuations give rise to the unbinding of dislocation pairs in the 2D VL according to the Kosterlitz-Thouless theory at a temperature $T_M = c_{66}a_0^2 s/4\pi k_B$. For Bi:2212 this amounts to 20-30 K when $\lambda_{ab} \approx 140$ nm is assumed.[32]

Because of the anisotropic penetration depth and GL coherence length, layered superconductors have anisotropic critical fields H_{c1} and H_{c2} as well. The determination of H_{c1} is problematic because of surface and demagnetization effects and pinning, that of H_{c2} because of thermal fluctuations. The result is that up to now for the HTS no truly reliable values of these parameters are available, e.g. estimates from different experiments on Y:123 single crystals using the GL theory, yield $-\mu_0 dH_{c2}/dT$ at T_c between 1.6 T/K from magnetization[44,45] to 7 T/K from the Ettinghausen effect.[46] As to the penetration depth λ_{ab} of Y:123, direct measurements yield $\lambda_{ab}(0) = 140$ nm[47] from which follows $\mu_0 H_{c1}(0) \approx 40$ mT in the c direction and about 7.5 mT in the ab plane. Here we used $\kappa = 70$. A similar value will hold for Bi:2212 in the c direction, but for the ab planes the larger Γ leads to a value of about 0.7 mT. In order to make estimates of pinning energies and critical currents it is necessary to decide on the value for ξ_{ab}. Rather arbitrarily, we will henceforth assume $\xi_{ab}(0) = 1.5$ nm for Y:123 and 2.3 nm for Bi:2212 with corresponding values of $B_{c2}(0) \equiv \mu_0 H_{c2}(0) = 100$ T and 44 T, and $B_c(0) = 1$ T and 0.3 T, respectively.

2.2 Pin mechanisms in HTS

The elementary pin interaction of a single defect with a single vortex or with a VL has been computed for several potential pin mechanisms, e.g. for twin planes in Y:123 with orientations (110) and $(1\bar{1}0)$[25] see also Ref. 48, for oxygen vacancies[49] in general, and for dislocation loops or stacking faults.[50] These are potentially the strongest pinning centers in bulk single crystals. In sintered bulk materials small precipitates of Y_2BaCuO_5[51] or CuO[52] may be effective; their elementary interaction is only roughly known.

Recently, STM studies of sputtered Y:123 films revealed a large density of screw dislocations along the growth direction.[53] The pin interaction can be estimated by considering the core of the dislocation as a nonconducting cylinder of diameter s. This assumption is supported by the fact that the resistance and T_c of Y:123 are very sensitive to disorder. Further, it seems reasonable to take the pitch of the screw dislocation as the diameter of the disordered core region. For parallel vortices the interaction can be estimated from the cross-section for electron scattering making use of Thuneberg's formalism,[54] see also Ref. 25. One obtains per unit length in the single vortex limit

$$f_{pl} \approx (B_c^2/\mu_0)pD\xi_0/\xi \qquad (1)$$

where ξ_0 is the BCS coherence length, p the scattering probability ($p = 1$ for a nonconducting defect) and D the diameter of the dislocation core, $D \approx s$.[55]

Another pin mechanism one should consider is the surface roughness created by the screw dislocations. If one models the roughness by $\delta d \cdot \sin(2\pi x/L_s)$ the pinning force per unit length for small fields is roughly

$$f_{pl} \approx 2\pi\epsilon_0(\delta d/d)/L_s \tag{2}$$

with the line energy $\epsilon_0 = (\Phi_0^2/4\pi\mu_0\lambda^2)\ln\kappa$, L_s the mean distance between the dislocations, d the film thickness and $2\delta d$ the average thickness variation. Note that f_{pl} represents the maximum force. To obtain the bulk pinning force one still has to sum over the actual forces which depend on the mutual positions of the defects and the vortices taking into account the vortex deformations, next Section.

As mentioned above, the order parameter is modulated by the crystal structure being small in the area between the superconducting layers. Consequently, if the field is along the ab planes, the energetically favorable configuration is that with the vortex cores between the CuO_2 double layers. To move the cores across the layers needs a large driving force close to the ultimate value set by the depairing current. This kind of pinning caused by the crystal structure is denoted as intrinsic.[34,35]

2.3 Summation models

The summation of the actual pinning forces of a realistic defect system has been a long standing problem which could be solved for a system of point pins with typical dimension smaller than ξ. It is assumed that each pin only gives rise to elastic strains and that the dislocation density of the VL is small enough to be ignorable. Even for weak pinning the positional order in the VL breaks down, given that the system is sufficiently large. Correlated regions determined by the transverse and longitudinal correlation lengths R_c and L_c, with volume $V_c = R_c^2 L_c$, may be defined in which the collective effect of the pins gives rise to a net force $F_c = (n_p V_c <f^2>)^{\frac{1}{2}} \equiv (WV_c)^{\frac{1}{2}}$. The average depinning force density follows from $F_p = F_c/V_c$ and leads to a critical current density $J_c = F_p/B$. This theory of collective pinning (CP)[56] has recently been reviewed,[6,25,57] [see also preceding paper by Ivlev]. Experiments have shown[58] that the CP theory applies to a 2D VL, but that for stronger pinning centers or for a soft VL, plastic deformations determine the disorder. In 3D the theoretical predictions are only valid in the amorphous limit defined by $R_c \approx a_0$. In layered superconductors and field normal to the layers the decoupling reduces the longitudinal correlation length L_c by a factor $\Gamma^{\frac{2}{3}}$ and J_c is enlarged by a factor $\Gamma^{\frac{1}{3}}$. Substituting experimental J_c values for Bi:2212 it can be shown that $L_c \approx s$, so that a 2D CP behavior can be expected.

For more extended defects in general only rough estimates can be made depending on the shape, orientation and nature of the defect and the vortex-defect coupling. A special example of such a defect is a flat grain boundary or a twin plane. Decoration experiments[59] showed that the vortices are attracted to the twin planes while between the planes the VL is very disordered. Assuming the effect of the twin planes to be predominant, a simple direct summation procedure seems appropriate, although limited to fields parallel to the planes and a geometry of mutually perpendicular twin planes.[60] Strong pins may disrupt the VL giving rise to a granular VL consisting of bundles separated by edge dislocations. This might be the case for the Y:123 films, see Eqs. (1) and (2). In such a situation the direct summation applies too.

For each of the pin mechanisms considered in the preceding section the situation at low fields is quite simple, because all vortices will be trapped by these strong pins. J_c follows from

$$J_c = f_{pl}/\Phi_0 \ . \tag{3}$$

In Section 3.2.4 we will further address this issue and compare with experimental results.

2.4 High temperature properties

2.4.1 Thermally activated depinning

So far, it has been assumed that thermal fluctuations are not important. However, close to T_c this assumption is not justified and one has to consider the effect of thermal depinning.[10] The reason is that both the elastic and the pinning energies scale with B_c^2/μ_0 so that near T_c they will be of order $k_B T$. Moreover, also a scaling factor $(1-b)^p$, with $b = B/B_{c2}$ and p either 1 or 2 related to the reduction of the order parameter, occurs which broadens the thermal depinning regime to part of the $B_{c2}(T)$ transition line. Thermal fluctuations of both the order parameter and its phase effectively lead to fluctuations of the vortices about their equilibrium positions. The influence on the pinning can be illustrated by considering a single vortex and a point pin. The pin potential has a typical range ξ. The thermal displacements then lead to an averaging of the pin potential over an area of order $<u^2>_T$. When this area is of order ξ^2 the pinning force decays $\propto \exp((-T/T_d)^3)$ in 3D[10] and $\propto (T/T_d)^{-5/2}$ in 2D at the depinning temperature T_d.[32] In larger fields similar considerations apply for a VL, but T_d is now determined by $<u^2>_T \approx a_0^2$.

2.4.2 Depinning or melting?

The above condition for depinning is very similar to the condition for melting. In the single vortex limit for which $\xi \ll a_0$, depinning may take place below the melting temperature, whereas for large fields probably the opposite happens. It should be mentioned, however, that strong pinning may still change these results quantitatively. When the pinning energy is larger than the elastic energy, the vortices move independently in their pin potentials and the squared average displacement is inversely proportional to the curvature of the pin potential, the Labusch parameter α_L.[21,22] Depinning now occurs when $k_B T_d^s \approx U_p$ and $U_p = F_p a_0^2 L_c^s \xi \leq (B_c^2/\mu_0)\xi^2 L_c^s$. (Superscripts s and w denote the strong- and weak-pinning case). When the pinning is weak the VL has short range order over distances large compared to a_0 and the displacements are determined by $k_B T/a_0 c_{ef}$ where $c_{ef} \approx (c_{44}c_{66})^{\frac{1}{2}}$ is the effective elastic constant of the VL. The depinning condition in this case can be written as $k_B T_d^w \approx c_{ef}\xi^2 L_c^w$. One can further analyze these conditions in order to compare the depinning temperatures. In case of strong pinning a temperature increase weakens the effect of the pins so that the correlation in the VL grows and eventually the configuration goes over into the weak-pinning case. The melting temperatures in both cases are expected to be almost equal. See also Section 3.2.6.

2.4.3 Flux creep in the vortex glass phase

It can be argued that a high-temperature vortex-liquid state is not a superconducting state, as the application of a current causes a resistance. On the other hand, the magnetization is diamagnetic. At low temperatures the vortices freeze into a disordered state caused by pinning. A new name has been introduced for this phase, the vortex-glass phase, because it lacks conventional superconducting order (off-diagonal long range order of the order parameter), but it has "Edward-Anderson spin-glass like order."[4,5] The vortex-glass phase is characterized by a true zero-resistance for $J = 0$. An infinite system would show a second order phase transition from a vortex-glass to a high-temperature vortex fluid at the glass-transition line $T_G(B)$. Transport coefficients are predicted to display critical behavior about the transition line, e.g. $E(J, T = T_G) \propto J^{(z+1)/(d-1)}$ with z the dynamical critical exponent and d the dimensionality. The latter should be ≥ 3, because conceptually a 2D vortex glass is not expected to exist.

In the vortex-glass phase thermal fluctuations in the presence of a driving force give rise to hopping of flux bundles over a distance u_h from one metastable configuration to another separated by an energy barrier U. The hopping frequency is given by $\nu_h = \nu_0 \exp(-U/k_B T)$ with ν_0 a microscopic attempt frequency. The net effect is a diffusion-like motion of the VL with an average velocity $v = \nu u_h$ in the direction of the driving force which is either caused by a transport current, a flux density gradient, or a flux-line curvature. The resistance generated by the creep process is given by

$$\rho \approx \rho_f \exp(-U/k_B T) \qquad (4)$$

where ρ_f is the flux-flow resistivity. The activation barrier depends on B and T, and generally also on J. For $\vec{J} = \nabla \times \vec{B}/\mu_0$ one obtains with $\nabla \times \vec{E} = -\partial \vec{B}/\partial t$, and using $\vec{E} = \rho(J)\vec{J}$, a nonlinear diffusion equation in $\vec{B}$,[61,62]

$$\partial \vec{B}/\partial t = -\mu_0^{-1} \nabla \times \{\rho(J)\nabla \times \vec{B}\} \qquad (5a)$$

or $\vec{J}$,

$$\partial \vec{J}/\partial t = \mu_0^{-1}(\nabla \cdot \nabla)(\rho(J)\vec{J}) , \qquad (5b)$$

results which describe the flux-density decay with time and position inside the superconductor. For simple geometries, e.g. a slab in parallel field, and $\rho \neq \rho(J)$, a linear equation describing the small-current behavior results,[63]

$$\partial B/\partial t = D_0 \partial^2 B/dx^2 \quad (6)$$

with $D_0 = \rho/\mu$ and ρ from Eq. (4). In this thermally assisted flux-flow (TAFF) limit a simple relation is obtained for the irreversibility lines as measured by several techniques,[64,65] see below. Note, that one should solve for $B(x,t)$ and average over the sample cross section in order to obtain the decay of the magnetization with time or the ac susceptibility before one can compare with experiments. The decay in the TAFF limit is exponential. The general behavior of χ' and χ'' can be expressed in terms of one parameter a/λ_{ac}, as shown in Fig. 1 for a slab and a cylinder. Here λ_{ac} is the penetration depth of the ac field,[49] and a the half-thickness or the radius of the sample. The peak in χ'' will occur when $\lambda_{ac} \approx a$. Since

$$\lambda_{ac} = (D_0/\pi\nu)^{\frac{1}{2}} \quad (7)$$

this condition leads to a relation between field, temperature, and frequency ν of the ac field which defines the irreversibility line. Below this line the VL oscillates in its pin potential, which offers a method to measure the Labusch parameter α_L and the range of reversible collective motion of the VL in the pin potential.[66,67] The combination with TAFF is worked out in Ref. 49; the effect is small.

In the limit of large current densities $J \leq J_c$ and for a simple geometry Eq. (5) leads to the well-known flux creep result[68,69] with a logarithmic decay regime for J in constant field:[70,139]

$$J(t) = J_c[1 - (k_B T/U)\ln(t/t_0)], t_0 \ll t \ll t_{cr} \quad (8)$$

where $t_{cr} \approx t_0 \exp(U/k_B T)$ is a cross-over time to exponential decay, and t_0 a macroscopic scaling time determined by the condition that the normal-state skin depth equals the sample size, i.e. $t_0 \approx (\mu_0 a^2/\rho_n)$ rather then ν_0^{-1} as is often assumed.[71] From Eq. (8) it follows that

$$S \equiv -d\ln(J)/d\ln(t) = [\ln(t_{cr}/t)]^{-1} \simeq k_B T/U . \quad (9)$$

Note, however, that S is not equal to the experimental decay parameter

$$S_{\exp} \equiv -M_0^{-1} dM(t)/d\ln(t) = [\ln(t_{cr}/t_i)]^{-1} = (k_B T/U)/\ln(t_0/t_i) \quad (10)$$

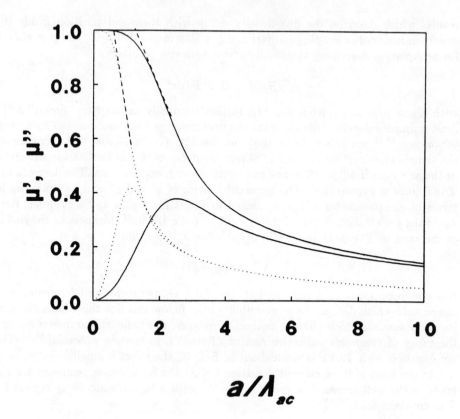

Fig. 1 The general behavior of the ac permeability $(1+\chi)$ for a slab (dotted lines) and a cylinder (solid lines) as follows from Eq.(6). In both cases a small ac field is added to a much larger dc field, both parallel to the surface. λ_{ac} denotes the ac-field penetration depth (7) and a is either the half-thickness or the radius of the sample.[49]

where $M_0 = M(t_i)$ and t_i is the time the measurement begins. Here it has been assumed that J is uniform so that $M \propto J$. This model can be extended to cases where U depends on J via the shape of the pin potential[69,72] and one may try to deconvolute for a distribution of energy barriers.[73]

In the preceding paragraphs the hopping distance and the size of the flux bundle L_b were assumed to be independent of J. This might be justified for $J \approx J_c$, or in the special case that the TAFF model holds (see below), but for smaller values of J both u_h and L_b have to grow in order to conserve energy in the hopping process, i.e. the energy gained from the external force should be equal to the elastic deformation energy of the VL. The thermally activated motion of an elastic object in a random potential has been treated for a VL in the collective creep theory.[10,11,74,32] The key result of this theory is

$$U(J) = U_c(J_c/J)^\alpha \tag{11}$$

which leads to

$$J(t) \simeq J_c/[1 + (\alpha k_B T/U_c)\ln(t/t_0)]^{1/\alpha} \tag{12}$$

with $\alpha = (2\zeta - 1)/(2 - \zeta)$ and ζ the wandering exponent, determined by the dimensionality of the medium and the hopping vector. Typically, for a VL, $\alpha \approx 1$. As L_b determines whether local or nonlocal elastic moduli are relevant, α also depends on the current regime leading to a more general result $U(J) = U_i(J_i/J)^{\alpha i}$ for $J_{i-1} < J < J_i \leq J_c$. In a 3D VL α changes from 1/7 to 3/2 to 7/9 with decreasing current density.[11,75,76] In 2D the values are 9/8 and 1/2, but in 2D, plastic creep of VL defects takes over at low J.[32,74] For plastic or dislocation mediated creep the energy barrier does not depend on J any more which means that the TAFF limit would apply. This is also expected for a 3D vortex liquid just above the glass-transition temperature where it probably behaves as a plasto-elastic medium.[12]

The logarithmic decay rate S, which follows from both Eqs. (8) and (10), will at low temperature increase with T since both U and U_c will be almost constant. Closer to the melting temperature the pinning and thus t_{cr} will decrease. At this point S will level off and eventually decrease for still higher T. The relation between U_c and the collective pinning potential $U_p = F_p V_c \xi$ depends on the size of the hopping bundle which consists of about $(c_{11}/c_{66})^{\frac{1}{2}}$ correlated regions of volume V_c.[10] Hence, $U_c \gg U_p$. Finally, by measuring ρ and using Eq. (4) it seems straightforward to determine the activation barrier. However, U at high temperature strongly depends on T, which may show up as a very large prefactor "ρ_f."

Thus far, the pinning has been treated as being random and rather uniform on the scale of a_0. It is well known that pinning in Nb_3Sn can be better described in terms of a percolation network of weak-pinning channels amidst strong pinning islands.[77] This model can be further worked out for creep making use of effective-medium theory.[78] The dependence of U on T and B follows from the nature of the energy distribution function. In the next chapter experiments will be discussed which illustrate the above mentioned pinning and creep features.

3. FLUX PINNING AND CREEP EXPERIMENTS

3.1 Experimental techniques

A great variety of experimental techniques are available to obtain information about pinning, critical currents, vortex mobility, VL disorder, etc. In this Section a review plus some discussion of typical characteristics of these techniques is given.

3.1.1 Flux line decoration

Decoration with magnetic smoke has been shown to be a very powerful technique to study VL configurations in small fields. For Y:123 it demonstrated that vortices are attracted to twin planes, but that other pinning defects also contribute considerably to the VL disorder.[59] It was also seen that the VL is anisotropic both with the field along c and in the ab planes.[79] For the Bi:2212 a totally new flux-line configuration with vortex chains was recently observed for field orientations within ≈ 25 degrees of the CuO_2 layers.[30] With the field in the c direction the VL disorder was investigated, providing evidence for the existence of a hexatic vortex-glass with short positional order, $R_c \approx 4a_0$, and a much weaker decay of the orientational order.[80] At high temperatures the smearing of the vortex images evidenced the strong influence of thermal fluctuations.[81] The limitation to low fields, because of the size of the magnetic particles and the decreasing amplitude of the local-field modulation when $a_0 < \lambda$, is the major drawback of this method.

A method to observe the penetration of a flux front, which thus has time resolution but only a few μm spatial resolution, makes use of the Faraday effect in an optically active thin layer. Such a layer, e.g. (EuS,EuF_2)-mixture or GdGa-garnet, is usually deposited on top of a superconducting sample.[82]

3.1.2 Magnetization and ac techniques

The measurements of magnetization loops and the ac response to small ripple fields added to large dc fields have been extensively used to determine J_c, the Labusch parameter α_L, the range of the pin potential r_f, and flux density profiles.[67,83] In the latter case Hall-probe experiments have been shown to give nice results as well.[84] At present, the magnetization is usually obtained by means of a vibrating sample magnetometer (VSM), a SQUID, or by a mechanical technique like the Faraday balance or a torque set-up. The main difference is that the former methods probe the average flux density in a pick-up coil, while the mechanical ones determine the total magnetic moment. The result is not necessarily the same because of demagnetizing effects, which certainly in the case of the platelet-like single crystals of the HTS cannot be ignored.[85] In this respect it is important to think in terms of current distributions rather then flux density gradients, since the flux-line curvature may carry an appreciable fraction of the screening currents. An advantage of the mechanical techniques is that the average current density is obtained for fully penetrated flux irrespective of demagnetizing effects. A disadvantage which is relevant for flux creep measurements is that the field should be extremely stable.[86] Moreover, for torque experiments the sample should have a platelet-like geometry and should make an angle with the applied field.

To determine $J_c(B)$ and the reversible magnetization curve $M_{rev}(H)$, one often exploits the Bean model[87] which assumes a uniform J_c. This is appropriate for large-κ superconductors and $H \gg H_{c1}$, but for small fields the model is not very reliable. Some extensions can be made, e.g. the B dependence of J_c giving rise to a positional dependence as well can be accounted for by a Taylor expansion.[88] When analyzing data after increasing the field in a zero-field cool-down procedure, one should be aware of the partial flux penetration below a field commonly denoted as H^*. For anisotropic superconductors J_c will also be anisotropic. For Y:123 this effect should be taken into account when $H \perp c$.[89] However, for extremely anisotropic materials like Bi:2212 one should analyze data taken in any field orientation except that which is exactly parallel to the ab planes, as if they were measured in a field $H_\perp$ equal to the projection of the applied field along the c axis.[39]

To determine J_c via the magnetization one has to know the relevant size of the sample, say a, as $J_c \propto \Delta M/a$. Especially for large "single" crystals of Y:123, it has been reported[90,91] that intragrain boundaries or variations in the density of oxygen vacancies give rise to an anomalous behavior of the $J_c(B)$ and $M(H)$ curves (fish tails). At low fields the sample diameter is the relevant size, switching to some smaller size, typically the distance between developing weak links, when the field is increased.

The above techniques are presently often used to determine the irreversibility (or mobility) line $T_{irr}(B)$. When the magnetization is measured the criterion is the field or temperature where ΔM increases above the resolution of the equipment. Apart from being rather ambiguous, this criterion also depends on the rate at which the data are taken. Therefore, ac techniques are preferable. When cooling down through the irreversibility line the dissipation goes through a maximum which can be used as a sharp definition of T_{irr}. Among these ac techniques are ac susceptibility,[92] ac-field transmission,[93] mechanical oscillator,[94] vibrating reed,[95] and ultrasonic attenuation.[96]

3.1.3 Resistive measurements

J_c in thin films is usually determined from a voltage-over-distance criterion, e.g., rather ambiguously, $1\mu V$/cm. Much more information can be obtained from the entire I-V curve: whether it is ohmic, the flux-flow resistance, and for the study of the glass transition the observation of a power-law dependence is essential. It has also become common practice to measure ρ vs. T in order to determine the in-field transition width and activation barrier for flux creep from an Arrhenius plot: $\ln(\rho)$ vs. T^{-1}.[97] As noted above, the explicit T dependence of U may end up in the prefactor ρ_f, see Eq. (4). Also when ρ is measured vs. H this can happen. The apparent prefactor can be as much as 12 orders of magnitude larger than ρ_n.[98]

Interesting novel features have been observed by measuring $\rho(T)$ or $\rho(H)$ for different angles between field, current and crystal orientations. For Bi:2212 it demonstrates the "$H_\perp$ effect,"[99,100] for Y:123 the effect of twin planes could be nicely probed.[101]

3.2 Pinning by natural defects

3.2.1 General features

To illustrate the state of the art an overview of $J_c(B)$ curves for wires and films of the Y and Bi HTS is given in Fig. 2, see Ref. 102 for references. For comparison, data of conventional materials are shown as well. The performance of Y:123 films is very promising,[103] both at 4.2 K and 77 K close to the estimated optimal values.[104] Y:123 wire, however, struggles with the weak-link and grain-boundary problem,[105] therefore Bi: 2212 wires and tapes seems to offer the best perspectives, especially with respect to applications above 20 T.[106] That is, at low temperatures, $T < 20$ K, for at 77 K and $B > 1$ T the J_c collapses. It should be noted that the J_c's for the $H \parallel ab$ configuration can be more than an order of magnitude larger than with $H \perp ab$.[107] It has been argued that this reduction is related to the large anisotropy.[39,108] A striking demonstration in support of this idea is given in Fig. 3.[108] The drop in resistance at 5 T clearly shifts to lower T for larger Γ. A recent confirmation has been seen in a Mo$_3$Ge/Ge multilayer model-system.[109]

From the very beginning it has been observed that the pinning force in the HTS in field goes to zero far below H_{c2}.[8,110] This phenomenon has been connected to the existence of an irreversibility line. An interesting property of the $F_p(B)$ curves near this line is the scaling which resembles the well-known Kramer scaling law for flux shear,[77] namely $F_p \propto (1 - b_0)^2$, with $b_0 = B/B_0$ instead of B/B_{c2}. B_0 is the field at which $F_p = 0$ and can be interpreted as the dc irreversibility field. This scaling has been noticed in Y:123,[111] but recently also in Bi:2212.[112] At low T the data display a linear decrease of F_p with field, see Fig. 4 for comparison. So it seems that the quadratic behavior is a high temperature feature. One may speculate about whether it reveals a shear behavior or is evidence for a melting transition.

3.2.2 Grain boundaries and twin planes

Grain boundaries (GB) in conventional superconductors are predominant pinning centers. In Y:123 they rather seem to act as weak links, which is not too surprising in view of the sensitivity to disorder and the short coherence length of

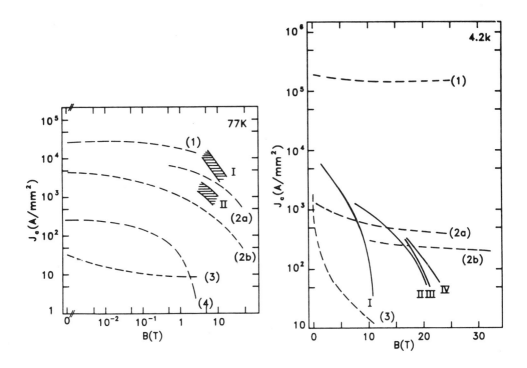

Fig. 2 (a) Typical critical current densities as functions of field for various conventional and high-temperature superconductors at 4.2 K: (1) Y:123 film on MgO substrate, (2a) Bi:2212 Ag-sheathed tape, (2b) Bi:2212 Ag-sheathed wire, (3) Y:123 film on Ag tape, (I) NbTi, (II) Nb$_3$Sn, (III) V$_3$Ga, and (IV) (Nb,Ti)$_3$Sn. Source: Nikkei Superconductors, Japan, February 1990.
(b) Comparison of current densities of various HTS at 77 K and conventional materials at 4.2 K. Note the change to logarithmic scale of B. (I) NB$_3$Sn, (II) NbTi, (1) Sputtered Y:123 film, (2a) CVD deposited Y:123 film, $H_\parallel$ ab-planes, (2b) same as (2a), $H \parallel c$, (3) Y:123 wire, (4) Bi:2212 tape, $H \parallel ab$. Source: Nikkei Superconductors, Japan, December 1989.

YBCO. In a series of elegant experiments on artificially grown, bicrystalline thin films the zero-field properties of GB have been systematically investigated.[113] The films were c-axis oriented with the GB also along this direction. The dependence of J_c at 4.2 K on the misalignment angle ϑ, as depicted in Fig. 5, shows a steep decrease between 5° and 15° leveling off to a value 50 times smaller than the J_c of 8×10^{10} A m^{-2} at $\vartheta = 0$. High-resolution electron-microscope pictures of a

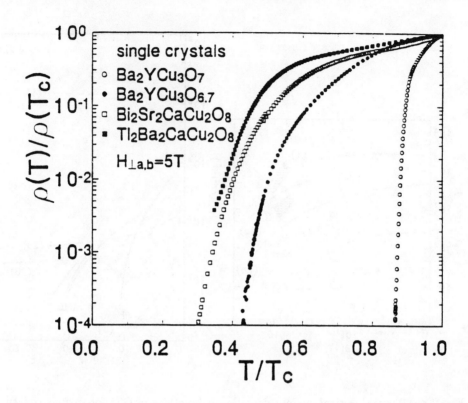

Fig. 3 The resistive transitions in a magnetic field of 5 T for HTS single crystals of different anisotropies.[108] A clear correlation is seen between the drop of $\rho(T)$ and the values of γ.[15-17]

10° GB only revealed a small amount of disorder along the GB. The temperature dependence of J_c through the GB turned out to be consistent with SNS-type weak-link behavior. For large-angle GB J_c was inversely proportional to a three-half power of the boundary resistance.[114]

The compositional changes at and near GB were studied by Babcock et al. on 91% dense sintered YBCO samples.[115] They also looked at the 77 K field dependence of J_c and the pinning properties of high-angle GB in bulk bicrystals. For a 90° GB, i.e. mutual perpendicular c axis, and 22° GB with common c axis, clear evidence was found for flux-pinning controlled $J_c(B)$ which differed by less than a factor of three from the single crystal J_c up to 7 T. No weak-link behavior was observed. For the 90° GB, by J_c at 5 T was as large as 10^7 A m^{-2}.

A special kind of GB in Y:123 are the twin planes (TP) with (110) and ($1\bar{1}0$) orientations. Their pinning potential and force have been computed for $H \parallel c$[25]

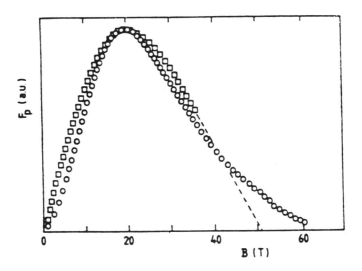

Fig. 4 The magnetic-field dependence of the pinning force F_p at 20.4 K (□) and 70 K (○) for $YBa_2Cu_3O_7$. The data for 70 K have been scaled in such a way that the maxima coincide with the data at 20.4 K.[110] The broader peak usually indicates linear decay.[77]

using the electron scattering formalism.[54] The result depends linearly on a scattering parameter p for which 0.4 has been assumed to obtain reasonable agreement with experimental $F_p(B)$ curves. For GB this value of p is reasonable, but TP do not seem to affect the resistance, so that a much smaller value is more appropriate for TP, in agreement with more recent $F_p(B)$ data on TP- free single crystals[116] which did not show a distinct decrease of J_c. It should be mentioned that the effect of TP is only to be expected if H is within a small angle of the TP orientation.[117] This is indeed observed, both in single crystals and high textured thin films[101,118,119] with the field along the ab planes and the driving force parallel to the TP, see Fig. 6. For thin films this pin mechanism may give a relevant contribution to the overall J_c because of the well defined orientations. For YBCO wires it is hard to believe that TP pinning will ever play an important role, except if one considers the TP as an aggregation of other (point) defects, e.g. oxygen vacancies.

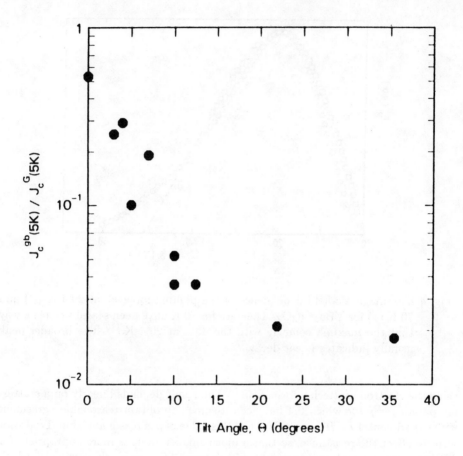

Fig. 5 Plot of the ratio of the grain-boundary critical current density to the average value of the critical current density in the two grains at 4.2-5 K vs the misorientation angle in the basal plane.[113]

3.2.3 Oxygen vacancies

Oxygen vacancies in the CuO_2 double layers pin most effectively because the pin interaction is caused by the local change of the GL coherence length and it is proportional to the maximum value of the order parameter. In the HTS the order parameter between the superconducting layers is either much smaller (Y:123) or almost zero and uniform (Bi:2212).[19] Therefore pinning must be due to defects in the superconducting layers. The pinning force between a single vacancy of diameter D and a single vortex is computed in Ref. 49, again using Thuneberg's electron scattering formalism.[54] At $t = T/T_c < 0.6$ one gets

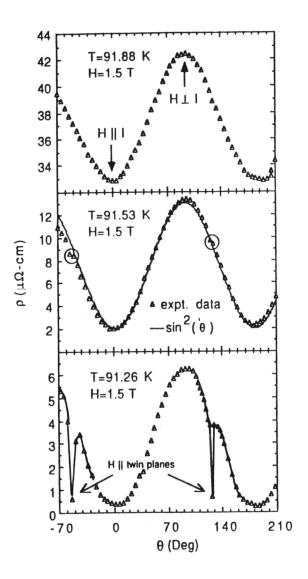

Fig. 6 Angular dependence of the (flux flow) resistivity for three temperatures just below T_c in a Y:123 single crystal with one twin plane orientation. The applied field (1.5 T) is parallel to the ab planes and is rotated with respect to the current direction defining ϑ, top panel. Flux flow in the c direction should obey a $\sin^2(\vartheta)$ behavior, see middle panel. The circles indicate the onset of pinning due to twin planes when the field orientation coincides. At slightly lower T this pinning effect is very distinct, see lower panel.[101] The sudden disappearance with increasing temperature indicates a steep reduction of the shear modulus when the VL melts.

$$f_p \approx [10/(1+t)^4](B_c^2/\mu_0\xi)\sigma_{tr}\xi_0 \tag{13}$$

where $\sigma_{tr} = \pi D^2/4$. Taking for D the diameter of an O^{2-} ion, 0.29 nm, and $B_c = 0.3-1.0$ T, a maximum force at $T = 0$ is computed between 0.6 and 8×10^{-13} N. Note that this value does not depend on $\xi(0)$ (= $0.74\xi_0$), but only on $B_c(0)$. Assuming that the distance between the defects is smaller than the range of the pinning force, i.e. $r_f \approx \xi$ in the single vortex limit, the collective effect should be estimated.

1. For the Bi compound in fields $B_{2D} < B < 0.1 B_{c2}$ the 2D vortices are individually pinned and one finds for the total force F_v per 2D vortex

$$F_v \approx (0.9 n_\square \pi \xi^2 f_p^2)^{\frac{1}{2}} \tag{14}$$

where $n_\square$ denotes the density of vacancies in the double layer. The pin energy follows from $U_v \approx F_v\xi$. Putting in numbers for the parameters[49] one obtains $U_v \approx 35$ K. F_p is computed by dividing F_v by the volume of the vortex segment, i.e. $a_0^2 s$ (s was the distance between the superconducting planes). This gives $J_c = F_p/B \approx 5\times 10^{10}$ A m^{-2}. In Section 3.4.2 the meaning of these results will be further discussed.

2. In case of Y:123 the 2D vortices are not decoupled and one should make an estimate for a vortex line of length L_c given by[25]

$$L_c \approx [\Phi_0^3 B r_f^2/(\Gamma^2 \mu_0^2 \lambda^4 W)]^{\frac{1}{3}} \tag{15}$$

with $W \equiv 0.5 n_\square f_p^2 \pi \xi^2/(a_0^2 s)$ the pinning strength, and Γ was the anisotropy parameter. Substituting $n_\square = 3.5\times 10^{17}$ m^{-2} (this number was found for Bi:2212[49] and means there is one vacancy per 80 oxygen atoms in each CuO_2 layer), $s = 1.2$ nm, $\xi = 1.5$ nm, $\lambda = 140$ nm, $\Gamma = 29$ and $f_p = 8\times 10^{-13}$ N, one gets $W \approx 0.18 B/[T]N^2 m^{-3}$ and $L_c \approx 6$ nm. Using $F_p = (W/L_c a_0^2)^{\frac{1}{2}}$ and $J_c = F_p/B$ one obtains $J_c \approx 1.2\times 10^{11}$ A m^{-2} in reasonable agreement with experimental values at low temperatures. For the pin energy one finds $U_p = F_p L_c a_0^2 \xi \approx 140$ K. This means that according to this model the creep barrier U at $T = 0$ is about 140 K. At higher temperatures the pinning will decrease due to thermal fluctuations yielding larger flux bundles and larger energy barriers. In addition, the effect of the compression modulus should be taken into account which increases the size of the activated bundle with a factor $(c_{11}/c_{66})^{\frac{1}{2}}$.[10] Therefore, this result means that pinning by oxygen vacancies can account for the observed J_c's and U's in Y:123 single crystals.

The justification for taking the limit of single-vortex pinning follows from an estimate of the elastic shear energy $U_{el} \approx 3c_{66}(\xi/a_0)^2 a_0^2 L$ for a vortex line element L. This can be expressed as $U_{el} \approx 0.12(\epsilon_0 L/\ln(\kappa))B/B_{c2}$. It turns out that for the HTS $\epsilon_0 = 10^3 \ln(\kappa)$ K/nm. Substituting $L = s$ and $L = L_c$ one gets $U_{el} = 180 B/B_{c2}$ K and $U_{el} \approx 7.6\times 10^2 B/B_{c2}$ K, respectively. A comparison with the pin energies U_v and U_p shows that they both exceed U_{el} in the field regime we consider.

Finally, it should be mentioned that other nonconducting defects with $D < \xi$ act as equivalent to vacancies. The pinning force of conducting defects, however, will be much smaller because of the proximity effect.[120] One may thus use the above considerations to estimate an optimal pinning force by taking $D \approx \xi(0)$ and $n_\square \approx 0.1\xi(0)^{-2}$. With this density of defects one may hope that T_c does not decrease due to proximity coupling but that an increase of J_c by an order of magnitude is still possible. More important is that U will increase by roughly the same amount which would reduce the creep process at 4.2 K considerably.

3.2.4 Screw Dislocations

In Section 2.2 two pin interactions due to screw dislocations in Y:123 films were proposed, namely the core interaction and the effect of thickness variations. The respective pinning forces per unit length were given by Eqs. (1) and (2). We now compare these forces with the possible effect of oxygen vacancies discussed above. Note, however, that $n_\square$ is not known from a direct measurement. Assuming that oxygen vacancies would give $J_c \approx 10^{11}$ A m^{-2} we estimate the line force **on each vortex** to be 2×10^{-4} N/m. This should be compared to the line force of the screw-dislocation core as determined from Eq. (1): $f_{pl} = 1.3 \times 10^{-3}$ N/m, and the line force caused by the surface roughness, Eq. (2). Assuming $L_s = 300$ nm, $d = 100$ nm and $\delta d = 5$ nm[53] the latter is estimated to be $f_{pl} = 6 \times 10^{-5}$ N/m. The corresponding J_c values following from Eq. (3) are $J_c = 6 \times 10^{11}$ A m^{-2} and 3×10^{10} A m^{-2}, respectively. In addition, we should investigate the suggestion put forward by Hawley et al.[53] that grain boundaries related to the screw dislocations may cause the large J_c's in films. Using the expressions derived in Ref. 25, we estimate this contribution to be almost two orders of magnitude too small and therefore irrelevant.

It follows from the above analysis that the interaction of the screw dislocation cores, the effects of surface corrugation, and oxygen vacancies are of roughly the same order of magnitude. At small fields the core mechanism is probably predominant since all vortices will be trapped by dislocation cores, while oxygen vacancies may take over at large fields. The combined effect provides a good explanation for the large J_c's reported in the literature. However, a definite conclusion cannot be given as long as crucial experimental evidence for the correlation between defect density and J_c is missing.

Till now only pin breaking has been considered. A quite different mechanism that might play a role here is depinning by VL shear.[77] This means that weakly pinned regions of the VL start to flow along strongly pinned areas. When both the screw dislocations and the valleys between them act as such strong pins, this effect might occur. Upon increase of the driving force, percolation paths of weak pinning regions would develop, bridging the width of the film. The force to be exceeded is the flow stress. Suppose the percolation path length is L_p, the width of this path is

$(q/2)a_0$ (being q VL planes), W the width of the film, and $\tau = \alpha c_{66}$ the flow stress, the resulting critical shear force density $F_s = J_s B$ is according to Pruijmboom et al.[77]

$$F_s = (4\alpha L_p/qW)c_{66}/a_0 \qquad (16)$$

In practice the prefactor will be of order unity yielding $J_s \approx c_{66}/(a_0 B) = (\Phi_0 B)^{\frac{1}{2}}/(16\pi\mu_0\lambda^2)$. Thus, we estimate $J_s \approx 3.7 \times 10^{10}(B/[\text{T}])^{\frac{1}{2}}$ A m^{-2}. Since this value is an order of magnitude smaller than the experimental value at low T and B being $J_c = 5 - 6 \times 10^{11}$ A m^{-2},[53] and because the $B^{\frac{1}{2}}$ dependence is not observed, the shear mechanism seems to be ruled out.

Very recently, new experimental data from Mannhart et al.[121] became available, see Fig. 7. The $J_c(B)$ data of the sample with a screw-dislocation density

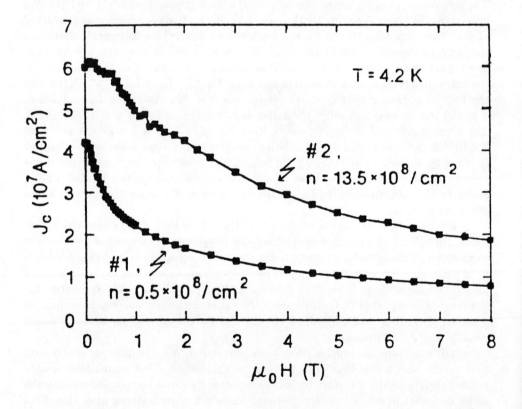

Fig. 7 $J_c(H)$ at 4.2 K of two characteristic YBCO films with screw dislocation densities of 0.5×10^8 cm^{-2} and of 13.5×10^8 cm^{-2}. The magnetic field was applied parallel to the c-axis, perpendicular to the transport current and to the sample surface.[121]

$n = 13.5 \times 10^8$ cm^{-2}, corresponding to $L_s \approx 300$ nm, show a behavior that can be well understood in terms of the above mechanisms. Suppose that up to a field $B_{occ} = \Phi_0 n$ all vortices can be accommodated by the screw dislocations yielding $J_c(B < B_{occ}) = 6 \times 10^7$ A cm^{-2}. Above this field, the access vortices will be mainly pinned by a background mechanism supposedly provided by the oxygen vacancies leading to a high-field level of about 10^7 A cm^{-2}. The observed gradual decrease of J_c will be related to the growing number of access vortices in combination with a constant voltage criterion for J_c. The field dependence of the flux-flow resistivity should give additional information by which this model can be tested.

The dislocation density would give $B_{occ} = 28$ mT, while from Fig. 7 one would rather estimate a value of 0.5 T. The density of screw dislocations is apparently not large enough and an equally strong pin mechanism is required to explain the discrepancy. It turns out that the STM picture also reveals the presence of a much larger density of edge dislocations parallel to the c-axis in agreement with other reports.[122] If these edge dislocations extend all the way down through the YBCO films the pinning force will be very similar to that of the dislocation cores. Further experiments are needed to test these ideas in more detail.

3.2.5 Intrinsic pinning

Till now we only considered pinning for the $H \parallel c$ configuration. As we argued in Ref. 39, situations in which the field makes an angle ϑ with the c axis can be understood by assuming that only the c component $H_\perp = H \cos \vartheta$ determines the pinning properties, except when H is parallel to the ab planes, i.e. within an angle Γ^{-1}. Evidence for such behavior in Bi:2212 has been found by several groups.[123] Outside this narrow range resistance tails seem to be independent of the Lorentz force,[99,124] which makes sense if the only relevant field component is $H_\perp$. Within the "locking angle" and below the 3D-2D crossover temperature the vortices remain strictly parallel to the ab planes[33] and the pinning is expected to increase sharply due to intrinsic pinning.[34,35] Evidence for intrinsic pinning is obtained from a sharp peak in J_c,[125] or a sudden drop in the resistance,[99] but a deviation from the $\cos \theta$ behavior[100] may also be an indication.

To describe the behavior of $J_c(\vartheta)$ in epitaxial Y:123 films, shown in Fig. 8a,[125] one has to assume that the pancake vortices are pinned by twin planes[34] which leads to a $(\cos \vartheta)^{-\frac{1}{2}}$ dependence that fits well to the low temperature and high field results. However, very recently, similar experiments on epitaxial films of Bi:2212 by Schmitt et al.[126] (Fig. 8b) showed clear evidence for the $\cos \vartheta$ dependence of $J_c(\vartheta)$ and scaling of J_c with $H \cos \vartheta$ for all $T < 70$ K. In practice, only the pinning in the CuO$_2$ planes will be relevant.

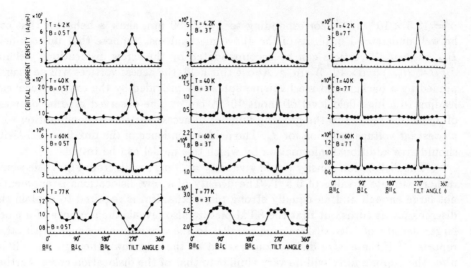
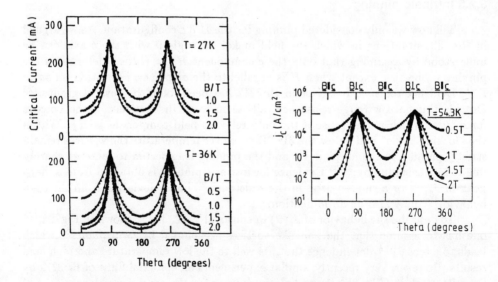

Fig. 8 (a) Rotation diagrams by Roas et al.[125] of J_c in thin films of Y:113 at various fields and temperatures. For low T and large B the curves can be well described by the intrinsic pinning model of Tachiki and Takahashi.[34]
(b) Rotation diagrams by Schmitt et al.[126] of J_c in thin films of Bi:2212 at various temperatures and fields. The lines drawn through the data points are obtained assuming J_c only depends on the normal component $H\cos\vartheta$, of the field, as suggested in Ref. 39.

3.2.6 Maximum critical currents

From the considerations in Section 3.2.4 one can easily derive the optimal value of J_c. Assuming that every vortex is pinned over its full length one gets $J_c < B_c^2 \xi_0 D/(\mu_0 \Phi_0 \xi) \approx 6 \times 10^{11}$ A m^{-2} at low temperature and $\approx 2 \times 10^{10}$ A m^{-2} at 77 K. The corresponding pin energy per unit length is $U_{pl} \leq B_c^2 D \xi_0/\mu_0 \approx 1.3 \times 10^{11}$ K/m. For a 100 nm thick film this would yield a pin energy of 1.3×10^4 K at low T and 1.2×10^3 K at 77 K. In the latter case U/kT is small enough to give rise to thermally activated dissipation leading to a somewhat smaller J_c than just computed. Depending on the density of line pins the above values do not depend on B. As soon as all line pins are occupied, however, J_c begins to decrease to the value dictated by the shear mechanism or by a background pinning mechanism.

It is clear that for Bi:2212 the quasi-2D nature above B_{2D} leads to much smaller pin energies, i.e. with $d = 1.5$ nm one obtains $U_p \leq 200$ K. This would already give considerable creep effects at 10 K in view of the general observation that flux creep is seen for $U_p/kT \leq 25$.

Apart from the reduction related to the explicit temperature dependence of the superconducting parameters, pinning is also reduced due to thermal fluctuations of the order parameter[10] and by the decrease of the order parameter when B goes to B_{c2}. The combination of both effects gives rise to a crossover line in the $B-T$ plane at which pinning steeply decreases, even when at $T = 0$ all vortices were optimally pinned. Below the crossover line the effective pinning has to become much weaker than it would be in the absence of thermal fluctuations. Consequently, one always passes from a strong- to a weak-pinning regime when the $B-T$ phase diagram is traversed towards the $B_{c2}(T)$ line. In the weak-pinning regime the melting theory supposedly describes the VL behavior, see also Section 2.4.2.

A rough estimate of the upper boundary for pinning can be obtained as follows. Suppose the effective pin energy is given by $\tilde{U} \leq U_p(0,0)(1-b)(1-t^2)^2$ with $U_p(0,0) \approx 10^{11}(d/[\text{m}])$ K as derived above. One readily obtains for the relative squared average displacement of a vortex $<u^2>/a_0^2 \approx (kT/\tilde{U})(\xi/a_0)^2$. When this quantity is of order 10^{-2} the effect of pinning is assumed to disappear, say at the $t_d(b_d)$ line. This line then will lie below the line determined by

$$(1-b_d)(1-t_d^2)^2/(b_d t_d) \approx 10^{-10}(T_c/d)/[K/m] . \qquad (17)$$

For conventional superconductors and a 10μm thick sample this rough estimate leads to $t_d \approx b_d \geq 0.95$, while for a 10$\mu$m thick Y:123 sample $b_d \geq 0.7$ at $t_d = 0.95$ results, i.e. $B_d \geq 6$ T. For Bi:2212 above B_{2D} $d = 1.5$ nm should be substituted and we get for $t_d = 0.25$ the result $b_d \geq 0.025$. These estimates seem to mimic the actually observed features rather well.

3.3 Artificial pinning centers

Of crucial importance for applications is to know whether the depinning line can be shifted by increasing the defect density. Probably the answer is yes, but not much, except, of course, in cases where pinning was weak to begin with. The overall impression that emerges is that there exists a borderline above which pinning disappears and that the position of this line depends mainly on the anisotropy.

To investigate this question one has to change the pin density in a controlled fashion, e.g. by substitutions or melt-growth processing,[127,128] neutron[129,130] or ion irradiation,[131-133] or radiation damage by fission products.[134] Ideally, the superconducting parameters do not change, otherwise one should scale the results; usually, in a first approximation, to reduced temperatures, but a true comparison becomes less reliable when the damage is too large. Because of the large effect of anisotropy and grain boundaries, these studies should be carried out on single crystals and highly textured or epitaxial films. Space and time does not allow us to further discuss results of the numerous reports which recently appeared and from which only a fraction is quoted above.

3.4 Flux creep

3.4.1 Introductory remarks

A great number and a rich variety of experiments on flux creep have been carried out since it became clear[7] that this effect has a much larger impact than it used to have in conventional superconductors. In two review papers[135,136] the results of the early years of HTS are summarized. In most reports the conventional creep theory is exploited to obtain using Eqs. (4) and (8-10) the temperature or field dependence of the creep barrier U. This theory can be extended to account for deviations from the Bean model.[137] Near the irreversibility line one can also use the expressions of the TAFF model.[64] Evidence which allows this model to be used is a logarithmic shift of the maximum in the ac losses with the probing frequency, or linear I-V characteristics at small current densities. The power-law I-V behavior as observed in Ref. 138 is instead evidence for the vortex-glass transition at T_g. This issue is still seen as controversial, though.[139]

3.4.2 Collective creep experiments

In this section we would like to move away from the irreversibility line in order to concentrate on the low temperature features of the creep process. We think, the necessary physical information can be obtained from an extensive study of the $U(J)$ dependence as a function of field.[140] We will choose the Collective Creep Theory (CCT), Section 2.4.3., as the theoretical framework describing the relevant physics. In this framework the decay parameter S attains the following two limiting forms:

$$S \approx (k_BT/U_i)\{1 - \alpha_i(k_BT/U_i)\ln(t/t_0)\}, t \ll t_{cr,i} \qquad (18a)$$

$$S = [\alpha_i \ln(t/t_0)]^{-1}, t \gg t_{cr,i} \qquad (18b)$$

with $t_{cr,i} \equiv t_0 \exp(U_i/k_BT)$. An interesting first attempt to analyze experimental results on a variety of Y:123 samples in terms of these expressions has been given in Ref. 13; see also Ref. 62. It is not clear, however, that all these data were obtained in the "long time" limit (18b) as was assumed in Ref. 13. The crossover time depends explicitly on the sample size and quality (via ρ_n) and on the exponential factor. Taking $a = 10^{-5} - 10^{-3}$ m and $\rho_n = 10^{-6} - 10^{-8}\Omega m$ t_0 ranges between 10^{-10}s and 10^{-4}s. Quite often, the experimental $U/k_BT \approx 23$ [140] leading to t_{cr} between 1 s and 10^6s. So, for relatively large, high-quality, single crystals experiments will rather refer to the short time limit (18a). Recent relaxation experiments on Bi:2212 single crystals[141] in fields below 20 mT and temperatures of 50 and 60 K yield $\alpha = 0.6$ and 0.4 and $t_0 = 10^{-6}$s and 10^{-4}s, respectively. This is considered to be in reasonable agreement with the predictions of CCT for the case of 3D disorder. Similar results for a field of 100 mT have been obtained in Ref. [142].

Strictly speaking, as mentioned in Section 2.3, the VL can be considered to be an elastically distorted medium only in 2D. The theoretical behavior of $U(J)$ for this case at low temperatures is sketched in Fig. 9, the general trend being $U = U_i (J_i/J)^{\alpha_i}$ see Eq. (11). When $J \approx J_c$ we have $U \approx U_p\{(J_c/J) - 1\}$ with U_p the pin energy, and a "bundle" size of one vortex, $L_b \approx a_0$. In the regime $J_2 < J < J_1 \leq J_c$, the bundle grows, $a_0 < L_b < \lambda$, but remains small enough to allow a nonlocal elastic description. The power α_1 is equal to 9/8. For a further decrease of J below $J_2 \approx J_c(a_0/\lambda)^{8/5}$, the bundle becomes large enough ($L_b > \lambda$) to use the local elastic modulus $c_{11}(0)$. The power changes to $\alpha_2 = \frac{1}{2}$. Expressions for U_1 and U_2 can be found in Ref. 32. Since this form of $U(J)$ diverges at low J any defect-mediated creep effect with finite activation barrier will be predominant at the lowest current densities, $J < J_3$. In this regime ohmic behavior of ρ in the diffusion equation (5b) leads to the TAFF limit.

Experiments on Bi:2212 in fields above B_{2D} may reveal 2D collective creep behavior. However the very small, but finite, Josephson coupling may give rise to a transition to 3D behavior when the bundle size surmounts the Josephson length $R_J = \Gamma^{\frac{1}{2}}s \gtrsim 100$ nm.[32] This would occur about J_2 since $\lambda \approx 140$ nm. Supposing oxygen vacancies are the predominant pinning centers, we found in Section 3.2.3 $U_p = 35$ K and $J_c = 5 \times 10^{10}$ A m^{-2}, seemingly in contrast to experiment. However, in terms of CCT one can easily understand that such a large J cannot be observed in magnetization experiments with a large time constant. The small U_p would give rise to a fast decay so that one moves up along the $U(J)$ curve in Fig. 9 until a steady state is obtained depending on the sweep rate of the field. Much larger J values have indeed been observed below 3 ms in short-time flux profile measurements on Y:123 samples.[143]

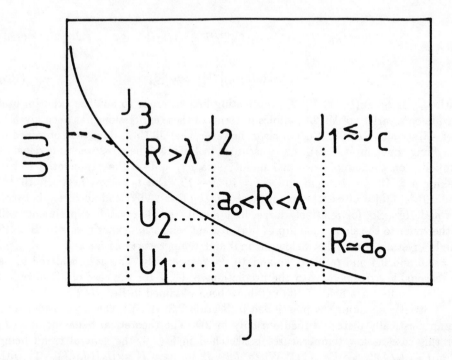

Fig. 9 Schematic behavior of the creep barrier vs J for collective creep of a 2D (straight) vortex lattice. The drawn line represents the expected power-law behavior in different current regimes. The dashed line denotes the transition to plastic creep when $J \to 0$.

In Fig. 10a the $U(J)$ data determined from magnetic relaxation experiments on a Bi:2212 single crystal in a field of 2 T ($\| c$) is displayed.[144] The measurements were carried out at temperatures of 5, 7, 9, 11, 14 and 17 K for a fully penetrated flux profile. A small temperature dependence has been accounted for. Details are given elsewhere.[144] The $M(t)$ data were elaborated according to a procedure suggested by Maley et al.[140] A slightly different procedure leading to similar results has recently been suggested in Ref. 145. When plotted on a double-logarithmic scale, Fig. 10c, the results can be compared to the 2D-CCT. The crossover from $\alpha = \frac{9}{8}$ to $\alpha = \frac{1}{2}$ should occur at about 5×10^9 A m^{-2}. The drawn lines denote $U = 20$ K $\times (J_c/J)^{\frac{2}{8}}$ for $J > J_2$ and $U = 70$ K $\times (J_2/J)^{\frac{1}{2}}$ below J_2, i.e. the extrapolations to J_c would give $U_p = 20$ K in reasonable agreement with the expected U_p of 35 K. The transition to the plastic creep regime, which would yield $U(J) \propto \ln J$, is not yet observed. It actually should occur below 10^7 A m^{-2}.[49]

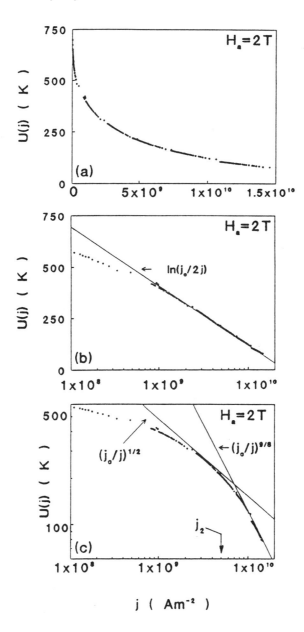

Fig. 10 Effective activation barrier $U(J)$ for a Bi:2212 single crystal in a field of 2 T ($\parallel c$) determined from magnetic relaxation experiments at 5, 7, 9, 11, 14 and 17 K using the analysis suggested in Ref. 140 on a linear (a), semi-logarithmic scale (b), and double-logarithmic (c). Lines denote: $U = 120$ K $\times \ln(J_c/J)$ (b); and $U = 20$ K $\times (J_c/J)^{\frac{9}{8}}$ and 250 K $\times (J_2/J)^{\frac{1}{2}}$ (c). J_c is taken to be 5×10^{10} A m^{-2}.

In Fig. 10b U is plotted vs log J. The line denotes $U = 120$ K $\times \ln(J_c/J)$. Evidence for such a current dependence in Y:123 has been reported by Zeldov et al.[146] According to Ref. 6, the relevant activation barrier in the vortex glass model cannot grow faster than the logarithm of a thermal length scale probed by the current density, $L_J \propto T/J\Phi_0 s$. At low current densities $U(J)$ increases slower than logarithmic implying that it may go to a constant as $J \to 0$, consistent with TAFF behavior. We conclude that so far the data are not conclusive regarding the flux-creep models, although both qualitatively agree with a collective glass-like behavior.

3.4.3 Numerical answers to creep questions

As for the comparison between experiments and theory a few key questions remain to be addressed. The general nonlinear diffusion equation for J (5b) with $\rho(J)$ given by $\rho(J) \approx \rho_f \exp(-U(J)/k_B T)$ can be solved in the slab geometry by separation of variables provided that $U(J) = U_p \ln(J_c/J)$.[147] If for other expressions of $U(J)$ the separation of variables would work as well, i.e. $J(t, x) = J(t) \cdot \tilde{J}(B(x))$, it would mean that sample geometry and demagnetization effects are not relevant for the time dependence of M, as they would give rise to a geometry-dependent prefactor only.

Numerical solutions of the equation for J have been obtained in infinite slab and cylinder geometries.[71] Several dependences of U on J were explored: $U = U_p(1 - J/J_c)$, $U = U_p \ln(J_c/J)$ and $U = U_p(J_c/J)^\alpha$. Different field dependences of J_c were investigated: J_c is field independent[87] and $J_c = J_{c0}/(B_0 + B)$,[148] as well as different initial conditions referring to a non-fully penetrated flux profile in increasing field, fully penetrated profile in both increasing and decreasing fields, and (constant) current densities larger or smaller than J_c. A great advantage is the time window of 20 decades that can be easily dealt with.

A number of features are found common to all models incorporating a $U(J)$ dependence. In the stage of incomplete flux penetration, $t < t^*$, the magnitude of the shielding current drops abruptly to zero at a well-defined flux front. At $t = t^*$ the fronts from the two sides meet in the sample center and a different relaxation regime is entered. For field-independent J_c, the flux profile remains linear at all times. For the field-dependent J_c the solution quickly evolves to the appropriate parabolic profile, $B \propto (x - x_0)^{\frac{1}{2}}$, in spite of the departure from a linear flux profile. It thus seems that at all times the solution tends to organize itself into the "critical" state appropriate to the input $J_c(B)$ dependence, i.e. the asymptotic solution for $M(t)$ is independent of the initial conditions.

This is demonstrated in Fig. 11 where the calculated $M(t)$ for $J_c = J_{c0}/(B_0 + B)$ and a variety of initial states in the infinite slab has been plotted. The parameters $U/k_B T = 2, 4, 7$ and 10, $\alpha = \frac{9}{8}$, $B_0 = 1$ T, $J_{c0} = 1.5 \times 10^{10}$ A m^{-2} and $\rho_f = 50 \times 10^{-8} \Omega m$ were chosen to represent the Bi:2212 compound in the 2D limit for large J. The solutions for $t > t^*$ (curved lines in Fig. 11) are well described by $M(t) = M^*[1 + (k_B T/U) \ln(\alpha t/t_0)]^{-1/\alpha}$; when $t < t^*$ the nearly straight lines are

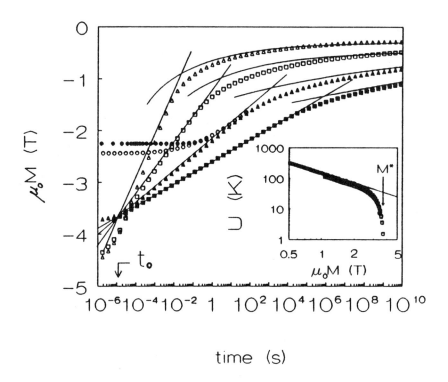

Fig. 11 Calculated time evolution of the magnetization in an infinite slab in parallel field for $U/k_BT = 2(\triangle)$, $4(\square)$, $7(\circ,\bullet,\blacktriangle)$, and $10(\blacksquare)$ and $U(J) = 35$ K $(J_c/J)^{\frac{9}{8}}$ with $J_c = J_{co}/(B_0 + B)$, $B_0 = 1$ T and $J_{co} = 1.5 \times 10^{10}$ A m^{-2} representative for the Bi:2212 compound at low temperatures. Different initial values of M correspond to different initial flux profiles, where open (closed) symbols refer to a linear (square-root) flux distribution. Curved and nearly straight lines represent the collective creep predictions for complete and incomplete flux penetration, see text. The inset shows a plot of the calculated data collapsed onto a universal curve following the method suggested in Ref. 140.

described by $M(t) = A + C[1 + (k_BT/U)\ln(\alpha t/t_0)]^{1/\alpha}$ with A and C constants. M^* is the magnetization corresponding to $J = J_c(B)$ in the entire sample. It can be shown that both limits express the same time-dependent decay of $J(t)$. The result therefore suggests that in general, for $t > t^*$, the ansatz of separable variables is justified.[76,147]

From the above results, it follows that it should be possible to superimpose all the curves in Fig. 11 on a single curve. This is illustrated by the inset: calculated data for various ratios of U/k_BT were analyzed according to the method of Ref. 140.

A plot of $-T\ln(dM/dt)+cT$ versus $-\mu_0 M$, with c an appropriate constant, causes the data to coincide on a single curve corresponding to the input dependence $U(J) = 35$ K $\times (M^*/M)^{\frac{9}{8}}$ denoted by the drawn line. It is thus found that the long-time solutions for both current-density and induction profiles are given by the *product* of the magnetization in the "critical state" corresponding to the $\tilde{J}_c(B(x))$ dependence and a scaling function describing the time evolution of the *local* current density. This should facilitate the data analysis of magnetic relaxation measurements on single crystals of platelet geometry where the field distribution in the critical state is a complicated function of the spatial coordinates.[85]

ACKNOWLEDGMENTS

I would like to thank Drs. J. Smith and M. Maley and the Exploratory Research and Development Center at Los Alamos National Laboratory for their hospitality.

Stimulating conversations with many colleagues are gratefully acknowledged. In particular, I would like to thank Valerii Vinokur and Kees van der Beek for many helpful, clarifying and pleasant discussions.

REFERENCES

1. J. E. Evetts, Physics World **3**, 24 (1990).
2. D. R. Nelson, Phys. Rev. Lett. **60**, 1973 (1988).
3. E. H. Brandt, Phys. Rev. Lett. **63**, 1106 (1989).
4. A. Houghton, R. A. Pelcovits and A. Sudbø, Phys. Rev. **B40**, 6763 (1989).
5. M. P. A. Fisher, Phys. Rev. Lett. **62**, 1415 (1989).
6. D. S. Fisher, M. P. A. Fisher and D. Huse, Phys. Rev. **B43**, 130 (1991).
7. Y. Yeshurun and A. P. Malozemoff, Phys. Rev. Lett. **60**, 2202 (1988).
8. A. P. Malozemoff, T. K. Worthington, Y. Yeshurun, F. Holtzberg and P. H. Kes, Phys. Rev. **B38**, 7203 (1988).
9. D. Dew-Hughes, Cryogenics **28**, 637 (1988).
10. M. V. Feigelman, V. M. Vinokur, Phys. Rev. **B41**, 8986 (1990).
11. M. V. Feigelman, V. B. Geshkenbein, A. I. Larkin, and V. M. Vinokur, Phys. Rev. Lett. **63**, 2303 (1989).
12. V. M. Vinokur, M. V. Feigelman, V. B. Geshkenbein and A. I. Larkin, Phys. Rev. Lett. **65**, 259 (1990).
13. A. P. Malozemoff and M. P. A. Fisher, Phys. Rev. **B42**, 6784 (1990).
14. W. E. Lawrence and S. Doniach, in Proceedings of the Twelfth International Conference on Low Temperature Physics, Kyoto, 1970, ed. E. Kanda (Kigaku, Tokyo, 1971) p.361.

15. D. E. Farrell, S. Bonham, J. Foster, Y. C. Chang, P. Z. Jiang, K. G. Vandervoort, D. L. Lam and V. G. Kogan, Phys. Rev. Lett. **63**, 782 (1989).
16. D. E. Farrell, R. G. Beck, M. F. Booth, C. J. Allen, E. D. Bukowski and D. M. Ginsberg, Phys. Rev. **B42**, 6758 (1990).
17. D. E. Farrell, C. M. Williams, S. A. Wolf, N. P. Bansal and V. G. Kogan, Phys. Rev. Lett. **61**, 2805 (1988).
18. D. E. Farrell, J. P. Rice, D. M. Ginsberg and J. Liu, Phys. Rev. Lett. **64**, 1573 (1990).
19. L. N. Bulaevski, Zh. Eksp, Teor.Fiz **64**, 2241 (1973) [Sov. Phys. JETP **37**, 1133 (1973)].
20. A. Sudbø and E. H. Brandt, Phys. Rev. **B43**, 10482 (1991).
21. R. Labusch, Crystal Lattice Defects **1**, 1 (1969).
22. A. M. Campbell and J. E. Evetts, Adv. Phys. **21**, 199 (1972).
23. R. Schmucker and E. H. Brandt, Phys. Stat. Sol.(b) **79**, 479 (1977).
24. H. Träuble and U. Essmann, J. Appl. Phys. **39**, 4052 (1968).
25. P. H. Kes, J. van den Berg in: Studies of High Temperature Superconductors, Vol. 5, A. Narlikar, ed. (NOVA Science Publishers, New York, 1990) p. 83.
26. V. G. Kogan and L. J. Campbell, Phys. Rev. Lett. **62**, 1552 (1989); L. J. Campbell, M. N. Doria, and V. G. Kogan, Phys. Rev. **B38**, 2439 (1988).
27. A. Sudbø and E. H. Brandt, Phys. Rev. Lett. **66**, 1781 (1991).
28. V. G. Kogan, Phys. Rev. **B38**, 7049 (1988); Z. Hao and J. R. Clem, Phys. Rev. **B43**, 7622 (1991).
29. A. I. Buzdin and A. Yu. Simonov, JETP Lett. **51**, 191 (1990); V. G. Kogan, N. Nakagawa and S. L. Thiemann, Phys. Rev. **B42**, 2631 (1990); A. I. Buzdin, S. S. Krotov and D. A. Kuptrov, Physica **C175**, 42 (1991).
30. C. A. Bolle, P. L. Gammel, D. G. Grier, C. A. Murray, D. J. Bishop, D. B. Mitzi and A. Kapitulnik, Phys. Rev. Lett. **66**, 112 (1991).
31. J. R. Clem, Phys. Rev. **B43**, 7837 (1991).
32. V. M. Vinokur, P. H. Kes and A. E. Koshelev, Physica **C168**, 29 (1990).
33. D. Feinberg, C. Villard, Phys. Rev. Lett. **65**, 919 (1990).
34. M. Tachiki and S. Takahashi, Sol. State Comm. **70**, 291 (1989).
35. B. I. Ivlev and N. B. Kopnin, Phys. Rev. Lett. **64**, 1828 (1990); J. Low Temp. Phys. **80**, 161 (1990); G. Blatter, B. I. Ivlev and J. Rhyner, Phys. Rev. Lett. **66**, 2395 (1991).
36. D. I. Glazmann and A. E. Koshelev, Physica **C173**, 180 (1991); Sov. Phys. JETP **70**, 774 (1990).
37. S. Chakravarty, B. I. Ivlev and Ya. N. Ovchinnikov, Phys. Rev. Lett. **64**, 3187 (1990); Phys.Rev. **B42**, 2143 (1990).
38. S. N. Artemenko, I. G. Gorlova and Yu. I. Latyshev, Phys. Lett. **A138**, 428 (1989).
39. P. H. Kes, J. Aarts, V. M. Vinokur and C. J. van der Beek, Phys. Rev. Lett. **64**, 1063 (1990).
40. B. I. Ivlev and N. B. Kopnin, Europh. Lett. **15**, 349 (1991).
41. J. R. Clem and M. W. Coffey, Phys. Rev. **B42**, 6209 (1990).
42. M. C. Marchetti and D. R. Nelson, Phys. Rev. **B41**, 1910 (1990).

43. D. R. Nelson and P. Le Doussal, Phys. Rev. B42, 10, 113 (1990).
44. U. Welp, W. K. Kwok, G. W. Crabtree, K. G. Vandervoort, and J. Z. Liu, Phys. Rev. Lett. 62, 1908 (1989).
45. Z. Hao, J. R. Clem, M. W. McElfresh, L. Civale, A. P. Malozemoff, and F. Holtzberg, Phys. Rev. B43, 2844 (1991).
46. T. T. M. Palstra, B. Batlogg, L. F. Schneemeyer and J. V. Waszczak, Phys. Rev. Lett. 64, 3090 (1990).
47. S. Sriahar, D.-H. Wu, and W. Kennedy, Phys. Rev. Lett. 63, 1873 (1989); Y. J. Uemura, L. P. Le, G. M. Luke, B. J. Sternlieb, W. D. Wu, J. H. Brewer, T. M. Riseman, C. L. Seaman, M. B. Maple, M. Ishikawa, D. G. Hinks, J. D. Jorgensen, G. Saito and H. Yamochi , Phys. Rev. Lett. 66, 2665 (1991); M. Weber et al., Supercond. Sci. Technol. 4, 403 (1991).
48. T. Matsushita et al., Jpn. J. Appl. Phys. 26, L1524 (1987); M. M. Fang et al., Phys. Rev. B37, 2334 (1988).
49. C. J. van der Beek and P. H. Kes, Phys. Rev. B43, 13032 (1991).
50. E. J. Kramer, Phil. Mag. 33, 331 (1976); C. S. Pande, Appl. Phys. Lett. 28, 462 (1976).
51. M. Murakami, M. Morita, K. Doi, K. Miyamoto, Jpn. J. Appl. Phys. 28, 1189 (1989).
52. S. Jin, T. H. Tiefel, S. Nakahera, J. E. Graebner, H. M. O'Bryan, R. A. Fastnacht and G. W. Kammlott, Appl. Phys. Lett. 56, 1287 (1990).
53. M. Hawley, J. D. Raistrick, J. G. Beery, and R. J. Houlton, Science 251, 1587 (1991). Ch. Gerber, D. Anselmetti, J. G. Bednorz, J. Mannhart and D. G. Schlom, Nature 350, 279 (1991).
54. E. V. Thuneberg, Cryogenics 29, 236 (1989); J. Low Temp. Phys. 57, 415 (1984).
55. An efficiency factor accounting for the layered structure is incorporated in the condensation energy. I am indebted to John Clem for pointing this out.
56. A. I. Larkin, Yu. N. Ovchinnikov, J. Low Temp., Phys. 34, 409 (1979).
57. E. H. Brandt and U. Essmann, Phys. Stat. Sol.(b) 144, 13 (1987).
58. P. H. Kes and C. C. Tsuei, Phys. Rev. Lett. 47, 1930 (1981); Phys. Rev. B28, 5126 (1983); R. Wördenweber, P. H. Kes and C. C. Tsuei, Phys. Rev. B33, 3172 (1986); R. Wördenweber and P. H. Kes, Phys. Rev. B34, 494 (1986); Cryogenics 29 321 (1989).
59. G. L. Dolan, G. V. Chandrashekhar, T. R. Dinger, C. Feild and F. Holtzberg, Phys. Rev. Lett. 62, 827 (1989); L. Ya. Vinnikov, L. A. Gurevich, I. V. Grigoryeva, A. E. Koshelev and Yu. A. Osipyan, J. Less. Comm. Met. 164&165, 1271 (1990).
60. P. H. Kes, A. Pruijmboom, J. van den Berg and J. A. Mydosh, Cryogenics 29, 228 (1989).
61. E. H. Brandt, Z. Phys. B80, 167 (1990).
62. R. Griessen, in Concise Encyclopedia of Magnetic and Superconducting Materials, ed. J. E. Evetts (Pergamon, 1991), to be published.
63. D. Dew Hughes, Cryogenics 28, 675 (1988).

64. P. H. Kes, J. Aarts, J. van den Berg and C. J. van der Beek, Superc. Sci. Technol. 1, 242 (1989).
65. A. Gupta, P. Esquinazi and H. F. Braun, Phys. Rev. Lett. 63, 1869 (1989).
66. K. Yamafuji, F. Fujiyoshi, K. Toko and T. Matsushita, Physica C159, 743 (1989).
67. A. M. Campbell, J.Phys. C: Solid St. Phys. 4, 3186 (1971).
68. P. W. Anderson and Y. B. Kim, Rev. Mod. Phys. 36, 39 (1964).
69. M. R. Beasley, R. Labusch and W. W. Webb, Phys. Rev. 181, 682 (1969).
70. C. W. Hagen, R. Griessen and E. Salomons, Physica C157, 199 (1989).
71. C. J. van der Beek, G. J. Nieuwenhuys and P. H. Kes, to be published in Proceedings MMS-HTS-III, Kanazawa, 1991.
72. Y. Xu, M. Suenaga, A. R. Moodenbaugh and D. O. Welch, Phys. Rev. B40, 10882 (1989); R. Griessen, Physica C172, 441 (1991).
73. C. W. Hagen and R. Griessen, Phys. Rev. Lett. 62, 2857 (1989).
74. M. V. Feigelman, V. B. Geshkenbein and A. I. Larkin, Physica C167, 177 (1990).
75. T. Nattermann, Phys. Rev. Lett. 20, 2454 (1989).
76. K. H. Fischer and T. Nattermann, Phys. Rev. B43, 10372 (1991).
77. E. J. Kramer, J. Appl. Phys. 44, 1360 (1976); D. Dew-Hughes, Phil. Mag. B55, 459 (1987); A. Pruijmboom, P. H. Kes, E. van der Drift and S. Radelaar, Phys. Rev. Lett. 60, 1430 (1988); Appl. Phys. Lett. 52, 662 (1988); Cryogenics 29, 232 (1989); C. J. G. Plummer and J. E. Evetts, IEEE Trans. Magn. Mag. 23, 1179 (1987); W. H. Warnes and D. C. Labalestier, IEEE Trans. Magn. Mag. 23, 1183 (1987).
78. A. Gurevich, Phys. Rev. B42, 4857 (1990); A. Gurevich, H. Küpfer and C. Keller, Superc. Sc. Technol. 4, 91 (1991).
79. G. J. Dolan, F. Holtzberg, C. Feild and T. R. Dinger, Phys. Rev. Lett. 62, 2184 (1989).
80. C. A. Murray, P. L. Gammel, D. J. Bishop, D. B. Mitzi and A. Kapitulnik, Phys. Rev. Lett. 64, 2312 (1990); D. G. Grier, C. A. Murray, C. A. Bolle, P. L. Gammel, D. J. Bishop, D. B. Mitzi and A. Kaptulnik, Phys. Rev. Lett. 66, 2270 (1991).
81. R. N. Kleiman, P. L. Gammel, L. F. Schneemeyer, J. V. Waszcza, and D. J. Bishop, Phys. Rev. Lett. 62, 2331 (1989).
82. M. R. Kobischka, N. Moser, B. Gegenheimer and H. Kronmüller, Physica C166, 36 (1990); M. V. Indenboom, N. N. Kolesnikov, M. P. Kulakov, I. G. Naumenko, V. I. Nikitenko, A. A. Polyanskii, N. F. Vershinin and V. K. Vlasko-Vlasov, Physica C166, 486 (1990).
83. R. W. Rollins, H. Küpfer and W. Gey, J. Appl. Phys. 45, 5392 (1974).
84. R. Riegler and H. W. Weber, J. Low. Temp. Phys. 15, 431 (1974).
85. M. Däumlung and D. C. Larbalestier, Phys. Rev. B40, 9350 (1989).
86. I. A. Campbell, L. Fruchter and R. Cabanel, Phys. Rev. Lett. 64, 1561 (1990).
87. C. P. Bean, Phys. Rev. Lett. 8, 250 (1962).
88. W. A. Fietz and W. W. Webb, Phys. Rev. 178, 657 (1969); P. H. Kes, C. A. M. van der Klein and D. de Klerk, J. Low Temp. Phys. 10, 759 (1973).

89. E. M. Gyorgy, R. B. van Dover, K. A. Jackson, L. F. Schneemeyer, and J. B. Waszczak, Appl. Phys. Lett. **55**, 283 (1989); F. M. Sauerzopf, H. P. Wiesinger, and H. W. Weber, Cryogenics **30**, 650 (1990).
90. H. Küpfer, I. Apfelstedt, R. Flükiger, C. Keller, R. Meier-Hirmer, B. Runtsch, A. Turowski, U. Wieck and T. Wolfs, Cryogenics **29**, 268 (1989).
91. M. Däumling, J. M. Seuntjes and D. C. Larbalestier, Nature **346**, 332 (1990).
92. J. van den Berg, C. J. van der Beek, P. H. Kes, J. A. Mydosh, M. J. V. Menken and A. A. Menovsky, Superc. Sc. Technol. **1**, 249 (1989); J. H. P. M. Emmen, G. M. Stollman and W. J. M. de Jonge, Physica **C169**, 418 (1990).
93. T. K. Worthington, Y. Yeshurun, A. P. Malozemoff, R. Yandrofski, F. Holtzberg and T. Dinger, J. de Physique **C8**, 2093 (1989).
94. P. L. Gammel, L. F. Schneemeyer, J. V. Waszczak and D. J. Bishop, Phys. Rev. Lett. **61**, 1666 (1988).
95. A. Gupta, P. Esquinazi and H. F. Braun, Phys. Rev. **B39**, 12271 (1989); H. Safar, C. Durn, J. Guimpel, L. Civale, J. Luzuriaga, E. Rodrigez, F. de la Cruz and C. Fainstein, Phys. Rev. **B40**, 7380 (1989).
96. P. Lemmens, P. Frönig, S. Ewert, J. Pankert, G. Marbach and A. Comberg, Physica **C174**, 289 (1991).
97. T. T. M. Palstra, B. Batlogg, L. F. Schneemeyer, R. B. van Dover and J. V. Waszczak, Phys. Rev. Lett. **61**, 1662 (1988).
98. T. R. Chien, T. W. Jing, N. P. Ong, and Z. Z. Wang, Phys. Rev. Lett. **66**, 3075 (1991).
99. Y. Iye, S. Nakamura, and T. Tamegai, Physica **C159**, 433 (1989); Physica **C174**, 227 (1991).
100. H. Raffy, S. Labdi, O. Laborde, P. Monceau, Phys. Rev. Lett. **66**, 2515 (1991).
101. W. K. Kwok, U. Welp, G. W. Crabtree, K. G. Vandervoort, R. Hulscher and J. Z. Lui, Phys. Rev. Lett. **64**, 966 (1990).
102. H. Krauth, K. Heine and J. Tenbrink, in Proceedings ICMC'90 Topical Conf. on Materials Aspects of HTS, eds. H. C. Freyhardt, R. Flükiger and M. Peuckert, (DGM Informationsgesellschaft, Oberursel, 1991) 29.
103. J. W. Ekin, K. Salama and V. Selvamanickam, Nature **350**, 26 (1991).
104. T. Matsushita, M. Iwakuma, Y. Sudo, B. Ni, T. Kisu, M. Funabi, M. Takeo and K. Yamafuji, Jpn. J. Appl. Phys. **26**. L1524 (1987).
105. S. Jin and J. E. Graebner, Materials Sc. Eng. B, preprint.
106. K. Sato, T. Hikata and Y. Iwasa, Appl. Phys. Lett. **57**, 1928 (1990).
107. K. Watanabe, N. Kobayashi, H. Yamone, T. Hirai and Y. Muto, ICMC'90, 965 (1991), see Ref. 100.
108. T. T. M. Palstra, B. Batlogg, R. B. van Dover, L. F. Schneemeyer and J. V. Waszczak, Phys. Rev. **B41**, 6621 (1990); Phys. Rev. **B43**, 3756 (1991).
109. W. R. White, A. Kapitulnik and M. R. Beasley, Phys. Rev. Lett. **66**, 2826 (1991).
110. P. H. Kes, Physica **C153-155**, 1121 (1988); J. van den Berg, C. J. van der Beek, P. H. Kes, J. A. Mydosh, A. A. Menovsky and M. J. V. Menken, Physica **C153-155**, 1465 (1988). K. Kadowaki *et al.*, Physica **B155**, 136 (1989).

111. J. N. Li, F. R. de Boer, L. W. Roeland, M. J. V. Menken, K. Kadowaki, A. A. Menovsky and J. J. M. Franse, Physica **C169**, 81 (1990); V. M. Pan *et al.*, Cryogenics **29**, 392 (1989).
112. R. Rose, S. B. Ota, P. A. J. de Groot, and B. Jayaram, Physica **C170**, 51 (1990).
113. D. Dimos, P. Chaudhari, J. Mannhart, and F. K. LeGoues, Phys. Rev. Lett. **61**, 219 (1988); J. Mannhart, P. Chaudhari, D. Dimos, C. C. Tsuei and T. R. McGuire, Phys. Rev. Lett. **61**, 2476 (1988).
114. R. Gross, P. Chaudhari, M. Kawasaki, and A. Gupta, Phys. Rev. **B42**, 10735 (1990).
115. S. E. Babcock and D. C. Larbalestier, Appl. Phys. Lett. **55**, 393 (1989); S. E. Babcock, X. Y. Cai, D. L. Kaiser and D. C. Larbalestier, Nature **347**, 167 (1990).
116. U. Welp, W. K. Kwok, G. W. Crabtree and K. G. Vandervoort, Appl. Phys. Lett. **57**, 84 (1990); B. M. Lairson, S. K. Streiffer and J. C. Bravman, Phys. Rev. **B42**, 10067 (1990).
117. G. Blatter, J. Rhyner, and V. M. Vinokur, Phys. Rev. **B43**, 7826 (1991).
118. E. M. Gyorgy, R. B. van Dover, L. F. Schneemeyer, A. E. White, M. M. O'Bryan, R. J. Felder, J. V. Waszczak and W. W. Rhodes, Appl. Phys. Lett. **56**, 283 (1989). J. Z. Liu, Y. X. Jia, R. N. Shelton and M. J. Fluss, Phys. Rev. Lett. **66**, 1354 (1991). B. Janossy, R. Hergt and L. Fruchter, Physica C **170**, 22 (1990).
119. Y. Iye, S. Nakamura, T. Tamegai, T. Terashima, K. Yamamoto, and Y. Bando, Physica **C166**, 62 (1990).
120. E. J. Kramer and H. C. Freyhardt, J. Appl. Phys. **51**, 4930 (1980); T. Matsushita, J. Appl. Phys. **54**, 281 (1983); A. A. Golub *et al.*, Fiz. Nizk. Temp. **10**, 258 (1984) [Sov. J. Low Temp.Phys. **10**, 133 (1984)].
121. J. Mannhart, D. Anselmetti, J. G. Bednorz, Ch. Gerber, K. A. Müller and D. G. Schlom, Proceedings of the International Workshop on Critical Currents, Cambridge, July 1991, to be published in Superc. Sci. Technol.
122. V. Pan *et al.*, Proceedings of the International Workshop on Critical Currents, Cambridge, July 1991, to be published in Superc. Sci. Technol.; S. K. Streiffer, B. M. Lairson, C. B. Eom, B. M. Clemens, J. C. Bravman and T. H. Geballe, Phys. Rev. **B43**, 13007 (1991).
123. S. Hayashi, K. Shibutani, I. Shigaki, R. Ogawa, Y. Kawata, V. Maret, K. Kitahama and S. Kawai, Physica **C174**, 329 (1991). H. Raffy. S. Labdi, O. Laborde and P. Monceau, Physica B **165& 166**, 1423 (1990).
124. K. C. Woo, K. E. Gray, R. T. Kampwirth, J. H. Kang, S. J. Stein, R. East, and D. M. McKay, Phys. Rev. Lett. **63**, 1877 (1989).
125. B. Roas, L. Schultz, and G. Saemann-Ischenko, Phys. Rev. Lett. **64**, 479 (1990).
126. P. Schmitt, P. Kummeth, L. Schultz, and G. Saemann-Ischenko, Phys. Rev. Lett. **67**, 267 (1991).
127. R. Wördenweber, K. Heinemann, G. V. S. Sastry and H. C. Freyhardt, Superc. Sci. Techn. **2**, 207 (1989). S. H. Whang, Z. X. Li, D. X. Pang, M. Suenaga,

D. O. Welch, S. Jin and T. H. Tiefel, Physica **C168**, 185 (1990). Y. Xu and M. Suenaga, Phys. Rev. **B43**, 5516 (1991).

128. D. Shi, M. S. Boley, U. Welp, J. G. Chen and Y. Liao, Phys. Rev. **B40**, 5255 (1989); S. Jin, T. H. Tiefel, S. Nakahara, J. E. Graebner, H. M. O'Bryan, R. A. Fastnacht and G. W. Kammlott, Appl. Phys. Lett. **56**, 1287 (1990); Z. J. Huang, Y. Y. Xue, J. Kulik, Y. Y. Sun and P. H. Hor, Physica **C174**, 253 (1991); M. Murakami, S. Gotoh, H. Fujimoto, K. Yamaguchi, N. Koshizuka and S. Tanaka, Supercond. Sci.Techno. **4**, S43 (1991).

129. H. W. Weber and G. W. Crabtree, to be published in Studies of High Temp. Superconductors, A. V. Narlikar ed. NOVA Science Publishers, New York, Vol. 9, (1991).

130. R. B. van Dover, E. M. Gyorgy, L. F. Schneemeyer, J. W. Mitchell, K. V. Rao, R. Puzniak and J. V. Waszczak, Nature **342**, 55 (1989); W. Schindler, B. Roas, G. Saemann-Ischenko, L. Schultz and H. Gerstenberg, Physica **C169**, 117 (1990); F. M. Sauerzopf, H. P. Wiesinger, W. Kritscha, H. W. Weber, G. W. Crabtree and J. Z. Liu, Phys. Rev. **B43**, 3091 (1991).

131. E. L. Venturini, J. C. Barbour, D. S. Ginley, R. J. Baughman and B. Morosin, Appl. Phys. Lett. **56**, 2456 (1990); R. B. van Dover, E. M. Gyorgy, A. E. White, L. F. Schneemeyer, R. J. Felder and J. V. Waszczak, Appl. Phys. Lett. **56**, 2681 (1990); L. Civale, A. D. Marwick, M. W. McElfresh, T. K. Worthington, A. P. Malozemoff, F. H. Holtzberg, J. R. Thompson and M. A. Kirk, Phys. Rev. Lett. **65**, 1164 (1990).

132. B. Roas, B. Hensel, S. Henke, S. Klaumnzer, B. Kabius, W. Watanabe, G. Saemann-Ischenko, L. Schultz and K. Urban, Europhys. Lett. **11**, 669 (1990); F. Rullier-Albenque, A. Legris, S. Bouffard, E. Paumier, and P. Lejay, Physica **C175**, 111 (1991).

133. H.-W. Neumüller, G. Ries, W. Schmidt, W. Gerhuser and S. Klaumnzer, Supercond. Sci. Technol. **4**, S370 (1990); H.-W. Neumüller, G. Ries, W. Schmidt, W. Gerhuser, and S. Klaumnzer, Journ. of Less-Comm.Met. **164&165**, 1351 (1990); B. Roas, B. Hensel, G. Saemann-Ischenko and L. Schultz, Appl. Phys. Lett. **54**, 1051 (1989).

134. R. L. Fleischer, H. R. Hart, Jr., K. W. Lay and F. E. Luborsky, Phys. Rev. **B40**, 2163 (1989); F. E. Luborsky, R. H. Arendt, R. L. Fleischer, H. R. Hart, Jr., K. W. Lay, J. E. Tkaczyk and D. Orsini, J. Mater. Res. **6**, 28 (1991).

135. A. P. Malozemoff in: Physical Properties of High Temperature Superconductors I, D. M. Ginsberg ed. (World Scientific Publishers, Singapore, 1989).

136. C. W. Hagen and R. Griessen in: Studies of High Temperature Superconductors, Vol.3, A. V. Narlikar ed. (NOVA Science Publishers, New York, 1990) p.159.

137. P. Chaddah and K. V. Bhagwat, Phys. Rev. **B43**, 6239 (1991).

138. R. H. Koch, V. Foglietti, W. J. Gallagher, G. Koren, A. Gupta and M. P. A. Fisher, Phys. Rev. Lett. **63**, 1511 (1989); P. L. Gammel, L. F. Schneemeyer and D. J. Bishop, Phys. Rev. Lett. **66**, 953 (1991); H. K. Olson, R. H. Koch, W. Eidellath and R. P. Robertazzi, Phys. Rev. Lett. **66**, 2661 (1991).

139. P. Esquinazi, Solid State Commun. **74**, 75 (1990).
140. M. P. Maley, J. O. Willis, H. Lessure and M. C. McHenry, Phys. Rev. **B42**, 2639 (1990).
141. P. Svedlindh, C. Rossel, K. Niskanen, P. Norling, P. Nordblad, L. Lundgren and G. V. Chandrashekhar, Physica C **176**, 336 (1991).
142. D. Shi and M. Xu, Phys. Rev. **B** (1991).
143. C. Keller, H. Küpfer, R. Meier-Hirmer, U. Wiech, V. Selvamanickam and K. Salama, Cryogenics **30**, 410 (1990).
144. C. J. van der Beek, P. H. Kes, M. P. Maley, M. J. V. Menken and A. A. Menovsky, to be published in Proceedings MMS-HTS-III, Kanazawa, 1991.
145. B. M. Lairson, J. Z. Sun, T. H. Geballe, M. R. Beasley and J. C. Bravman, Phys. Rev. **B43**, 10405 (1991).
146. E. Zeldov, N. M. Amer, G. Koren, A. Gupta, M. W. McElfresh and R. J. Gambino, Appl. Phys. Lett. **56**, 680 (1990).
147. V. M. Vinokur, M. V. Feigelman, and V. B. Geshkenbein, submitted to Phys. Rev. Lett.
148. Y. B. Kim, C. F. Hempstead and A. R. Strnad, Phys. Rev. **129**, 528 (1963).

DISCUSSION

D. S. Fisher: When you showed your U(J), and you said that there was a $J^{-1/7}$ or something behavior at small J, and that might be like 3D ... I don't understand why from this picture you would get something as small as that from 3D.

P. H. Kes: Well, okay. You see that's straight, and the power turns out to be 1/7, and 1/7 is one of the powers derived in the 3D collective creep ...

D. S. Fisher: For which regime?

P. H. Kes: ... for the vortex string regime, I would say.

D. S. Fisher: But that's not 3D, that's sort of a 1D regime.

P. H. Kes: Well, but for a string. I mean, 2D means that the pancake vortices actually are decoupled. If the bundle grows, the coupling between the pancakes becomes stronger, so the bundle actually becomes a 3D object. But the power for 3D objects of this size is not 1/7.

D. S. Fisher: Yes, but if one goes to the real 3D limit at small currents, the minimum power you can have is 4/5, or something like that, if I got the calculation right.

P. H. Kes: 2/3, I think.

D. S. Fisher: It's 4/5. Well, it's a question of what bounds one believes.

P. H. Kes: I won't stress this too much.

D. S. Fisher: The fact that this is getting to be a small power there is really not surprising. You would expect the power to go down a lot once you get to a regime where dislocations and things are coming in, and it's more likely to be crossing over to vortex glass behavior.

D. R. Nelson: It's on its way to being a constant.

P. H. Kes: I just want to mention that this is straight, and 1/7.

D. R. Nelson: I'm still a little confused about this decoupling field in the bismuth, which is about one tesla at liquid nitrogen temperatures. I don't see how that could be correct at zero temperature. In the absence of disorder I would expect a lattice of pancakes, a three-dimensional lattice of pancakes – with a weak interplanar coupling to be sure, but I don't see why that would be two-dimensional unless there were thermal fluctuations around that were sufficiently strong to decorrelate the layers. So why wouldn't it be temperature-dependent, this decoupling field?

P. H. Kes: Well you can simply say that it tells you that the shear energy of vortices in one plane becomes more important than the tilt energy of vortices in different planes.

D. R. Nelson: I guess I wouldn't call that decoupling. I would say that as long as that shear energy – which is admittedly small – is small compared to $k_B T$, there will be correlations along the z-axis of these pancakes.

P. H. Kes: It's a disorder kind of decoupling.

D. S. Fisher: I think a better term for it might be that it is the field above which the discreteness of the layers is important. Below that field, the discreteness of the layers doesn't matter.

D. R. Nelson: Yes, but even if the discreteness does matter the pancakes can be decoupled, and you're not necessarily able to treat them as decoupled layers.

D. S. Fisher: Right.

P. H. Kes: They're only decoupled on small length scales. If you look at the coupling of a larger area, then they start to become coupled again.

D. R. Nelson: I would worry about this coupling even in small regimes, when I'm below this temperature.

V. M. Vinokur: It just means that the interaction between vortices lying in one layer is more important than magnetic interaction between vortices in different layers.

D. R. Nelson: Like graphite, and I wouldn't say that graphite is independent 2D layers.

D. C. Larbalestier: Peter, I have two questions actually.

First one is, I'm confused about all the U values because what I heard you say at the beginning was that really it's all point pinning. And you calculated I think a U in BSCCO of about 35 K, but I'm looking here at an extrapolation to 500 K, so . . .

P. H. Kes: Okay, yes. It's good you mentioned that because I forgot to tell that.

D. C. Larbalestier: I think that's also rather similar to what Peter Gammel tried . . .

P. H. Kes: Well this has to do with the size of the bundle. When the bundle grows, the energy of the bundle – the barrier over which it has to hop – also grows. And that's just what is expressed by this power-law dependence. So when you only consider . . . in this case, when J is J_c, you have to consider only one vortex. It's the single-vortex limit. Then the pinning energy is just the one I computed: 35 K. But when the bundle grows, it becomes a much larger value. And actually what I wanted to point out here (I forgot that): it's funny that you always see, looking at these data, that $U/k_B T$ is always of the order 20 to 25, which actually is related to the time window of your experiment.

D. C. Larbalestier: So really what you want to emphasize is that it is a collective effect, but it's collective within the plane and not coupled between planes.

P. H. Kes: Collective in the plane. Yes. Yes. By the way, if you take the inverse of $U/k_B T$ it's 4/100 or 5/100, it's very close to this value Alex Malozemoff and Matthew Fisher found, for the S.

D. C. Larbalestier: The second point that I wanted to raise really relates to, I guess, the question of vacancies in the CuO_2 planes, because it's clear that what you have done is to work back from the experimental data of J_c.

P. H. Kes: Well, yes. We're back in the sense that we didn't know one number, and that's the density of these vacancies.

D. C. Larbalestier: The only thing I want to say is I'm not sure there's any hard evidence for vacancies in the CuO_2 planes. Now in bismuth I'm frankly just ignorant, I don't know what's going on, but in YBCO the argument I think I would make against the question of vacancies in CuO_2 planes is that people have done careful thermogravimetric analysis, and when you do, as you know – and of course it's very difficult to fully oxygenate single crystals – you can oxygenate polycrystalline material up to 7.0. So presumably, in polycrystalline material anyway, you really have oxygenated to the stage where the vacancies which exist are negligible. Now mostly we think that those vacancies exist in the chains. That is where the oxygen goes in and out.

P. H. Kes: Yes, of course, of course. It's a large number. But you only need one percent vacancies in the CuO_2 planes. That's not that much, and I learned from chemists that these perovskites are a sort of Gruyère. There are a lot of these vacancies.

D. C. Larbalestier: Okay. Well, you may be right and you may be wrong. But all I would say is that one should perhaps admit an element of doubt here.

P. H. Kes: Yes, yes.

A. P. Malozemoff: I want to ask about the maximum J_c you were quoting – I guess it's 2×10^6 at 77 K, but there's been a lot of data in the literature particularly showing 5×10^6 and even up to 8×10^6. Is that a problem?

P. H. Kes: Maybe. Let's see.

D. R. Nelson: Sounds good to me.

A. P. Malozemoff: What is the accuracy of your estimate?

P. H. Kes: It depends a little bit on T_c, what T_c you take. I think I took 92 K. I usually take 90 K. It's reasonably close to T_c where this $(1-\frac{T}{T_c})^{5/2}$ starts to become important. It might be a problem; I don't know. I have to go check it with other data.

B. I. Ivlev: I'd like to make some comments on spiral forest problem. We have calculated the following: Suppose we have the empty spiral, and the vortex of this spiral is parallel to this spiral axis. Even in the absence of the T_c decreasing in the spiral region, taking into account only the topological properties of this spiral defect, one can obtain that there is an attraction between the spiral and the vortex. And this attraction potentially is large enough. The critical current in order to release this vortex is of order, more or less, very near to the Ginzburg-Landau depairing current. So there are both effects: your effect, and that topological effect. And both effects are of the same size.

P. H. Kes: Right.

D. S. Fisher: Also on the screw dislocation. If you have a very strongly anisotropic material like BSCCO, or maybe the thallium ones are even more so, and you have screw dislocations with 500 or 1000 Å spacing between them – whether or not one can get to that regime is unclear – there's a novel mechanism by which you can take a vortex and you can split it up onto two screw dislocations or several screw dislocations. And if you do that, the kind of pinning energies you can get are more like those from cutting a hole in the film of size the penetration length, rather than the core size. You get much bigger than core pinning energies, because to pull a half vortex off a screw dislocation costs you an enormous amount of energy. You have to make a planar defect to do that. Whether one can ever get to this regime is unclear, but I think it might be possible if one can grow systems with lots of screw dislocations in BSCCO or thallium. It might be very useful.

D. R. Nelson: If I understand it, you're saying that the vortices are no longer precisely quantized because they're . . .

D. S. Fisher: Right.

D. R. Nelson: . . . going around the spiral staircase and they don't have to come back to themselves.

B. I. Ivlev: It should quantize, should be quantized.

D. S. Fisher: No, it doesn't have to be. If it's anisotropic enough and the screw dislocations are not too far apart . . . in BSCCO, the number's around 1000 Å. If they were 1000 Å apart, the vortices don't have to be quantized.

B. I. Ivlev: No, they're de-quantized. But there is some additional quantum number.

D. S. Fisher: No, they get de-quantized. The vortex is split up into two dislocations, or three if there's enough of them. But it's not quantized on an individual screw dislocation. It's different from the case you analyzed.

B. I. Ivlev: It should be there. Why?

D. S. Fisher: I'll explain later.

M. P. Maley: On another point, the weakness of the pinning seems to be very highly correlated with the degree of anisotropy, and yet both our measurements and I think some of Ken Gray's measurements seem to show that, for instance, in thallium-2223 and thallium-2212, that the irreversibility line, even in reduced units, is higher in temperature for the thallium compounds than it is for the BSCCO, while thallium presumably has a higher mass anisotropy ratio. Do you have any feeling for why that would be true or what else is?

P. H. Kes: No, I have no explanation for that. Except for a very dull one, is that it may be the sample properties, the sample quality. I don't like to mention that as a possibility. I don't know.

K. E. Gray: I think the anisotropy is greater in BSCCO. I mean, our measurements certainly would indicate that.

M. P. Maley: The anisotropy is larger in the BSCCO compounds than it is in the thallium compounds?

K. E. Gray: At least in our measurements.

D. S. Fisher: You recorded large anisotropies?

P. H. Kes: 50 000 is the largest one measured in bismuth.

K. E. Gray: The thallium's about 100 000 . . . . You get better and better crystals, and maybe it will get higher.

P. H. Kes: A Japanese group measured it on three crystals, or more even, and they found numbers for the square root of Γ as large as 220 . . .

P. L. Gammel: To get 50 000 you must be close to infinite, taking into account the mosaic spread of the crystal.

M. P. Maley: This is a really important problem for applications, because if you are really stuck with intrinsic anisotropy as being the fundamental limit to pinning and critical current densities at high temperatures . . . it's also important . . . I mean, if you can improve the crystals and increase the anisotropy, doesn't that seem to say that sort of introducing a lot of stacking faults and a lot of other things in these things might decrease the anisotropy and increase the pinning, improving the operation at higher temperatures?

P. H. Kes: Yes, on a very short length scale it would certainly have some influence. Well, I think to increase the pinning force or the critical current density is certainly to decrease the distance between the copper oxygen layers. And I recently heard from Kitazawa that – I think it's possibly Hitachi – was able to make wire of the thallium compound, the 1-2-2-3, which has a smaller distance and has a much higher irreversibility line, so it seems to work.

M. P. Maley: This is a wire without weak links?

D. C. Larbalestier: There are some weak links, but there is a large magnetization hysteresis well above 77 K.

X-D Wu: My question is about the pinning centers. All the numbers you estimated for the pinning centers look as on the same order. So nobody knows which was really the pinning center. At the same time, you also mentioned the stacking faults are weak pinning centers. But in most of high-T_c films – if you look at cross-sectional TEM's of those samples, most of the defects you see are the stacking faults. Cannot the defects, stacking faults, be pinning centers, which do not exist in single crystals?

P. H. Kes: No, I won't say that. They also exist in single crystals.

X-D Wu: No, they do not exist in single crystals.

P. H. Kes: Well, okay. For instance, a single crystal of 2-2-1-2 certainly will have some second phases – 2-2-0-2 – which probably is a stacking fault.

X-D Wu: That's mainly the insertion in the 2212-phase.

P. H. Kes: No, very few. One in 500 or so layers, typically. Yes, the pinning mechanism ususally is a little different for stacking faults and comes out to be much smaller than the Thuneberg-kind of mechanism. On the other hand you can say a stacking fault gives rise to disorder and electron scattering. Then you would be back in business. I didn't consider that yet. But usually the density of stacking faults is much less than the density of oxygen vacancies I assumed here.

X-D Wu: The other question is about single crystals. And if you think about it, the surface of those single crystals are larger. Most of the single crystals have steps, because of the way you grow those single crystals. But J_c is low. It means that the steps do not act as pinning centers as the surface step you see in thin films. However, after neutron irradiation, J_c goes up in the crystals.

P. H. Kes: Well, it depends on what the spacing is between these steps. If it's of the order of the vortex lattice spacing then the pinning becomes large again. If the distance is very wide, then you have to divide by a long distance, and then the effect is very small. So just a few surface steps, it's not such an important effect, I would say.

M. P. Maley: It could give rise to very large surface currents, but when you take the whole critical current of the crystal and divide by the thickness then you won't come up with a large number.

X-D Wu: But after neutron irradiation the J_c in the crystal goes up by a couple of orders of magnitude.

P. H. Kes: No, no no. Not that much, I would say. A factor of five, maybe ten. It depends on how large the critical current is to begin with. If you start with nothing, yes.

5. Survey of Applications and Summary
Chair — Jim Smith

Robert B. Hammond and Lincoln C. Bourne
Superconductor Technologies Inc.,
Santa Barbara, California

Survey of Potential Electronic Applications of High Temperature Superconductors

1. INTRODUCTION

In this paper we present a survey of the potential electronic applications of high temperature superconductor (HTSC) thin films. During the past four years there has been substantial speculation on this topic.[1] We will cover only a small fraction of the potential electronic applications that have been identified. Our treatment is influenced by the developments over the past few years in materials and device development and in market analysis. We present our view of the most promising potential applications.

Superconductors have two important properties that make them attractive for electronic applications. These are a) low surface resistance at high frequencies, and b) the Josephson effect. These properties can be exploited to enable unique electronic circuits; and with the advent of the high temperature oxide superconductors, practical operation at temperatures near 77 K. As mentioned, a wide variety of applications have been considered by others. Here we cover those applications which, in our view, have the best market potential. These are passive microwave devices, magnetic sensors, logic circuits, and computer interconnect boards. We include also a brief discussion of Josephson microwave mixers, infrared detectors, and 3-terminal devices. We would like to add that it is always difficult to predict the future in electronics products and markets. So, the future for HTSC electronic applications is not at all clear. And it is likely that the largest markets for HTSC electronics have not yet been identified.

2. PASSIVE MICROWAVE DEVICES

HTSC passive microwave devices exploit the very low surface resistance in HTSC thin films at microwave frequencies. Thin film resonators, filters, and delay lines are examples of such devices. Superconducting devices are attractive for two reasons. First, with respect to conventional devices, they can provide a comparable performance with a greatly reduced size and weight. Orders of magnitude reduction in volume and weight are, in principle, possible. Second, higher performance than conventional devices is possible. The following discussion summarizes several of the most attractive passive microwave device applications for high temperature superconductors. We focus on applications at ~77 K and above. This is because the requirement for cooling is the major hurdle to developing successful products. At ~77 K and above, cooling is greatly simplified for both product development and in actual applications. Circuit development is simplified because one can use liquid nitrogen. For products, cooling to these temperatures is much simpler and cheaper than cooling to lower temperatures. Cooling can be provided by liquid nitrogen, single-stage mechanical cryocoolers, or by efficient Joule-Thomson systems.

2.1 Microwave Properties of HTSC Thin Films

The relevant properties of HTSC materials for passive microwave applications are the frequency- and power-dependent surface impedance. $YBa_2Cu_3O_7$ (YBCO) and $Tl_2Ba_2CaCu_2O_8$ (TBCCO) have both demonstrated excellent microwave properties at 77 K. The low-power surface reactance is determined by the superconducting penetration depth, which for high quality films has a BCS-like temperature dependence with a value of[2] $\lambda < 2000$Å at 0 K. The low-power surface resistance is 0.2 to 0.4 mΩ at 10 GHz and 77 K, and is approximately proportional to the square of the frequency.[2,3] Figure 1 demonstrates the surface resistance vs. temperature for a film of $Tl_2Ba_2CaCu_2O_8$. At 10 GHz, this is about 50 times better than copper. At 1 GHz the surface resistance is more than 1 000 times better than copper.

The power dependence of the surface resistance at 77K is excellent in high quality films, showing no increase up to microwave magnetic fields above 50 gauss and microwave current densities well above 1 000 000 A/cm^2 (Figure 2).

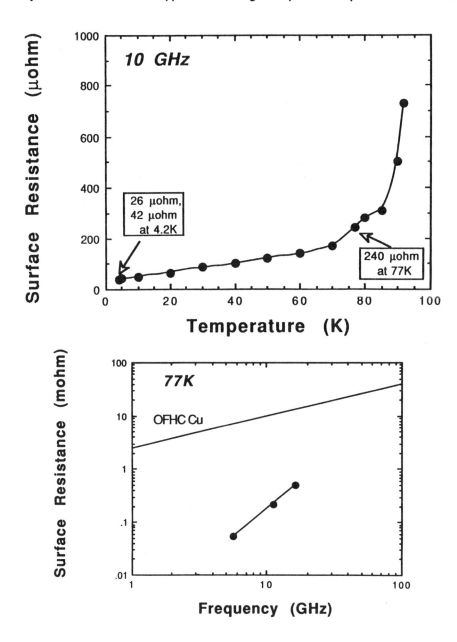

Fig. 1 Surface resistance data taken on thin-film $Tl_2Ba_2CaCu_2O_8$ samples grown at Superconductor Technologies, Inc. (a): Surface resistance vs. temperature at 10 GHz, (b): Surface resistance as a function of frequency. The data points are derived from the first three harmonics of a 5 GHz resonator. The line drawn through the data points has a slope of f^2.

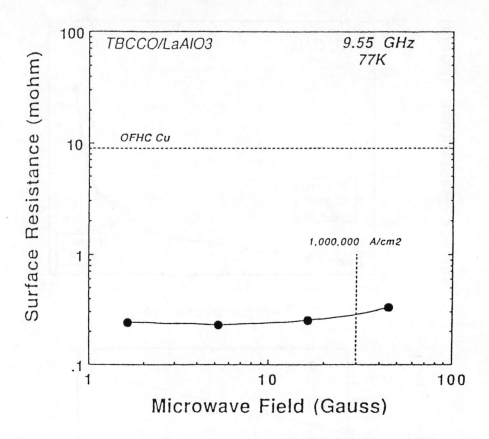

Fig. 2 Surface resistance as a function of microwave field for HTSC film and high purity copper at 77 K (from G. V. Negrete, McD. Robinson and R. B. Hammond, "Microwave Applications of High Temperature Superconductivity," Technical Digest, Electro '91, New York, April 16-18, 1991).

2.2 Resonators and Oscillators

Very pure microwave signal sources are required for many applications including instrumentation, communications, and radar. The noise characteristics of the signal source are usually determined by the characteristics of the resonator used for reference stabilization. Figure 3 shows the signal phase-noise performance attainable at 10 GHz using several different types of resonators for stabilization. Curve (1) is an 80 MHz quartz crystal stabilized source multiplied to 10 GHz. Curve (2) is a dielectric resonator stabilized oscillator (DRO) operating directly at 10 GHz. The use of ultrahigh Q microwave resonators can improve the predicted phase noise performance, as shown in curves (3) and (4) of Figure 3, for resonator Q's respectively of 300 000 and 1 500 000.

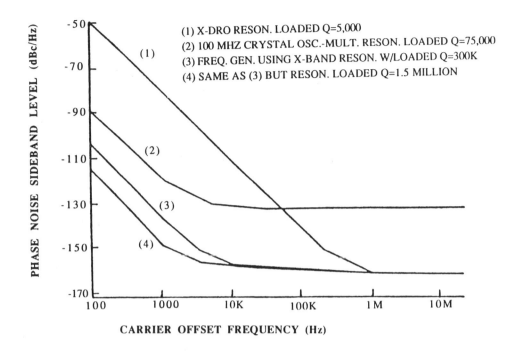

Fig. 3 Phase noise as a function of resonator Q, after reference 4.

Resonators with ultrahigh Q's can be obtained using cooled sapphire dielectric resonators inside an HTSC enclosure. Such a resonator is shown schematically in Figure 4. The cylindrical piece of sapphire is supported between two HTSC films by a low loss foam material. The sapphire and foam have dielectric constants of e_{r1} and e_{r2}, respectively. Unloaded Q's of up to 650 000 have been measured at 77K and 10.5 GHz, using TBCCO thin films to enclose the sapphire cylinder.[5]

Another type of HTSC resonator is attractive for applications that require high performance and significant reductions in size and weight. High Q's are obtained in stripline-type HTSC resonators, which can be used in oscillators and also as the building block for compact filters (see the next section.) Figure 5 shows the measured unloaded Q plotted vs. device power for a HTSC microstrip resonator. The structure, shown in the inset to the figure, is a 7-mm × 1-mm HTSC film above a 1-cm × 1-cm HTSC film used as the ground plane. The structure has a fundamental resonance at 5.1 GHz and a Q of 33 000, which is some 50 times higher than that of an equivalent structure made with Cu at 77K. This type of simple resonator offers higher Q's and much smaller size than can be obtained with copper cavities operated at 77 K.

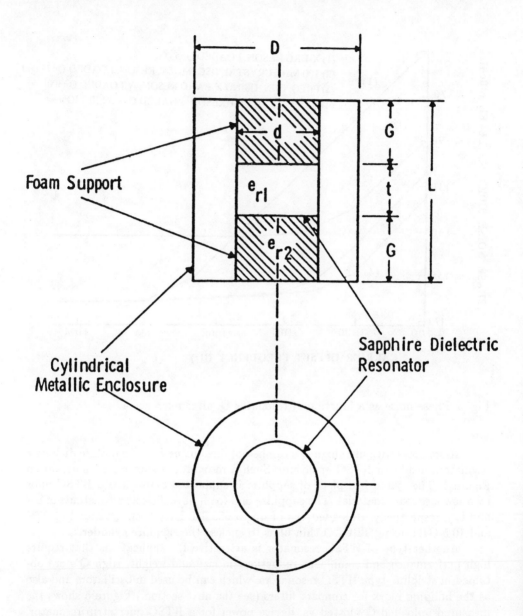

Fig. 4 Diagram of a cylindrical dielectric resonator, after reference 4. The sapphire dielectric is supported by two low-loss foam disks. In the HTSC version of this device, superconducting thin films are used for the endwalls of the resonator.

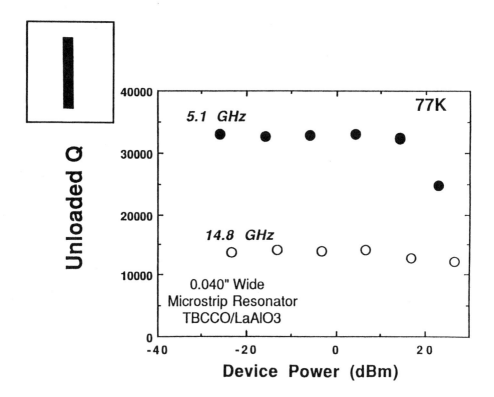

Fig. 5 Unloaded Q vs. device power for a microstrip $Tl_2Ba_2CaCu_2O_8$ resonator fabricated at Superconductor Technologies, Inc. Data is shown for the first and third harmonics. The inset shows the geometry of the resonator, represented by the black bar on a 1 cm^2 substrate of $LaAlO_3$.

2.3 Filters

In the previous section we discussed single-pole resonant structures. Multi-pole resonant structures are commonly used to provide a frequency response tailored to a specific circuit/system need. In conventional microwave devices, there is a direct relationship between Q and volume. The larger the volume, the higher the Q obtainable.[6] The lower surface resistance of HTSC films means that high Q can be obtained in a much smaller volume. Filters that pass a narrow band of frequencies require high-Q resonators, and must now be made with large volume structures. Large-volume filters are commonly used in sensitive microwave receivers.

Figure 6 shows model calculations of the loss in a 100 MHz bandpass filter centered at 10 GHz.[7] Losses in three different Cu structures are shown. Cu microstrip,

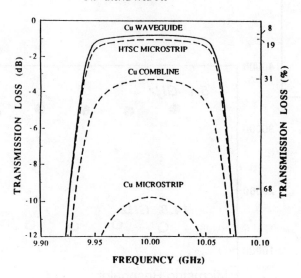

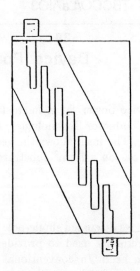

Fig. 6 Model calculations of loss in a 100 MHz bandpass filter centered at 10 GHz (from reference 7). The performance of an HTSC microstrip filter (diagrammed to the bottom of the figure) is compared with three types of Cu structures.

the structure shown in the inset, is impractical for applications because of the 10 dB loss. Waveguide is commonly used, and provides 1 dB of loss. The microstrip structure can be used, however, if it is made with HTSC films, and gives a loss that is comparable to Cu waveguide. HTSC filters, in principle, can be used to replace bulky waveguide filters now used in satellites for microwave communications.[8] These filters constitute most of the weight and volume of the satellites made today. Superconducting stripline-type filters also offer the potential for significantly lower manufacturing cost, because they are produced using photolithographic techniques rather than complex precision machining.

2.4 Delay Lines

Delay lines are used in microwave systems to provide a simple delay of a pulsed microwave signal by many nanoseconds. Frequently signals are stored in this manner to give the receiver time to process the signal accurately or simply to make a decision as to whether or not to process it. More complex analog signal processing is performed with arrays of delay lines with weighted taps. These arrays can perform real-time integral transforms (Fourier, correlation, matched filter, etc.). Figure 7 is a schematic diagram of an analog signal processor.[9] The time delays t_1, t_2, t_3, etc. are provided by microwave delay lines.

In current microwave systems, delays of tens to hundreds of nanoseconds are commonly provided by very high quality Cu coaxial cables. These are heavy and bulky. Superconductors offer the promise of performing the same function with orders of magnitude reductions in weight and volume.

One example of an attractive HTSC delay line has been developed at STI. Figure 8 shows a photograph of this device. It provides 30 nanoseconds of delay for frequencies from dc up to several GHz. The losses are lower than achieved in 0.135" coaxial cable, with a volume and weight of only 1% of the cable.

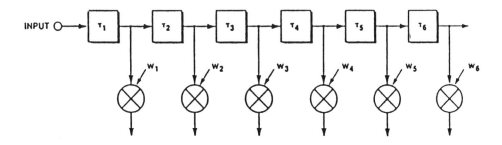

Fig. 7 Schematic diagram of an analog signal processor, with time delays provided by low-loss microwave delay lines (after reference 9).

Fig. 8 Photograph of a 30-ns wide-band delay line fabricated from a $Tl_2Ba_2CaCu_2O_8$ film at Superconductor Technologies Inc. The substrate, a wafer of $LaAlO_3$, is 1.2" square.

2.5 Antennas

Superconductors offer the potential to improve efficiency in phased array antennas in radio frequency and microwave systems.[10,11] At rf frequencies, superconductors permit reductions in antenna size through the use of superdirective beamforming techniques combined with superconductive matching networks. Figure 9 shows the efficiency of a simple dipole antenna for three values of the surface resistance, ranging from 1 microhm to 10 milliohm. For a particular value of efficiency, reducing surface resistance by a factor of 100 can permit a factor of 10 reduction in dipole length. This, of course, permits much smaller antennas. Lower surface resistance in combination with superdirective beamforming techniques can reduce the size of phased array antennas significantly. Figure 10 shows the improvements in directive gain achievable from superdirective beamforming. The low losses in HTSC allow the practical implementation of superdirective beamforming in radio frequency systems.

At millimeter wave frequencies, the losses in normal metals introduce significant degradation in the received signals. Superconductors can permit dramatic improvements in receiver sensitivity by lowering the losses at the receiving antenna. Millimeter-wave phased-array antennas will become more attractive because of the major reductions possible in losses in the feed networks.

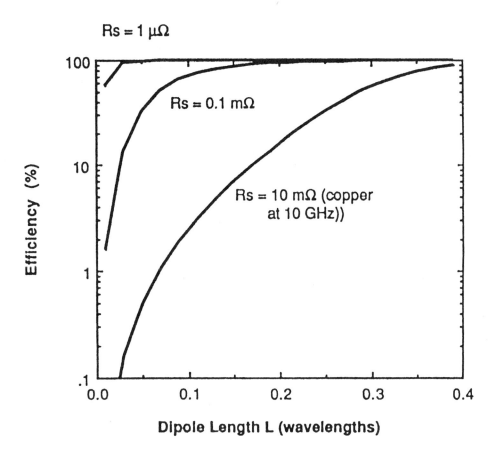

Fig. 9 Efficiency of a dipole antenna for three different values of surface resistance (from reference 12).

3. DIGITAL CIRCUITS

3.1 Latching Logic

Over the past 20 years there has been significant effort expended in developing superconducting logic circuits for computer applications. IBM had a major program in this area that was discontinued in 1983. Since then, several large Japanese

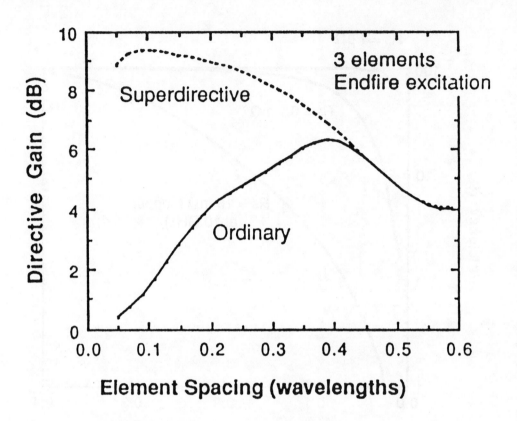

Fig. 10 Directive gain vs. element spacing, demonstrating the advantage of superdirective beamforming at narrow element spacings (from reference 12).

computer manufacturers, including Fujitsu, Hitachi, and NTT, have continued work on Josephson logic. Josephson circuits with high clock speeds and low power dissipation have been demonstrated at a high level of circuit complexity. One example is Fujitsu's 8-bit Nb Josephson digital signal processor,[13] which has an internal clock cycle of 1 GHz and a power dissipation of 12 mW. These circuits require superconducting tunnel junctions, and are now most commonly made with Nb technology and operate at 4 K. However, latching logic is not attractive for applications at ~77 K, since at these temperatures there is little advantage over semiconductor circuits in terms of speed and power dissipation. In addition, latching circuits require junctions with hysteretic I-V characteristics, which have not yet been demonstrated in any of the HTSC materials that have T_c's above 77 K. Thus, the availability of high T_c superconductors does not provide any new applications opportunities for superconducting latching logic. However, there is another type of superconducting logic, described in the following section, which does have promise for HTSC electronics.

3.2 Flux Quantum Logic

The first flux-quantum logic circuit was conceived by Goto et al., [14] and named the quantum flux parametron. Since then, Likharev and coworkers[15] have invented rapid single flux quantum (RSFQ) logic. RSFQ circuits do not require tunnel junctions, and can be built with non-hysteretic (i.e., SNS or weak-link) Josephson junctions. Junctions of this type have been demonstrated in HTSC materials by a number of laboratories.

In contrast to superconducting latching logic, RSFQ logic does have the potential to outperform semiconductor circuits at temperatures near 77 K. Simple RSFQ circuits have been demonstrated at clock speeds of over 100 GHz,[15] and are expected to reach speeds of over 300 GHz. Another advantage lies in power dissipation: RSFQ circuits operate at 100-500 nW/gate at 4 K, with an expected increase in power of about an order of magnitude for operation at 40-50 K.[16] In contrast, fast GaAs circuits dissipate on the order of 1 mW/gate.

Despite the significant promise of RSFQ logic, very little in the way of circuit development has been done as yet, and actual demonstrations have been made with only the simplest of logic elements. To assess the promise of the technology, substantially more circuit development will be necessary. The potential for this approach to logic is difficult to assess at this time because it is radically different from conventional approaches. It requires complex circuits, and these circuits need to be optimized to establish the trade-offs between error rates, operating speed, and temperature.

These circuits may become attractive for a variety of digital applications, including A/D converters, shift registers, and small/fast digital processors. The most promising near-term application appears to be A/D converters for infrared focal plane arrays, which require processors with low heat dissipations. Practical applications of flux-quantum logic are probably at least 5 to 10 years away. To summarize, there is evidence that RSFQ logic can be capable of high speed and extremely low power dissipation. However, not enough is known about RSFQ logic design to determine if the theoretical advantages can be achieved in a real system.

An additional concern for superconducting digital circuits is that nearly all applications require dense memory circuits as well as logic circuits. The memory circuits must be compatible with the logic in terms of voltage, current and speed. Further innovative work is needed to develop fast memory that is compatible with superconducting logic. This may require superconductor/semiconductor hybrid approaches, or entirely new all-superconducting circuit concepts.

4. DIGITAL SYSTEM PACKAGING (MULTICHIP MODULES)

The central processing unit in a high performance computer may contain up to a few hundred integrated circuit chips, with up to a few hundred input/output wires per chip. As IC's become faster, the limitations on computer speed increasingly come from the IC interconnections, instead of from the IC's themselves. The best computer packaging allows chips to be placed close together, demanding a very high density of interconnection wires. High performance computer interconnect boards are called multi-chip modules (MCM), as illustrated in Figure 11. MCM's are a key element in modern computers. The technology for MCM's is as sophisticated, complex and expensive as the technology for the integrated circuits themselves. IBM now manufactures an MCM for its mainframe computers that contains 64 layers of copper wiring embedded in a low-dielectric-constant glass ceramic.[18]

As computers improve in performance, more and more layers of copper interconnect wiring will be required, which will make the MCM's increasingly more difficult to build. Superconductors offer potential advantages in MCM wiring. The low ohmic losses in superconductors permit, in principle, the use of very small cross section wiring compared to what is required with copper. Superconducting wires with 1 micron widths might be used, whereas the resistance of normal metal wires begins to become prohibitive at line widths of less than 20 microns. Superconducting wires could thus reduce the number of wiring layers by orders of magnitude, and potentially make the MCMs much simpler to fabricate.

There are many significant challenges to making HTSC MCM's with such fine wiring.[19] Current densities above 1 000 000 A/cm^2 are required, thus necessitating the highest quality films. The films must be made in large area (>2" to 8") and with an extremely low defect density to support the very high density of narrow

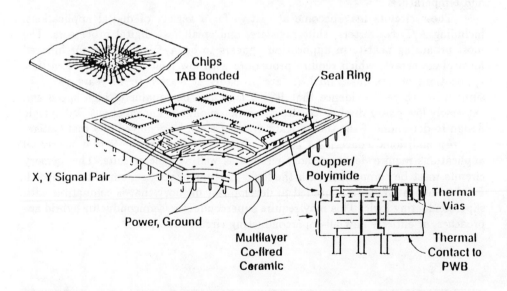

Fig. 11 Schematic diagram of a multi-chip module (from reference 17).

lines. Micron-scale patterning with sub-micron registration must be done over large areas, requiring processes that are not yet available. Micron-scale interconnect vias must be developed. And finally, low dielectric constant insulating layers must be made that are compatible with the HTSC films.

The materials challenges for HTSC MCM's are difficult, but the long-term payoffs for this technology could be truly significant.

5. MAGNETOMETRY - HTSC SQUID APPLICATIONS

5.1 Introduction

Superconducting quantum interference devices (SQUIDs) provide the best sensitivity of any technique for magnetic field detection at low frequencies over small areas. Significant development work has been done with SQUIDs for medical research and for detection of submarines and mines, and research has been conducted on a wide array of other possible applications. However, the only significant commercial application of SQUIDs to date has been in susceptibility measurements of magnetic materials and geological samples.

The advent of high-temperature superconductivity enables SQUIDs to operate in liquid nitrogen, thus allowing for smaller, simpler SQUID systems with much longer hold times. In this section we will discuss some areas where HTSC SQUIDs may find practical applications.

5.2 HTSC SQUID Technology

There are two types of superconducting components in a SQUID magnetometer: the SQUID itself, and a superconducting flux transformer coil, which is used to pick up magnetic signals and couple them into the SQUID. Although the coils carry only tiny signal currents, and the limiting noise sources are traditionally in the SQUIDs, the coils need to be fabricated from high-quality superconductor material, to avoid noise from flux motion in defects. For planar coils, this will require multi-layer processes with good crystal quality, as have been demonstrated[20] with *in situ* $YBa_2Cu_3O_7$ films, using $SrTiO_3$ for insulated cross-overs. Three-dimensional coils will probably require bulk wires with low noise levels.

The SQUID itself is formed from two Josephson junctions. SQUIDs using natural grain boundaries as junctions have shown noise levels low enough at 77K for some applications.[21] An all-HTSC SQUID magnetometer with a sensitivity of 3.1 $pT/\sqrt{Hz}$ at 10 Hz and 38 K was demonstrated[22] with a $YBa_2Cu_3O_7$ coil coupled to a $Tl_2CaBa_2Cu_2O_8$ grain-boundary SQUID. A more recent effort has given a result of 0.6 $pT/\sqrt{Hz}$ at 10 Hz and 77 K.[23] There appear to be no fundamental reasons preventing the sensitivity level from being increased beyond this value by several orders of magnitude. Although systems requiring ultimate sensitivity will almost undoubtedly continue to be made from LTSC materials, there is a large range in sensitivity between the best niobium-based SQUID systems (1 fT) and competing technologies, such as fluxgate magnetometers (10 pT). Many applications requiring intermediate sensitivities could be built with present-day HTSC SQUID technology, and the substitution of liquid nitrogen for liquid helium will allow engineering solutions that are not allowed for LTSC SQUIDs, including long hold times and very short separations between SQUIDs and room-temperature samples.

5.3 Susceptibility Measurements

The only truly commercial SQUID market is in susceptibility instrumentation. Two U.S. companies make these instruments: 2-G Enterprises, primarily for testing of geological samples, and Quantum Design, for measurements of magnetic materials. The market for these instruments is perhaps $10 million per year, with little prospects for rapid growth. Instrument prices are typically a few hundred thousand dollars for a complete measurement system, with the SQUID itself providing only a tiny fraction of the total cost. Switching from liquid helium cooling to liquid nitrogen cooling would make these instruments somewhat cheaper and easier to use, but the small market size does not provide a strong driver to develop HTSC SQUIDs.

5.4 Medical Instrumentation

It has been estimated that in the U.S. alone, more than $100 billion dollars is spent annually on health care for neurological disorders, including epilepsy, cardiac arrythmia, stroke, Alzheimer's and Parkinson's diseases, and psychiatric disorders. Existing tools for noninvasive analysis of the human body include magnetic resonance imaging and computer-aided x-ray tomography, both of which give structural but not functional information. Positron emission tomography gives information about metabolic activity, but the time response is on the order of minutes. SQUID magnetometry offers a completely noninvasive, real-time analysis of electrical activity in the body,[24] and this unique capability has attracted a great deal of development activity.

At least four companies have built prototype SQUID instruments for medical research: Siemens, Philips, CTF Systems, and Biomagnetic Technologies. These instruments have arrays of SQUIDs with as many as 100 channels, and are designed to measure magnetic fields primarily in the brain and heart. The current development issues are cost (typically a few million dollars per facility), signal analysis, and patient throughput. Since liquid helium cooling is not a limiting problem for these systems, they do not currently provide a strong driver for development of HTSC SQUIDs. (As an example, the requirement for liquid helium cooling in nuclear magnetic resonance imaging machines did not hinder commercialization, once there was a demonstrated clinical need). The first use of HTSC in these systems will probably be as pickup coils to transfer magnetic signals from the head to LTSC SQUIDs, since an array of nitrogen-cooled coils can be placed closer to the scalp than can helium-cooled coils. The difficulty here will be to build three-dimensional HTSC coils that are quiet enough to handle signal levels of a few fT, since the signal must be picked up near the scalp and then transferred to a liquid helium environment. It will thus be necessary to either form epitaxial thin films into 3D structures, or to develop low-noise bulk wires or thick films.

If the cost of medical SQUID systems could be brought down to the hundred thousand dollar level, the market would be expected to expand rapidly, since many hospitals would then be able to afford the systems for routine monitoring. At this price level, the difference in cost between helium and nitrogen cooling would be significant, leading to a stronger justification for all-HTSC systems.

One medical application that has been suggested for a simple HTSC SQUID system is a fetal heart monitor. The standard monitoring techniques in present-day clinical use rely on ultrasonic reflections, with the heartrate obtained from any of a variety of heart movements, such as the opening of a valve. Perinatologists would prefer to monitor the distinctive electromagnetic heart pulse, but the electrocardiogram of the fetus's heart is difficult to obtain because of interference from the mother's heartbeat. The magnetic signal of the fetal heartbeat, however, is typically a few pT outside the mothers' abdomen in the third trimester,[25] and could thus be detected with HTSC SQUID magnetometry.

5.5 Sensors for Navy Applications

Historically, military research has provided a strong impetus for SQUID systems development. The Naval Coastal Systems Center in Panama City, Florida, has conducted work on magnetic anomaly detection of mines and submarines. HTSC SQUIDs have potential applications here, because the background noise levels at the ocean surface are within the range of HTSC SQUID systems. Nitrogen cooling here would offer a number of advantages: liquid nitrogen is available on Navy ships, the dewar design would be simplified, and hold times would be much longer. The availability of nitrogen-cooled systems would also allow for new basing modes for SQUID sensors.

Another potential military SQUID use would be for magnetic antennas for extremely low frequency (ELF) communications. The current submarine communication system relies on a long towed cable as an E-field antenna. The antenna is uniaxial, so the submarine can travel only in a limited directional range while receiving signals. The directional problem could be solved with a compact triaxial H-field antenna, but the necessary sensitivities are on the order of 10 fT, thus requiring SQUID sensors. A LTSC SQUID antenna system was demonstrated[26] in 1977, but the program was not continued because of the difficulty of handling liquid helium in a real operating environment. Interest in SQUID ELF antennas could be revived by HTSC, since a liquid nitrogen sensor would be compatible with fleet requirements. The development of a towed HTSC buoy with a 10 fT sensitivity would, however, be a very challenging task.

5.6 Other SQUID applications

Many possible SQUID applications have been suggested and in some cases attempted, such as geophysical prospecting, borehole magnetometry in oil wells, or location of ferrite-tagged materials, such as buried plastic piping. Non-destructive evaluation (NDE) of materials has been one active field of research, including studies of metal fatigue in strained bars[27] and aging of nuclear reactor materials,[28] but commercial applications have not yet been developed. One problem is that real structures have complex geometries, leading to extreme difficulties in imaging and discrimination. In addition, the commercial NDE community tends to use only conservative and inexpensive technologies because of cost and liability concerns. However, if simple, inexpensive, nitrogen-cooled SQUID systems were available, novel applications would undoubtedly result.

6. MICROWAVE MIXERS

The lowest noise mixers available at millimeter and sub-millimeter frequencies are superconducting-insulating-superconducting (SIS) quasi-particle mixers, which are currently used in radio astronomy.[29] Superconducting mixers will operate only up to the gap frequency, which is about 1 THz for LTSC materials. An ideal SIS HTSC junction should have a bandwidth larger than this by an order of magnitude, assuming BCS behavior. Although there have been no reports of true SIS junctions in HTSC materials with T_c's above 77K, mixing has been demonstrated in controlled grain-boundary Josephson junctions in $Tl_2BaCa_2Cu_2O_8$ at frequencies of up to 94 GHz and at temperatures of up to 100 K.[30] Although it is unlikely that these weak-link junctions will give better noise performance than the best semiconductor mixers, they offer the advantage of low power operation. For example, a major problem for millimeter-wave converters lies in providing a stable signal at high frequencies with sufficiently high power to drive a mixer. A minimum Local Oscillator (LO) drive level of -5 dBm is required for a biased Schottky diode, whereas weak-link HTSC junctions can operate at power levels that are lower than this by several orders of magnitude. HTSC junctions therefore have promise as active devices in millimeter-wave systems, especially when integrated with low-loss passive HTSC elements.

7. INFRARED DETECTION

Since their discovery, thin film HTSC materials have been seriously considered for applications in infrared detection.[31] Detection mechanisms include direct detection and transition-edge bolometry. Devices based on direct detection rely on the breaking of Cooper pairs by direct absorption of infrared photons. The pair breaking is measured from changes in the penetration depth, or by changes in potential with the superconductor in the voltage state. Although there has been no clear demonstration that the performance of these devices can surpass that of semiconductor detectors, there continues to be a substantial amount of research in this area.

Transition-edge bolometers depend on the increase in thermal temperature from the absorption of infrared radiation. A superconducting microbridge with a small thermal mass is held at the transition temperature, so that the electrical resistance of the bridge is a strong function of temperature. A design analysis[32] of HTSC transition-edge bolometers showed that these devices have the potential to achieve a performance level competitive with any other type of bolometric detector operating above 77 K, for wavelengths greater than 20μm. The analysis estimated that noise equivalent powers for HTSC transition-edge bolometers should be in the range of $(1\text{-}20) \times 10^{-12}$ W/$\sqrt{\text{Hz}}$.

8. THREE-TERMINAL DEVICES

The marriage of superconducting and semiconducting materials to form novel three-terminal devices is a concept that has been explored for many years. The hope has been to produce a superconducting transistor that would offer advantages over semiconductor-based transistors in the parameters of speed, power dissipation and noise. The advent of HTSC materials raises the prospect of developing such devices with operating temperatures above 77 K. One device that shows promise and has been examined in detail theoretically is the Josephson field effect transistor. In principle, voltage gain in this device should be possible[33] at 77 K. The device would require short gate lengths, of less than 1000 Å, and high carrier mobilities, of greater than 100 000, in the semiconductor. The materials challenges here are significant, and as yet there have been no demonstrations of devices such as these.

9. SUMMARY

Let us conclude with a brief summary of the HTSC electronic applications that are most likely to develop in the next 10 years.

HTSC passive microwave devices are likely to start appearing in applications during the next 3 to 5 years. The first applications will be specialized filters and delay lines for military use, and low-noise HTSC resonator stabilized oscillators for instrumentation applications.

HTSC SQUID sensors will probably appear in applications within the next several years, with military and medical needs driving the technology development. Other uses will include measurements of magnetic materials and geological samples, as well as non-destructive evaluation of materials.

During this decade HTSC transition-edge infrared bolometers will be used for space astronomy, and also are likely to begin to be used in laboratory instruments.

Also during this decade, HTSC technology will be developed for digital applications. The earliest and most probable use will be in packaging for cryogenic semiconducting computers, e.g. multi-chip modules. In addition, HTSC flux quantum logic circuits will see considerable research and development. By the end of the decade, we may see the first applications of these circuits.

ACKNOWLEDGMENTS

We gratefully acknowledge discussions of SQUID devices with Professor F. Wellstood at the University of Maryland, Dr. Ted Clem at the Naval Coastal Systems Center, and with Dr. Robert Dinger at China Lakes Naval Weapons Center.

REFERENCES

1. K. K. Likharev, "Progress and Prospects of Superconductor Electronics," Superconductor Science Technol. **3**, 325 (1990).
2. D. E. Oates, A. C. Anderson, D. M. Sheen, and S. M. Ali, "Stripline Resonator Measurements of Z_s Versus Hrf in $YBa_2Cu_3O_{7-x}$ Thin Films," IEEE Trans. On Microwave Theory And Tech. **39**, 1522 (1991).
3. R. B. Hammond, G. V. Negrete, L. C. Bourne, D. D. Strother, A. H. Cardona, and M. M. Eddy, "Epitaxial $Tl_2CaBa_2Cu_2O_8$ thin films with low 9.6 GHz surface resistance at high power and above 77 K," Appl. Phys. Lett. **57**, 825 (1990).
4. M. M. Driscoll, J. T. Heynes, S. S. Horwitz, R. A. Jelen, R. W. Weinert, J. R. Gavaler, J. Talvacchio, G. R. Wagner, K. A. Zaki, and X.-P. Liang, "Cooled, Ultra-High Q, Sapphire Dielectric Resonators for Low Noise, Microwave Signal Generation," IEEE 45th Annual Symp. on Freq. Control, 29-31 May 1991, Los Angeles.
5. Private communication from George R. Wagner (Westinghouse Science & Technology Center).

6. R. V. Snyder, "All The World Is a Filter," IEEE MTT-S Newsletter, (1990).
7. Private communication from Ming-Jong Shiau (TRW Space Communications Division, Electronic Systems Group).
8. S. H. Talisa, M. A. Janocko, J. Talvacchio, and C. Moskowitz, "Present And Projected Performance Of High-Temperature Superconducting Filters," IEEE 1991 Intl. Microwave Symp.; and S. H. Talisa, M. A. Janocko, C. Moskowitz, J. Talvacchio, J. F. Billing, R. Brown, D. C. Buck, C. K. Jones, B. R. McAvoy, G. R. Wagner, and D. H. Watt, "Low- and High-Temperature Superconducting Microwave Filters," IEEE Trans. On Microwave Theory And Tech. 39, 1448 (1991).
9. W. G. Lyons, R. S. Withers, J. M. Hamm, A. C. Anderson, P. M. Mankiewich, M. L. O'Malley, and R. E. Howard, "High-Tc Superconductive Delay Line Structures And Signal Conditioning Networks," IEEE Trans. Mag. 27, 2932 (1991).
10. R. C. Hansen, "Antenna Applications of Superconductors," IEEE Trans. On Microwave Theory And Tech. 39, 1508 (1991).
11. R. C. Hansen, "Superconducting Antennas," IEEE Trans. On Aerospace And Electronic Systems 26, 345 (1990).
12. R. J. Dinger, D. R. Bowling, and A. M. Martin, "A Survey of Possible Passive Antenna Applications of High-Temperature Superconductors," IEEE Trans. On Microwave Theory And Tech. 39, 1498 (1991).
13. S. Hasuo, S. Kotani, A. Inoue, and N. Fujimaki, "High-Speed Josephson Processor Technology," IEEE Trans. Mag. 27, 2602 (1991).
14. See, for example, E. Goto, T. Soma, Y. Harada, H. Nakane, N. Miyamoto and U. Kawabe, "Basic Operation of the Quantum Flux Parametron," IEEE Trans. Mag. 23, 3801 (1987).
15. For a recent review, see K. K. Likharev and V. K. Semenov, "RSFQ Logic/Memory Family: A New Josephson-Junction Technology for Sub-Terahertz-Clock-Frequency Digital Systems," IEEE Trans. Appl. Superconductivity 1, 3 (1991).
16. Private communication from John X. Przybysz (Westinghouse Science & Technology Center).
17. T. A. Lane, F. J. Belcourt, and R. J. Jensen, "Electrical Characteristics of Copper/Polyimide Thin-Film Multilayer Interconnects," IEEE Trans. Components, Hybrids, Manuf. Technol., Vol. CHMT-12, 577 (1987).
18. Solid State Tech., Jan. 1991.
19. D. Herrell and C. Hilbert, "The Role of Superconductivity in Electronics Packaging," in "Electronics Packaging Forum," ed. J. E. Morris, vol. 1, p. 117 (Van Nostrand Reinhold, New York, 1990).
20. F. C. Wellstood, J. J. Kingston, M. J. Ferrari and J. Clarke, "Superconducting Thin-Film Flux Transformers of $YBa_2Cu_3O_{7-X}$," Appl. Phys. Lett. 57, 1930 (1990).

21. R. H. Koch, W. J. Gallagher, B. Bumble and W. Y. Lee, "Low-Noise Thin-Film TlBaCaCuO DC SQUIDs Operated at 77 K," Appl. Phys. Lett. **54**, 951 (1989).
22. A. H. Miklich, F. C. Wellstood, J. J. Kingston, J. Clarke, M. S. Colclough, A. H. Cardona, L. C. Bourne, W. L. Olson and M. M. Eddy, "High-Tc Thin-Film Magnetometer," IEEE Trans. Mag. **27**, 3219 (1991).
23. A. H. Miklich, J. J. Kingston, F. C. Wellstood, J. Clarke, M. S. Colclough, K. Char, and G. Zaharchuk, "Sensitive $YBa_2Cu_3O_{7-x}$ Thin-Film Magnetometer," Appl. Phys. Lett. **59**, 988 (1991).
24. A good review of medical SQUID work can be found in S. J. Williamson and L. Kaufman, "Biomagnetism," J. of Magnetism and Magnetic Materials **22**, 129 (1981).
25. V. Kariniemi, J. Ahopelto, P. J. Karp and T. E. Katila, J. Perinat. Med. **2**, 214 (1974).
26. J. R. Davis, R. J Dinger and J. A. Goldstein, "Development of a Superconducting ELF Receiving Antenna," IEEE Trans. on Antennas and Propagation, Vol. AP-25, p. 223 (1977).
27. H. Weinstock, "A Review of SQUID Magnetometry Applied to Nondestructive Evaluation," IEEE Trans. Mag. **27**, 3231 (1991).
28. G. Donaldson, S. Evanson, M. Otaka, K. Hasegawa, T. Shimizu and K. Takaku, "Use of SQUID Magnetic Sensor to Detect Aging Effects in Duplex Stainless Steel," Brit. J. NDT **32**, 238 (1990).
29. A review of superconducting devices for receivers is given by P. L Richards and Q. Hu, "Superconductive Components for Infrared and Millimeter-Wave Receivers," Proc. IEEE **77**, 1233 (1989).
30. J. P. Hong, T. W. Kim, H. R. Fetterman, A. H. Cardona and L. C. Bourne, "Millimeter Wave Mixing from Deliberate Grain-Boundary Weak Links in Epitaxial $Tl_2CaBa_2Cu_2O_8$ Films," Appl. Phys. Lett. **59**, 991 (1991).
31. A recent review of superconducting infrared dectectors is given by S. A. Wolf, U. Strom, and J. C. Culbertson, "Visible and Infrared Detection Using Superconductors," Solid State Tech., April 1990, p. 187.
32. P. L. Richards, J. Clarke, R. Leoni, Ph. Lerch, S. Verghese, M. R. Beasley, T. H. Geballe, R. H. Hammond, P. Rosenthal, and S. R. Spielman, "Feasibility of the High T_c Superconducting Bolometer," Appl. Phys. Lett. **54**, 283 (1989).
33. A. W. Kleinsasser and T. N. Jackson, "Prospects for Proximity Effect Superconducting FETs," IEEE Trans. Mag. **25**, 1274 (1989).

DISCUSSION

J. Smith: How many filters have you sold?

R. Hammond: I don't have a good number. I can tell you . . .

J. Smith: What is the order?

R. Hammond: Oh, tens. Tens. We sell lots of other things too. Actually our biggest business right now is in selling patterned and metallized films to customer design.

X-D Wu: What do you think about the future of American high-T_c small companies, like STI, Conductus, American Superconductors. Because in this country, most investors are always looking at the short term return, and we all have been in business almost three or four years. What do you think about your future? (Laughter.)

R. Hammond: We just had a Board of Directors meeting last week, so I'm really close to this issue. The reason that companies like STI and Conductus, and American Superconductors were started a few years ago was because there was expectation that there were large markets that would develop, and they would develop in a relatively short time frame. And here we are, four years later, and it really hasn't happened. The markets haven't "exposed" themselves. (Laughter.)

I still believe that they are out there. And I guess I don't know how this field is going to develop. But, the microwaves is going to happen; I think that we're going to do that. But it's going to happen slowly. It's not going to provide the kind of growth that our investors want. And you know, we're going to make a viable business out of it. But something that has not been identified yet, I think, is going to have to come along to get the kind of growth that the venture capitalists want. I don't know if that answers your question.

D. C. Larbalestier: Can I just comment on that. Intermagnetics General is a small company formed 25 years ago, and they are holding their 25th anniversary, I think, this Thursday and Friday. And they've now just lost their small business status. So, essentially it is possible. I mean, they were formed in the early days of low-T_c technology . . .

R. Hammond: Well I think that, you know, I guess this will happen with High T_c. I mean, there's real business out there and this technology's going to develop, and I think there's a real industry that's going to grow around it. It's just not clear how quickly it's going to grow, at this point.

P-J Kung: I have three questions. The first one is, for materials evaluation, did you observe any electromigration phenomenon? And the second question is, could you say something on the device packaging, the current research on device packaging of high-T_c superconductors?

R. Hammond: Well, electromigration we never really looked for. I can't say much more than that. In terms of the packaging: packaging is a real challenge, and we're working on it a lot. And there's a bunch of issues; materials compatibility, basically. The superconductors . . . like with vacuum packaging, you need to use elevated temperatures, you need to use vacuum, you need to use epoxies, and various kinds of materials. And the process temperature has to be compatible. So far we've had no fundamental problems. I think that these superconductors . . . we're going to be able to package them okay. I think those packaging fabrication issues, those are all solvable problems.

P-J Kung: And the last question is, right now, do you have any research on, say you put a second heat sink – for example, a semiconducting diamond thin film – with superconductor, together, and then get some benefits from there.

R. Hammond: I'm not sure I follow you. You mean getting a benefit from cooling the semiconductors, as well as the superconductors?

P-J Kung: Yes.

R. Hammond: We do see that in microwave circuits. Our applications in microwaves will, I think, be facilitated by the fact that semiconductors operate much better at 77 K than they do at room temperature. For instance, microwave amplifiers, microwave mixers, they all work better.

A. M. Goldman: Everybody talks about multichip modules in the context of pulsed propagation for short pulses in superconductors. Have you looked at power dissipation? I mean, maybe it's easier to make the ground planes and the power leads.

R. Hammond: Oh it is. Yeah, I think that it will happen first in the power planes, and strontium titanate's ideal because it's probably the highest-dielectric-constant practical material at 77 K. So superconductors on strontium titanate for power planes, that will be where it happens first.

A. P. Malozemoff: Is your technology mostly a thallium technology? Or, how do you evaluate thallium, for instance, against yttrium barium copper oxide?

R. Hammond: Well that's a subject of a whole talk. We have YBCO capability – we make first-rate YBCO films – and it's been a constant issue with us as to which film to go with. We choose thallium for all the same reasons we started out. Thallium has lower loss at 77 K. It's got the other advantage: we can put high-quality films on both sides of the substrate on two inches, uniform. Doing that with YBCO has not been demonstrated. Because of the process we have we can do it with thallium pretty straightforwardly. So, that's basically why we choose thallium.

I think YBCO has strengths. I mean, YBCO has a number of strengths, and I'm not committed to one material or the other in my own mind. We're going to pursue these applications, and the applications are going to tell us which film is the right one. And for instance, this issue of excess noise. We may find that thallium, because of this weaker pinning, has flux noise in it and it makes it, say, unacceptable for the stable oscillator application, for the really high performance applications. So we don't know yet how that's going to come out. My expectation now is YBCO will have some applications and thallium will have some applications, and it's not sure how that will shake out. Time will tell.

R. E. Muenchausen: Just to add a little bit to that. I disagree [with your assertion] that it hasn't been demonstrated that you can deposit YBCO double-sided, epitaxial-quality, [thin films] which below 77 K have a lower surface resistance [than thallium superconducting films]. Furthermore you showed a slide where you had the Sterling cooler, implying that for some applications at least you're not necessarily stuck with 77 K cryogenics.

R. Hammond: Oh, that's absolutely true. For the applications, I think lower temperatures are probably okay, eventually. Near term applications will be at 77 K because liquid nitrogen is there, and because the development is going to be a lot faster. When you're doing circuit development, and you've got the choice of whether you can put this thing in a mechanical cooler or a liquid helium cryostat, or you can put it in a liquid nitrogen dunk, I mean, there's no comparison. And the rate at which you can do the development, early on, is going to make a difference. And in terms of the double-side YBCO films, I would love to measure one. So, if you have one, I'd love to measure it.

X-D Wu: One more question Bob. What do you think about the problems involved in developing devices? Do you think it's materials issues, or device issues, or packaging issues – of course, let's forget marketing, okay – what do you think are the most important issues right now?

R. Hammond: All the above, but I'd say number one is microwave engineering, doing things like these filter designs. Developing a filter topology and a filter design capability. I mean, developing the software so you can do the designs, as well as developing the topologies that'll work. That's number one.

Number two is the cryogenic packaging. Cryogenic packaging for these types of applications is just non-existent, and so we're developing that in collaboration with some other companies.

The third is the materials. Materials are going to require continual development, and exactly what is going to be emphasized in that development is going to come out of the need of the applications. And that's an evolving story, and it's going to continue to evolve in the next few years. Things like this excess noise issue: don't know. Things like the non-linear property. Even though we do pretty well up to ten milliwatts, we still see some non-linearities so we may have to optimize the films for that. We may have to go to different substrates, for example.

David C. Larbalestier
University of Wisconsin
Madison, Wisconsin 53706

Development of High Temperature Superconductors for Magnetic Field Applications

The key requirement for magnetic field applications of high temperature superconductor (HTS) materials is to have conductors with high transport critical current density available for magnet builders. After 3 or 4 years of being without any such object, conductor makers have had recent success in producing simple conductor prototypes. These have permitted the construction of simple HTS magnets having self fields exceeding 1 tesla at 4K. Thus the scientific feasibility of making powerful HTS magnets has been demonstrated. Attention to the technological aspects of making HTS conductors for magnets with strong flux pinning and reduced superconducting granularity is now sensible and attractive. However, *extrinsic* defects such as filament sausaging, cracking, misaligned grains and other perturbations to long range current flow must be controlled at a low level if the benefit of *intrinsic* improvements to the critical current density is to be maintained in the conductor form. Due to the great complexity of HTS materials, there is sometimes confusion as to whether a given sample has an intrinsically or extrinsically limited critical current density. Systematic microstructure variation experiments and resistive transition analysis are shown to be particularly helpful in this phase of conductor development.

1. INTRODUCTION

The much-discussed technological applications of high temperature superconductors (HTS) to magnetic field technologies such as motors, generators, transmission lines, magnetic energy storage devices and train and ship propulsion systems all require high critical current densities (J_c). In fact the key requirement is for a high *transport* critical current density (J_{ct}) in a conductor form. A high J_c means of order 10^5 A/cm^2 and the qualifier transport means that the current be measured over macroscopic dimensions, thus crossing grain boundaries or other interfaces which produce the superconducting granularity which so strongly limits the J_{ct} of polycrystalline 123 compounds. The requirement of a conductor form establishes that the superconductor be parallel with a good normal metal conductor, that it be fabricable in long lengths at reasonable cost and that it be capable of being wound into a magnet while retaining these qualities.[1-3] Nothing like this existed during the first 3 or 4 years after Bednorz and Müller's discovery. However, the development of 2212-phase Bi-Sr-Ca-Cu-O (BSSCO)[4] and 2223-phase BSCCO phase[5] oxide-powder-in-tube (OPIT) conductor prototypes showed that J_{ct} values of 10^4-10^5-A/cm^2 were possible when *genuinely polycrystalline* conductors were encased in a silver sheath. It was the development of these prototype conductors that today makes it realistic to invest important resources directed at making the technologies mentioned above practically feasible.

In this article I want to discuss some of the issues that arise in developing a technological understanding of the transport current of HTS materials. I do not discuss mechanical or fabrication properties because much of the knowledge on these topics is incomplete and empirical. A technological understanding of the J_c benefits enormously from a good scientific basis, even though scientific understanding is not absolutely necessary for technological progress. Indeed, the scientific and technical approaches to the problem may be quite different, as may the two communities who study the problem. Some interesting examples of this were found in the development of high current density in Nb-Ti, the material from which virtually all superconducting magnets are presently built, and this paper will accordingly recall some of the lessons learned in developing high J_c in conductors of this material.

One important difference of emphasis can be easily illustrated by contrasting the scientific and technological approach to the J_c. As argued elsewhere,[1] there is a tendency to treat J_c as if it were as well-defined a parameter as the superconducting transition temperature, (T_c), is. In this simple view, current flow is dissipation-free below a well-defined J_c and dissipative above J_c. Flux creep and thermal activation phenomena make this approach conceptually wrong, as companion papers in this volume discuss.[6-9] However, discussion is undoubtedly simplified if a single value of J_c is chosen as the basis for a comparison of theory and experiment. Starting from a fundamental view of the J_c, the tendency is to identify the possible fluxon-microstructure interactions, calculate the energy of such interactions, estimate the length scale of the interactions in order to derive the elementary pinning force, sum all such forces over the different types of defects existing in the microstructure according to the most appropriate fluxon summation scheme and then, having determined the correct field dependence of the pinning functions from a knowledge of the pinning interaction and the appropriate thermodynamic parameters of the superconductor (e.g. $H_c, \lambda, \kappa, \xi$), finally predict J_c (H). Such calculations, if truly *ab initio* are seldom accurate to better than a factor of 5 or 10, as might be expected from the many steps involved in the calculation. Conductor users and producers generally take a different approach. By virtue of the fact that the wire is being produced, they assume that the flux pinning is high enough and superconducting granularity restricted enough that the transport current density is usefully high. Their problem is to maintain J_{ct} near the level permitted by the microstructure, without interruptions of long-range current flow by irregularities of cross-section, cracks, misaligned grains or other defects which produce local variations of J_c. Their focus of attention is necessarily quite different from the flux pinner and tends to focus on fabrication and utilization issues.

Particular focus to the difference in the two viewpoints is given by the fact that their tolerable uncertainty scales are vastly different. As already stated, a first principles scientific calculation of J_c (J_{ct} is seldom explicitly distinguished from J_c in this approach) may well be considered good, if accurate to within a factor of 5 or 10 (particularly if the field or temperature dependence is accurately predicted), while a technological uncertainty in J_{ct} of even 50% for an established product would be a commercial disaster. Thus most present development of Nb-Ti conductor neglects flux pinning in favor of controlled materials or fabrication parameter variation studies. For HTS conductors the best approach is still unclear, since the parameter space inhabited by the materials themselves, as well as present prototype conductors, is so large that it is frequently debated what controls the properties of any given sample. The present uncertainties in J_{ct} for apparently well-made prototype silver-sheathed BSCCO conductors are still quite large, of order at least 2 and perhaps as high as 5 or more.

This paper will explore some of the reasons for discrepancies such as the above, developing a viewpoint complementary to the more physics-based ones of Kes,[6] Fisher,[7] Nelson[8] and Kapitulnik.[9]. This paper will explore some microstructural issues from the viewpoint of the need to design experiments which bring the *macro*structure, as well as the *micro*structure under explicit control. We take the view that the development of HTS magnetic field applications is underway and that such conductor development issues, as well as the fundamental ones of flux pinning, flux creep, phase transitions of the vortex lattice, granularity and other basic issues of fluxon dynamics all need to be understood if the technological applications that so many have dreamed of since the first discoveries of high temperature superconductors are actually to come to fruition.

2. INTRINSIC AND EXTRINSIC CONTROL OF THE CRITICAL CURRENT DENSITY

In the development of optimized Nb-Ti conductors, a key step was the explicit recognition that, although the upper limit to the J_c is determined by *intrinsic* elementary fluxon-microstructure pinning interactions (f_p) and by the summation of these elementary interactions into the bulk pinning force (F_p), in practice the transport J_{ct}, defined as the critical current (I_c) divided by the cross-sectional area (A), is subject to a second limit introduced by *extrinsic* factors which produce larger-scale, frequently macroscopic variations of the critical current.[10] Figures 1-4 illustrate this point, both for a Nb-Ti and for a Bi-Sr-Ca-Cu-O (BSCCO) conductor. Fig. 1 shows the nanometer scale α-Ti precipitates within an optimized Nb48wt%Ti filament that provide the strong elementary flux pinning interaction which makes a high J_c possible. These precipitates have a ribbon-like morphology and are highly anisotropic in shape. By measuring the precipitate volume fraction, size (in particular, the thickness) and separation and by correlating these quantitative microstructural parameters to the electromagnetic properties, it was possible to prove[11,12] that it is these α-Ti particles which determine the flux pinning and that the particles exert their maximum effect when thinner than and separated by less than a coherence length (ξ). Given that $\xi(4K)$ is 5nm for Nb47wt%Ti, this sets the natural length scale needed for flux pinning studies of Nb-Ti. This is of order 1nm or 0.2ξ, a scale which is fortunately accessible to existing transmission electron microscopes.

The extrinsic, more macrostructural-scale limits are provided by many factors.[10] One of the simplest to understand is shown by Fig. 2. A filament which has a periodically varying filament cross-section is said to be sausaged. Such a filament has an I_{ct} which is extrinsically limited by the smallest local cross-section. If the current in such a necked region exceeds the local critical current, dissipation occurs due to flux flow within the superconductor or by resistive current transport across the matrix as the excess current transfers to a neighboring filament. Sausaging has many causes and for this reason can be difficult to understand and control. It is the most common limit on performance of all practical conductors.[10,13,14]

Similar phenomena certainly exist in HTS materials too, but much less is explicitly understood about both intrinsic and extrinsic factors. Fig. 3 provides a high magnification view of a misaligned pair of grains in a silver-sheathed 2223-BSCCO conductor prototype. The magnification is larger than that used in Fig. 1 to reveal the flux pinning microstructure of Nb-Ti and the anisotropic, atomic-scale planar nature of the microstructure is clear. In principle, the microstructural defects responsible for flux pinning and/or superconducting granularity ought to be visible. However, transmission microscopes are not very sensitive to individual atoms, particularly if they are the important, but low atomic number, oxygen atoms. By analogy to the 123 compounds,[15] there is a tendency to assign the flux pinning of BSCCO to oxygen vacancies.[6,16] However, microscopy cannot yet give the sort of

Development of High Temperature Superconductors for Magnetic Field Applications   **467**

Fig. 1 Transverse cross-section micrograph of a region in a Nb47wt% Ti filament, optimized by multiple heat treatment and draw cycles to produce a high J_c. The critical flux pinning sites are the ribbons of normal α - Ti which form $\sim 25\%$ by volume of the microstructure (Courtesy P. J. Lee, University of Wisconsin)

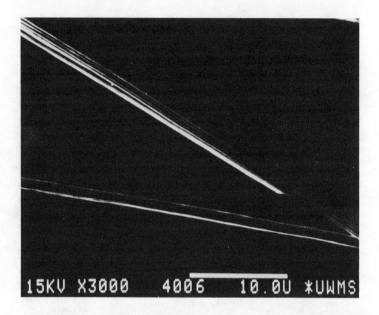

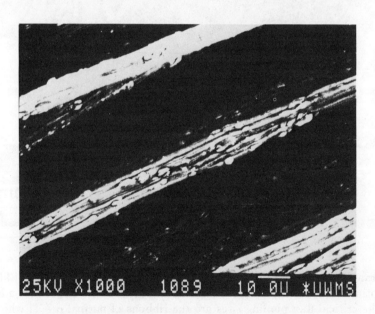

Fig. 2 Scanning electron micrographs of exposed Nb-Ti filaments in a) Nb-barrier clad filament of uniform cross-section and b) a barrier-free filament in which Cu-Nb-Ti nodules have instigated sausaging. The interfilament Cu has been removed by etching. (From Warnes and Larbalestier, ref. 13)

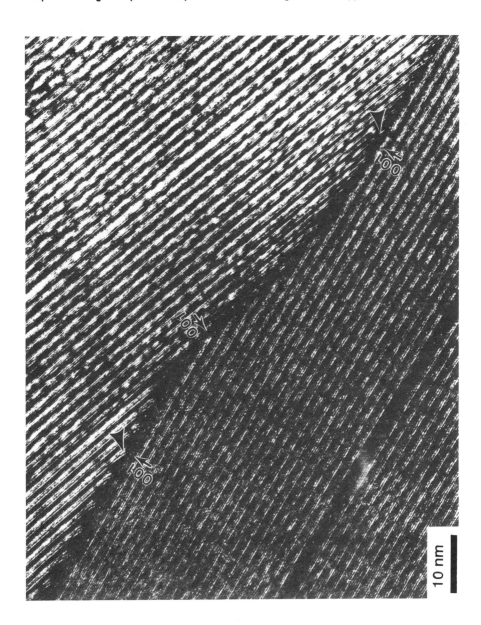

Fig. 3 Longitudinal section micrograph of two abutting 2223-BSCCO grains found in a silver-sheathed tape. The c axis of each grain is normal to the layer structure of each grain. Some evidence of lattice interpenetration and facetting is seen at the grain boundary. (Courtesy Yi Feng, University of Wisconsin)

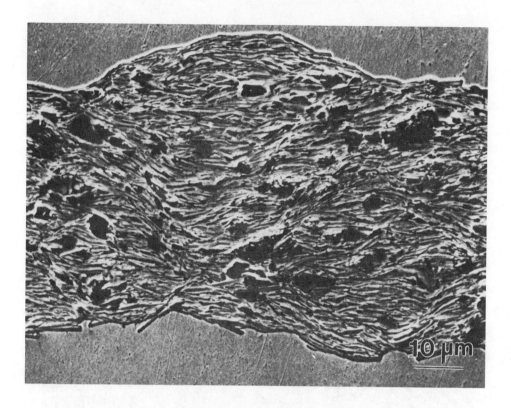

Fig. 4 Longitudinal section of the grain structure in Ag-sheathed 2223-BSCCO tape. An irregular section has been chosen to show how the local alignment tends to follow that of the silver sheath. Grain alignment is also interrupted by undissolved second phase particles. An apparent shear band on the left produces an aligned band lying at about 45° to the tape axis, which is horizontal. (Courtesy Y. High, University of Wisconsin)

quantitative picture of the defect structure exemplified by Fig. 1, either for oxygen or for cation defects of vacancies. Figs. 1 and 3 are strikingly different, in that the Nb-Ti microstructure is clearly heterogeneous and two phase, while that of BSCCO appears single phase and almost featureless, except on the periodic scale of the crystal structure itself. Nor is it yet possible to deduce from TEM grain boundary images whether the grain boundaries have depressed superconducting properties and exhibit superconducting granularity or whether they are fully coupled and exhibit no granularity. Thus we already see a major difference between Nb-Ti and HTS materials: microstructural examination can define the magnitude and origin of the defect microstructure in Nb-Ti but, so far, there are very few cases where this has been possible for HTS materials. A crucial consequence of the difference between the two materials is that much experimentation with HTS materials occurs "blind," because the consequences of changing the processing cannot be explicitly defined by microstructural examination.

By contrast, at the present early stage of HTS conductor development, the imaging of extrinsic defects is rather easy. As Fig. 4 shows, there are many possibilities that must be taken into account when considering possible extrinsic limits on J_{ct}. Fig. 4 is a macroscopic view on a scale of about $50 \times 100 \mu m$ of a silver-sheathed 2223-BSCCO tape. In this case the grain structure has been made visible by etching.[17] The macrostructure is seen to be anything but uniform: the sheath thickness is not uniform, the grain alignment varies both through the thickness of the tape and along the tape axis. Insulating second phase particles also disrupt the overall cross-section and the local grain alignment. In contrast to the almost perfect microstructure of Fig. 3, the macrostructure is clearly highly defective.

In conductors, rather than scientific single crystal samples, it is almost invariably the extrinsic aspects of the microstructure illustrated in Figs. 2 and 4 that control J_{ct}. Apparently well-planned experiments frequently do not have a logical explanation unless this is appreciated. These technological issues do not make it any less important to understand the basic flux pinning or granularity issues that are central to so many scientific studies of HTS. What is important is to recognize that both aspects be treated, generally simultaneously.

When do the extrinsic factors become important and how large are their magnitude? In Nb-Ti conductors the extrinsic factors can be easily assessed by measuring the resistive transition. It has long been common to fit the resistive transition to a power law, where the voltage V is related to the current I by the resistive transition index n ($V \propto I^n$). Fig. 5 shows that the index n has a characteristic dependence on magnetic field. In the intrinsic limit, n(H) increases steeply from a value of order 10 to one of more than 100, as H decreases from about 0.1 H_{c2} to 0.9 H_{c2}. This progressively changes to a flattened plateau-like characteristic as the J_{ct} becomes extrinsically limited. The intrinsic to extrinsic transition for Nb-Ti conductors occurs as the wire is drawn to progressively finer sizes. Deformation instabilities, provoked by a variety of microstructural causes, produce the sausaging visible in Fig. 2 and this can be clearly correlated to the flattened n-H characteristic seen at small sizes, where filament sausaging is extensive.[13,14] The characteristic can be correlated to the Coefficient of Variation (COV) in the filament cross-sectional area, where the COV is defined by the standard deviation of the measured filament cross-section (σ) divided by the mean cross-sectional filament area (A) and thus to the J_{ct} defined by I_{ct}/A.[13,14]

The magnitude of the effects can be obtained by treating the resistive transition more quantitatively. As originally suggested by Baixeras and Fournet[18] and then experimentally studied by Warnes and Larbalestier,[13] the curvature of the resistive transition contains information about the distribution of local critical f(I') within the conductor:

$$V(I) = A \int_0^I (I - I') f(I') dI' \qquad (1)$$

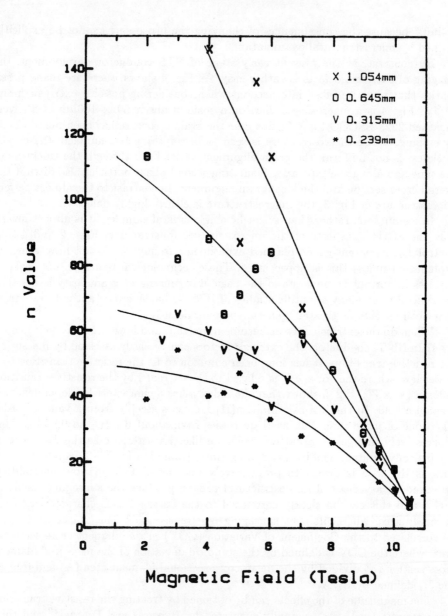

Fig. 5 Resistive transition index (n) vs. H characteristics for a high J_c multifilament Nb47wt% Ti wire as it is drawn from 1.054 to 0.239mm diameter. As the wire is drawn the filaments become irregular in cross-section and the resistive transition ($V \alpha I^n$) broadens. At large sizes the filaments are uniform and flux pinning controls the critical current; at small sizes the smaller necked cross-sections impose an extrinsic limit on I_c. (Data from Warnes and Larbalestier, ref. 68)

where V(I) is the voltage produced at a given current I and A is a factor appropriate to the particular dissipative processes occurring within the conductor when I is greater than I'. By differentiating Equation 1 twice with respect to the current, one obtains:

$$d^2V/dI^2 = A\,f(I) \qquad (2)$$

Two typical distribution plots are shown in Fig. 6. In the case of smooth filaments the I_c distribution is sharp and the full width of the distribution at half maximum (FWHM) divided by the average critical current of the distribution $<I_c>$ is of order 15%. When the COV of the filament cross-section is large, FWHM/$<I_c>$ can easily reach 50%. Nb-Ti conductors having such properties are practically unusable.

Few similar measurements have yet been made on BSCCO conductors. One indication that the resistive transition characteristics may be more complex than for Nb-Ti is shown in an investigation by Heine et al.[19] They compared the n values of Ag-sheathed YBCO and BSCCO conductors (Fig. 7). Their n(H) characteristics are rather different from those seen for Nb-Ti in Fig. 5. The two HTS conductors have n values which fall sharply in only weak millitesla fields, then tending to a plateau value of less than 10. The initial fall is presumably associated with the field-induced destruction of some Josephson-coupled current paths, particularly in the YBCO conductor. The low plateau values may be associated with the locally variable cross-section which is active in passing transport current. One component is the real longitudinal variation of cross-section which is still very common in prototype conductors (Fig. 3); a second contributor may be a variation in the active cross-section because of variability in the local grain-to-grain connectivity associated with local grain misalignments. This variable can operate in both radial and length-wise directions. Flux creep, particularly in the BSCCO tape, should also contribute to the shape of the transition, indeed making a power law fit of the resistive transition inappropriate. The essential point is that multiple factors can operate and that all need to be identified and brought under control. Certainly the resistive transitions of present BSCCO conductors tend to be very broad, as compared to Nb-Ti conductors of even moderate quality.

The best $J_{ct}(0T,77K)$ values for the 2223 conductors are developed by multiple pressing and reaction cycles which tend to produce an aligned grain structure which is largely 2223 phase.[5,20,21,22] However, nominally identical thermal treatments applied to rolled samples produce J_{ct} values which factors of 2 to 5 less than the present maximum of 50,000 A/cm^2.[5] Is it an *intrinsic* flux pinning or an *extrinsic* macrostructural variation which produces such a big difference? Given the evidence of cross-sections like Fig. 3, it appears likely that the controlling factors often lie in differences of crack density, grain alignment and second phase distribution. Thus controlled flux pinning enhancement experiments will only achieve full success when the extrinsic factors limiting the J_{ct} are identified and controlled.

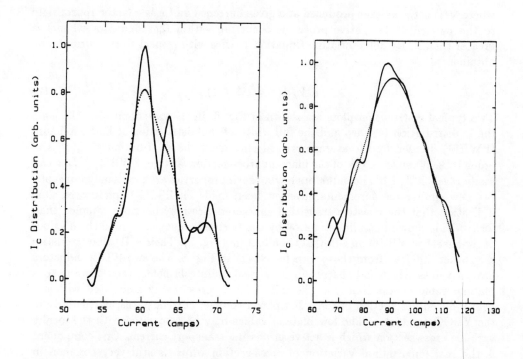

Fig. 6 Critical current distribution plots for a) barrier clad composite such as that shown in Fig. 2a and b) sausaged barrier-free composite such as that shown in Fig. 2b. The 2 curves for each plot represent different degrees of smoothing of the second differential (d^2V/dI^2) data points. (From Warnes and Larbalestier, ref. 13)

3. FLUX PINNING

A high level of flux pinning is very desirable, since this is the only way to develop a high bulk current density. Granularity and extrinsic factors can only reduce the intrinsic flux pinning current density. Broadly speaking, it appears that the 123 compounds have high flux pinning even at 77K, while the development of greater flux pinning in the more anisotropic 2212 and 2223 BSCCO compounds is still needed. Thus flux creep and low flux pinning are more of a problem for the BSCCO compounds than for the 123 compounds. Unfortunately, granularity of polycrystalline samples is severe in the 123 compounds and $J_{ct}(H)$ is typically 2 to 4 orders of magnitude below the flux pinning values.[23] For BSCCO, the flux pinning and transport J_c can be rather similar, thus showing that granularity need

Development of High Temperature Superconductors for Magnetic Field Applications

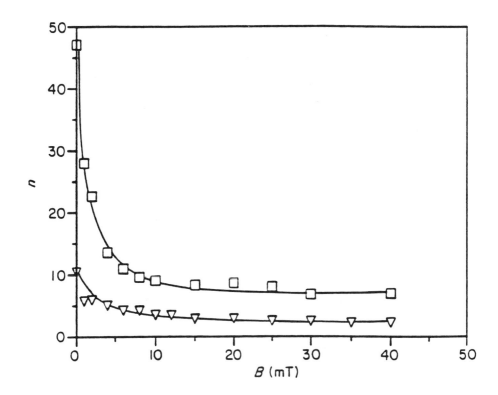

Fig. 7 Resistive transition index (n) vs. field (H) at 77K for Ag-sheathed YBCO and 2212-BSCCO conductors the upper curve is for YBCO, the lower for BSCCO. (Data of Heine et al. ref. 22)

not be a problem in HTS materials.[19] One of the key goals of present research is to understand the specific factors that produce such contrasting behavior, with a view towards engineering the best features of each into one or both of these material classes.

Flux pinning and flux creep are covered in detail in the companion paper by Kes,[6] where calculations of pinning by various microstructural defects observed in YBCO and BSCCO are presented. These calculations are useful for establishing the general magnitude of the pinning that is feasible for HTS materials. However, such calculations are subject to at least three important uncertainties. One is that few experiments exist where the density of a given proposed pinning site has been varied in a controlled way. Such experiments are vital if the choice of pinning site is to be verified and if the magnitude of the pinning interaction (and its temperature and field dependence) is to be checked. A second issue is the choice of appropriate thermodynamic parameters. Almost all elementary pinning interactions have the thermodynamic critical field $H_c(T)$ as the parameter which scales the strength of the elementary pinning interaction:

$$\delta E = -\mu_o \, \delta \, H_c^2(T) V_i / 2 \tag{3}$$

where δE is the change in fluxon energy as an interaction volume (V_i) of fluxon occupies a pin having a difference in critical field $\delta H_c(T)$ from the matrix. In general, of course, the most favorable pinning sites are not superconducting and $\delta H_c(T)$ then equals the full $H_c(T)$ of the superconducting matrix.

Respectable investigations of the properties of single crystals yield values of H_c which are not yet all consistent with each other.[6,24-26] Thus the zero temperature extrapolations for $H_c(0)$ range from 0.7 to 1.55T for YBCO. For 2212-BSCCO a value of 0.35T is frequently chosen, although no explicit variation of $H_c(0)$ as T_c varies from about 80 to 95 K is yet available.[6,24] Since $f_p(=\delta E/\xi)$ is proportional to $H_c^2(T)$, any uncertainty in H_c leads to significant imprecision in the calculated pin strength f_p. A third difficulty lies in identifying the correct elementary pinning force summation scheme. In a strong pinning LTS system such as Nb-Ti, full summation can be explicitly verified.[12] The J_c values of HTS materials, at least the 123 compounds, are sufficiently high that full summation of f_p also seems intuitively reasonable. However, collective pinning summations have found more favor than full summation schemes.[6] This may be because there is as yet no general agreement on how to treat the experimental difficulties of taking J_c data over wide ranges of temperature and field space without introducing significant uncertainty factors associated with granularity and flux creep.[27,28] The net result of these uncertainties is that *a priori* calculations of the flux pinning in HTS systems are remarkably hard to verify. For example, early flux pinning assessments tended to emphasize pinning by twins.[29] When experiments varying the twin density failed to show any systematic dependence of the J_c on the twin density,[30] it became clear that other defects or mechanisms must control the J_c. A similar situation appears probable in so far as the pinning effect of growth dislocations in thin films is concerned.[31,32] As Jin *et al.*[33] have reported, there is in fact no positive correlation between J_c and dislocation density that can yet be made. The true reason for much of this uncertainty is probably that the coherence lengths of HTS materials are so short that almost any defect can pin flux.[34] Whether any one particular defect in fact controls the pinning can only be addressed by experiments that systematically vary that particular defect density. Unfortunately, HTS microstructures cannot, in general, yet be controlled to the extent required.

The lesson of conventional low temperature superconductors (LTS) is that J_c is maximized by providing both a strong elementary pinning interaction and a high density of interactions.[11,12] Optimized Nb-Ti represents a very interesting case, because the bulk pinning force F_p is optimized at a precipitate thickness t considerably smaller than that which optimizes the elementary pinning force f_p. In fact F_p is optimized when t is 0.2ξ (Fig. 8) and when the separation of pinning centers is about ξ. This clearly corresponds to a very dense pinning situation, in

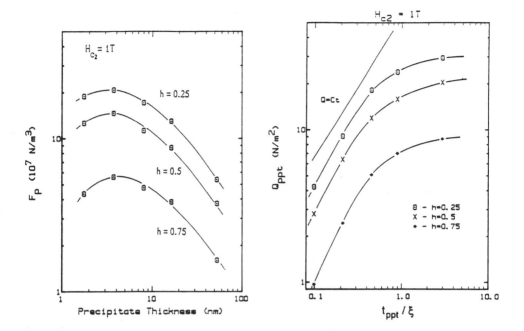

Fig. 8 The α-Ti precipitate thickness (t) dependence of the bulk pinning force (F_p) and the measured elementary pinning force Q_{ppt}, defined by F_p divided by the precipitate density. The data are given for $H_{c2}=1T$ because the peaks in F_p and Q_{ppt} are particularly evident close to T_c. ξ is ~ 18nm for the above plots. F_p is observed to peak at t_{ppt}, ~ 3.5nm (~ $\xi/5$), while Q_{ppt} peaks at ~ t_{ppt}, as expected for core pinning calculations. (Data of Meingast and Larbalestier, ref. 12)

which the density of pinning sites (assuming that each α-Ti precipitate acts as an independent pinning site) exceeds the density of fluxons over almost all temperature and field space.

Can such a situation be developed for HTS materials? An interesting analysis is provided by Hylton and Beasley.[35] They asked what pin site density would be required to explain the very high J_c (4K,0T) values (~ 10^7 A/cm^2) measured on the best thin films. Their conclusion was that strong pins need to be spaced ~5 nm apart. This spacing is very comparable to Nb 47wt%Ti: the important difference is that the pins are visible in Nb-Ti but have not been identified in thin films. The lack of microstructural evidence for dense pinning centers in YBCO thin films led us to consider alternative explanations for the high J_c values of thin films. One characteristic of the best films is that their thickness (d) is comparable to or less than the penetration depth λ. In this case a substantial fraction of the current can be due to shielding currents, whose magnitude is limited by depairing rather than flux pinning. The conclusions of Stejic et al.[36] are that J_c tends to the depairing current density $J_d = H_c/\sqrt{3}\lambda$ for d< λ and to a limit of $2J_d(\lambda/d)$ for d> λ. In the thick film limit, however, J_c becomes dominated by flux pinning rather than depairing.

This prediction has been experimentally checked on thin films of Nb47wt%Ti having a *low* density of pinning centers. For a $\lambda/2$ thick film, it was possible to develop J_{ct} which reached $J_d/3$, while a microstructurally similar 4λ thick film reached a limit almost an order of magnitude lower. Comparison can be made to the very best values attained in bulk-scale Nb-Ti conductors[12] where the high J_c values are certainly developed by flux pinning. In this case (Fig. 9) the maximum values are about a factor of 3 less than those seen in the $\lambda/2$ thick film. Thus we conclude that calculations which attribute all of the J_c of thin films to flux pinning are unlikely to be correct. This conclusion can also provide a plausible explanation of the contrasting normal and superconducting state behavior of high quality single crystals and thin films. Good single crystals have J_c values which are typically 1 or 2 orders of magnitude below thin films.[30] The natural hypothesis is to invoke the atomic scale defect density as the difference. Good thin films have lattice parameter anomalies that suggest some disorder.[30] However, such disorder is hard to reconcile with the low normal state resistivities of high J_c thin films, which are almost identical to those of good quality (but low J_c) single crystals.[37] Indeed, thin films and single crystals show quite different responses to small oxygen deficiencies: small deficiencies significantly raise the J_c of crystals[15] but depress the J_c of films.[38,39] Cation site disorder might produce strong pinning in films and could be compatible with the lattice parameter anomalies but this explanation appears incompatible with the similar normal state resistivity of films and crystals already quoted. Again, therefore, this emphasizes the need for experiments in which the microstructure is varied in a controlled manner. However, in the absence of agreement on what are the controlling microstructural parameters of thin films, we turn to bulk samples, where all dimensions are large compared to the penetration depth and concerns about the contribution of shielding currents to the measured transport or magnetization currents do not apply.

One set of interesting experiments on bulk materials are those on melt-processed quasi-single crystals of YBCO in which the diameter of insulating Y_2BaCuO_5 (211 phase) particles can be varied in a size range of order 0.1-10 μm and with a volume fraction of order 10-30%. The accompanying paper by Murakami[40] describes the properties of such samples. Excluding irradiation methods, such dispersions of fine 211 particles, when small in size, have produced the highest J_{ct} values yet seen in bulk materials, of order 10^5 A/cm^2 and 5×10^4 A/cm^2 in fields of 0 and 1T at 77K. The dependence of J_c on added volume fraction of 211 particles has been demonstrated, as has the dependence on size of the particles. For example, no explicit addition of 211 particles (it should be noted, however, that even nominally stoichiometric 123 samples have about 20% of 211 phase) produces J_c values almost 10 times lower than those quoted above.[15,40] A dependence of J_c on inverse particle diameter in the range 1-10μm for samples containing 10% of added 211 phase has been

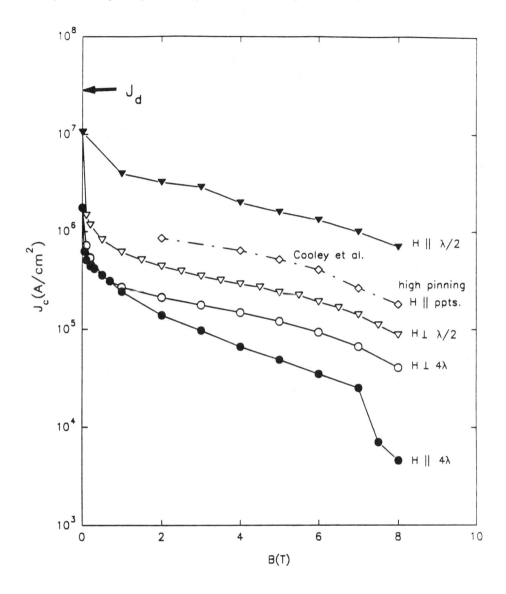

Fig. 9 Transport critical current density of two thin films and one bulk sample of Nb47wt%Ti. The two thin films are single phase with a low flux pinning center density, while the bulk sample is a rolled tape having a microstructure similar to that of Fig. 1, with the exception that the precipitates are highly aligned by the rolling process. This sample has the highest J_c ever achieved for a bulk Nb-Ti sample. The two thin films have thicknesses of one half and four penetration depths, respectively, and were measured with the field parallel to and perpendicular to the film plane. (Thin film data courtesy of G. Stejic, University of Wisconsin: bulk data from Cooley et al., ref. 69)

demonstrated. Such dependencies demonstrate that the significant pinning centers have been identified and that the system is useful for modelling flux pinning. By comparison to the Nb-Ti benchmark quoted earlier, we see that the volume fraction of pinning center is high (about 20% in Nb47wt%Ti, as compared to 20-40% in YBCO), that the properties of the precipitate phase are appropriately different from the matrix (α-Ti is normal, while 211 is insulating), but that the size of the 211 particles is still far from optimum (the smallest particles are $0.1\mu m$ in diameter and are still 2 orders of magnitude larger than the coherence length).

A complementary view of the flux pinning possibilities of a much smaller-sized and higher density defect population in 123 compounds is provided by oxygen vacancies. The systematic effects of changing δ (in the composition $YBa_2Cu_3O_{7-\delta}$) over the approximate range 0.05 to zero were first explored by Daeumling et al.[15] Fig. 10 shows that the magnetization hysteresis of good single crystals continuously declines as δ declines. A characteristic signature of the deficiency is the intermediate field minimum in the hysteresis, which is the result of the more weakly superconducting O-deficient regions being driven normal by the increasing field. In good crystals containing little or no mosaic sub-structure, diffusion is very slow at the temperatures below 400°C needed to produce δ of 0.02 or less.[41] Thus this behavior is common and we were able to show that these vacancies are the dominant pinning centers in good single crystals, as well as in melt-textured crystals containing large (>5nm) 211 particles.[15] The behavior is reversible when oxygen is taken in and then discharged.[42]

An interesting question is how the oxygen vacancies are distributed, whether as individual lattice defects or as ordered regions. For large values of δ (e.g. $\delta > 0.2$), there is abundant evidence of ordering into superstructures, particularly the $O_{6.5}$ or OII phase,[43-48] but this has not been possible to observe for $\delta < 0.1$. However, ordering at small δ seems very likely on many grounds. For example, the trends of the a and b lattice parameters do not show any smooth extrapolation to the tetragonal (i.e. a=b) value for any value of $\delta < 0.4$.[43] These various considerations stimulated Vargas and Larbalestier[49] to consider the consequences of flux pinning if the oxygen vacancies order themselves by a spinodal decomposition of $YBa_2Cu_3O_{7-\delta}$ into O_7 and $O_{6.5}$ regions.

One characteristic of a spinodal decomposition is that there is a characteristic wavelength of the precipitation. From observations at larger values of δ this can be measured to be of order 25nm.[45,50] Knowing the equilibrium $\delta(T)$ relationship, one can then calculate the concentration and size of OII domains as a function of temperature. A typical oxygenation temperature is 425°C, yielding a δ of 0.037, an OII volume fraction of 0.074 and diameter (on a spherical approximation) of 2.8nm. An interesting consequence of such an oxygenation treatment is that it produces a very high density of pinning centers (the flux line lattice (FLL) spacing, ϕ_0/B, is 22nm at 5T) with a size comparable to a fluxon diameter ($\sim$3nm). Thus the oxygen defect microstructure is much more comparable in size and spacing to that seen in Nb-Ti (Fig. 1) than to that produced by the 211 particle dispersions discussed above. Unfortunately the OII phase is itself superconducting (its bulk T_c value is 60K) and its elementary pinning force is thus reduced.

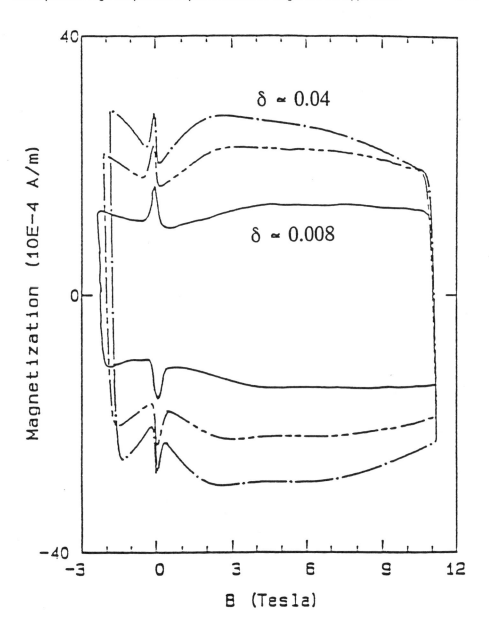

Fig. 10 Magnetization hysteresis (M) vs. field (H) for a flux grown single crystal of $YBa_2Cu_3O_{7-\delta}$. As δ diminishes from ∼0.04 to ∼0.008, the hysteresis continuously declines. Data were taken at 30K. (Data courtesy of J. Seuntjens, ref. 42)

The flux pinning due to the core interaction with the OII domains can be calculated using standard models:[49]

$$\delta E = -\frac{1}{2}\mu_o \left[(H_{cOI})^2 - (H_{cOII})^2\right] V_i \; [J] \qquad (4)$$

The elementary pinning force is then given by:

$$f_p = \delta E/\xi_{ab} [J/m] \qquad (5)$$

An upper limit to the bulk pinning force F_p is given by the full summation of the elementary pinning interactions, i.e. $F_p = nf_p$ [N/m^3], where n is the number density of interactions. The predicted area density of OII domains exceeds the fluxon density $n_{fl} = B/2\phi_0$ [m^{-2}] over the whole field range B < 10 tesla ($n_{fl} \leq 2 \times 10^{15}$ m^{-2}) so that full summation then occurs over all fluxons, n_{fl}. Thus F_p can be calculated from:

$$F_p = \frac{\pi B \mu_o \left[(H_{cOI})^2 - (H_{cOII})^2\right] r^2 \nu_{OII}}{4\phi_0 \, \xi_{ab}} \; [N.m^{-3}] \qquad (6)$$

and the critical current density is calculated using $J_c = F_p/B$. Values of H$_{cOI}$(0)= 1.55T, H$_{cOII}$(0)=0.38T(0 0.38 T and ξ_{ab}(0)=1.6 nm from single crystal measurements are used. We assume that H$_c$(T) follows H$_c$(0)(1-t^2) where t = T/T$_c$, and ξ_{ab}(T) is calculated from $(2\pi\phi_0/H_{c2}(T))^{1/2}$.

Such calculations appear to fit the data well in the low temperature limit at 4K (Fig. 11). Indeed, the predicted dependence of F$_p$ on the radius r and volume fraction of the O$_{II}$ phase (F$_p \propto r^2 \nu_{OII}$) is observed experimentally.[51] Thus when the annealing temperature is reduced from 425°C to 375°C, J$_c$ (4.2K, 3T) falls from 3.5 × 10^6 A/cm^2 to 2.4 × 10^5 A/cm^2, as compared to the predicted decline from 2.6 × 10^6 A/cm^2 to 2.4 × 10^5 A/cm^2. The temperature dependence is not well-fitted by the simple model, however. Flux creep and single crystal granularity[15] cannot be ignored in deducing J$_c$ values from magnetization measurements. The whole problem of how to treat the temperature dependence is complex. For example, Kes[6] has calculated the pinning due to uncorrelated single oxygen vacancies using the quasi-particle scattering model. Such a model has a strong temperature dependence almost identical to that of Figure 11, *without* needing to take account of flux creep or granularity. This all points to the fact that multiple models can be applied to HTS pinning. With no decisive reason to choose one or the other, systematic microstructural variation experiments are crucial.

Few systematic experiments have yet been performed on BSCCO. Oxygen vacancies in the CuO$_2$ planes have been suggested as the principal pinning centers,[6,16] although there is no evidence for such vacancies either in BSCCO or YBCO. One set of systematic experiments have been performed in 2212-phase BSCCO by Nomura and Chiang.[52] They found that cation (i.e. Bi, Cu, Sr or Ca) vacancies in the 2212-phase of order 10^{23} cm^{-3} could be achieved. J$_c$ and the flux creep rate were

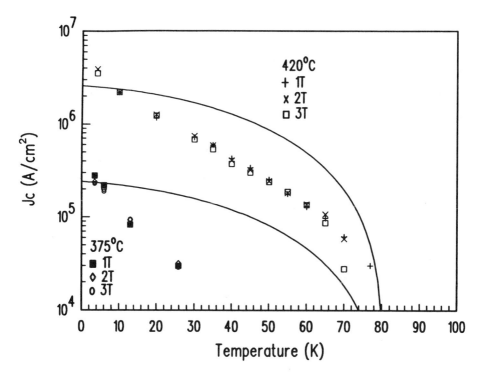

Fig. 11 Predicted J_c (see equation 6) versus oxygenation condition for single crystals of $EuBa_2Cu_3O_{7\delta}$ oxygenated at 420°C and 375°C. The low temperature limit of J_c is well-predicted by the model but the temperature dependence is not (see text). (Data courtesy of J. L. Vargas, University of Wisconsin, ref. 49, 51).

both strongly dependent on the cation vacancy concentration, the improvements reaching almost an order of magnitude in both J_c (60K,1T) and creep rate for the high defect concentrations.

Controlled microstructural variation experiments of the sort described above are not easy to perform or to interpret because of the rather imperfectly known phase relationships of HTS materials and because of the limitations of electron microscopy for determining the true density of atomic-scale defects. Thus, developing a verifiable scientific understanding of naturally occurring flux pinning centers in HTS materials may be slow. Artificial pinning centers, such as those introduced by heavy ion irradiation, are particularly interesting in this context, since they allow the formation of line pins, whose size and density should be susceptible to quantitative measurement.[53]

4. SUPERCONDUCTING GRANULARITY

It is well-established that high angle grain boundaries (HAGB) impede the transport of current. The reason for this is that many grains are only weakly coupled across the grain boundaries, this leading to a lowered and strongly field-dependent depression of the supercurrent across the boundary. The most explicit demonstration of this was provided by the artificial thin film bicrystal experiments of Dimos et al.[54] By investigating a variety of tilt and twist misorientations, they concluded that Josephson junction behavior was seen when the misorientation across the boundary exceeded about 5°.

These thin film bicrystal results have been central to setting the framework within which to understand the role of grain boundaries in limiting current flow in HTS materials. The Josephson junction character of HAGB can explain the low and strongly field-dependent J_c of polycrystalline 123 compounds very well. On the other hand, there are important situations which were not addressed by the experiments. For example, few measurements were made in a magnetic field (and then only in weak fields). Even the weakly coupled bicrystals had J_c values exceeding 10^5 A/cm^2 (0T, 4.2K), values 1 to 2 orders of magnitude higher than seen in bulk-scale bicrystals. Given the previous discussion concerning the role of depairing and flux pinning currents in determining the J_{ct} of thin films, it is not surprising that to rely only on the *magnitude* of the J_{ct} is not an infallible guide to the character of the coupling across the boundary.

In work at Wisconsin, we have tried to develop a deeper understanding of the character of the behavior of grain boundaries by measuring the electromagnetic properties of bulk-scale bicrystals and then determining the grain boundary microstructure and composition of the *same* grain boundary.[55-57] Thus we hope to develop the first direct microstructural description of the grain boundary electromagnetic properties for HTS materials. Our analysis of the properties of HAGB is somewhat more complex than that used for the thin film bicrystals, because it assesses the character of the superconducting coupling across the boundary by analyzing the $J_c(H)$ characteristic of the boundary. In such a description, the boundaries are described as being fully coupled (i.e. flux pinning controls the J_{ct} across the boundary), weakly coupled (i.e. the boundary acts as a Josephson junction) or resistive. The goal is to understand why the boundaries have such different properties. So far some 20 naturally grown bicrystals have been studied. An overview of their properties is shown in Table 1.

Naturally grown bicrystals may have different properties from the artificially prepared bicrystals of Dimos et al.[54] They grow while slowly cooling from the melt, thus allowing opportunities for abutting crystals to rotate and to seek low energy orientations.[58] This was not the case for the thin film bicrystals, because the misorientation was imposed by the misorientation of the underlying SrTiO$_3$ bicrystal which serves as the substrate. There is a definite tendency for certain crystal misorientations to be favored in the bulk-scale bicrystals. The favored misorientations

Table 1.
Properties of bulk-scale YBCO Bicrystals
(Larbalestier et al., ref. 56)

BiCrystal[1]	Character	T_c (R=0) (K)	J_{ct} (0, 77K) (A/cm^2)	J_{ct}(0, 4.2K) (A/cm^2)	$\rho_n d$ ($\Omega\mu m^2$)	$I_c R_n$ (μV)
C3	FP	93	3000	nm^3	—	—
C6	FP	95	3300	nm	—	—
C10-1	FP	90.5	1700	nm	—	—
C10-2[4]	JJ → JJ	85.1 → 92	57 → 342	1000 → nm	6.5 → 2.6	14 → 35
C14	FP	92.5	850	nm	—	—
C22...37[2]	FP	90	370	nm	—	—
C22	JJ	84.7	210	4000	0.92	23
C26	JJ	87.3	170	900	1.0	9
C28[4]	JJ → JJ/FP	81.5 → 93	20 → 590	1100 → 4400	10.0 → 2.0	—
C33	JJ	90	269	nm	1.2	—
C40	JJ	32.5	0	273	26.4	98
C43	R	<4.2	0	0	16.5	—
C45	R	<4.2	0	0	89	—
P3	FP	92.5	1750	nm	—	—
P15-1	FP	89	33	nm	—	—
P15-2	R	<4.2	0	0	—	—
RAH[5]	JJ	42	0	14	71.4	10

tend to be those having low Σ, that is misorientations for which the Coincidence Site Lattice (CSL) formed by the interpenetration of the two lattices has a relatively small volume. The ratio of the CSL cell volume to the crystal unit cell is denoted Σ and this ratio is generally held to be significant up to values of at least 50. $\Sigma 5$, $\Sigma 17$ and $\Sigma 41$ misorientations were all found to be favored for YBCO bicrystals.[59] The structural units which form the boundary are also important: grain boundaries were not amorphous and their particular structure clearly can be expected to determine the local superconducting coupling.

To support this statement, several pieces of evidence can be presented. For example, two bicrystals of nominally identical misorientation (C10 - i.e. parallel [001] c axes with a 10° rotation about the [001] axis and P15 - i.e. crystal II rotated 90° about the common b axes of both crystals, crystal II then again being rotated 15° about the c_{II} axis) had contrasting character. In each case, one was flux pinning and one a Josephson junction (Table 1). The P type misorientations are interesting because they were not studied in the thin film bicrystal studies and because they cause the basal (001) planes of the two crystals to be normal to each other. In principle, therefore, the coupling across such a boundary might be expected to be weak because one direction is the short coherence length ($\xi_c \sim 0.3$nm). Figure 12 shows the J_c(H, 77K) characteristics of 2 high angle grain boundaries (P3, C14), as compared to one low angle grain boundary (C3) and a single crystal. The essential point is that, even though the magnitude of the J_c varies from one sample to another, all have a J_c(H) characteristic which smoothly diminishes towards zero at 7-8T, field values which are quite similar to the irreversibility fields H* observed in epitaxial thin films and single crystals.

What microstructural feature is it that permits strong coupling across the highly misoriented P type boundaries? Unfortunately, we were not able to perform TEM on this particular bicrystal but the P0 misorientation has been studied in MOCVD films by Gao et al.[60] This misorientation occurs in predominantly c-axis films; a small proportion of grains grow with an a/b axis normal to the substrate. Chan et al.[50] have shown that high zero field J_c values can be obtained across such boundaries. High resolution transmission electron microscopy (HREM) on such films[60] showed that the macroscopically planar (001) boundary actually facets into steps which lie alternately along (001) and (013). Across the (001) facet the CuO_2 planes in the two abutting crystals do not contact, since they have a T orientation. Gao et al. made the reasonable suggestion that it is the (013) facets which provide the strongly-coupled flux pinning connection shown in Figure 12.

Reasonable though this suggestion is, it cannot be accepted without question. A significant problem is that the (013) orientation is found in step-edge Josephson junctions, where clearly the boundary is weakly coupled. Of course, one can postulate that Josephson behavior occurs by contamination of the step-edge junction boundary which is much easier than for a bulk-scale bicrystal, since the film is in close proximity to the substrate. Such arguments, while plausible, merely reinforce how many variables there are in these systems and how vital it is to build a parallel microstructural and electromagnetic understanding of grain boundaries.

A natural question is to ask what it is that goes wrong at Josephson junction grain boundaries? Figure 13 shows an HREM image of a C25 ($\sim 2°$ away from the $\Sigma 17$ CSL which occurs at 28°) grain boundary. What is most striking about this image is the perfection of the atomic structure right up to the boundary. Observable displacements of the atoms (in fact it is the cations which are imaged) occur only within a thin layer right at the boundary, the thickness of which appears comparable to or less than ξ_{ab}(~ 1.5nm at 4K). Whatever produces the weak coupling of this boundary occurs on a very fine scale.

Transmission microscopy is, unfortunately, not very sensitive to light elements. Oxygen is of course a key element for HTS compounds and its electromagnetic effects can be large, as Figure 14 shows. In this case a C27.5° bicrystal was studied (this lies within 0.5° of the $\Sigma 17$ CSL). When leads were attached, the crystal lost a little oxygen because it was heated in air to $\sim 250°$C. Its J_c(H) behavior showed

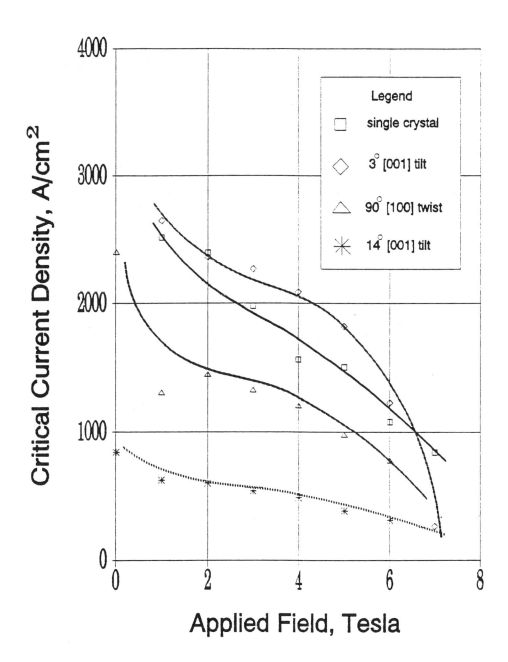

Fig. 12 J_c(H) characteristics at 77K of two C type bicrystals (Parallel (001) axes), one P type bicrystal (c axes at right angles) and one single crystal. All have flux pinning, rather than Josephson junction characteristics. (Data from Babcock et al., ref. 55)

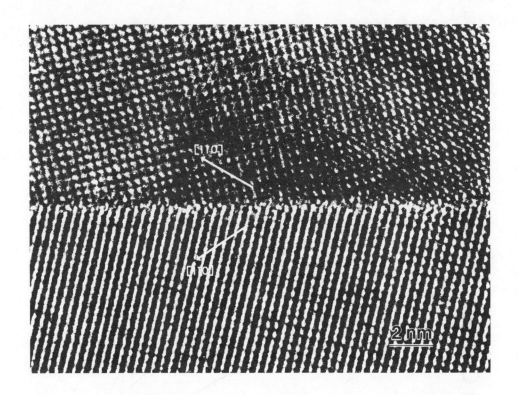

Fig. 13 High resolution electron micrograph of a 26° [001] tilt boundary which acts as a Josephson junction. (Courtesy of N. Zhang, University of Wisconsin, ref. 65)

a steep fall in only weak fields, behavior characteristic of a Josephson junction. After re-oxygenating for 50 hrs at 420°C, the characteristic had markedly changed, now exhibiting a double step characteristic of parallel weak and strong coupling paths.[56] A further oxygenation of 72 hrs at 420°C raised J_c(77K,7T) by two orders of magnitude. These results directly show the important role that oxygen can play in changing both the magnitude and character of current transport across a high angle grain boundary. Recent analysis of the structure of grain boundaries within a CSL framework suggests that the local oxygen content could be a direct function of misorientation.[61]

Little direct information about the properties of BSCCO grain boundaries is yet available. However, the very striking property of polycrystalline BSCCO is that it is possible to achieve J_{ct} exceeding 10^5 A/cm^2 to fields exceeding 20T at 4.2K and J_{ct} exceeding 10^4 A/cm^2 at fields of ~0.5T at 77K.[5] Polycrystalline 123 compounds seldom achieve more than 10^2 A/cm^2 under these conditions.[61] Is the grain

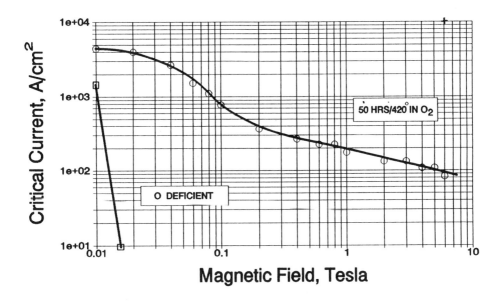

Fig. 14 $J_c(H)$ characteristics for 28° [001] tilt boundary in slightly oxygen-deficient condition, as oxygenated for 50hrs at 240°C and then after a further 72hrs at 420°C (7T point only). (Data from Larbelestier et al., ref. 56)

boundary structure of 2212 - or 2223 - BSCCO inherently more strongly coupled than that of the 123 compounds? We do not know the answer to this question, since bicrystal experiments have not yet been performed on BSCCO. A new view of the high J_{ct} values in BSCCO has recently been provided by Bulaevskii et al.[62] and Malozemoff.[63] The microstructural basis of this model is that BSCCO grains tend to grow as plates which align themselves with mutually parallel c axes (see Figure 4), such that neighboring plates overlap each other, as in a brick wall. This grain morphology is not developed in 123 compounds. The basis of the model is that the current transfers around the "bad" (001) tilt boundaries by traversing the large area planar (001) twist boundaries. This current is a Josephson current. The key points that make c axis current transport possible across the boundary are (i) that the CuO_2-CuO_2 plane spacing does not increase across such a grain boundary because the boundary forms at the double Bi-O layers without increasing the plane separation; (ii) that the length (L) to thickness (D) aspect ratio of the grains is very large; (iii) that the c axis [001] current is a Josephson current and (iv) that the characteristic size of the Josephson junctions is very small, the characteristic dimensions being set both by the insulator thickness (d ~ 1nm) between the CuO_2 planes and by the spatial variation of coupling (r_0) within the grain boundary. This latter dimension is not very well-defined: it may originate in grain boundary dislocations, incommensurate distortions of the Bi-O layers or other yet to be defined parameters. The predictions of the model are in general accord with the behavior of BSCCO tapes. The model predicts that the J_c will first drop at a field $H_1 = \phi_0/Ld$, stabilize at a plateau and then exhibit a second fall off to zero at a field H_2, where $H_2 = \phi_0/r_0 d$. BSCCO tapes indeed exhibit a plateau in J_c which starts at about 1 T, a value compatible with grain lengths of order $2\mu m$ and an insulator thickness of 1 nm. The second fall-off beyond the plateau has not yet been seen in BSCCO; however, it is intriguing to note that it has been seen for a c-axis aligned $DyBa_2Cu_3O_7$ sample which has the plate-like grain morphology of BSCCO. In this case the fall off occured at ~ 25T at 4.2K,[64] thus yielding a value of 80nm for r_0, a not unreasonable value nor one which should be impossible to verify.

The contrasting behavior of 123 and BSCCO polycrystalline materials is of enormous technological significance. Whether HAGB in BSCCO and YBCO are fundamentally different in their properties is not yet known, nor is it known whether the brick wall model provides the correct explanation of the excellent high field J_{ct} of BSCCO. If so, this is hypothesized as being a direct consequence of the grain boundary being a sub-divided Josephson junction. Just as in the flux pinning boundaries, the local atomic structure of the boundary should control the properties and thus the grain boundary structure needs to be identified. In this regard, the recent demonstration by Lelay et al.[65] that grain boundaries of the short coherence length lead and tin molybdenum sulfides are *not* weak-linked is a provoking result. This adds further complexity to the issue of whether short coherence length superconductors are inherently granular, as suggested by Deutscher and Mueller.[66]

5. SUMMARY

Factors important to the development of high transport critical current density of polycrystalline HTS materials have been described and reviewed. High flux pinning within the grains can be produced in 123 compounds by observable defects. Neglecting irradiation methods, the highest J_c values have been obtained with 0.1-$1\mu m$ diameter insulating 211 particles. Oxygen defects having a diameter of order 3nm can produce a finer and denser dispersion of pinning centers, but such defects have a small elementary pinning force and tend to produce intra-grain granularity and are thus not desirable pinning centers. Optimum flux pinning would in principle combine the best features of these defect classes. Flux pinning in epitaxial 123 films is still not well understood, but the high J_{ct} values of films having thicknesses of order a penetration depth or less have a significant contribution from shielding currents and therefore J_c should not all be attributed to pinning. Whatever level of flux pinning is possible within grains, the transport current of HTS materials is controlled by the currents which can cross grain boundaries. Transport across high angle grain boundaries of the 123 compounds remains difficult but some high angle grain boundaries do have a flux pinning character which responds strongly to oxygenation condition. Much less is understood about the significant microstructural defects of BSCCO: flux pinning and granularity appear less strong and less marked, respectively, than in 123 compounds. The more 2 dimensional nature of BSCCO produces a plate-like grain morphology which aligns more easily than does 123. This may permit high critical current densities to cross grain boundaries even if they are weak linked.

ACKNOWLEDGMENTS

I am grateful to many colleagues for discussions, in particular S. E. Babcock, X. Y. Cai, L. D. Cooley, H. Edelman, Y. Feng, E. E. Hellstrom, P. J. Lee, A. P. Malozemoff, G. Stejic, A. Umezawa and J. L. Vargas. The support of our work by DARPA, Electric Power Research Institute, the Department of Energy and the National Science Foundation is gratefully acknowledged.

REFERENCES

1. D. C. Larbalestier, Physics Today **44**, 74 (1991).
2. D. C. Larbalestier, G. Fisk, B. Montgomery, D. Hawksworth, Physics Today **35**, 25 (1986).
3. S. J. Dale, S. M. Wolf, T. R. Schneider, "Energy Applications of High Temperature Superconductivity," Electric Power Research Institute Report No. ER-6682 (1991).
4. K. Heine, N. Tenbrink, M. Thöner, Appl. Phys. Lett. **55**, 2441 (1989).
5. K. Sato, T. Hikata, H. Mukai, M. Ueyama, N. Shibuta, T. Kato, T. Masuda, M. Nagata, K. Iwata, T. Mitsui, IEEE Trans. Magn. **27**, 1231 (1991).
6. P. H. Kes, "Vortex Pinning and Creep Experiments," article in this volume "Phenomenology and Applications of High Temperature Superconductors," Eds. Kevin Bedell, M. Inui, D. Meltzer, J. R. Schrieffer and S. Doniach, Addison Wesley, NY 1991.
7. D. S. Fisher, "Phase Transitions and Transport in Anisotropic Superconductors with Large Thermal Fluctuations," ibid ref. 6.
8. D. R. Nelson, "Correlations and Transport in Vortex Liquids," ibid ref. 6.
9. A. Kapitulnik, "Dynamics of the Vortex State in High Temperature Superconductors," ibid ref. 6
10. D. C. Larbalestier, "IEEE Transactions on Magnetics," MAG **21**, 257 (1984).
11. C. Meingast, P. J. Lee, D. C. Larbalestier, J. Appl. Phys. **66**, 5962 (1989).

12. C. Meingast, D. C. Larbalestier, J. Appl. Phys **66**, 5971 (1989).
13. W. H. Warnes, D. C. Larbalestier, Cryogenics **26**, 643 (1986).
14. Y. High, P. J. Lee, J. C. McKinnell, D. C. Larbalestier, To appear in Adv. in Cryogenic Eng. **38** (1992).
15. M. Daeumling, J. Seuntjents, D. C. Larbalestier, Nature **346** (6282) 332 (1990).
16. E. M. Chudnovsky, Phys. Rev. Lett. **65**, 3060 (1990).
17. Y. Fend, K. E. Hautanen, Y. E. High, D. C. Larbalestier, R. Ray, E. E. Hellstrom, S. E. Babcock, Physica C to appear 1992.
18. J. Baixeras, G. Fournet, J. Phys. Chem. Solids **28**, 1541 (1967).
19. J. Tenbrink, K. Heine, H. Krauth, Cryogenics **30**, 422 (1990).
20. Y. Yamada, B. Obst, R. Flükiger, Supercond. Sci. Technol. **4**, 165 (1991).
21. M. Wilhelm, H. W. Neumüller, G. Ries, To be published in Physica C (1992) ibid ref. 27.
22. K. Heine, J. Tenbrink, H. Krauth, M. Wilhelm, To appear Adv. in Cryo. Eng. **38** (1992).
23. H. Küpfer, I. Apfelstadt, R. Flükiger, C. Keller, R. Meier-Hirmer, B. Runtsch, A. Turowski, U. Weich, T. Wolf, Cryogenics **28**, 650 (1988).
24. D. H. Kim, K. E. Gray, R. T. Kampwirth, J. C. Smith, D. S. Richeson, T. J. Marks, J. H. Kang, J. Taluacchio, M. Eddy, Physica C **177**, 431 (1990).
25. Z. Hao, J. R. Clem, M. W. McElfresh, L. Civale, A. P. Malozemoff and F. Holtzberg, Phys. Rev. **B43**, 2844 (1991).
26. K. G. Vandervoort, G. W. Crabtree, Y. Fang, S. Yang, H. Claus, J. W. Downey, Phys. Rev. **B43**, 3688 (1991).
27. A. P. Malozemoff, "Flux Creep in High Temperature Superconductors," To appear Physica C 1991 in proc. of M^2S - HTSC.
28. M. P. Maley, J. O. Willis, H. Lessure and M. C. McHenry, Phys. Rev. **B42**, 2639 (1990).
29. P. H. Kes, J. van den Berg in "Studies of High Temperature Superconductors," Vol. 5, A. Narlikar, ed. (NOVA Science Publishers, New York, 1990) 83.
30. B. M. Lairson, S. K. Streiffer, J. D. Bravman, Phys. Rev. **B42** 10067 (1990).
31. M. Hawley, J. D. Raistrick, J. G. Beery and F. J. Houlton, Science **251**, 1587 (1991).
32. Ch. Gerber, D. Anselmetti, J. G. Bednorz, J. Mannhart and D. G. Schlom, Nature **350**, 279 (1991).
33. S. Jin, G. W. Kamlott, S. Nakahara, T. H. Tiefel, J. E. Graebuer, Science **251**, 1587 (1991).
34. M. Tinkham, Helv. Physica Acta **61**, 443 (1988).
35. T. L. Hylton, M. R. Beasley, Phys. Rev. **B41**, 11669 (1991).
36. G. L. Stejic, R. Joynt, D. K. Christen, D. C. Larbalestier, "Depairing and Depinning Critical Currents in Superconducting Thin Films," submitted to Phys. Rev.
37. T. Ho, H. Takago, S. Ishibashi, T. Ido, S. Uchida, Nature **350**, 596 (1991).
38. B. M. Lairson, J. L. Vargas, S. K. Strieffer, D. C. Larbalestier, J. C. Bravman, To appear Physics C (1992) ibid ref. 27.
39. D. K. Christen, R. Feenstra, To appear in Physica C (1992) ibid ref. 27.

40. M. Murakami, "Flux Pinning Enhancement by $YBa_2Cu_2O_5$ Inclusions in Melt Processed YBCO Superconductors," ibid ref. 6.
41. S. J. Rothman, Phys. Rev. **B44**, 2326 (1991).
42. J. Seuntjens, Phys Rev., PhD thesis, "Understanding Weak Links Through the Processing of $ReBa_2Cu_3O_{7-\delta}$," University of Wisconsin - Madison (1991).
43. R. Cava, A. W. Hewat, E. A. Hewat, B. Batlogg, M. Marezio, K. M. Rabe, J. J. Krajewski, W. F. Peck, Jr. and L. W. Rupp, Jr., Physica **C165**, 419 (1990).
44. R. Beyers, B. T. Ahn, G. Gorman, V. Y. lee, S. S. Parkin, M. L. Ramirez, K. P. Roche, J. E. Vasques, T. M. Gur and R. A. Huggins, Nature **340**, 619 (1990).
45. J. Reyes-Gasga, T. Krekels, G. Van Tendeloo, G. Van Landuyt, S. Amelinckx, W. H. M. Bruggink and H. Verweij, Physica **C159**, 831 (1989).
46. A. G. Khachaturyan and J. W. Morris, Phys. Rev. Lett. **61**, 215 (1988).
47. D. de Fontaine, G. Ceder and M. Asta, Nature **343**, 544 (1990).
48. L. T. Wille, A. Berera and D. de Fontaine, Phys. Rev. Lett **60**, 1065 (1990).
49. J. L. Vargas, D. C. Larbalestier, "Flux Pinning by Ordered Oxygen-deficient Phases in Nearby Stoichiometric $YBa_2Cu_3O_{7-\delta}$ Single Crystals," Submitted Appl. Phys. Lett.
50. C. H. Chen, D. J. Werder, L. F. Schneemeyer, P. K. Gallagher and J. V. Waszczak, Phys. Rev. **B38**, 2888 (1988).
51. J. L. Vargas, D. C. Larbalestier, Work in progress.
52. S. Nomura, Y. M. Chiang, Appl. Phys. Lett. **58**, 768 (1990).
53. L. Civale, A. D. Marwick, Mk. W. McElfresh, T. K. Worthington, A. P. Malozemoff, F. H. Holtzerg, J. R. Thompson, M. Kirk, Phys. Rev. Lett. **65**, 1164 (1990).
54. D. Dimos, P. Chaudhari and J. Mannhart, Phys. Rev. **B44**, 4038 (1990).
55. S. E. Babcock, X. Y. Cai, D. L. Kaiser and D. C. Larbalestier, Nature **347**, 6289 (1990) 167.
56. D. C. Larbalestier. S. E. Babcock, X. Y. Cai, M. B. Field, Y. Gao, N. F. Heinig, D. L. Kaiser, K. Merkle, L. K. Williams, N. Zhang, "Electrical Transport Across Grain Boundaries in Bicrystals of $YBa_2Cu_3O_{7-\delta}$," To appear Physics C (1992) ibid ref. 27.
57. S. E. Babcock, in "Structure - Properties Relationships for Interfaces," ed. J. W. Walter, Materials Park: ASM International, to appear in August 1991.; S. E. Babcock, N. Zhang, Y. Gao, X. Y. Cai, D. L. Kaiser, D. C. Larbalestier, K. L. Merkle, Proc. 2nd Tokai University Int'l Workshop on Superconductivity, October 1991. To appear J. Adv. Science (1992).
58. D. L. Kaiser, F. Holtzberg, B. A. Scott and T. R. McGuire, Appl. Phys. Lett. **51**, 1040 (1987).
59. D. A. Smith, M. F. Chisholm and J. Clabes, Appl. Phys. Lett. **53**, 2344 (1988).
60. Y. Gao, G. Bai, D. J. Lam, K. L. Merkle, Physica C **173**, 487 (1991).
61. J. Seuntjents, D. C. Larbalestier, J. Appl. Phys. **67**, 2007 (1991).
62. L. N. Bulaevskii, J. Clem, L. I. Glazman, A. P. Malozemoff, submitted to Phys. Rev. (1991).

63. A. P. Malozemoff, "Models of Current Density in Bi-based High Temperature Superconducting Tapes," Proc. of NY State Symposium of High Temp. Superconductivity, to be published.
64. Y. Zhu, H. Zhang, H. Wang, M. Suenaga, J. Mater. Res. 6, 2507 (1991).
65. L. Lelay, T. Willis, D. C. Larbalestier, "Fully Connected Chevrel Phase Polycrystalline Compacts Made by hot Isostatic Processing," to appear Appl. Phys. Lett. Feb. 1992.
66. G. Deutscher, K. A. Müller, Phys. Rev. Lett. 59, 1745 (1988).
67. W. H. Warnes, D. C. Larbalestier, Proc. Int. Symp. on Flux Pinning and Electromagnetic Properties of Superconductors, (Ed. T. Matsushita) Matsukama Press, Fukuoka, Japan, p. 156 (1986).
68. L. D. Cooley, P. D. Jablonski, P. J. Lee, D. C. Larbalestier, Appl. Phys. Lett. 58, 2984 (1991).

DISCUSSION

D. K. Finnemore: It doesn't matter whether the oxygen defect is on the chains or in the planes. If the oxygen defect is on the chain, it gives you a ripple in the pair potential in the planes, and it just simply doesn't matter.

Second point: I'll bet you, any bottle of Scotch or Irish Whiskey that you like, that H_c of bismuth is not that low compared to yttrium if the T_c's are the way they are.

K. G. Gray: Let me just make a comment directly on that. The number that you quoted from me was from trying to fit a very elementary pinning model to some data at low temperatures, and we stated very clearly that there were many assumptions and that this was just to show that we were in the right ballpark. It's not meant to be a realistic ...

D. C. Larbalestier: Well I certainly don't want to pick on you or anybody else here, but what I did try and do here was to point out that there are various numbers out there, all of good provenance, and we need to agree on a set of numbers and not just choose the number that happens to fit our particular calculation.

K. G. Gray: This wasn't fit to calculation, it was to show that the calculation was somewhat reasonable.

D. S. Fisher: Two comments. One on this question of H_{c2}, and ξ and so on. Now ξ can be well defined near T_c, even though H_{c2} isn't. However, it's not clear how useful the quantity is, because the vortex lines are fluctuating all over the place, and what their core size is is dependent on whether you take a snapshot or you let them fluctuate around a bit. But at low temperatures, ξ is perfectly well defined – presumably H_{c2} is well-defined there, even though it's much too big to measure

– and to find it one ought to think in terms of what measurements are there that give ξ or things like condensation energies more directly at low temperatures.

The other comment was on the question of distributions of critical currents. If I take a very long, very thin wire then clearly I can talk about critical current in different sections of it and I can model what goes on in the whole wire in terms of those. But if I take something which is relatively thick, and I'm measuring the properties of that, in the absence of thermal fluctuations – which, down at four degrees, or maybe one degree at least, are not very important – there really should be a sharp critical current in the system. That critical current is some percolation-like process by which little local regions where vortices start trying to move around, and then they get shorted by the current; the currents go around them and make other regions; they make other regions unstable . . . eventually the whole thing goes. But there still will be a sharp critical current. The I-V curve near that critical current will not just be the simple form and just come out like a straight line, it'll come up with some curvature, away from that point. So one really has to distinguish between whether what one is thinking of is really a long wire, with distributions of critical currents in the usual sense along that wire, or something where one is measuring the I-V curve in a bulk material and wants to relate that to things that are going on locally, and the bulk critical current is determined by collective effects.

D. C. Larbalestier: Well I take the point you're making but I'm not sure I agree with you, for the following reason. First of all if you take a percolation model, and inherently, where you're accepting that not all of the cross-section is active – and in a percolation model, I think it's therefore reasonable that the active cross-section varies as a function of length. And indeed, even in short samples, i.e. thin films, those experiments in the SEM – I guess by the IBM and Bellcore groups – clearly showed that even in good epitaxial films in zero field that the critical current is not uniform from place to place. As you scan across with the electron beam you can see imaging of regions which become locally . . .

D. R. Nelson: He agrees with you.

D. S. Fisher: Locally there are critical currents which vary; there is a distribution of those. But to get from there to what you measure in a bulk sample, in an I-V measurement, is something which is very non-trivial and I don't think the I-V curve that comes out has anything to do with the local distribution of critical currents.

A. P. Malozemoff: There's another aspect to that problem, and that's just the experimental indication that in fact there's a huge difference between the niobium-titanium data – your n-values that you showed – and what is seen typically in the high-temperature superconductors.

An interesting example is the work of Tenbrink et al. – the Vakuumschmelze group – reported in the last Applied Superconductivity Conference, where they showed

n-values ($V \propto I^n$) as a function of field, and they were completely flat at about a value of 25. And Matt Fisher and I have earlier pointed out [PRB **42**, 6784 (1990)] that these n-values in a variety of the yttrium barium copper oxide materials are typically coming out in the range of 30 over a wide range of conditions. So there's something very different going on in the high-temperature superconductors, simply empirically, as compared to the low-temperature superconductors.

And of course we think this difference is related to flux creep and in particular to the version of flux creep which comes either from collective pinning theory or from the vortex glass theory, with these funny U(J) dependences. And that's the other point I wanted to comment on: that as far as I could understand your modeling of J – that Vargas plot – I don't see you taking flux creep into account, and that is a huge effect. It seems now to be the case that one can nicely model the temperature dependences of J_c if one includes flux creep, and that makes a very big difference. Actually, Mike McHenry and Marty Maley were showing that in some of their work in one of the posters. We've also been doing modeling of this type, and it works a lot better than the kind of modeling that you've been struggling to do there without that factor.

D. C. Larbalestier: Well let me make two comments. The first of them is that this change here [referring to plot of n-value versus magnetic field] corresponds to something like, if you look at the standard deviation of the filament cross-section over the average, it's about three percent up here, and as you come down here it comes to about ten percent. Nb_3Sn conductors, as mostly made today, have characteristics like this, and in fact frequently plateau off with n-values something like 20-30. And so if you looked at those over a small range of field, you would see exactly what Tenbrink and people saw for bismuth: that n was about 20-30, and it was independent of field. Cross-sectional area variations of the conductors that are made of bismuth are certainly on that order of five to ten percent or so. And so whether or not flux creep exists as well, this explanation here is equally valid for looking at those sorts of conductors and I think needs to be taken into account. That fits entirely into the framework of what happens in Nb_3Sn.

D. S. Fisher: But you're not measuring one filament, you're measuring something with a huge number of filaments. So it's not clear the relevance of the properties of one filament . . .

D. C. Larbalestier: Well, but you're measuring down at a level where you're picking up, perhaps in Nb_3Sn, flux flow within the individual filament, because the matrix is rather resistive between the filaments – in niobium titanium almost certainly, current transfer between the filaments. But in either case, you are looking at the onset of dissipation and I think we know that in all of these conductors we have significant variations in cross-section as a function of length. That's true for the bismuth, and Nb_3Sn, and Nb-Ti.

So far as the flux creep argument is concerned, I think you've raised a good point: that we need some way of getting back to some time-independent measure of the critical current density. What we find, however, is that these numbers – up to about 77 K – are not very time dependent . . . if we change the sweep rate in the VSM by factors of four, then the numbers up to about 70 K change by less than ten percent or so. Another reason for thinking that flux creep effects are not major is that, if you look at the scaling curves as a function of temperature from 70 K down to 10 [K], they superimpose rather well.

P. Kes: Two short comments. I agree with the interpretation of Doug Finnemore. My choice of oxygen vacancies in the copper-oxygen planes: they act as pinning centers, whereas the other ones will give rise to modulation of the order parameter in the planes which then act as pinning centers as well, but more like a precipitate where the T_c is a little bit changed.

As to the last point about B_c and H_{c2}. There was a discussion on Thursday that you could get H_{c2} from fit or scaling of fluctuations to magnetization data, and the question was can you do that as well for the bismuth compound. I called yesterday, and we did that. It doesn't work so well. The fit only works above T_c, and you get, really, deviations below T_c. Anyway, if you plot the T(H) values you get out of this analysis, you get a value for the slope of a half tesla per kelvin, which is pretty small.

M. McHenry: I'd just like to make one comment about the creep data at high temperatures. I think the fact that you are seeing a slow creep rate is just a reflection of the fact that most of the creep has happened before you ever started the experiment, and this nonlinear U(J) formalism would explain that temperature dependence of what we call the magnetization current – which is a strongly relaxed current – quite well.

D. C. Larbalestier: I'd like to have a copy of your latest there.

G. W. Crabtree: I just want to comment on Peter's comment. That is, this scaling technique that was being discussed on Thursday is specifically for the high-field regime, where you should have lengths limited by the radius of the lowest Landau level and it could be that in the bismuth compounds you need to go to higher fields to see that. I mean, there are possibly some reasons why it shouldn't work.

M. P. Maley: You alluded to this granularity problem, and one of the things I think that's striking is the fact that with the yttrium barium copper oxide system, which seems to have acceptable pinning at high temperatures, no one has defeated the granularity problem in a process that's amenable to making wires and tapes. It appears – particularly with the work that Murakami's done with melt processing – that you can actually kind of get around or minimize the weak links. But do you see any prospects for working on this problem, after studying all these grain boundaries? All of the continuous processing wires and tapes so far are just full of weak links, and they fail at about 100 gauss.

D. C. Larbalestier: Well, I certainly don't think it's time to quit. The point [of] the bi-crystal work we're doing is that, from a cation point of view, when you look at a Josephson junction boundary and you compare that to a flux-pinning boundary – there's no difference. The atomic planes are continuous, they're very well ordered right up to the boundary, whatever is happening at the boundary is localized to within a nanometer. On the other hand, electron microscopy doesn't tell you, and you can't detect, what's happening to oxygen – and we have demonstrated this strong sensitivity to oxygen. And oxygen has not explicitly, really, been a control parameter in very much YBCO work. Much, I think, depends on what happens in bismuth. If bismuth works well then interest in the problem of YBCO grain boundaries will just tend to go away.

M. P. Maley: It's really striking that even if you prepare a polycrystalline BSSCO material, like Jin did recently, that the critical currents are just enormously higher, just through the weak links, than they are in the yttrium barium copper oxide. So there's a real qualitative different between the two systems.

D. C. Larbalestier: Right.

M.Tinkham
Harvard University
Cambridge, MA 02138

CONFERENCE SUMMARY

This summary will begin with short remarks, trying to recall some of the spirit of the presentations of each of the speakers during the first day, with no attempt at detail or completeness, given the need for a 20:1 compression relative to the original talk. I hope these idiosyncratic recollections do not infuriate the speakers too much! Since the speakers on the second day presented such interlocking topics, I shall simply try to present some sort of consensus report, to which I shall add some comments of my own. The two talks preceding this Summary on the final day dealt with the prospects for applications; since I have had no chance to attempt to prepare a proper report on these, I shall say only a few words about those presentations.

Day One: Materials and Junctions

On the first day, we started off with a splendid keynote address by Professor Tanaka, in which he presented his grand vision of the energy revolution which was needed to support the information society which he anticipates. He showed us some nice pictures of magnetic levitation, emphasizing the fact that this was based on flux pinning in the high-temperature superconductors, rather than on the Meissner effect. There has been a certain progression in MagLev. You first lift goldfish, then you lift people, and I believe I heard mention of a hundred-ton flywheel, which was going to be levitated in an underground cavern for energy storage; so this is actually going someplace. On the lighter side, Murakami told us about levitating sake barrels, so the applications are quite broad! Professor Tanaka also commented on the need for superconducting microwave filters to manage bandwidth for this communications and information society, and he mentioned in particular that he was waiting for and expecting a 150°K superconductor to really make these things work in a large way. That is a challenge to us all, particularly to those of you who consider yourselves to be proper materials scientists, as opposed to physicists like me who also attend meetings like this. In other words, I wouldn't *think* of trying to make a 150° superconductor; I wouldn't know where to start!

Next, we heard from Aharon Kapitulnik, whose report emphasized the nice Stanford experiments on a synthetic multi-layered system. This is a model system for high-T_c materials, in which the superconducting layers are made of amorphous Mo-Ge, separated by germanium insulating layers. It shows a resistive transition resembling that of YBCO or BSCCO depending on the thickness of the insulating barrier. In discussing this data, he distinguished a temperature T^*, which was not terribly well-defined, but above which the resistance versus temperature seemed to be the same whether you had one layer or a whole bunch of layers, and below which the resistance dropped more rapidly and *did* depend on the number of layers, implying effective coupling between the layers. The implication is that above this T^* the resistance is dominated by universal fluctuations, while below T^* the resistance is thermally activated, and depends more on details of pinning and coupling energies.

Kapitulnik also told us quickly about some BSCCO experiments, in which the resistance in the c-direction was found to have a semiconducting-sort of temperature dependence. If this dependence becomes mixed by some mechanism into the resistance as nominally measured in the a-b plane, it could account for some anomalies that are observed in that quantity.

George Grüner then spoke about the microwave properties of various superconductors. One thing I particularly appreciated was that he confirmed a peak in σ_1 just below T_c. This is a feature which ought to be there in BCS superconductors, but which had not been confirmed directly before. Of course, there's a lot of doubt as to whether the HTSC are BCS-like superconductors, so he was looking mostly at classic superconductors, to make sure he got the right answer there before addressing the HTSC. But a question seemed to be left open as to whether the peak is really a BCS coherence factor effect as described in the books, or whether it is due to something else, still unknown. Grüner also showed us that the surface resistance in thin films of HTSC is down well below that of copper, and is approaching the BCS limit for relatively high temperatures, i.e., $T > 70$ K, where the quasiparticle resistance is still large. However, the surface resistance falls more slowly than expected from BCS quasiparticles at lower temperatures, and he was less sure about the residual resistance at lower temperatures, blaming it on things like magnetic flux, which is very plausible, but leaves open the question of whether these losses can be reduced in practice.

Later in the proceedings, Grüner told us about the superconducting properties of "Bucky Balls," after drawing on his classical mathematical knowledge to suggest calling these "Eulerenes" instead of "Fullerenes," on the basis that Euler pointed out that you needed the right number of pentagons as well as hexagons to make soccer-ball-like structures long before Buckminster Fuller popularized them. Anyway, he was telling us that the H_{c2}'s of these doped variants on the organic material C_{60} looked like 50 tesla, from which he inferred coherence lengths of 26 Å. Palstra questioned the quantitative reliability of this inference, wondering if Pauli limiting was affecting this measured H_{c2}, so that maybe the estimated ξ wasn't totally settled yet. The temperature dependence of the penetration depth in the Bucky Balls reported by Grüner seemed to be pretty BCS-like; this suggested that the gap didn't have any nodal points or lines to give anomalous numbers of low-lying quasiparticles which would make the penetration depth rise more rapidly with increasing T at low temperatures than predicted by BCS.

Continuing down the list, we heard from Murakami about his marvelous developments using inclusions of (nonsuperconducting) 2-1-1 in 1-2-3 YBCO to increase the pinning, and these are really quite impressive figures. He managed to raise the critical current by a factor of ten, by putting in a 25 percent admixture of 2-1-1. Similarly, he raised the irreversibility line by about a factor of tw-o in magnetic field, at ten percent content of 2-1-1. The critical current was raised by a factor of two by going from ten micron to one micron grain size, confirming the importance of having these inclusions finely subdivided. In a nice final touch, he showed with decoration experiments that the flux really does prefer to pass through these pinning sites, by observing that that's where the nickel particles were attracted. This result doesn't surprise us, but I always like experiments which confirm that your presumption is really true, and does provide the correct way to think about a phenomenon.

As expected, John Clarke gave one of his brilliant overviews about the prospects for Josephson junctions made from high temperature superconductors. He covered so much ground in that hour that I can't even give an outline here! First he talked about the many efforts to make good fabricated junctions of these difficult materials. He considered four categories of approaches, each of which had four subcategories. One I think he liked especially was the bi-epitaxial junction, being developed at Conductus, but he did think several other methods also were very promising. Next he emphasized that, although it was fine to make these junctions out of HTSC, they could not really be used unless you can also make HTSC flux transformers, crossovers, vias and other things, which add greatly to the fabrication complexity. Despite these complexities, his group has overcome these problems sufficiently well that they were able to make a working "flip-chip" SQUID device. This utilizes components fabricated on two separate chips, one of which is then flipped over and placed on top of the other to get magnetic coupling. When refrigerated in a simple thermos flask of liquid nitrogen, and shoved against a graduate student's chest, excellent magnetocardiograms were obtained. This device has a sensitivity of one picotesla per root Hertz, which seemed to be good enough to see the heartbeat pretty well. The big challenge to be overcome is the 1/f noise, which is still quite large, so that the rollover to the flat white noise region occurs at relatively high frequencies. Once you get up to these frequencies, the white noise seems to be pretty near the theoretical value, which means that things are working as expected from standard SQUID theory and experience. One simply must learn how to beat down the 1/f noise. No doubt he is working on tricky methods to accomplish that, which we'll hear about at the next conference!

I found Doug Finnemore's talk fascinating, because he asked some rather basic questions, such as, "Are H_{c1} and H_{c2} meaningful? Are they measurable?" Of course the problem with measuring H_{c1} is the surface barrier which makes it hard to really tell at what field you can first put flux in equilibrium into the interior of a bulk superconductor. Similarly, the fluctuation rounding, which we'll hear more about, causes the definition of H_{c2} to not be a crystal clear experimental quantity, either. He discussed these things in a very useful way, stirring up lots of debate, and I think we even managed to get Daniel Fisher to concede that H_{c2} was a worthwhile semiquantitative concept despite being poorly defined.

After raising these interesting conceptual questions, Finnemore gave us a peek into the black arts of metallurgy, melting, and annealing of multi-phase composite systems. I will freely admit that I had never had any notion of how tricky these materials were to make until this presentation, and I can not really produce any sort of summary version of its complexity and remaining poorly resolved issues.

Day Two: Vortex Motion

On the second day, the vortex lattice and its pinning and motion were discussed from various perspectives by David Nelson, Peter Gammel, Daniel Fisher, Gabe Aeppli, Boris Ivlev, and Peter Kes. What I have tried to do here is first to produce a set of statements on which I think there was general consensus. This is not to be minimized; on many of these subjects there was not such general consensus a year ago, so I think this represents a significant step forward. After that, I shall discuss in more detail the resistive regime near T_c, reviewing some of the arguments about scaling.

The first general consensus statement is that it is meaningful to make a phase diagram of the general sort presented here by Nelson and by Fisher. In these, there is an $H_{c2}(T)$ line based on the mean field concept of H_{c2}, but which is agreed to be fuzzy because of the prominence of fluctuations, and not a precisely determined experimental quantity. Inside this "line" one goes continuously into a vortex fluid regime, after which there is supposedly a *sharp* line, provided one is at *zero current*, at which one goes from the vortex fluid into the vortex glass or vortex lattice or entangled vortex solid, depending on the model. The exact nature of the solid seems open to question, but the idea that there should be a rather well-defined line seems to be getting better established by experimental evidence. Of course, below H_{c1} is the Meissner region, which should be resistanceless and not very controversial.

In an anisotropic 3D material like YBCO, the liquid-solid transition line is fairly close to the $H_{c2}(T)$ "line," whereas in a more highly-layered material like BSCCO the fluid region extends much further down in temperature. There are important questions of dimensional crossovers, which I shall skip over since I'm trying to be very general, and simply focus on the fact that the existence of distinct liquid and solid regimes (at zero current) seems to be reasonably well accepted by many people, together with a fuzzy crossover at H_{c2}. Other "crossover lines" may be useful, and people introduce them to indicate things like the onset of activation, or hexatic order, or interplanar coupling. Well, the onset of activation is rather continuous when kT gets comparable with characteristic energies, so that's obviously a qualitative line. Hexatic order can be seen directly in some decoration experiments, but its effect on R(T) curves is less well established; some think they see it, others do not. These all are concepts for which some workers put "lines" in the H-T plane, but a consensus is not well established.

If we give up the restriction to J=0, which is of course a rather impractical regime, and go to a finite current, then I think it was said without controversy that, in principle, there is always dissipation; however, it may be immeasurably small on laboratory time scales. There now seems to be general agreement that the barrier energy for activation of a dissipative process often can be thought to vary as an inverse power of the current, as predicted in the collective pinning model of the Russian school. When that is so, then as the current goes to zero the exponential activation rate goes to *zero* even at finite temperatures, but for any finite value of current the activation rate is always a nonzero quantity except at T = 0. This is why one has to make a distinction, between zero current as a theoretical abstraction in which you have these nice sharp phase-boundary lines, and a finite-current case where you have to ask how big these activation energies really are and how much dissipation there really is.

In this connection I was rather impressed by the report here in the poster session of the Los Alamos work of Maley's group on the analysis of their flux creep measurements in terms of a current-dependent barrier energy. They were able to unfold this data to yield activation energies which seemed, as Peter Kes showed in his presentation, to fit in quite well with the theoretical expectations, except for a possible question about an inverse 1/7 power at the lowest current end.

A final general comment that I think we would all agree on is that we probably need new types of experimental measurements to actually discriminate amongst the many theoretical models which are roughly consistent with all of the above general statements. The presently available types of experiments, which have led to our consensus and about which I will be saying more below, have already been pushed very hard, while a new approach might yield qualitatively fresh results. For example, David Nelson has frequently urged experimenters to put down a bunch of voltage contacts at one micron intervals or less, and measure the velocity profile of the fluxon distribution near a discontinuity in a current-carrying sample. The intervortex viscosity in his entanglement model would lead to a velocity healing length with a characteristic temperature dependence near the solid transition, which should be distinct from that resulting from other models such as the vortex glass model. Clearly that would be a great experiment, but it has yet to make the transition to the real world. Nonetheless, experiments of that sort seem possible of success, and are being attempted by various groups.

The idea of using noise measurements as a probe of these various models was another idea which received a fair amount of discussion here. Such experiments have already been informative in the low flux density region of isolated vortex hopping. Further progress will probably wait upon theorists coming up with predictions from these various models of collective fluxon motion so as to give the experimenters an idea of what sort of signature they might hope to find.

The linear resistance:

At this point, I would like to focus more on one specific subsection of this subject, one in which I've dabbled myself, and that is the question of the *linear resistance* as a function of temperature. The various regimes are indicated in the sketch shown in Fig. 1. This is purely schematic, with a vertical scale that is totally nonlinear and represents data stretching over many orders of magnitude in addition to zero. The circles identify three regions, each of which has to be approached in a different way.

The region down near the putative vortex glass, or other solid, transition temperature is one in which there have been very nice experiments by Koch et al.,[1] Worthington et al.,[2] Gammel et al.,[3] and others. These have shown that the linear resistance does appear to be zero — not just something exponentially small as in the TAFF model, but *zero* — below a definite temperature, and thereafter rise with some high ($\sim$3-6) power or critical exponent of (T - T_g). In discussing his data, Peter Gammel made the sage observation that a thing which is going up with a large power is hard to tell from something that is coming down with a large exponential. You are limited by how wide a range of voltage you can track within the sensitivity of your instrument. Needless to say, all these fits of data to determine T_g involve extrapolations to zero voltage, because no one, even at Bell Labs, can measure *zero* voltage; you have to have a finite signal. Despite these difficulties, I think the available data offer quite convincing support for the hypothesis of a sharp onset of linear resistance at some sharply defined (glass) temperature as opposed to the earlier view that the temperature variation would go as $e^{-U/kT}$ all the way down, and vanish only at T=0. I must reemphasize, however, that this sharp transition exists only in the limit as J approaches zero. For all but the most sensitive measurements, the transition to zero resistance remains blurred by the measuring current.

In addition to these dc resistance measurements, there is recent data by Olsson et al.[4] (IBM) on the ac impedance in the critical region. These measurements show that right at T_g, the complex impedance displays both the frequency dependence and phase expected from the vortex glass model with critical exponents similar to those found in the dc resistance measurements. To me, this appears to be a nice confirmation, but others at this conference have cautioned that this data is only preliminary.

CONFERENCE SUMMARY 505

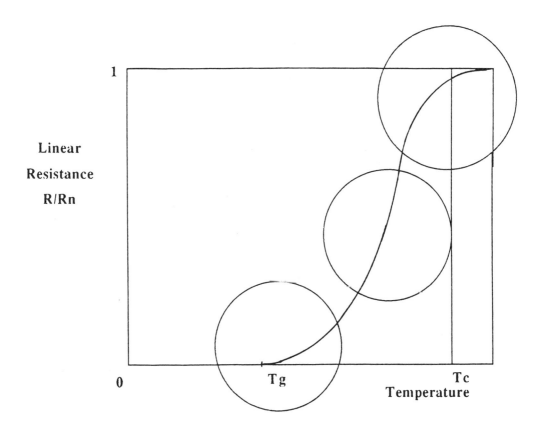

Fig. 1 Schematic plot (on a *very* nonlinear vertical scale) of the *linear* resistance vs. temperature. The circles identify three regions with different characteristics. Resistance is thought to be zero below T_g and rise as $(T-T_g)^n$ (with $n \gg 1$) above T_g. Resistance near T_c is also thought to show scaling properties, perhaps depending on the variable $(T-T_c)^{3/2}/H$. The behavior in the intermediate region depends more strongly on details of pinning, interlayer coupling, etc.

The other region of Fig. 1 in which some reasonably general remarks can be made is that near T_c where fluctuation effects dominate. As sketched in the Figure, the fluctuations cause a depression of the resistance already above T_c, followed by an ever more rapid loss of resistance below $T_c(H)$. There is no discontinuous behavior at T_c, which as noted above is blurred by fluctuations. It was in this regime near T_c that I showed[5] in 1988 that at least the top 90% of the resistive transition of YBCO as measured by Iye could be fitted surprisingly well by a simple analytic function of the variable $[(T-T_c)^{3/2}/TH]$. In that paper I used the specific scaling function [a modified Bessel function $I_0(x)$] found by Ambegaokar and Halperin for phase evolution in the "tilted washboard potential" describing an overdamped Josephson junction, arguing that it should also qualitatively describe the driven thermally activated motion of fluxons past one another in what is now recognized as a fluxon liquid. The comparison of this model with Iye's data is reproduced here as Fig. 2. Since this model takes account only of fluctuations of the phase of the order parameter and not of its magnitude, it did not describe the reduction of R above T_c by fluctuations; that was built into the curves by simply putting in "by hand" the experimentally observed fluctuation rounding in zero field, and using the model to describe only the further, field-induced broadening of the resistive transition. Although the fit of this simple one-parameter model to the data was quite impressive, Palstra quickly showed that if the resistive transition was followed further decades down, it dropped with a stronger exponential than that found from fitting the top two decades in the resistive transition, indicating that the excellent apparent agreement was somewhat fortuitous.

DAVID NELSON: Mike, can I ask a question? On that viewgraph that you just took off, the 60 kilogauss data of Iye: do I see a shoulder there?

Yes; it is not as clearly revealed as in more recent and more precise data, but it was already there.

This general approach has recently been advanced substantially by the Argonne group, as reported in a preprint by Welp et al.,[6] and represented in discussions at this conference by George Crabtree. With his permission, I reproduce some of their data as Fig. 3. What is plotted is not the resistance but the fluctuation conductivity σ_{fl}, which they found by subtracting out the extrapolated normal state conductivity to find the extra conductivity due to the onset of superconductivity. The insets in the Figure show this fluctuation conductivity as a function of temperature at different fields. One sees it going up at lower temperatures; the lower the field, the quicker the superconducting effects set in. What they observed was that if they scaled this σ_{fl} data by multiplying it by $H^{1/3}/T^{2/3}$, and then plotted it against the variable $[T-T_c(H)]/(TH)^{2/3}$, all of these curves fell upon a single, universal-appearing curve. [Obviously this scaling variable is equivalent to that used in my paper, which is simply the 3/2 power of it.] This universal function does not have any singular behavior at this nominal T_c; that is simply a parameter in this way of plotting the data. However there is some subtlety here. When they fit the data, they use a $T_c(H)$ which they assume decreases linearly with H as expected in a mean-field $H_{c2}(T)$ curve, and the variations of $H_{c2}(T)$ they found were generally consistent with the results of other approaches to the determination of this admittedly inexactly defined quantity. This use of a mean-field concept was questioned by Daniel Fisher, who favored using the unshifted T_c as the reference point; further work is needed to clarify this issue. However, one should bear in mind that the downward shifts in $T_c(H)$ are quite small compared to the field-induced

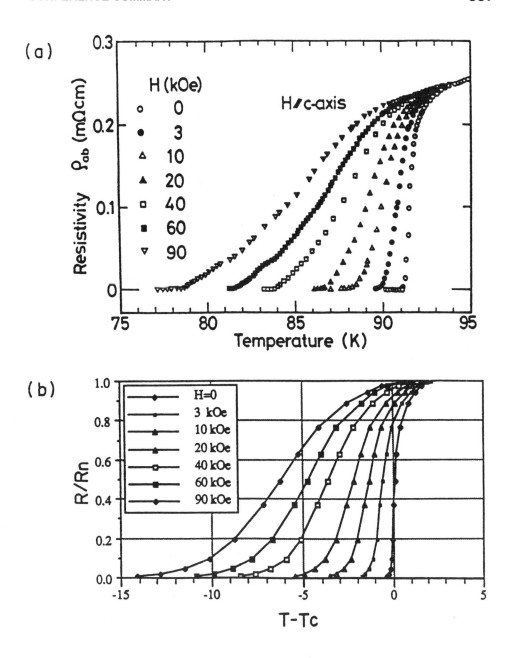

Fig. 2 Comparison of YBCO resistive transition data of Iye et al. with model predictions, reproduced from Ref. 5. (a) $\rho_{ab}(T)$ of a YBCO crystal for various values of H along the c-axis as reported by Iye et al. (b) Curves computed from simple scaling model, in which R/R_n is a specified function (I_0^{-2}) of the scaling variable $A(1-t)^{3/2}/2HT$.

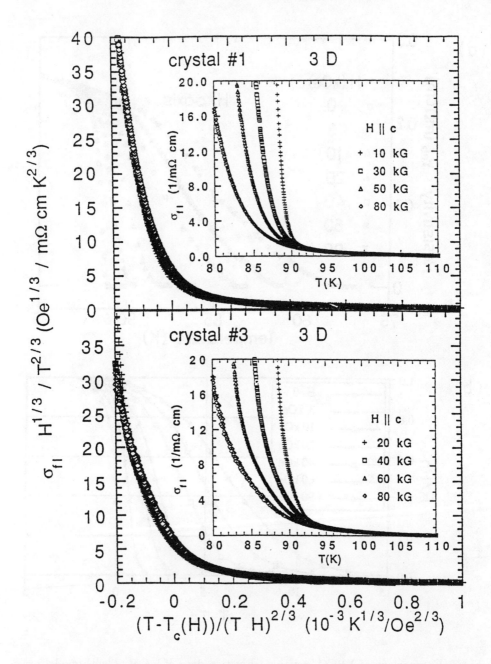

Fig. 3 Three-dimensional scaling of the fluctuation conductivity of two crystals of YBCO. The insets show the temperature dependence of the fluctuation conductivity before scaling. From preprint by Welp *et al.*, (Ref. 6) with permission.

transition broadening in moderate fields, so the distinction may not be crucial to the value of the approach. For example, in the fits in my earlier paper, for simplicity I ignored the field-induced decrease of T_c.

One way of using a scaling model to analyze data which I found useful was to make the sort of pseudo-phase-diagram shown in Fig. 4. Here one plots curves of constant resistance in the H-T plane, for example, at half the normal resistance, one

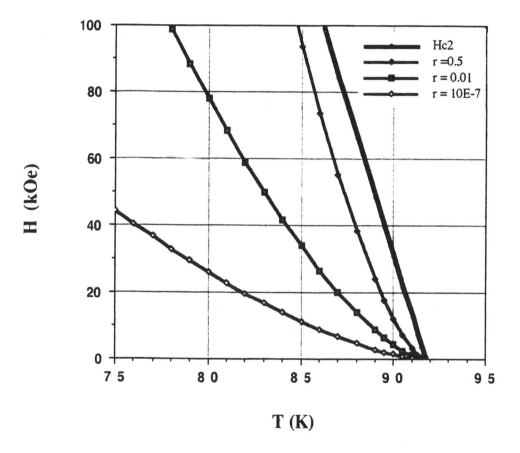

Fig. 4 Contours of constant reduced resistance in the H-T plane, for YBCO crystal with magnetic field along c-direction, as given by scaling model of Ref. 5. Curves for $r = R/R_n = 1/2$ and 0.01 are in good agreement with data of Iye et al., shown in Fig. 2. Curve for $r = R/R_n = 10^{-7}$ is a model extrapolation which *qualitatively* agrees with Palstra resistive data or measured position of irreversibility line or glass temperature. Position of "fuzzy" $H_{c2}(T)$ line as estimated from equilibrium magnetization measurements is also indicated.

percent, and 10^{-7}. Iye included a plot of his data for half- and zero-resistance in this format. These data are fitted almost within the experimental precision by the result of my simple scaling model, if one relabels as 1% resistance the curve he labeled as "zero" resistance, since the sensitivity of the data appeared to be only about 1%. According to this model, for constant ρ/ρ_n, H should scale with $(T-T_c)^{3/2}$, with a coefficient given by the Ambegaokar-Halperin dependence. The curve labeled 10^{-7} is an extrapolation using this same specific scaling function $[I_0(x)]^{-2}$ down to a resistance level low enough to begin to act like a superconductor. The resulting curve is not too far from (though not in quantitative agreement with) Palstra's later measurements made with sufficient sensitivity to reach this resistance level, nor from the position of the irreversibility lines found in other types of experiments. If one used the more general kind of scaling that Crabtree is proposing, one might get an even wider range of quantitative fitting, but in any case scaling down from T_c can hardly be expected to hold all the way down to T_g! None the less, it is reassuring that the form of the $T_g(H)$ dependence found by Koch et al., is similar to the curve at which R becomes immeasurably small (as was pointed out by Koch at the time), which in turn is at least qualitatively very similar to the curve predicted by this simple scaling down from T_c to a very small resistance. This means that one can "piece together" the two expansions, about T_c and T_g, without any big disparity where they meet.

DANIEL FISHER: Mike, in which regime is the pinning important?

I am not sure. I'm just plotting data. Actually, this is not all data; the curve at 10^{-7} is what my model gave as an extrapolation of Iye's data at higher resistance levels. But there's nothing in this model which is so specific as to deal with the amount of pinning.

Having emphasized the apparent usefulness of a scaling with the variable $[T-T_c(H)]^{3/2}/(TH)$, I would now like to make a few personal remarks about the rationalization of such a form. In my 1988 paper, as originally submitted, I had simply accepted the observation of Yeshurun and Malozemoff[7] that a thermal activation energy which varied as $(T-T_c)^{3/2}/H$ could account for the observed irreversibility line. I then argued that for higher values of T the same thermally activated flux motion process would give rise to measurable resistance which could account for the broadening of the resistive transition reported by Iye, which was considered unexplained and surprising at the time. One of the referees requested that, in addition to showing how well this hypothesis worked, I should also provide a more explicit rationale for this form of activation energy. In response, even though I had previously considered this form as an empirical "given," I produced the argument which appears in the published version. It is based on a crude estimate of the energy barrier to push a segment of length ξ of one fluxon past its neighbors, taking no account of the energies stemming from the long-range connectedness of the fluxon. This energy scaled with the condensation energy per unit volume $(H_c^2/8\pi)$ times

($\xi\Phi_o/H$), the volume of a fluxon of length ξ. With the standard mean-field temperature dependences, this product scales with $(1-t)^{3/2}/H$ near T_c, where $t \equiv T/T_c$, and gave an impressive fit to the resistive transition, as noted above.

Because the origin of these characteristic activation energies continues to spark considerable discussion, I am going to use this opportunity to record a few additional thoughts on the subject. Resistance in a superconductor results from phase slippage. In a one-dimensional superconducting filament, this can occur near T_c by a fluctuation which takes the magnitude of the order parameter down near zero, as discussed by Langer-Ambegaokar and McCumber-Halperin. They showed that the activation energy for this process could be written as proportional to $(H_c^2/8\pi)$ times ξ times the cross-sectional area of the filament. This is the same as the energy discussed in the previous paragraph if this cross-sectional area is Φ_o/H, the area associated with one fluxon. But what physical basis is there for such an association?

I suggest the following possible point of view. Near T_c, with the normal state as a reference, the important thing is how much phase coherence of the order parameter ψ can be maintained to increase the diamagnetism, produce extra *conductance* (i.e. reduce the resistance), etc. [This contrasts with the situation at low temperatures, where the zero resistance state is the reference, and *resistance*-producing motions of the fluxons are the appropriate focus of attention.] Near the second-order transition at $T_{c2}(H)$ the order parameter is small and the structure of superconducting ψ can be described approximately by the linearized Ginzburg-Landau equation. This has many degenerate solutions, including the Abrikosov state, but a simpler one is a Gaussian with cylindrical symmetry about the field direction, and a particular radius r_o such that $\pi r_o^2 H \approx \Phi_o$. This suggests considering a little "rod" of superconductivity of radius r_o. [An alternative view is that this is also the area of the compact region *between* vortices in the Abrikosov lattice.] What is the energy which is trying to maintain the phase coherence along such a rod? This has a universal answer, namely Φ_o times the critical current I_c ($=J_c A=J_c\Phi_o/H$), which is equivalent to the Langer-Ambegaokar-McCumber-Halperin result for a filament, or to the standard result for the phase coupling energy in a Josephson junction. Using standard mean-field interrelations, $J_c \propto H_c/\lambda \propto H_c^2\xi \propto (1-t)^{3/2}$, so the energy which is enforcing phase coherence along this rod of superconductivity scales with $(1-t)^{3/2}/H$. It then seems plausible that the strength of the superconducting coherence should be a function of the ratio of this phase integrity energy to the thermal energy kT, suggesting a scaling with a dimensionless variable proportional to $(1-t)^{3/2}/HT$. [Of course this plausibility argument is too simple to pick up the prefactors (fractional powers of H and T) which Crabtree *et al.* use to prescale the conductance, diamagnetism, and other quantities.] Evidently, this scaling variable based on phase decoupling by amplitude fluctuations of the order parameter is the same as the one obtained in Ref. 5 by considering fluxon segments moving past one another. The latter is another way to decouple the long range phase coherence that is more important when the superconducting order parameter is strongly established. This equivalence suggests that the basis for the relevance of this variable is more general than suggested by either of these simple physical arguments.

Let us now return to the new data from Argonne, interpreted using this scaling variable. They have very nice data on the reversible magnetization, i.e., the diamagnetism near T_c, which is an *equilibrium* property. The raw magnetization data at different H values look quite different because of the nonlinearity of the diamagnetic response, but when they replot the data against the same scaling variable (together with prefactors), all of the experimental data fall together on a single universal curve. They showed the same sort of data collapse for two other physical quantities. I found these new results a very fascinating bit of experimental information which lends support to the usefulness of some kind of a universal fluctuation/scaling picture for the behavior near T_c, before all of the more detailed things to do with pinning and layer structure really set in.

Day Three: Applications

On the third day, we had two surveys of applications of the high temperature superconductors: the electronic applications reviewed by Bob Hammond and the high current applications by David Larbalestier. With these talks still ringing in our ears, I really have had no opportunity to prepare any detailed summary of them. However, it is clear that in both of these areas very impressive progress has been made toward reducing these technologies to commercial practice. Unfortunately, it is also clear that some problems still remain to be cleared up with further work before all the potential opportunities can be fully realized. Because many of the important issues are quite specific, detailed, and material-sensitive, I can only recommend consultation of the written versions of these talks, when they become available, for all the details.

I would now like to turn to my final item of business, which is to thank the organizers for providing the framework for this spirited discussion. As far as I know no physical casualties have resulted, perhaps mental ones! In any case, it's been a very stimulating meeting and I look forward to reading this book that you all are writing.

ACKNOWLEDGMENT

This work was supported in part by NSF grant DMR-89-12927 and by the Harvard MRL grant DMR-89-20490.

REFERENCES

1. R. H. Koch et al., Phys. Rev. Lett. **63**, 1511 (1989).
2. T. K. Worthington et al., Cryogenics **30**, 417 (1990); also, Phys. Rev. **43**, 10538 (1991).
3. P. L. Gammel et al., Phys. Rev. Lett. **66**, 953 (1991).
4. H. K. Olsson et al., Phys. Rev. Lett. **66**, 2661 (1991).
5. M. Tinkham, Phys. Rev. Lett. **61**, 1658 (1988).
6. U. Welp, S. Fleshler, W. K. Kwok, R. A. Klemm, V. M. Vinocur, J. Downey, and G. W. Crabtree, preprint.
7. Y. Yeshurun and A. P. Malozemoff, Phys. Rev. Lett. **60**, 2202 (1988).

DISCUSSION

D. R. Nelson: Just a minor comment. You plotted critical field on one of your transparencies, and I presume that was determined from the resistivity in some sense.

M. Tinkham: Critical field?

D. R. Nelson: H_{c2}.

M. Tinkham: Those were Daniel's plots, those phase diagrams . . .

D. R. Nelson: No, there was another plot, which I thought had H_{c2}. Let me pose the question in the following way: H_{c2}, it seems to me, this approximate place that's fuzzily defined, ought to best be defined in terms of Meissner, the onset of the Meissner effect. And if you do it in terms of resistivity, you're going to get answers that depend on whether you have twin planes and the density of other things which are not intrinsic features of the material.

M. Tinkham: Well that's why in Finnemore's talk he emphasized that approach, but even that has its complications because of fluctuation effects. Daniel?

D. S. Fisher: Just a comment on all these scaling plots, and so on. You know there was a paper of Kapitulnik, et al. a couple of years ago, showing that all kinds of things scaled with the resistivity if you picked out various points reasonably far down the resistivity curve to choose, and everything scaled quite nicely with H going as $(T_c - T)^{4/3}$.

M. Tinkham: I think that Kapitulnik likes the 4/3 power instead of the 3/2.

D. S. Fisher: Right. But there is actually a theoretical reason to believe that there's a wide regime, which is maybe ten degrees wide in temperature, and then goes up to relatively large fields, perhaps 100 kilogauss fields. If the pinning is small scale, even with the pinning included everything should scale nicely in this regime.

M. Tinkham: Yes, I didn't say the pinning wasn't in there. I just didn't say anything about pinning.

D. R. Nelson: But surely it's a valid experimental thing, Daniel, to try 4/3 and 3/2 and see what does the best . . .

D. S. Fisher: Right. But I don't think there's a mystery as to why it should look like that. There is basically only one energy scale in that regime and the pinning ends up being effectively independent of temperature in that regime also, basically because ν is 2/3 and so you'd really expect everything to scale nicely there.

D. R. Nelson: But I would also comment that the beautiful data collapse with the 3/2 exponent – which may well *not* be theoretically justified – isn't proof that 3/2 is right. So we have to take the data collapse there with a grain of salt, and near the vortex glass transition as well.

M. Tinkham: They did try another power for 2D instead of 3D, and the fit wasn't quite as good.

Contributed Papers

Langevin Dynamics Simulations of Large Frustrated Josephson Junction Arrays

N. Grønbech-Jensen,* A. R. Bishop and P. S. Lomdahl
*Stanford University, Stanford, California 94305-4090,
Los Alamos National Laboratory, Los Alamos, New Mexico 87545

Long-time Langevin dynamics simulations of large ($N \times N, N = 128$) 2-dimensional arrays of Josephson junctions in a uniformly frustrating external magnetic field are reported. The results demonstrate: (i) Relaxation from an initially random flux configuration as a "universal" fit to a "glassy" stretched-exponential type of relaxation for the intermediate temperatures $T(0.3T_c \lesssim T \lesssim 0.7T_c)$, and an "activated dynamic" behavior for $T \sim T_c$; (ii) A *glassy* (multi-time, multi-length scale) voltage response to an applied current. Intrinsic dynamical symmetry breaking induced by boundaries as nucleation sites for flux lattice defects gives rise to *transverse* and noisy voltage response.

The Hamiltonian for the JJA takes the form (see Ref. [1] and references therein)

$$\mathcal{H} = -E_0 \sum_{i,j} [\cos(\theta_{ij} - \theta_{i-1j} - A_{i-1j,ij}) + \cos(\theta_{ij} - \theta_{ij-1} - A_{ij,ij-1})], \quad (1)$$

where θ_{ij} is the phase of the superconducting island with the discrete coordinates (i,j) of the lattice, and $A_{ij,kl} \equiv (2e/\hbar c)\int_{ij}^{kl} \mathbf{A}\cdot d\mathbf{l}$ is the integral of the vector potential from island (i,j) to a neighboring island (k,l). The $A_{ij,kl}$ summed around a plaquette obeys the following relation: $A_{ij,kl} + A_{kl,kl-1} + A_{k-1l-1,ij-1} + A_{ij-1,ij} = 2\pi f$, where the frustration $f = Ha^2/\Phi_0$ is a constant giving the average number of flux quanta $\Phi_0 = hc/2e$ of the external magnetic field H through the area a^2 of each plaquette of the array. We also introduce the fractional charge q_{ij}, obtained as the gauge invariant phase sum around the ij'th plaquette:

$$q_{ij} = \frac{1}{2\pi} \sum_{ij'\text{th plaquette}} (\theta_{kl} - \theta_{mn} - A_{mn,kl}) \bmod \pi. \quad (2)$$

This quantity exhibits a checkerboard pattern for the ground state of the frustration $f = \frac{1}{2}$ used in this report. It is therefore sometimes convenient to introduce the staggered order parameter $\tilde{q}_{ij} = (-1)^{i+j} q_{ij}$. The dynamical equations become[2]

$$\ddot{\theta}_{ij} = \sum_{kl:\text{neighbors}} [\sin(\theta_{kl} - \theta_{ij} - A_{ij,kl}) - \eta(\dot{\theta}_{kl} - \dot{\theta}_{ij})] + \lambda_{ij}(t), \quad (3)$$

where the thermal noise is introduced in a classical Langevin sense: $\langle\lambda_{ij}(t)\rangle = 0$, $\langle\lambda_{ij}(t)\lambda_{kl}(t')\rangle = 2\eta G_{ij,kl}^{-1} T\delta(t-t')$, where the temperature T is normalized to E_0/k_B, k_B being the Boltzmann constant. Time is normalized to $\tau = (C_0\hbar/2eI_0)^{1/2} G_{ij,kl}$ is the discrete Green's function for the square lattice, the normalized dissipation is given by $\eta = \frac{1}{R}(\hbar/2eC_0 I_0)^{1/2}$ with R the normal resistance of the junctions, and C_0 is the capacitance between a superconducting island and the ground plane.

(i) Relaxation studies with periodic boundary conditions ($\eta = 1$):
The following gauge invariant quantity is defined: $C(t) = \frac{1}{N^2}\sum_{i,j} q_{ij}(q_{i+1j} + q_{i-1j} + q_{ij+1} + q_{ij-1})$. This quantity is $C = -1$ for the $f = \frac{1}{2}$ ground state and $C = 0$ for the random flux configuration. The initial conditions in θ_{ij} are random in all cases and normalized times up to 10^4 are included. The results are shown for various temperatures T in Fig. 1. Here it is important to note that the transition temperature T_c for $f = \frac{1}{2}$ is $T_c \simeq 0.45$: for $T > T_c$ long-range flux order and superconductivity are lost. $C(t)$ is displayed for a selection of temperatures illustrating the various time-dependencies we have observed: viz. (a) $T \ll T_c$ and short times : Here, flux creep is found (not shown) with $C(t) \sim \ln t$.[2] The "asymptotic" value of C first decreases as T is increased from $T \ll T_c$ and then increases again ($T \gtrsim T_c$). This can be understood as trapping into a metastable (after the initial relaxation) flux configuration at very low T because of the uniform frustration; as T is increased, thermal

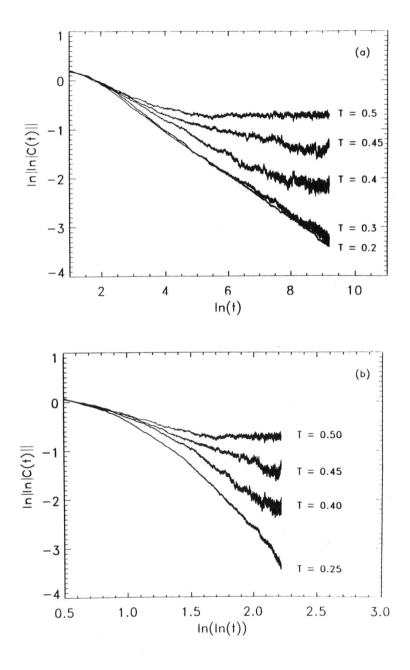

Fig. 1

tunneling over the frustration barriers is allowed and C approaches closer to its $f = \frac{1}{2}$ ground state value ($C \to -1$); at higher T ($\geq$ frustration pinning energy) thermal randomization occurs ($C \to 0$ for $T \gg T_c$). (b) *Intermediate* $T(0.3T_c \lesssim T \lesssim 0.7T_c)$. Here, after the initial rapid relaxation, we observe (Fig. 1a) an excellent "universal" fit to a "glassy" stretched-exponential type of relaxation.[3] Specifically, Fig. 1b shows the fit to $C(t) \sim \exp[-(t/\tau)^\beta]$, for which we find $\beta \simeq 0.45$, in good agreement with studies of other multi-time-scale systems.[3] (c) $T \sim T_c(\sim 0.45)$. Here critical effects dominate. An "activated dynamic" behavior of the form $C(t) \sim \exp[-(\ln(t/\tau))^\delta](\delta > 1)$ has been proposed.[4] As seen in Fig. 1b, this is in good agreement for $T \sim T_c(\approx 0.45)$, with $\delta \sim 0.9$. (d) $T > T_c$. Here the flux lattice melts and $C \to 0$ for $T \to \infty$.

(ii) Driven system with open boundary conditions ($\eta = 5$):
Figure 2a shows the $V_x - J_1$ characteristic with uniform edge driving for open boundary conditions. We see little difference from periodic boundary conditions[1] at this macroscopic level, with $I_c \approx 0.35$. Examination of the noise spectrum $S_{V_x}(\omega)$,[2] reveals the presence of a window of *noisy* response for $I_c \leq I \lesssim 0.55$. This multi-time scale response is understood[1] in terms of irregular domain wall separations, shown in the staggered charge $\tilde{q}_{ij}$ plot representation of the insets in Fig. 2a. With periodic boundary conditions,[1] there is obviously no transverse voltage V_y generated for any I. However, with open boundaries, Fig. 2b shows that a *large* transverse V_y is exclusively associated with the chaotic regime of driving current. Indeed we see that V_y is largest in the most noisy driving regime and vanishes (within our numerical resolution) in the regular regimes. Furthermore the transverse power spectrum $S_{V_y}(\omega)$ supports this scenario.[2] Examination of the $\tilde{q}_{ij}$-plots in Fig. 2a (inserts) clearly demonstrates the nature of the longitudinal symmetry breaking. As shown in the inserts of Figs. 2a and 2b, in the regular regime exact symmetry is maintained about the middle of the array: thus V_y for the upper and lower half arrays *exactly* cancel, even though there is considerable voltage in each half. However, in the noisy regime this upper-lower symmetry is broken, distinct $\tilde{q}_{ij}$-distributions develop (Fig. 1), and the upper- and lower-half contributions to V_y no longer cancel (insets of Fig. 2b), leading to a finite, noisy transverse voltage response. Furthermore, as suggested in the $\tilde{q}_{ij}$-plots of Fig. 2a, and as we have observed through careful examination of the time evolution, the asymmetry between upper and lower half planes develops by local flux defects, defined with respect to the domain wall pattern, nucleating (differently) at the upper and lower edges and propagating into the interior to form the mesoscopic structures shown in Fig. 2a.

We thank the Los Alamos Advanced Computing Laboratory for generous support and for making their facilities available to us. This work was performed under the auspices of the U.S. Department of Energy.

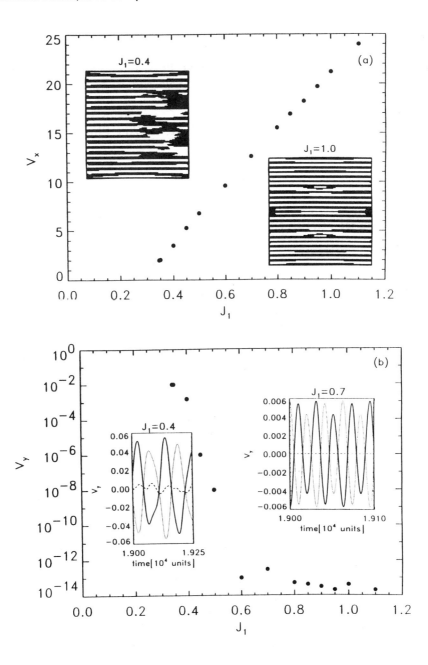

Fig. 2

REFERENCES

1. F. Falo, A. R. Bishop and P. S. Lomdahl, Phys. Rev. B **41**, 10983 (1990).
2. N. Grønbech-Jensen, A. R. Bishop, F. Falo and P. S. Lomdahl, preprints (1991).
3. I. M. Lifshitz and V. V. Slyozov, J. Phys. Chem. Solids **19**, 35 (1961).
4. See, e.g. C. Dekker, A. F. M. Arts and H. W. de Wijn, Phys. Rev. B **40**, 11243 (1989).

Simulations of vortex lattice melting: evidence for two melting transitions[1]

S. Ryu, S. Doniach, A. Kapitulnik*, G. Deutscher**
*Stanford University, Stanford, CA 94305
**Tel-Aviv University, Ramat-Aviv, 69978, Israel

Thermal fluctuations of the vortices induced by an applied magnetic field are believed to play an important role in the high temperature superconductors (HTSC), due to strong anisotropy of the Ginzburg Landau effective mass and the high transition temperature.[2] The cuprates are all strongly type-II materials ($\kappa \cong 100$) with high anisotropy (effective mass ratio at large as 100) and very short coherence lengths (typically $\xi(0) \simeq 10\text{Å}$). Thus, instead of forming a rigid Abrikosov lattice, each flux line is expected to execute considerable lateral fluctuations normal to the direction of the applied magnetic field.

Our simulations are based on a model where the vortex cores in each plane are treated as explicit degrees of freedom in a Monte Carlo evaluation of the partition function. We consider a stack of superconducting layers each of thickness d, interlayer spacing a and dimensionless interlayer coupling strength g following the Lawrence-Doniach model, and concentrate on the case where the field is applied perpendicular to the planes.

For a separation $\delta R = |R_i^z - R_i^{z+1}|$ of vortices in adjacent planes much smaller than a length scale defined by $r_g \equiv \xi_{ab}/\sqrt{g}$, the interaction can be shown to be approximately quadratic in the separation of the cores, while for a larger separation, a 'string' of vortex with core running between the planes provides[3] the necessary phase-healing by 2π over a length scale given by r_g. For separation of the vortices beyond $2r_g$, the interplanar Josephson coupling between two-dimensional vortices may be written in the following way

$$\begin{cases} \frac{\phi_0^2}{8\pi^3 \lambda_J \lambda} [1 + \log(\lambda/a)]\{|R_z - R_{z+1}| - 2r_g\} & \text{for } |R_z = R_{z+1}| > 2r_g \\ \frac{\phi_0^2}{8\pi^3 \lambda_J \lambda} [1 + \log(\lambda/a)]r_g \left(\frac{|R_z - R_{z+1}|^2}{4r_g^2} - 1\right) & \text{otherwise} \end{cases}$$

where $\lambda_J = r_g\lambda/a$. The vortices in the same plane are assumed to interact with the usual $K_0(|R_i^z - R_j^z|\lambda)$ form:

$$\delta F_{in-plane}(R_i - R_j) = (\phi_0^2/8\pi^2\lambda^2)K_0(|R_i^z - R_j^z|/\lambda_{ab}) .$$

In performing the Monte-Carlo simulation, we followed the usual Metropolis algorithm on a system of 64 (or 16) flux lines and 32 layers (total maximum of 2048 particles) confined to a 256 by 256 grid-space. Boundary conditions were periodic in the xy-plane and free in the z-direction. Since the in-plane interaction $K_0(\delta R/\lambda)$ is long ranged for a moderately high vortex density $H_{c1} \leq H \ll H_{c2}$, periodic boundary conditions also necessitate proper treatment of the potential by image charges. For each value of the flux density, we find the optimum number of image charges (up to 200) and include their effect.

Throughout the simulation, the areal density of the system is fixed at a value determined by a selection of the length scale corresponding to a grid unit. For example, choice of $\Delta x = \alpha\lambda$ gives a magnetic field intensity of $B = n\phi_0 = 2.07 \times 10^{-7}(Gcm^2)/\sqrt{3}a_0^2(cm^2) = (0.145/\alpha^2)G$ with $a_0 = 64\Delta x, \lambda = 2000$ Å. We chose to study a model with parameters similar to those of the $Bi_2Ca_1Cu_2O_8$ system for which there are some interesting experimental results available. The other length related parameter, $r_g = \xi_{ab}/\sqrt{g}$, is fixed by the choice of values for ξ and g. Taking $\kappa = 100$ and $\sqrt{g} = 1/50$ as is appropriate for BiSrCaCuO gives $g_g/\Delta x = 0.5/\alpha$. Therefore, we vary the effective density of the flux lines by varying the length-scaling factor α while keeping the total number of flux lines and the number of grid cells in the simulation fixed. The energy is scaled with $d\phi_0^2/8\pi^2\lambda^2 \equiv T_0$ where $\lambda^2 \approx \lambda_0^2(1 - T/T_c)$ was used. Using the value of 1.2×10^6 Å for $\lambda(0)^2/d$, we have $T_0 \approx 1000°K$ placing our results in a reasonable range in the phase diagram.

While slowly warming up the system, we measure the average of each component of the internal energy and energy fluctuations. The mean square deviation of each particle position from its equilibrium position ($\delta R_{in-plane} \equiv \langle |\vec{R}_{i,z} - \vec{R}_{i,z}^{av}|^2\rangle^{1/2}$) was measured to evaluate the Lindemann number for melting. Another important quantity, especially for a system with many layers is the 'end to end' distance fluctuation. The degree to which a flux line wiggles is measured by the quantity $|d\delta R/dz| \equiv \langle |\vec{R}_{i,z} - \vec{R}_{i,z+1}|^2\rangle^{1/2}$. The collective structure of the vortices was measured by the three dimensional structure factor along with the hexatic order parameter $\psi_6 = \langle \frac{1}{z_i} \sum_{j=1}^{z_i} e^{i6\theta_{ij}(r_i)}\rangle$ where θ_{ij} is the bond-angle between nearest neighbors. The pair correlation functions for the two types of order were also measured. The procedure was repeated with and without flux line cutting.

The system is found to be melted over most of the phase diagram which spans five decades in density.[1] We identified two phase transition curves from our simulation. The lower temperature curve was determined by observing the disappearance of the in-plane translational order monitored by $S(\vec{q} = \vec{G}_1)$. The second curve is from a similar analysis of the "hexatic" order parameter. We observe that these two lines are significantly separated from each other in the low field regime while they get closer in the high field limit. Thus, within the limitations of a finite sized sample, melting seems to occur through two steps.

We took a thorough look at the equilibrium configurations and found that the low temperature and high field regime exhibits in-plane order but the two-dimensional lattice planes wiggle as a whole. For fields greater than about 500 kG the simulation yields virtually independent layer crystals asymptotically approaching the 2D limit. Following Huberman and Doniach,[4] an upper limit for the melting temperature of an individual layer can be set to be $k_B T_m \leq (1/8\pi\sqrt{3})d\phi_0^2/16\pi^2\lambda^2$ which translates to $k_B T_m \leq 11.5°K$ in our case. Given the fact that the renormalization of the interaction pulls down[5] that limit to $5 \sim 9°K$, our result is in good agreement with the theoretical bounds.

As the temperature is raised across the melting lines in the intermediate regime, the system clearly goes through a triangular lattice phase, a disentangled liquid, and through an entangled liquid state. In the low field regime, the interlayer coupling is relatively stronger than the in-plane correlations and the system displays more rigid flux lines forming a very fragile lattice. In the very low field limit, the melting temperature shifts downward drastically giving rise to a reentrant behavior. This behavior is expected since the vortices are so far apart from each other that the in-plane interaction is in the exponential limit.[2]

The Lindemann constant as determined from the simulation is found to give a reasonable value of about 0.2 but deviates in both the large field and the low field limits. Since we have two different types of interaction in the system which scale differently with length (i.e. with changing density), it is reasonable to expect such a behavior. This finding may also suggest that the nature of the melting changes as one goes from the 2D-like high density (high field) limit to the 3D-like low density limit. The calculation by Houghton et al.[6] in the continuum-elastic theory had used the somewhat larger value of 0.4.

REFERENCES

1. S. Ryu, S. Doniach, G. Deutscher and A. Kapitulnik, to be published.
2. D. R. Nelson, Phys. Rev. Lett. **60**, 1973 (1988), D. R. Nelson and H. S. Seung, Phys. Rev. B **39**, 9153 (1989).
3. S. Doniach, in High Temperature Superconductivity, Proceedings edited by K. S. Bedell, D. Coffey, D. E. Meltzer, D. Pines and J. R. Schrieffer (Addison Wesley, Redwood City) 406 (1989).
4. B. A. Huberman and S. Doniach, Phys. Rev. Lett. **43**, 950 (1979).
5. D. S. Fisher, Phys. Rev. B **22**, 1190 (1980).
6. A. Houghton, R. A. Pelcovits and A. Sudbo, Phys. Rev. B **40** (1989).

Determination of the Sommerfeld Constant and the Electron-Phonon Coupling Strength in YBa$_2$Cu$_3$O$_{7-\delta}$

M. E. Reeves,* S. A. Wolf,* and V. Z. Kresin**
*Naval Research Laboratory, Washington, DC 20375
**Lawrence Berkeley Laboratory, Berkeley, CA 96720

We present estimates for the bare Sommerfeld constant (γ_B) and the strength of the electron-phonon coupling based on specific heat measurements of YBa$_2$Cu$_3$O$_{7-\delta}$. γ_B is determined from measurements made at high temperatures (300 to 700K). The strength of the electron-phonon coupling is determined by the ratio of the low- to the high-temperature values of the Sommerfeld constant.

The high temperature specific heat is determined from measurements of the enthalpy of YBa$_2$Cu$_3$O$_{7-\delta}$ between 300 and 700K. Here, we will present only the data; full details of the analysis are given in Ref. 1. As seen in Fig. 1, the high temperature specific heat agrees well with the lower temperature measurements.[2]

The electronic specific heat is determined by subtracting the lattice specific heat (solid line in Fig. 1) as calculated from the measured phonon density of states.[3] The difference data plotted in the inset of Fig. 2 are linear as expected from a Fermi liquid description of the charge carriers. However, the data do not extrapolate through the origin, which we attribute to lattice anharmonicity not accounted for by the dilatation correction. We forced the electronic specific heat to pass through the origin by assigning a small (5%) quartic anharmonicity to the low energy (< 20meV) modes of the lattice. The resulting lattice and electronic specific heats are plotted in Fig. 1 and in Fig. 2, respectively. This arbitrary assignment of a quartic anharmonicity is inconsistent with the temperature dependence of the neutron DOS[3] which shows a softening rather than a hardening of the modes with increasing temperature. Nevertheless, the slopes of the data in Fig. 2 provide a useful upper bound to γ_B of 40±5mJ/moleK^2.

Following the analysis of Knapp and Jones,[4] we have also determined the electron-phonon coupling constant from specific heat is renormalized by the electron-phonon coupling. Thus, the low temperature value of the electronic specific heat (γ_{LT}) is renormalized by a factor, $1 + \lambda$, over the γ_B.[5] In the two band model,[6] phonons are coupled more strongly to the plane carriers than to the chain carriers, so the relative contribution of the planes to the total density of states must be ascertained before λ can be determined. Following the results of the band structure calculations and specific heat measurements on oxygen reduced samples,[7] we assume that the total density of states can be written as $\gamma_B = \gamma_{ch} + \gamma_{pl}$ where $\gamma_{ch} = 2\gamma_{pl}$.

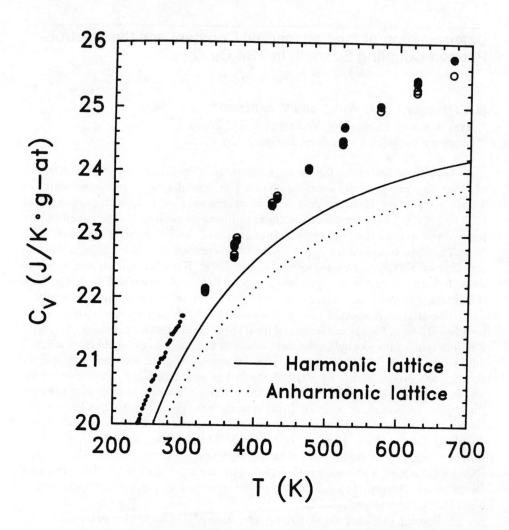

Fig. 1 The high-temperature specific heat of Y123 (large circles). The small circles are the specific heat data reported in Ref. 2. The solid and dotted lines represent the lattice specific heat.

In an applied magnetic field, the specific heat of a type II superconductor is enhanced by the normal electrons in the vortex cores. These contribute linear temperature dependence, and the Sommerfeld constant is related to the field dependent enhancement $(d\gamma/dH)$ by $\gamma_{LT} = H_{c2} d\gamma/dH$. Based on Ref. 8, we take 0.55 ± 0.05 mJ/moleK^{2}T as the value for $d\gamma/dH$, and use the angular average of H_{c2} (171T) as determined from measurements of YBa$_2$Cu$_3$O$_{7-\delta}$ single crystals.[8] The value of γ_{LT} is then determined to be 94 mJ/moleK^2.

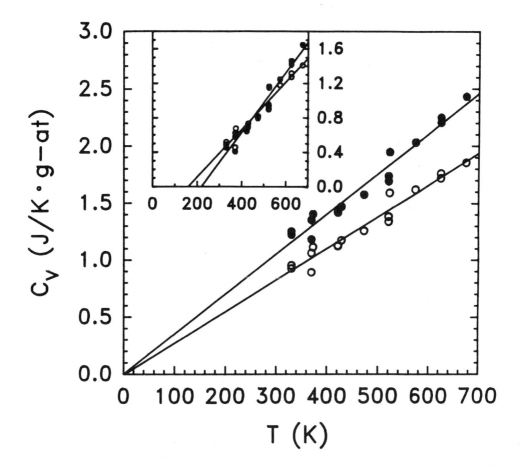

Fig. 2 The specific heat of Fig. 1 with the harmonic lattice (inset) and the partially anharmonic lattice subtracted. The open and closed circles represent two different determinations of the specific heat from the enthalpy data.

The planes are assumed to be strongly coupled and the chains are assumed to be weakly coupled so the relationship between the high- and low-temperature Sommerfeld constants can be written: $\gamma_{LT} = (1 + \lambda_{pl})\gamma_{pl} + \gamma_{ch}$. Substituting for γ_{pl} and γ_{ch}, we conclude that $\lambda_{pl} \geq 3$. Thus, the electron-phonon coupling is very strong for the plane carriers.

One of the authors (MER) acknowledges support from a National Research Council Post-doctoral Fellowship.

REFERENCES

1. M. E. Reeves, D. A. Ditmars, S. A. Wolf and T. A. Vanderah, preprint.
2. A. Junod et al., Physica C159, 215-225 (1989).
3. W. Reichardt et al., Physica B156-157, 897 (1989).
4. G. S. Knapp and R. W. Jones, Phys. Rev. B 6, 1761 (1972).
5. G. Grimvall, J. Phys. Chem. Solids 29, 1221 (1968).
6. V. Z. Kresin and S. A. Wolf, Physica C169, 476 (1990).
7. A. Junod et al., Physica C 162-164, 1401 (1989).
8. M. E. Reeves et al., Phys. Rev. B40, 4573 (1989); G. Panova et al., JETP Lett. (Suppl.) 46, S68 (1987); T. Sasaki et al., Physica C 156, 395 (1988).
9. U. Welp et al., Phys. Rev. Lett. 62, 1908 (1989); R. C. Morris et al., Phys. Rev. B 5, 895 (1972).

Evidence for Strong Electron-Phonon Coupling in Thermal Conductivity of $YBa_2Cu_3O_{7-\delta}$

Joshua L. Cohn and Stuart A. Wolf*
Terrell A. Vanderah**

*Naval Research Laboratory, Washington, DC 20375
**Naval Weapons Center, China Lake, CA 90124

The in-plane thermal conductivity (κ) in the high-T_c superconductors[1] is widely observed to increase abruptly for $T \leq T_c$. This behavior is consistent with a low carrier density and demonstrates the importance of phonon-carrier scattering in limiting the lattice heat conduction; the phonon mean free path (and hence κ) increases for $T \leq T_c$ due to the reduced scattering of phonons by carriers, as the latter condense into superconducting pairs. Here we discuss an analysis of this slope change in κ (T) data for $YBa_2Cu_3O_{7-\delta}$ (YBCO) crystals. A more extended discussion is presented elsewhere.[2]

Figure 1 shows κ (T) data for two YBCO crystals. The total thermal conductivity is a sum of carrier (e) and lattice (L) components, $\kappa = \kappa_e + \kappa_L$, and thus the first step in analyzing the slope change at T_c is to estimate the carrier contribution. We do so by applying the Wiedemann-Franz law, which states that $\kappa_e/\sigma T = L_0$, where σ is the measured electrical conductivity, and $L_0 = 2.45 \times 10^{-8}$ WΩ/K^2 is the Lorenz number. This yields an upper limit estimate for κ_e and we find at T=100K, $\kappa_e/\kappa \approx 0.31$ for both specimens.

The ratio of the thermal conductivities in the superconducting (s) and normal (n) states, κ^s/κ^n, is most readily compared with theory. We are interested in the quantity,

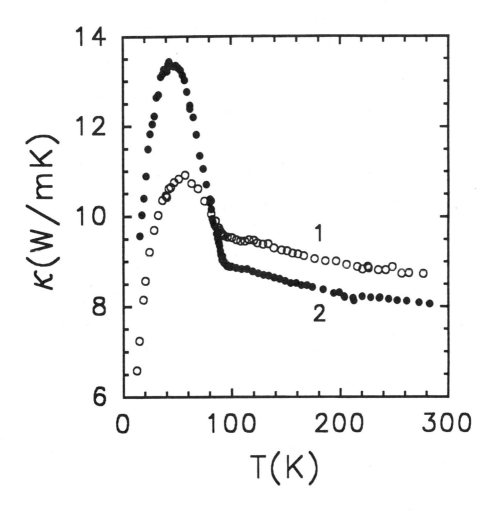

Fig. 1 Thermal conductivity versus temperature for two YBCO single crystals.

$$\frac{d}{dt}(\kappa^2/\kappa^n) = \eta \frac{d\alpha}{dt} + (1-\eta)\frac{d\beta}{dt} + \frac{d\ln\kappa^n}{dt}, \qquad (1)$$

where $t=T/T_c$ is the reduced temperature, $\alpha \equiv \kappa_e^s/\kappa_e^n, \beta \equiv \kappa_L^s/\kappa_L^n$, and $\eta = \kappa_e^n/\kappa^n$. The normal-state thermal conductivity, κ^n, is determined, for $T < T_c$, from the experimental data by smooth extrapolation of the normal-state data to several degrees below T_c. In Fig. 2 we plot κ^s/κ^n versus t near t=1. The slopes are $d(\kappa^2/\kappa^n)/dt|_{t=1} = -0.4$ (sample 1) and -1.1 (sample 2) as indicated by

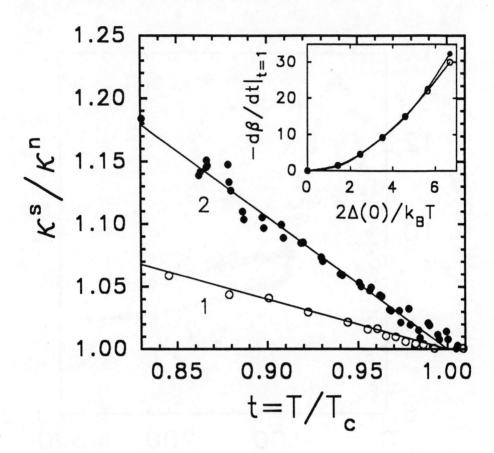

Fig. 2 Normalized thermal conductivity plotted versus reduced temperature. Inset: values of the reduced-temperature derivative of the normalized lattice thermal conductivity, $d\beta/dt|_{t=1}$, plotted versus gap size, as calculated from the BRT theory (refs. 5, 6). See reference 2 for more details.

the solid lines. The third term on the right-hand-side of (1) is quite small and thus the slope is determined by a competition between the first two terms which have opposite sign.

It is important to emphasize that the increase in κ_L for $T \leq T_c$ generally reflects the coupling of carriers to longitudinal acoustic phonons (presumed responsible for most of the lattice heat conduction) and these phonons may not be the most important for superconductivity. Thus the value of $d\alpha/dt|_{t=1}$ is more directly related to the strength of the carrier-phonon interaction relevant to superconductivity. Our approach is to calculate $d\beta/dt|_{t=1}$, and then to place a lower limit on $d\alpha/dt|_{t=1}$ using our data and Eq. (1). In the context of the existing theory of κ in superconductors, a value of $d\alpha/dt|_{t=1}$ substantially greater than 1.6 implies strong electron-phonon coupling.

We calculate $d\beta/dt|_{t=1}$ from the BRT theory[3] and its extension,[4] the latter including the effects of phonon scattering by defects and other phonons, as well as by carriers. The details of this procedure can be found in reference 2. Following reference 5 we take $\Delta = \chi\Delta^{BCS}$, where χ is a constant. In the inset of Fig. 2 we plot $-d\beta/dt|_{t=1}$ versus $2\Delta(0)/k_B T_c$, calculated using scattering parameters established by fitting to the normal-state κ_L data. We place a conservative lower limit on $d\beta/dt|_{t=1}$ by using the smallest gap size (for the Cu-O chains) reported in the literature, $2\Delta(0)/k_B T_c = 2.5$. This implies $d\beta/dt|_{t=1} \approx 5$ which, from Eq. (1), yields $d\alpha/dt|_{t=1} \approx 6$. A more reasonable picture[5] that assigns strong coupling to the planes and weak coupling to the chains, would imply a plane contribution, $d\alpha^{pl}/dt|_{t=1} \approx 20$. For comparison, $d\alpha/dt|_{t=1} \approx 9$ for lead.[6] Our main conclusion, $d\alpha/dt|_{t=1} \geq 6$, indicates that strong coupling characterizes some of the carriers in YBCO.

REFERENCES

1. C. Uher, J. Supercond. **3**, 337 (1990).
2. J. L. Cohn, S. A. Wolf and T. A. Vanderah, submitted to Phys. Rev. B.
3. J. Bardeen, G. Rickayzen and L. Tewordt, Phys. Rev. **113**, 982 (1959).
4. L. Tewordt and Th. Wölkhausen, Sol. St. Commun. **70**, 839 (1989).
5. V. Z. Kresin and S. A. Wolf, Physica C**169**, 476 (1990).
6. J. H. P. Watson and G. M. Graham, Can. J. Phys. **41**, 1738 (1964).

Copper Chloride, Revisited

N. V. Coppa, W. P. Beyermann,[1] J. D. Thompson, F. M. Mueller and J. L. Smith

Los Alamos National Laboratory, Los Alamos, New Mexico 87545

Exploration for new non-oxide high temperature superconductors (HTS) had led to the reconsideration of copper chloride as a candidate for superconductivity. In 1975, Chu et al.[2] demonstrated that copper chloride at room temperature exhibits a sharp drop in resistivity when under an imposed pressure of 40 kbar, (delta $\Omega > 7$ orders of magnitude). It was suggested that there was a progressive change from covalent to metallic bonding as the pressure was increased and then at still higher pressures, where the resistivity increased (55 kbar) there was a progression to ionic bonding. Later, Brandt et al.[3] showed that rapidly cooling CuCl under a hydrostatic pressure of 5 kbar, induced a transient diamagnetic response of about

$X = -1/4\pi$ at $\sim$100 K. Chu et al.,[4] using three different methods of preparation, showed there was a correspondence between the magnetic and resistive anomalies. It was proposed that the diamagnetic response was a manifestation of superconductivity and may originate at the Cu-CuCl interface, where Cu was the result of the disproportionation of 2 CuCl = Cu + $CuCl_2$ within the sample. Numerous accounts of superconductivity followed, but the superconducting properties observed never matured beyond an anomalous status of some samples. Batlogg and Remeika[5] prepared a definitively pure CuCl sample that showed no superconductivity. To our awareness, no further work has been reported on this system.

Since the discovery of HTS materials, an important idea has emerged regarding the discovery of other HTS materials: that pure systems can be made superconducting upon the introduction of impurities (doping), although the parent compound may exhibit no superconductivity itself. Furthermore, the addition or substitution of foreign ions into a parent lattice can simulate conditions such as pressure. Thus we have doped CuCl with fluorine in an attempt to stimulate the disproportionation reaction and to simulate the effect of pressure, chemically by the substitution of chlorine for fluorine.

Polycrystalline CuCl itself is translucent white and an insulator, but when reacted with CuF_2 forms a black material which conducts poorly at room temperature. A two point measurement showed the resistance was about 90 - 100 kΩ. Magnetization measurements as a function of temperature in a field of 1000 Oe reveal a small diamagnetic *kink* at 70 K in the otherwise paramagnetic behavior, (Fig. 1). Upon further doping with CuF_2, little or no change was observed in the magnitude of the feature at 70 K. Anhydrous copper fluoride (blue in color), when thermally decomposed in helium, undergoes a stepwise decomposition. Stoichiometric quantities of water are lost below 200 C, forming white CuF_2. Further decomposition at 650 C results in a mass loss corresponding to the loss of one fluorine per CuF_2. This black product, which is an insulator at room temperature, has a zinc blend structure and we believe it to be CuF. This phase exhibits antiferromagnetism with a Néel temperature of 80 K, (Fig. 2a). The temperature dependent magnetic response of CuCl is shown in Fig. 2b. Comparison of data in Figs. 1 and 2 suggests the 70 K kink found in the susceptibility of CuCl reacted with CuF_2 may result from antiferromagnetism of the second phase CuF. Substitution of Cu^{+1} with ions of equivalent atomic radii systematically shifts the 10 K magnetic susceptibility. The Bi^{+3} doped compounds become less paramagnetic, whereas, Ti^{+3} doped compounds become more paramagnetic, (Fig. 3).

It is apparent that the disproportionation of Cu^{+1} in CuCl may be stimulated by the addition of F^- which gives rise to the increased conductivity in the F^- doped samples. The magnetic anomalies observed in CuCl under pressure, by Chu and others, may be a result of inducing weak antiferromagnetism in those materials. On the face of it these conclusions appear disappointing. However, chemical doping to produce antiferromagnetism and electrical conductivity appears on the road to high temperature superconductors in the copper halides as we have seen in the

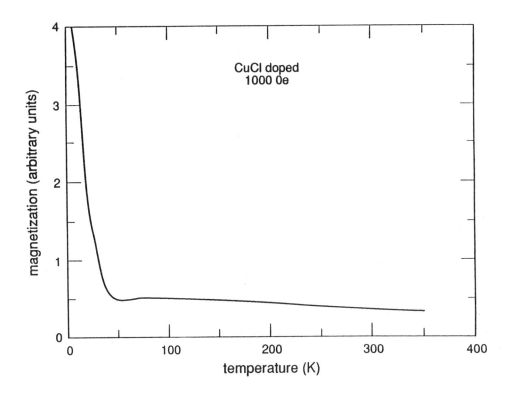

Fig. 1 Magnetization of CuCl doped with CuF_2 measured as a function of temperature in a field of 1000 Oe, atmospheric pressure. Note the small diamagnetic trend at 70 K.

copper oxides. We believe that although we have not seen any superconductivity here, that perhaps superconductivity will emerge in other chemically substituted copper halides as had been sought in the past. This research continues.

This work was supported by the Electric Power Research Institute, and under the auspices of the United States Department of Energy.

REFERENCES

1. Present address: Department of Physics, University of California, Riverside, CA 92521.
2. C. W. Chu, S. Early and R. E. Schwall, J. Phys. C **8**, L241, (1975).
3. N. B. Brandt, S. V. Kuvskinnikov, A. P. Rusakov and M. V. Semyonov, JETP Lett **27**, 33 (1978).
4. C. W. Chu, A. P. Rusakov, S. Huang, S. Early and C. Y. Huang, Phys. Rev. B **18**, 2116 (1978).
5. B. Batlogg and J. P. Remeika, Phys. Rev. Lett. **45**, 1126 (1980).

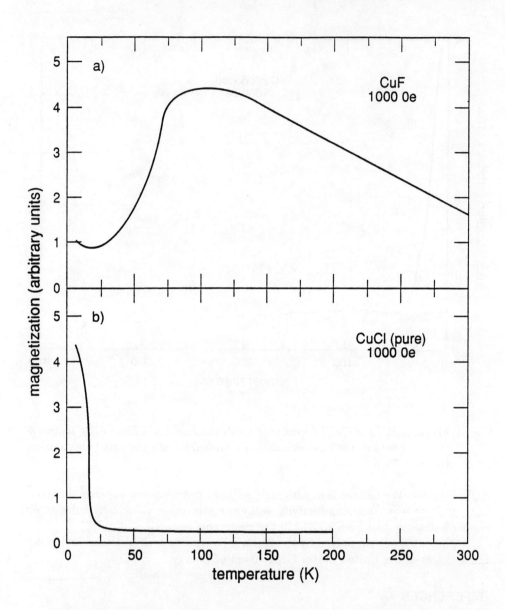

Fig. 2 (a) Magnetization of CuF measured as a function of temperature. (b) Magnetization of CuCl measured as a function of temperature, (both curves: in a field of 1000 Oe, atmospheric pressure).

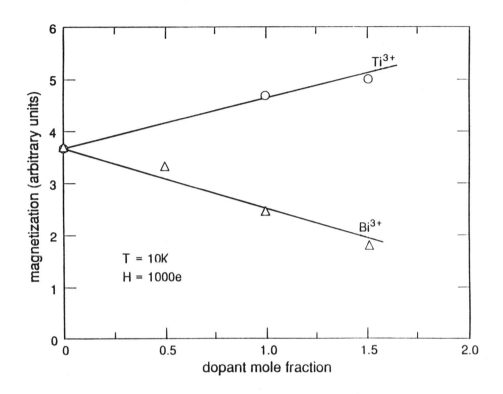

Fig. 3 Magnetic response of Bi^{+3} or Ti^{+3} doped CuCl/F at 10 K and 100 Oe. X-ray diffraction studies (not shown) show that the products retain the original zinc blend structure.

Flux Motion and Dissipation in High-Temperature Superconductors[*]

K. E. Gray and D. H. Kim

Materials Sciences Division, Argonne National Laboratory, Argonne, Illinois 60439, USA

The effects on flux motion and dissipation of interlayer coupling of the Cu-O planes along the c-axis are considered for the high-temperature superconductors (HTS). It is argued that for the *highly-anisotropic* HTS, the weak interlayer coupling plays a dominant role that can be described by *incoherent* Josephson tunneling between superconducting Cu-O bi- or tri-layers. In $YBa_2Cu_3O_7$, the layers are strongly coupled, presumably because the conducting Cu-O chains short circuit the Josephson tunneling, so that these effects are weak or missing.

Recently,[1] the effects of anisotropy and fluctuations on critical current densities, $J_c(T,H)$ and the field-induced broadening of resistivity transitions, $\rho(T,H)$, have been studied in high-temperature superconductors (HTS). Although the broadening looks similar for the applied field, H, oriented either parallel to the superconducting Cu-O layers ($H\|ab$) or parallel to the c-axis ($H\|c$), its width and the detailed shape of $\rho(T,H)$ are different. The explanations given below for the *highly-anisotropic* HTS differ in detail for the two cases, but have a crucial feature in common: they result from fluctuations affecting the Josephson coupling across the interlayer junctions.[2,3]

For $H\|ab$, the broadening is smaller: the absence of any measurable Lorentz-force dependence[1,3-5] in the highly-anisotropic HTS together with the anticipated intrinsic pinning of the insulating region between layers, questions explanations involving motion of vortices from the external field.[1,4] Various mechanisms can explain the weak or missing Lorentz-force dependence of $\rho(T,H)$ and $J_c(T,H)$ for $H\|ab$. Some relate to sample perfection: (a) meandering current paths in the *ab* plane due to poorly-coupled grains or other defects; fluctuation of the Josephson coupling, either (b) between Cu-O bi- or tri-layers when meandering current paths include a c-axis component,[3] or (c) between grains;[6] (d) misalignment of the sample, or (e) misorientation of individual grains with respect to the field;[7] and (f) field-induced granularity.[8] Another is intrinsic: (g) *field-induced* thermal excitation of vortex/anti-vortex pairs in the Cu-O planes.[9] The results on single crystals[5] and epitaxial films[3] of $Tl_2Ba_2CaCu_2O_x$ seem to adequately rule out explanations (a) and (c). The measured distributions of c-axis misorientations[1,3] are <0.3°, which rules out (e), while the need of a 3.8° sample misalignment[1] for (d) likewise precludes it. Recently, the degree of Lorentz-force dependence of $\rho(T,H)$ and $J_c(T,H)$ in HTS has been shown to depend on the interlayer spacing,[3] suggesting that the interlayer coupling, predominated by Josephson tunneling between neighboring Cu-O bi- and tri-layers, may be important for (b) or (g).

In the case of (b), these tunnel junctions would occur between isolated finite-area plates of neighboring Cu-O bi- or tri-layers, with defects in these layers causing a meandering of the current path between such plates.[3] Fluctuations of the relative phase across these junctions occur when kT exceeds $E_{cj}(T,H)$, the Josephson coupling energy between adjacent Cu-O multilayers, and this would result in a crossover to finite resistance and a reduction in the low-temperature $J_c(T,H)$, both of which effects are known to occur in thin-film Josephson tunnel junctions. For (g), we note that as kT exceeds $E_{cj}(T,H)$, there would be a crossover to isolated 2D superconducting layers, such that the thermal activation of vortex/anti-vortex pairs is greatly enhanced over the well-coupled, 3D system.

Josephson fluctuations were used to explain the observed Lorentz-force independence of the broadened $\rho(T,H)$ in *granular* NbN films[6] and of $J_c(T,H)$ in *granular* multilayers[10] of NbN with AlN. Motion of the external flux was suppressed by the relatively strong pinning, e.g., the insulating AlN layers, and a distinct crossover in $J_c(H)$ was observed between depinning of the *external* flux and Josephson fluctuations between grains.[10] Recent experiments[11] on discrete Josephson junctions, made with high-quality Nb films, confirm this conclusion. A broadened resistive transition, very similar to that of HTS materials, was observed in such junctions in fields, perpendicular to the film plane, up to 0.03 T. These measurements used a current density of 0.1 A/cm^2, for which the resistive transitions of the films were very sharp, indicating that the external flux was completely pinned in the electrodes. The dissipation was caused by self-field, Josephson vortices which are perpendicular to the applied field direction.

In zero field, E_{cj} is proportional[12] to the product of the superconducting order parameters on each side of the junction, ψ_a and ψ_b, divided by the normal-state resistance, R_N. For HTS interlayer junctions with an area A, $R_N = \rho_c s/A$, where ρ_c is the c-axis resistivity and s with cell size along the c-axis, so that $E_{cj} \sim FA$. This relation reflects the fact that fluctuations must produce self-field, Josephson vortex loops which cover the total junction area, while F accounts for the energy required per unit area. The activation energy measured on discreet Nb junctions[11] indicate that $A = \Phi_o/H$ at high fields, where Φ_o is the flux quantum. In zero field, A will be limited either by the sample dimensions or the inevitable presence of defects, even in single crystals and epitaxial films, to a value A_o. For the Nb junctions, A_o was found to be $\sim 1\mu m^2$, while the physical junction area is $\sim 8 \times 12\mu m^2$. We suggest that the effect of H may be to further limit the minimum size of the fluctuation-induced vortex loops to Φ_o/H, since they can then connect with the pinned, external-field vortices in the electrodes. This is analagous to dislocation-mediated shearing (melting) of crystal lattices.[13]

For HTS, the lack of intrinsic pinning for $H\|c$ implies that the broadening is due to thermally-activated flux motion. This broadening is fairly independent of sample quality, but depends strongly on the spacing between Cu-O bi- or tri-layers.[2] Thus, we suggest that for $H\|c$, thermally-activated decoupling of the Josephson-coupled superconducting phases causes the broadening by decoupling the magnetic-field-induced pancake-like[14] vortices in adjacent Cu-O layers. The resulting independent motion of vortices in adjacent layers, i.e., 2D behavior, greatly reduces the effectiveness of pinning compared to extended, 3D vortex *lines*. For the *highly-anisotropic* HTS, such a crossover from 3D to 2D vortices was found[2,3] for $kT \sim E_{cj}(H,T)$. In addition, at sufficiently low temperatures, the finite pinning strength, $E_p(H,T)$, of individual Cu-O multilayers was found[2] to be effective even in the 2D regime.

A finite dc resistance requires that the vortices are excited out of their potential wells of *both* energy barriers, so $k_BT=E_p(H^*,T)+2E_{cj}(H^*,T)$ was solved[2] for the crossover field, H^*. For $H\|c$, this model gives convincing fits[2] to measurements of resistive transitions for the Bi- and Tl-cuprates with realistic values for the parameters ρ_c, B_c and H_{c2}, providing that $E_{cj} \sim 1/H$, in agreement with the above Josephson-junction model. Mechanical oscillator experiments can also probe E_p and E_{cj} individually, since dissipation can also occur *without* vortices being excited out of their potential wells (i.e., when $kT>E_p$ and $2E_{cj}$, but $kT<E_p+2E_{cj}$). The two

loss peaks found in such experiments[15] on $Bi_2Sr_2CaCu_2O_x$ single crystals agree[3] surprisingly well with the Josephson model with substantially the same parameters as found resistively.[2]

Returning to the case of H|| ab, Josephson vortex cores divide the interlayer junctions into areas given by $\sqrt{A_o\Phi_o\lambda_c/\lambda_{ab}H}$, providing $A_o > \Phi_o\lambda_c/\lambda_{ab}H$, where λ_c and λ_{ab} are the c-axis and in-plane magnetic penetration depths, respectively, and thus $E_{cj} \sim 1/\sqrt{H}$. Experimentally, the activation energy which best fits[3] the resistive transitions in epitaxial $Tl_2Ba_2CaCu_2O_x$ films is (4800 [$KT^{0.54}$]) $k_B(1-t)/H^{0.54}$. This dependence is valid for H$\geq$0.25 T, which, together with a lower limit of λ_c/λ_{ab} ~350 from torque magnetometry,[16] implies that $\sqrt{A_o}$ must be $\geq$1.5 μm. One can also obtain $\sqrt{A_o}$ from the saturation of the 1/H dependence in the same films at low fields with H||c: although this occurs very near T_c and there is no independent measure of T_c, a lower limit of $\sqrt{A_o}$ ~1.5 μm is found. An important difficulty arises when fitting $\sqrt{A_o}$ to the experimental prefactor, 4800 [$KT^{0.54}$], with the other parameters of the Josephson model. Using λ_c/λ_{ab} ~350, we find $\sqrt{A_o}$ ~0.3 μm, in disagreement with the above estimates. Although this is ~800 unit cells and may be reasonable even for epitaxial films, a five-times-larger value of λ_c/λ_{ab} would be necessary to make the model quantatively compatible.

We note that the predicted vortex/anti-vortex pair creation energy[9] is $E_{cv} = \Phi_o^2 d_s/8\pi^2\lambda_{ab}^2$ ~(1400 [K]) $k_B(1-t)$ for $Tl_2Ba_2CaCu_2O_x$, where d_s is the Cu-O bilayer thickness and λ_{ab} is the in-plane magnetic penetration depth. For presently attainable H, E_{cv} is less than experiment, so we cannot choose between mechanisms (b) and (g). For larger H, (g) would predict a field-independent activation energy.

Although there is a quantatitive inconsistency in the detailed Josephson-coupling model for H||ab, the correlation of the degree of Lorentz-force dependence on interlayer spacing/coupling suggests that it is important in this case. Thus, for the highly-anisotropic HTS, the dissipation may be described by thermal fluctuations of the interlayer coupling, resulting in dissipation by Josephson vortices crossing interlayer junctions, rather than motion of the external field vortices. For H||c, thermal fluctuations of the interlayer Josephson coupling decouples the pancake vortices of the external field, leading to significantly greater dissipation than the well-coupled case at low fields and temperatures.

This research was done in collaboration with J.C. Smith, R. Holobof, M.D. Trochet and M. Eddy. This work is supported by the U.S. Department of Energy, Basic Energy Sciences-Materials Sciences under contract #W-31-109-ENG-38, and the National Science Foundation-Office of Science and Technology Centers under contract #STC8809854.

REFERENCES

1. D. H. Kim, K. E. Gray, R. T. Kampwirth and D. M. McKay, Phys. Rev. B**42**, 6249 (1990).
2. D. H. Kim, K. E. Gray, R. T. Kampwirth, J. C. Smith, D. S. Richeson, T. J. Marks, J .H. Kang, J. Talvacchio and M. Eddy, Physica C**177**, 431 (1991).
3. D. H. Kim and K. E. Gray, unpublished.
4. K. C. Woo, K. E. Gray, R. T. Kampwirth, J. H. Kang, S. J. Stein, R. East and D. M. McKay, Phys. Rev. Lett. **63**, 1877 (1989).
5. H. Iwasaki, N. Kobayashi, M. Kikuchi, T. Kajitani, Y. Syono, Y. Muto and S. Nakajima, Physica C**159**, 301 (1989).
6. D. H. Kim, K. E. Gray, R. T. Kampwirth, K. C. Woo, D. M. McKay and S. J. Stein, Phys. Rev. B**41**, 11642 (1990).
7. P. H. Kes, J. Aarts, V. M. Vinokur and C. J. van der Beek, Phys. Rev. Lett. **64**, 1063 (1990).
8. M. Daeumling, J. M. Seuntjens and D. C. Larbalestier, Nature **346**, 332 (1990).
9. S. Doniach and B. A. Huberman, Phys. Rev. Lett. **42**, 1169 (1979).
10. K. E. Gray, R. T. Kampwirth, D. J. Miller, J. M. Murduck, D. Hampshire, R. Herzog and H. W. Weber, Physica C**174**, 340 (1991).
11. D. H. Kim, K. E. Gray, J. H. Kang and J. Talvacchio, to be published.
12. P. W. Anderson, Lectures at Ravello Spring School, 1963; V. Ambegaokar and A. Baratoff, Phys. Rev. Lett. **10**, 486 (1963).
13. V. M. Vinokur, private communication.
14. J. R. Clem, Phys. Rev. B**43**, 7837 (1991).
15. C. Duran, J. Yazyi, F. de la Cruz, D. J. Bishop, D. B. Mitzi and A. Kapitulnik, preprint.
16. D. E. Farrell, R. G. Beck, M. F. Booth, C. J. Allen, E. D. Bukowski and D. M. Ginsberg, Phys. Rev. B**42**, 6758 (1990).

LIST OF ATTENDEES

Elihu Abrahams
Rutgers University
Physics & Astronomy Department
Piscataway, NJ 08855

Gabriel Aeppli
AT&T Bell Laboratories
600 Mountain Avenue
Murray Hill, NM 07974-2070

Al J. Arko
Los Alamos National Laboratory
P-10
Mailstop K764
Los Alamos, NM 87545

Kevin Bedell
Los Alamos National Laboratory
T-11
Mailstop B262
Los Alamos, NM 87545

Dermot Coffey
Los Alamos National Laboratory
CMS
Mailstop K765
Los Alamos, NM 87545

Bryan R. Coles
Imperial College of Sciences
  & Technology
Prince Consort Rd.
London SW7 282
UNITED KINGDOM

Alan R. Bishop
Los Alamos National Laboratory
T-11
Mailstop B262
Los Alamos, NM 87545

A. M. Boring
Los Alamos National Laboratory
CMS
Mailstop K765
Los Alamos, NM 87545

Laurence J. Campbell
Los Alamos National Laboratory
T-11
Mailstop B262
Los Alamos, NM 87545

John Clarke
University of California
Department of Physics
Berkeley, CA 94720

Sebastian Doniach
Stanford University
Applied Physics
Stanford, CA 94305-4090

Douglas K. Finnemore
Iowa State University
Department of Physics
Ames, IA 50011

Nicholas Coppa
Los Alamos National Laboratory
ERDC
Mailstop K763
Los Alamos, NM 87545

George W. Crabtree
Argonne National Laboratory
MSD 223
Argonne, IL 60439

Luke L. Daemen
Los Alamos National Laboratory
T-11
Mailstop B262
Los Alamos, NM 87545

Alen M. Goldman
University of Minnesota
Physics Department
Minneapolis, MN 55455

Kenneth E. Gray
Argonne National Laboratory
Bldg - 223
Argonne, IL 60439

George Gruner
University of California, Los Angeles
Department of Physics
405 Hilgard Avenue
Los Angeles, CA 90024-1547

P. Chris Hammel
Los Alamos National Laboratory
P-10
Mailstop K764
Los Alamos, NM 87545

Daniel S. Fisher
Harvard University
Lyman Lab.
Department of Physics
Cambridge, MA 02138

Zachary Fisk
Los Alamos National Laboratory
CMS
Mailstop K765
Los Alamos, NM 89545

Peter L. Gammel
AT&T Bell Laboratories
600 Mountain Avenue
Room 1D221
Murray Hill, NJ 07974

Marilyn E. Hawley
Los Alamos National Laboratory
MST-7
Mailstop K762
Los Alamos, NM 87545

Masahiko Inui
Los Alamos National Laboratory
CMS
Mailstop K765
Los Alamos, NM 87545

Boris Ivlev
University of Southern California
Department of Physics
Los Angeles, CA 90089-0484

Aharon Kapitulnik
Stanford University
Department of Applied Physics
Stanford, CA 94305-4085

Robert Hammond
Superconductors Technologies, Inc.
460 Ward Drive
Suite 7
Santa Barbara, CA 93111

Vladimir G. Kogan
Ames Laboratory
Physics Department
Ames, IA 50011

Pang-Jen Kung
Los Alamos National Laboratory
ERDC
Mailstop K763
Los Alamos, NM 87545

David C. Larbalestier
University of Wisconsin-Madison
College of Engineering
1500 Johnson Drive
Madison, WI 53706

Ross Lemons
Los Alamos National Laboratory
ERDC
Mailstop K763
Los Alamos, NM 87545

Li Luo
Los Alamos National Laboratory
CMS
Mailstop K765
Los Alamos, NM 87545

Fred M. Mueller
Los Alamos National Laboratory
CMS
Mailstop K765
Los Alamos, NM 87545

Peter H. Kes
Kamerlingh Onnes Laboratory
P.O. Box 9506
2300 RA Leiden
THE NETHERLANDS

Martin P. Maley
Los Alamos National Laboratory
ERDC
Mailstop K764
Los Alamos, NM 87545

Alexis P. Malozemoff
American Superconductor
149 Grove Street
Watertown, MA 02172

Michael McHenry
Los Alamos National Laboratory
ERDC
Mailstop K763
Los Alamos, NM 87545

David Meltzer
Department of Chemistry & Physics
Box 878
Southeastern Louisiana University
Hammond, LA 70402

Don M. Parkin
Los Alamos National Laboratory
CMS
Mailstop K765
Los Alamos, NM 87545

Dean E. Peterson
Los Alamos National Laboratory
ERDC
Mailstop K763
Los Alamos, NM 87545

Ross E. Muenchausen
Los Alamos National Laboratory
ERDC
Mailstop K763
Los Alamos, NM 87545

Masato Murakami
International Superconductivity
　Tech Center
Superconductivity Research Lab.
10-13, Shinonome, 1-Chome, Koto-ku
Tokyo 135, JAPAN

David Nelson
Harvard University
Department of Physics
Cambridge, MA 02138

Thomas T. M. Palstra
AT&T Bell Laboratories
Room 1B118
600 Mountain Avenue
Murray Hill, NJ 07974

Douglas Scalapino
Univ. of California, Santa Barbara
Department of Physics
Santa Barbara, CA 93106

J. Robert Schrieffer
Florida State University
1800 East Paul Dirac Drive, B223
Tallahassee, FL 32306

Richard N. Silver
Los Alamos National Laboratory
T-11
Mailstop B262
Los Alamos, NM 87545

David Pines
University of Illinois
1110 West Green Street
Urbana, IL 61801

David W. Reagor
Los Alamos National Laboratory
MEE-11
Mailstop D429
Los Alamos, NM 87545

T. Maurice Rice
ETH-Honggerberg
Theoretische Physik
8093 Zurich
SWITZERLAND

Shoji Tanaka
International Superconductivity
　Tech Center
Superconductivity Research Lab
10-13 Shinonome, 1-Chome
Koto-ku, Tokyo 135 JAPAN

Joe D. Thompson
Los Alamos National Laboratory
P-10
Mailstop K764
Los Alamos, NM 87545

Michael Tinkham
Harvard University
Physics Department
Cambridge, MA 02128

Stuart Trugman
Los Alamos National Laboratory
T-11
Mailstop B262
Los Alamos, NM 87545

James L. Smith
Los Alamos National Laboratory
ERDC
Mailstop K763
Los Alamos, NM 87545

Kendall Springer
Los Alamos National Laboratory
MEE-11
Mailstop D429
Los Alamos, NM 87545

John Wilkins
Ohio State University
Department of Physics
4100 Smith Lab.
174 West 18th Avenue
Columbus, OH 43210

Xindi Wu
Los Alamos National Laboratory
ERDC
Mailstop K763
Los Alamos, NM 87545

Valeri Vinokur
Argonne National Laboratory
MSD
9700 South Cass Avenue
Argonne, IL 60439

Eugene Wewerka
Los Alamos National Laboratory
ADCM
Mailstop A102
Los Alamos, NM 87545

Stuart A. Wolf
Head, Materials Physics Branch
Code 6340
Naval Research Laboratory
Washington, DC 20375-5000